Developments in Precambrian Geology 6

IRON-FORMATION: FACTS AND PROBLEMS

DEVELOPMENTS IN PRECAMBRIAN GEOLOGY

Advisory Editor B.F. Windley

Further titles in this series

1. B.F. WINDLEY and S.M. NAQVI (Editors)
   Archaean Geochemistry
2. D.R. HUNTER (Editor)
   Precambrian of the Southern Hemisphere
3. K.C. CONDIE
   Archean Greenstone Belts
4. A. KRÖNER (Editor)
   Precambrian Plate Tectonics
5. Y.P. MEL'NIK
   Precambrian Banded Iron-formations. Physicochemical Conditions of Formation

DEVELOPMENTS IN PRECAMBRIAN GEOLOGY 6

IRON-FORMATION FACTS AND PROBLEMS

Edited by

A.F. TRENDALL

Geological Survey, Department of Mines, Perth, W.A., Australia

and

R.C. MORRIS

Division of Mineralogy, CSIRO, Wembley, W.A., Australia

ELSEVIER, Amsterdam – Oxford – New York – Tokyo 1983

ELSEVIER SCIENCE PUBLISHERS B.V.
Molenwerf 1, 1014 AG Amsterdam
P.O. Box 211, Amsterdam, The Netherlands

Distributors for the United States and Canada:

ELSEVIER SCIENCE PUBLISHING INC.
52, Vanderbilt Avenue
New York, N.Y. 10017

Library of Congress Cataloging in Publication Data
Main entry under title:

Iron-formation, facts and problems.

(Developments in Precambrian geology ; 6)
Includes bibliographies and index.
1. Iron ores. I. Trendall, A. F. (Alec Francis)
II. Morris, R. C. III. Series.
QE390.2.I76I76 1983 553.3 83-1494
ISBN 0-444-42144-0 (U.S.)

ISBN 0-444-42144-0 (Vol. 6)
ISBN 0-444-41719-2 (Series)

Printed in The Netherlands

FOREWORD

The year 1973 marked publication of two milestone volumes dealing with Precambrian iron-formation: "Precambrian Iron-formations of the World", published as a special issue of the journal *Economic Geology* (Vol. 68, No. 7); and "Genesis of Precambrian Iron and Manganese Deposits", published by UNESCO (Earth Sciences, No. 9). These two volumes, together containing some 56 separate papers, are veritable storehouses of fact and theory. With the publication now of "Iron-Formation: Facts and Problems" it is appropriate to consider and appraise the advances of the past 10 years, as reflected in the 16 papers that comprise the present volume.

But first a general comment on the book as a whole. Readers who seek an orderly, internally consistent, and satisfyingly conclusive statement on the nature and significance of the rock generally known as iron-formation are likely to be disappointed; in fact they may be dismayed by the diversity of approach, the lack of consistent terminology (as stated with disarming candor in the Introduction, authors were not instructed in nomenclature; they were not even provided with an agreed-upon definition of the term iron-formation itself), and by the obvious individuality of conclusions. Indeed, some may feel that the volume has something in common with Stephen Leacock's legendary knight, who mounted his horse and rode off in all directions! But for the reader who is willing to cope with what in fact is a healthy diversity in actively evolving fields of research there is much here that will reward his patient effort. The issue of nomenclature, for example, is not simply a matter of semantics: introduction of a new family of terms (femicrite, and the like) introduced initially by Eric Dimroth, patterned after the widely used classification of limestone textures, and here used (and expanded) by Beukes, represents a broadening of approach, one that seeks to provide more adequate expression of physical factors operative during sedimentation and of the impress of diagenesis. Whether the analogy between the behavior of precipitates of strictly divalent elements (calcium and magnesium) and those of multivalent elements (iron and maganese) is valid may properly be questioned, but certainly no one can take issue with an attempt to make use of research results on chemical sediments other than iron-formation.

The advances in knowledge since publication of the 1973 compendia are evident in a number of ways. The descriptions of five of the great iron districts of the world — Lake Superior, Hamersley, Labrador Trough and its extensions around the Ungava craton, Transvaal-Griqualand West, and Krivoy Rog, which together contain at least three-fourths of the world's iron-formation, as a minimum constitute valuable up-dates of previous accounts; and several, notably those of Lake Superior, the Ungava belt of Canada, and of the Transvaal Supergroup of South Africa, are syntheses considerably superior to any elsewhere available.

Most of the topical papers, such as Klein's thorough review of major-element chemistry, mineralogy, and petrology of iron-formation assemblages, record non-spectacular but steady progress in the accumulation of basic data, and some, such as Perry's perceptive treatment of oxygen isotope trends, may have important implications concerning Precambrian environments and ocean waters. On the other hand, there appear to be few if any significant advances in iron-formation paleontology and paleoecology; indeed, as evident from the careful and thoughtful review by Walter and Hofmann, some previously cited evidence for a biologic role in iron-formation sedimentation may be open to question.

In a completely unregimented way, and expressed in this volume by highly individualistic and independent summations, a convergence in thought concerning basic factors involved in deposition of the major iron-formations has subtly evolved during the past decade or so. Though not yet at the level of complete consensus, there is now widespread acceptance, either implicit or explicit, of the assumption that a controlling factor was the particular composition of ocean waters during Archean and early Proterozoic time — a composition that differed significantly from that of later eras with respect to pH and oxidation potential. In a word, many now believe that the early oceans were major reservoirs for dissolved iron and silica, the ultimate source of which was diverse — volcanic, terrestrial, and even cosmic. This conclusion, of fundamental importance to concepts of earth's evolution, does not derive from theoretical models; it is, in the view of many, demanded by the evidence from the rocks themselves.

Much remains to be done before understanding is reached as to the mechanisms by which iron and silica were withdrawn from this hypothetical reservoir to form the great iron-formations. The various models offered in the present volume are generally incomplete and highly subjective; and I suspect that most authors would concur with Ewers' regretful comment (this volume): "I would have wished to have transferred more of this topic from areas of opinion and controversy into areas of reasonable certainty". But if one accepts the widely held view that a principal measure of research succes is the degree to which it stimulates further work, then the reports that comprise this volume are likely to be accorded high marks indeed.

HAROLD L. JAMES
Port Townsend, Wash.
U.S.A.
Dec. 7, 1982

CONTRIBUTING AUTHORS

R.YA. BELEVTSEV	Institute of Geochemistry and Physics of Minerals, Academy of Sciences of the Ukraine SSR, Palladin Prospect 34, Kiev 252068, U.S.S.R.
YA.N. BELEVTSEV	Institute of Geochemistry and Physics of Minerals, Academy of Sciences of the Ukraine SSR, Palladin Prospect 34, Kiev 252068, U.S.S.R.
N.J. BEUKES	Department of Geology, Rand Afrikaans University, Auckland Park, PO 524, Johannesburg 2000, South Africa.
J.A. BUNTING	Exploration Department, BHP Pty Ltd, 695 Burke Road, Camberwell, Victoria 3124, Australia.
P. CLOUD	Department of Geological Sciences, University of California, Santa Barbara, California 93106, U.S.A.
R. DAVY	Geological Survey. Department of Mines, Mineral House, 66 Adelaide Terrace, Perth Western Australia 6000, Australia.
W.E. EWERS	Unit 16, 128 Forrest Street, Peppermint Grove, Western Australia 6011, Australia.
B.J. FRYER	Department of Geology, Memorial University of Newfoundland, St. Johns, Newfoundland A1B 3X5, Canada.
A.D.T. GOODE	Exploration Department, BHP Pty. Ltd., 695 Burke Road, Camberwell, Victoria 3124, Australia.
G.A. GROSS	Geological Survey of Canada, 601 Booth Street, Ottawa, Ontario K1A OE8, Canada.
W.D.M. HALL	Broken Hill Proprietary Co. Ltd., 37 St. Georges Terrace, Perth, Western Australia 6000, Australia.
H.J. HOFMANN	Department of Geology, University of Montreal, Montreal, Quebec H3C 3J7, Canada.
H.L. JAMES	1617 Washington Street, Port Townsend, Washington 98368, U.S.A.

C. KLEIN — Department of Geology, 1005 East Tenth Street, Bloomington, Indiana 47405, U.S.A.

G.B. MOREY — Minnesota Geological Survey, 1633 Eustis Street, St. Paul, Minnesota 55108, U.S.A.

R.C. MORRIS — Division of Mineralogy, CSIRO, Private Bag, PO Wembley, Western Australia 6014, Australia.

E.C. PERRY, JR. — Department of Geology, Northern Illinois University, De Kalb, Illinois 60115, U.S.A.

R.I. SIROSHTAN — Institute of Geochemistry and Physics of Minerals, Academy of Sciences of the Ukraine SSR, Palladin Prospect 34, Kiev 252068, U.S.S.R.

A.F. TRENDALL — Geological Survey, Department of Mines, Mineral House, 66 Adelaide Terrace, Perth, 6000 Australia.

M.R. WALTER — Baas-Becking Geobiological Laboratory, PO Box 378, Canberra City, ACT 2601, Australia.

I.S. ZAJAC — Iron Ore Company of Canada, 100 Retty Street, Sept-Iles, Quebec G4R 3E1, Canada.

CONTENTS

Chapter 5. THE KRIVOY ROG BASIN
Ya.N. Belevtsev, R.Ya. Belevtsev and R.I. Siroshtan

Chapter 6. IRON-FORMATION IN FOLD BELTS MARGINAL TO THE UNGAVA CRATON
G.A. Gross and I.S. Zajac

Chapter 7. THE NABBERU BASIN OF WESTERN AUSTRALIA
A.D.T. Goode, W.D.M. Hall and J.A. Bunting

Chapter 9. THE PALAEONTOLOGY AND PALAEOECOLOGY OF PRECAMBRIAN IRON-FORMATIONS
M.R. Walter and H.J. Hofmann

Chapter 10. BANDED IRON-FORMATION — A GRADUALIST'S DILEMMA
P. Cloud

Chapter 11. DIAGENESIS AND METAMORPHISM OF PRECAMBRIAN BANDED IRON-FORMATIONS
C. Klein

Chapter 12. DISTRIBUTION OF BANDED IRON-FORMATION IN SPACE AND TIME
H.L. James

Chapter 13. CHEMICAL FACTORS IN THE DEPOSITION AND DIAGENESIS OF BANDED IRON-FORMATION
W.E. Ewers

Chapter 14. SUPERGENE ALTERATION OF BANDED IRON-FORMATION
R.C. Morris

Chapter 1

INTRODUCTION

A.F. TRENDALL

ORIGIN, PURPOSE, AND SCOPE OF THIS VOLUME

This book results from Project 132 of the International Geological Correlation Programme — "Basins of Iron-formation Deposition". The main objective of the project, which was established in 1975, was to encourage systematic parallel studies of the major basins of iron-formation deposition of all continents, so as to arrive at a more complete understanding of the origin and significance of this scientifically challenging and economically important rock type. As time went by it became clear that the small number, and scattered distribution, of specialised workers on iron-formation, were such as to make the normal activities of an IGCP project difficult to pursue. A suggestion from Dr Brian Windley in 1977 that the project should incorporate its results into a book on iron-formation was therefore accepted, and this volume represents the final outcome.

The structure of the volume is simple. In the first six chapters, accounts are presented of a number of major basins in which iron-formation is a significant component. Authors of these papers were asked not to focus on the details of the iron-formation itself, but instead to show how the iron-formation was related to the development of the basin as a whole. For these chapters well-preserved basins were chosen with wide regional extent and relative lack of metamorphism and deformation. It was intended that these first papers as a group should represent essentially an up-dated and greatly extended version of a short comparative review which had earlier been published on three of them (Trendall, 1968). Clearly, the amount of new information available since then had made such a task impossible for one author. To supplement these basin-by-basin descriptions the succeeding eight chapters give summaries of the present state of knowledge in various sub-disciplines of geology which have immediate relevance to the study of iron-formation.

Neither group of chapters pretends to be comprehensive. For the first group some depositional basins that should certainly have been included, such as that of Minas Gerais in Brazil, are unfortunately missing. But accounts of other iron-formations are not present simply because their poor exposure, intensity of metamorphism, or tectonic deformation, are such as to make interpretation of their depositional significance largely derivative from better-preserved examples. A complete descriptive catalogue of known iron-formations was not attempted.

A second way in which this volume is not comprehensive derives from problems of definition. None of the authors contributing to this volume was told what nomenclature to use; none was told what an iron-formation was! This deliberate initial policy decision was intended to produce, and has produced, a collection of papers whose total content itself reflects current concepts of the nature of iron-formation: most of the authors, regardless of their field of interest, accept the term "iron-formation" as designating principally a Precambrian sedimentary rock now consisting largely of silica (as "chert" or quartz) and iron oxides (as magnetite or hematite); they accepted, in effect, the Precambrian "oxide facies iron-formation" of James (1954) as a form of archetypal iron-rich sedimentary rock displaying in its most evident form the general genetic problems which are presented by all sedimentary rocks very rich in iron.

CLASSIFICATION AND NOMENCLATURE OF IRON-FORMATION

General

The main purpose of classification and nomenclature are to assist understanding and communication respectively. To the non-geological observer it must often seem that the force of this simple proposition has historically escaped geologists when they have approached the classification and naming of rocks; and no group of rocks better illustrates geologists' mishandling of nomenclature than the iron-rich sedimentary rocks. On the other hand, their immense diversity, and the continuum between different types, present a genuine challenge to classification, which must be met and overcome before a single rational nomenclature can be developed.

In the Foreword to the UNESCO-published volume resulting from the 1970 Kiev Symposium (see UNESCO, 1973) "the systematization and classification of the rocks of the chert-iron-manganese formations, the correlation of nomenclature of these rocks in different countries, the elaboration of a unified system of nomenclature for iron rocks in different regions of the world . . ." are presented as tasks to follow on from the Symposium. Unfortunately, the untidy nomenclatural situation existing in 1970 has only deteriorated through the intervening years.

Faced with the present confusion, the main purposes of this present contribution are to briefly review the existing English language classification and nomenclature in a historical and regional perspective to examine some of the special problems that have arisen in naming and classifying these rocks, and to make recommendations on future practices, and to discuss the relationship between English and Russian nomenclature.

Development of English-language usages

A major historical reason for the present variety in iron-formation nomenclature is that these rocks were studied and described in their main areas of occurrence with a high degree of independence. The term "iron-formation" was an early contraction from the "iron-bearing formation" of Van Hise and Leith (1911); however, it was first formally defined by James (1954) for the Lake Superior area as follows: a chemical sediment, typically thin-bedded or laminated, containing 15 per cent or more iron of sedimentary origin, commonly but not necessarily containing layers of chert". By the time of James' paper the term had come to replace, and embrace, the earlier "jaspilite" and "ferruginous chert" of, for example, Van Hise and Leith (1911). Gross's (1959) further definition consolidated the term and gave it an unchallenged position in North American literature; he included in the term "all stratigraphic units of layered, bedded, or laminated rocks that contain 15 per cent or more iron, in which the iron minerals are commonly interbanded with quartz, chert, or carbonate, and where the banded structure of the ferruginous rocks conforms in pattern and attitude with the banded structure of the adjacent sedimentary, volcanic, or metasedimentary rocks". The local Minnesota name "taconite", earlier used as a synonym of "ferruginous chert" by Van Hise and Leith (1911), is not at present in conflict with the term "iron-formation" since taconite is used strictly as an economic term, often in a mining context, for an iron-formation which can be viably extracted and beneficiated after fine grinding.

James (1954) did not at first make a clear nomenclatural distinction between the Precambrian iron-formations, as typified by those of the Lake Superior area, and the Phanerozoic "ironstones"; indeed he noted their similarity in iron content and mineralogy. However, he (James, 1966) later emphasized the difference between the two rock types, and recommended that this difference be reflected in nomenclature. Gross (1965, fig. 3) also initially included all iron-rich sedimentary rocks within the term "iron-formation" but later (Gross, 1980, fig. 1) accepted James' ironstone/iron-formation distinction and nomenclature.

The term "ironstone", or more commonly "banded ironstone", was applied early in South Africa to rocks which would have been called "iron-formation" in North America (e.g., Wagner, 1928). While Beukes (1973) found no difficulty in substituting "iron-formation" for "ironstone" in his review paper, it will clearly take time to see whether any South African acceptance of "iron-formation" specifically includes also the replacement of the older connotation of "ironstone" by that of James.

"Quartzite" has also been used in the South African literature, and in the form "banded hematite quartzite", or "BHQ", became firmly established in descriptions of Indian iron-formations (e.g., Krishnan, 1973).

Although the term "itabirite" has as its root an American Indian word (meaning black rock) adapted into Portuguese it has become firmly entrenched into the English-language literature not only on the iron-formations of the Minas Gerais area of Brazil, where it was originally applied, but also in areas such as Venezuela (Gruss, 1973) and West Africa (Gruss, 1973; Sims, 1973), to which it was carried by geologists familiar with the Brazilian rocks, or the literature on them.

In Australia the terms "jasper" (e.g., Feldtmann, 1921) and "jaspilite" (e.g., Ellis, 1939) were most commonly applied to iron-formations of the Western Australian Precambrian, but "banded iron-formation" or "BIF" is now generally used. Australia never developed a restricted local term comparable to "ironstone" (in its South African sense), "itabirite", or "BHQ".

In addition to the confusion caused by this local variation of names for the rock type itself, some difficulties of iron-formation nomenclature have been caused by two schemes of iron-formation classification and nomenclature introduced in North America. These are the "facies" classification of James (1954) and the "type" classification of Gross (1965).

James (1954), following extensive work on the iron-formations of the Lake Superior area, proposed their fourfold subdivision into four "facies" on the basis of the dominant original iron mineral: sulphide, carbonate, oxide, and silicate. James (1954, fig. 3) published a diagrammatic cross-section of a conceptual basin of iron-formation deposition in which the first three of these are presented as intergradational lateral equivalents deposited simultaneously in the deep, intermediate, and shallow parts of the basin respectively; and he related the different iron minerals characteristic of deposition at these different depths to parallel depth-related variations in Eh and pH. The silicate facies was believed to have a more complex, and less precisely depth-related, control.

Gross (1965) accepted James' facies classification as having palaeoenvironmental (depth) significance, and superimposed upon it an independent subdivision into four main "types" — Algoma, Superior, Clinton, and Minette. The defining criteria of each type are variable, but broadly comprise a set of parameters relating the lithology of the iron-formations to a conceptual tectono-sedimentological model. "Superior type" and "Algoma type" iron-formations, for example, are distinguished principally by differing thickness and lateral extent, differing associated rocks, and by a lack of evidence for volcanicity associated with the former. Gross's "type" classification was first evolved as a reflection of the variety of iron-formation types occurring in Canada, but was later (Gross, 1980) reaffirmed to cover most other major occurrences of iron-formation. Its merits are discussed further below.

The first concerted international attempt to resolve this regionally developed confusion between iron-formation, taconite, ironstone, quartzite, BHQ, itabirite, jaspilite and so on, came, as far as English-language literature is concerned, at the International Symposium on the Geology and Genesis of Precambrian Iron-Manganese Formations and Ore Deposits, held in Kiev in 1970, under the joint auspices of the Ukrainian Academy of Science, UNESCO, and the International Association for Geochemistry and Cosmochemistry. A five-man Ad hoc Committee on Nomenclature formed at that Symposium prepared an analysis of the nomenclature of the "banded ferruginous-cherty sedimentary rocks and their metamorphic equivalents", as used by the authors of the 38 papers presented at Kiev, with some supplementary comments by the committee. The resultant statement by the committee (referred to for convenience here as the "Kiev Symposium statement") appeared in three forms: in English as a letter to "Economic Geology" (Brandt et al., 1972a), concurrently in Russian (Brandt et al., 1972b), and finally again in English (UNESCO, 1973).

Each version of the Kiev Symposium statement is significantly different. The English texts of the 1972a and 1973 versions are essentially identical, but the latter incorporates also a supplementary table of English and Russian equivalents; that table does not appear in the 1972b version, but its content is included in additional text. Some text comment also appears in the 1972b version which is not present in either of the English-language (1972a, 1973) versions, and vice versa. Notwithstanding these differences, all three versions are content simply to expose the confusion of nomenclature at the time, to express the hope that "from this small beginning a coherent and internationally acceptable nomenclature for these rocks will eventually evolve", and to offer the sound counsel that "until it does, clear definitions of rock terms used in papers for international audiences, if only by reference to standard accessible publications, will prevent obscurities and misunderstandings".

The 1972a version of the Kiev Symposium statement was published specifically as a basis for further discussion at a subsequent symposium on "Precambrian Iron-formations of the World", held at Duluth in November 1972 under the joint sponsorship of the Society of Economic Geologists and the University of Minnesota. No explicit analysis of nomenclatural problems appears in the published proceedings of the Duluth Symposium (Econ. Geol., v. 68 (7)). However, the editorial adherence to the name iron-formation did at least establish the authority of this term more widely by comparison with such locally applied names as "itabirite", "BHQ", and "ironstone" (in South African usage).

Since 1973, the year which saw the publication of the papers presented at both the Kiev and Duluth Symposiums, the likelihood of early success in reaching a satisfactory international nomenclature for iron-rich sedimentary rocks has receded rather than increased, largely due to the appearance of a number of innovative suggestions which depart widely from all previously established nomenclature. Those of Dimroth (1975), Kimberley (1978), and Beukes (1980) are particularly notable.

Dimroth's (1975) nomenclature was initiated in an earlier paper (Dimroth, 1968) and applied in a description of the Sokoman Iron Formation of Labrador by Dimroth and Chauvel (1973). Arguing that the close similarity between the sedimentary textures and structures of cherty iron-formations, Minette-type iron-formations and limestones demonstrates that the mechanical processes responsible for the deposition of all these rock types were the same, Dimroth (1975) adapts the carbonate nomenclature of Folk (1962) for both textural description and palaeoenvironmental interpretation of iron-formation. In particular, terms such as micrite, oomicrite, biomicrite, and intramicrite are applied to corresponding components of iron-formation as femicrite, oofemicrite, biofemicrite, and intrafemicrite.

Kimberley (1978) introduced two new proposals. Firstly, troubled by the long-recognized anomaly that the term "iron-formation", either with or without a hyphen, has been applied both alone as a lithological descriptor and in combination as part of the name of a stratigraphic formation, Kimberley proposed that the term "ironstone" should be used to serve the former purpose, and "iron formation" (not hyphenated) the latter. "Ironstone", defined simply as a chemical sedimentary rock which contains over 15 per cent Fe, would thus directly replace the "iron-formation" of James (1954).

Secondly, Kimberley (1978) proposed a palaeoenvironmental classification and nomenclature of iron formations (i.e. mappable units composed mostly of ironstone as he used the term). In this system six acronyms — SVOP-IF, MECS-IF, SCOS-IF, DWAT-IF, SOPS-IF, and COSP-IF — derived from brief descriptions of depositional environments, were used to divide iron formations into six classes.

Most recently, Beukes (1980) proposed a comprehensive nomenclatural scheme that builds on Dimroth's earlier suggestion that carbonate rock nomenclature provides the best model for application to iron-formation. Beukes (1980) extended Dimroth's nomenclature to cover the full range of banded, granular, and intermediate textural types present in iron-formation of the Transvaal Supergroup of South Africa.

The first effort to develop a comprehensive nomenclature for the component bands of regularly banded iron-formation — BIF — was that of Trendall (1965). He proposed what appeared to be a hierarchy of three scales of band, named "macroband", "mesoband", and "microband", in order of decreasing thickness, for use in description of the BIFs of the Hamersley Group. Subsequent work on Hamersley Group BIFs (e.g., Ewers and Morris, 1980, 1981) has shown that these three terms, *as defined*, do not provide a sufficiently comprehensive set of names to describe the full range of band types present; in addition, there were logical shortcomings in the original definitions. As a consequence, Trendall et al. (in prep.) have suggested a revised scheme in which two sets of names are proposed. The "scale unit names" of the first set — metre band, decimetre band, centimetre band, millimetre band, and micron band — provide a simple and self-explanatory indication of the general order of scale of bands. A second set of hierarchical names is only loosely bound to actual band thickness, and is intended to reflect relationships between different types of band. It is based on the mesoband (which may be a decimetre band, a centimetre band, or a millimetre band) as the primary unit. Submesobands then form components within mesobands, while bands commonly comprising many mesobands are macrobands.

Further comments on the relative merits of these recent nomenclatural proposals appear beneath the following heading, together with implications for future use.

Difficulties and suggested resolutions

Nomenclatural systems for classes of rocks are adopted not because they are advocated with special force by particular authors, nor because an international committee has recommended along certain lines, but because their innate merits lead to their general acceptance after many years of practical application. For this reason I have some reluctance to make recommendations about nomenclature. At the same time, in the Introduction to a volume of this kind, which brings together a number of important contributions on iron-formation employing different approaches to nomenclature, there is a responsibility to do more than point out that this is so. Under this heading, therefore some comments on the nature of the difficulties involved in naming iron-formation are interspersed with personal views on their best resolution; these latter are clearly indicated by use of italic type.

Authors writing in English who seek to apply to iron-rich sediments a widely accepted and understood nomenclature which at the same time is logical and simple, face a number of difficulties, among them being:

(a) the ironstone/iron-formation problem;
(b) the "facies" problem;
(c) the nomenclature of components and textures;
(d) the avoidance of genetic implication.

The first of these is one of the simpler. Apart from its common application to near-surface concretionary iron-rich material, three quite different usages of "ironstone" have already been noted: it is the traditional South African name for iron-formation (e.g., Wagner, 1928), it is used to differentiate Phanerozoic from Precambrian iron-rich sediments (James, 1966), and it has been proposed as the lithological, as distinct from the stratigraphic, term for all such rocks (Kimberley, 1978). Kimberley's (1978) proposal appears to be based on the views that separate lithological and stratigraphical names are desirable for sedimentary rock types and that the "usage of 'iron-formation' as a lithologic term is inconsistent with the concept of a formation". Both views seem weakly founded. "Limestone" and "sandstone", among many others, effectively serve as both lithological and stratigraphic (formation) names. As far as the second point goes the word "formation" has well established antecedents for its application in such general senses as "any assemblage of rocks which have some character in common, whether of origin, age, or composition" (Lyell, 1858, quoted in Howell, 1960). Geologists need feel no constraint to avoid the use of "formation" in a general sense merely because of its subsequent restricted and codified use in formal stratigraphic nomenclature.

It is therefore recommended that the term "ironstone" be avoided wherever possible, that "iron-formation" be retained as the general lithological and stratigraphic term for iron-rich sedimentary rocks, in essentially the original sense of James (1954, not 1966). However, because the setting of a strict quantitative lower limit of iron content is arbitrary and restrictive, a suggested definition is "any sedimentary rock whose principal chemical characteristic is an anomalously high content of iron". The recommended abbreviation of iron-formation is IF, and of banded iron-formation, BIF.

The "facies" difficulty also appears to be a straightforward one. Although James (1954) specifically included the words "highly diagrammatic" in the caption to his fig. 3, its now classical cross-section became almost universally accepted as indicating a stratigraphically demonstrable and intergradational relationship between the four facies of iron-formation which he defined. The fact of the matter is that James (1954) never claimed to have observed such a relationship, and even stated that (James, p. 242) "it is doubtful if the pattern of precipitation indicated is ever actually obtained in nature . . .".

In view of the clear evidence from the Hamersley Basin that in some basins of iron-formation deposition abrupt changes from "oxide facies" to "silicate facies" (and vice versa) are basin-wide phenomena related to volcanicity rather than water depth, it seems wise to abandon the term "facies" in such combinations as "oxide facies iron-formation", insofar as the word is inevitably associated with a particular basinal model which is unlikely to be universally true. However, the four terms oxide iron-formation, carbonate iron-formation, sulphide iron-formation, and silicate iron-formation, remain good descriptive terms to indicate the dominant iron-bearing mineral present in particular iron-formations. Although they could potentially be applied stratigraphically they would be expected to be most useful as lithological terms, as applied by James (1954), for example, in his table headings.

The merits of Gross' (1965) "type" classification of iron-formations are best discussed by reference first to his Superior and Algoma types. Among other features typical of Algoma type Gross (1965, pp. 90—91) specified thin banding or lamination, with oolitic or granular textures absent or inconspicuous, lateral extent rarely more than a few miles, and an intimate association with various volcanic rocks. By contrast, granules and oolites were listed as a typical textural feature of Superior type iron-formations, together with a common lateral extent of hundreds of miles, and a close association with quartzite, black carbonaceous shale, conglomerate, dolomite, massive chert, chert breccia, and argillite. Good descriptions of the major iron-formations of the Hamersley, Cape Province and Transvaal basins were not available to Gross when these two types were proposed. He (Gross, 1973, 1980) subsequently classified the Hamersley Group iron-formations as Superior type. This could certainly be justified on the basis of their enormous lateral extent, but the "typical" granules and oolites are completely absent from these thinly banded iron-formations, which also have a closer association with volcanic rocks than with the sedimentary rock types listed by Gross. It is therefore not surprising that Dimroth (1975), regarded the Hamersley Group iron-formations as one of the best described examples of Algoma type. Kimberley (1978) has put forward other cogent objections to the type classification.

No suggestion is made here concerning future application of the terms Clinton type and Minette type: this volume has not added to knowledge of these rocks. However, it is recommended that the classification of iron-formations not falling into either of those categories into Algoma and Superior types is discontinued. This demonstrably subjective distinction is not only inadequate to accommodate, without distortion, many major iron-formations of the world, but imposes an artificial two-fold division not present in the complex spectrum of rocks it seeks to cover. Gross' (1965, p. 83) view that "the use of type names can thus be more misleading than helpful", is endorsed.

The last two nomenclatural difficulties listed above are harder to resolve and are to some extent inter-related, and are therefore discussed together. While it is easy to arrive at a clear recommendation that iron-formation (IF) and banded iron-formation (BIF) be applied to the exclusion of other alternatives, so little progress has been made on the nomenclature of components of iron-formation that only the passage of time will determine which, if any, current suggestions will be widely accepted.

For BIF, Trendall's (1965) band nomenclature has been adopted widely by others, but later revision has been necessary (Trendall et al., in prep.). However, these suggestions were devised mainly to indicate the scale and relationship of different bands; the question of what the material of different bands should be called was subordinate, and is not satisfactorily resolved. While "chert" is universally accepted as a name for quartz-rich bands there is no agreement on a name for iron-rich bands. Trendall's (1965) early suggestion of QIO (quartz-iron oxide) was later changed to "chert-matrix" (Trendall and Blockley, 1970), but this name is not entirely satisfactory, and later workers (e.g., Ewers and Morris, 1981) have tended to prefer the simple "iron-rich mesoband". Alternatives to refer to the actual material of such mesobands would presumably be the "femicrite" of Dimroth (1968) or "felutite" of Beukes (1980). However, either of these names involve the difficulty of genetic implication, femicrite implying that the material is genetically analogous in all respects with the micrite of limestone, and felutite having a similar connotation in respect to clastic silt. Nevertheless, Beukes' (1980) very detailed scheme is at present the only one specifically designed to cover comprehensively the full range of textures and their components within iron-formation.

A strong implication of genesis, or at least of genetic environment, is also present in Kimberley's (1978) palaeoenvironmental classification of iron-formations, referred to earlier; few, if any, iron-formations are at present sufficiently well understood to be classifiable with confidence into Kimberley's divisions, and it seems unlikely that his system will be widely adopted.

Relationship between English and Russian nomenclature

A detailed commentary on the nomenclature of iron-rich sedimentary rocks in the Russian language literature is not attempted here; some notes on this are available on request to the author. However, since this volume includes a paper on one of the classic Russian iron-formations a note is appropriate on the problems involved in finding the best equivalence between Russian and English nomenclature.

These problems come from two main sources. Firstly, Russian language nomenclature of iron-formation has been historically just as diverse and contentious as in English. Secondly, as in all translation, translators into English of terms applied to Russian iron-rich sediments must face the dilemma of choice between "correct", literal, translations of the words used, or terms most closely equivalent to those which would have been used and applied by English-speaking geologists to the same rocks. In the latter case the main judgement required is geological rather than linguistic, while in the former case the situation is reversed; in practice the choice often depends on the principal skill of the translator.

For Krivoy Rog the dilemma is particularly acute, since an early transfer of Van Hise and Leith's (1911) term "ferruginous chert" for iron-formation of the Lake Superior area to the Krivoy Rog region was incorrectly made in the form "rogovik" (Pyatnitskiy, 1925). "Rogovik", much used since that time in Krivoy Rog literature to denote banded iron-formation, or BIF, is most "correctly", but to English-speaking geologists quite incomprehensibly, rendered as "hornfels" (see UNESCO, 1973) or archaically and inaccurately, as "hornstone" (N. Rast, in Nalivkin, 1973). In Chapter 5 by Belevtsev et al. (p. 211) the second course has been followed, and the following equivalents between the nomenclature of the original Russian text and the final English one were adopted:

Russian	English
zhelezorudnaya formatsiya	iron-formation
zhelezistyy rogovik	oxide iron-formation
silikatnyy rogovik	silicate iron-formation
zhelezisto-silikatnyy rogovik	oxide-silicate iron-formation
dzhespilit	jaspilite

This system differs somewhat from the classification and nomenclature of Semenenko (1973, 1978) used by Mel'nik (1982).

ACKNOWLEDGEMENTS

For help of various kinds in connection with the production of this volume R.C. Morris and I gratefully acknowledge the help of the following: J.A. Morris, J.R. Gray, Ray Connolly, Nell Stoyanoff, Elizabeth Amann, Mrs Manjeet Kumar, and a number of anonymous scientists who kindly gave time to act as referees.

REFERENCES

Beukes, N.J., 1973. Precambrian iron-formations of Southern Africa. Econ. Geol., 68 (7): 960—1004.

Beukes, N.J., 1980. Suggestions towards a classification and nomenclature for iron-formation. Trans. Geol. Soc. S. Afr., 83: 285—290.

Brandt, R.T., Gross, G.A., Gruss, H., Semenenko, N.P. and Dorr, J.V.N., 1972a. Problems of nomenclature for banded ferruginous-cherty sedimentary rocks and their metamorphic equivalents. Econ. Geol., 67 (5): 682—684.

Brandt, R.T., Dorr, Dzh.V., Gross, G.A., Gruss, G. and Semenenko, N.P., 1972b. Problemy nomenklatury poloschatykh zhelezisto-kremnistykh osadochnykh porod i ikh metamorficheskikh ekvivalentov. In: Geologiya i genezis dokembriyeskikh zhelezisto-kremnistykh i margantsevykh formatsiy mira (Geology and genesis of Precambrian ferruginous-siliceous and manganiferous formations of the world). Naukova Dumka, Kiev, pp. 380—384 (in Russian).

Dimroth, E., 1968. Sedimentary textures, diagenesis and sedimentary environment of certain Precambrian ironstones. Neues Jahrb. Geol. Palaeontol., Abh., 130: 247—274.

Dimroth, E., 1975. Paleo-environment of iron-rich sedimentary rocks. Geol. Rundsch., 64 (3): 751—767.

Dimroth, E. and Chauvel, J.-J., 1973. Petrography of the Sokoman Iron Formation in part of the Central Labrador trough. Geol. Soc. Am. Bull., 84: 111—134.

Ellis, H.A., 1939. The geology of the Yilgarn Goldfield, south of the great eastern railway. West. Aust., Geol. Surv., Bull. 97.

Ewers, W.E. and Morris, R.C., 1980. Chemical and mineralogical data from the uppermost section of the upper BIF member of the Marra Mamba Iron Formation. CSIRO, Perth, Rep. FP23.

Ewers, W.E., and Morris, R.C., 1981. Studies on the Dales Gorge Member of the Brockman Iron Formation. Econ. Geol., 76 (7): 1929—1953.

Feldtmann, F.R., 1921. The geology and mineral resources of the Yalgoo Goldfield. Pt. 1, The Warriedar gold-mining centre. West. Aust., Geol. Surv., Bull. 81.

Folk, R.L., 1962. Spectral subdivision of limestone types. In: W.D. Ham (Editor), Classification of Carbonate Rocks. Mem., Am. Assoc. Pet. Geol., 1: 62—84.

Gross, G.A., 1959. A classification of iron deposits in Canada. Can. Min. J., 80 (10): 87—92.

Gross, G.A., 1965. Geology of iron deposits in Canada, vol. 1 — General geology and evaluation of iron deposits. Geol. Surv. Can., Econ. Geol. Rep. 22, 181 pp.

Gross, G.A., 1973. The depositional environment of the principal types of Precambrian iron-formations. In: Genesis of Precambrian Iron and Manganese Deposits. UNESCO, Paris, Earth Sciences, 9: 15—21.

Gross, G.A., 1980. A classification of iron formations based on depositional environments. Can. Mineral., 18: 215—222.

Gruss, H., 1973. Itabirite iron ores of the Liberia and Guyana shields. In: Genesis of Precambrian Iron and Manganese Deposits. UNESCO, Paris, Earth Sciences, 9: 335—359.

Howell, J.V. (Chairman), 1960. Glossary of Geology and Related Sciences. American Geological Institute, Washington D.C., 2nd Ed., 325+72 pp.

James, H.L., 1954. Sedimentary facies of iron-formation. Econ. Geol., 49: 235—293.

James, H.L., 1966. Data of geochemistry (6th ed.). W. Chemistry of iron-rich sedimentary rocks. U.S. Geol. Surv. Prof. Pap. 440-W.

Kimberley, M.M., 1978. Paleoenvironmental classification of iron formations. Econ. Geol., 73: 215—229.

Krishnan, M.S., 1973. Occurrence and origin of the iron ores of India. In: Genesis of Precambrian Iron and Manganese Deposits. UNESCO, Paris, Earth Sciences, 9: 69—76.

Mel'nik, Y.P., 1982. Precambrian Banded Iron-Formations, Physicochemical Conditions of Formation. Elsevier, Amsterdam, 310 pp.

Nalivkin, D.V., 1973. Geology of the U.S.S.R. Oliver and Boyd, Edinburgh, 855 pp.

Pyatnitskiy, P.P., 1925. Genetic relationships of the Krivoy Rog ore deposits. Vol. 1, Iron-formations and jaspilites. Trudy. Inst. Priklad. Mineralogii i Petrografii (Trans. Inst. Econ. Mineral Petrography), Kharkov, 17: 42 pp. (in Russian).

Semenenko, N.P., 1973. The iron-chert formations of the Ukrainian shield. In: Genesis of Precambrian Iron and Manganese Deposits. UNESCO, Paris, Earth Sciences, 9: 135—142.

Semenenko, N.P., 1978. Obshchaya kharakteristika zhelezisto-kremnistykh formatisiy Ukrainskogo shchita kak zhelezorudnoy bazy. In: N.P. Semenenko (Editor), Zhelezisto-kremnistye formatsii Ukrainskogo shchita. T.l (Iron-formations of the Ukrainian shield, vol. 1). Naukova Dumka, Kiev, pp. 7—41 (in Russian).

Sims, S.J., 1973. The Belinga iron ore deposit (Gabon). In: Genesis of Precambrian Iron and Manganese Deposits. UNESCO, Paris, Earth Sciences, 9: 323—334.

Trendall, A.F., 1965. Progress report on the Brockman Iron Formation in the Wittenoom-Yampire area. West. Aust., Geol. Surv., Annu. Rep., 1964: 55—65.

Trendall, A.F., 1968. Three great basins of Precambrian banded iron formation deposition: a systematic comparison. Geol. Soc. Am. Bull., 79: 1527—1544.

Trendall, A.F. and Blockley, J.G., 1970. The iron formations of the Precambrian Hamersley Group, Western Australia, with special reference to the associated crocidolite. West. Aust., Geol. Surv., Bull. 119, 366 pp.

Trendall, A.F., Morris, R.C., and McConchie, D., in prep., Classification of "bands" and "banding" in banded iron-formation (BIF).

UNESCO, 1973. Problems of nomenclature for banded ferruginous-cherty sedimentary rocks and their metamorphic equivalents. In: Genesis of Precambrian Iron and Manganese Deposits. UNESCO, Paris, Earth Sciences, 9: 377—380.

Van Hise, C.R., and Leith, C.K., 1911. The geology of the Lake Superior region. U.S. Geol. Surv., Mon. 52, 641 pp.

Wagner, P.A., 1928. The iron deposits of the Union of South Africa. Geol. Surv. S. Afr., Mem. 26, 268 pp.

Chapter 2

ANIMIKIE BASIN, LAKE SUPERIOR REGION, U.S.A.

G.B. MOREY

INTRODUCTION

The Lake Superior region (Fig. 2-1) contains numerous bodies of iron-formation that have been important sources of iron ore for over 130 years. The region has shipped 4.6 billion metric tons since mining started in 1848. In 1978, the region produced 75 million metric tons of ore or 89% of the total ore produced in the United States and 10% of the total produced in the world.

Iron ores have been produced in the region at one time or another from rocks ranging in age from the late Archean to Late Cretaceous. However, slightly more than 96% of the iron ore has been derived from strata of early Proterozoic (ca. 2500—1600 m.y.) age. These rocks still contain vast resources; it has been estimated that more than 271 billion metric tons of crude iron ore or 36 billion metric tons of iron-ore concentrate are recoverable from Minnesota, Wisconsin, and Michigan by present-day technological methods (Marsden, 1978a, b; Cannon et al., 1978).

The lower Proterozoic rocks occur in a broad intracontinental basin that underlies much of east-central and northern Minnesota, adjacent parts of Ontario, northern Wisconsin, and the northern peninsula of Michigan (Fig. 2-2). This basin has been informally termed the Animikie basin (e.g., Trendall, 1968; Sims, 1976) because most strata in it were once assigned to the so-called "Animikie Series" of James (1958). The term Animikie Series has now been abandoned as a lithostratigraphic descriptor, but the term Animikie basin has been retained for convenience.

In terms of surface exposures the Animikie basin occurs at the southern extremity of the Canadian Shield where it forms a major part of the so-called "Southern province" (Stockwell et al., 1970) or the "Hudsonian foldbelt" on the "Tectonic Map of North America" (King, 1969). However, if the subsurface geology is considered, the basin occurs near the center of the known Precambrian basement of the North American craton (Fig. 2-1).

Rocks of the Animikie basin crop out in an oval-shaped area having a major east-trending axis of about 700 km and a minor axis probably about 400 km, giving an area of about 220 000 km^2. The original basin may have been much larger inasmuch as parts of it have been removed by erosion, and other parts are covered by younger Proterozoic and Phanerozoic strata.

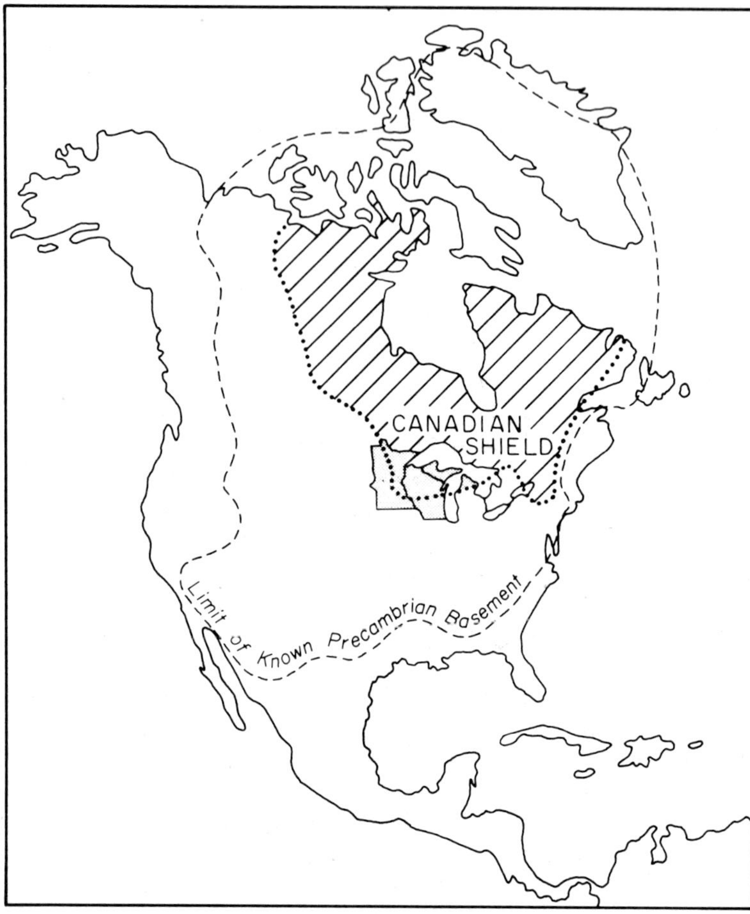

Fig. 2-1. General map of the North American continent showing the location of the Lake Superior region relative to the Canadian Shield and to known or inferred Precambrian basement rocks of the North American craton.

REGIONAL GEOLOGIC SETTING

Archean rocks of two contrasting types, which differ in age, rock assemblages, metamorphic grade and structural style (Morey and Sims, 1976), form the basement for the supracrustal rocks of the Animikie basin. Greenstone-granite complexes of late Archean (2750—2600 m.y.) age, which are typical of most of the southern part of the Superior province (Peterman, 1979), underlie the northern part of the basin. In contrast, migmatitic gneiss and amphibolite, in part about 3600 m.y. old, underlie the southern part of the basin. The type area for the gneiss terrane is the Minnesota River Valley

in southwestern Minnesota (Sims and Peterman, 1981); the type area for the greenstone-granite terrane is the Vermilion district of northeastern Minnesota (Morey, 1980).

The two Archean terranes are juxtaposed along a major crustal feature — the Great Lakes tectonic zone of Sims et al. (1980) — extending eastward more than 1200 km from central South Dakota to the Grenville front in eastern Ontario. The boundary was initiated in late Archean time when the two crustal segments were joined into a single, large continental block. Although the two segments were tightly juxtaposed thereafter, the boundary was the focus of major crustal movements during early Proterozoic time.

The Animikie basin was one of several basins that formed over and approximately parallel to the Great Lakes tectonic zone (Fig. 2-2). Much of the Animikie basin is filled with sedimentary rocks of clastic origin, and it contains nearly all of the commercially exploited iron-formations of the Lake Superior region as well as appreciable quantities of volcanic rocks, particularly in northern Michigan. In general the rocks in the basin record a complete transition from that of a "stable craton" to that of a "eugeosynclinal" environment (Bayley and James, 1973). However, the nature of the lithic fill varies considerably from place to place, and this variability appears to be related, at least in part, to contrasting kinds of Archean basement rocks (Sims, 1976).

The rocks of the Animikie basin are bounded on the south by a possibly coeval sequence of dominantly mafic to felsic volcanic rocks that form an east-trending belt across much of north-central Wisconsin (Fig. 2-2). Similar volcanic rocks also occur as a fault-bounded block in south-central Wisconsin. These volcanic sequences are poorly exposed and nowhere interlayered with strata of the Animikie basin; they have not yet been integrated with the better known succession of the basin proper.

Sedimentation and volcanism within the Animikie basin were terminated or closely followed by a tectono-thermal event, the Penokean orogeny of Goldich et al. (1961). The resulting deformation and metamorphism of the supracrustal rocks was most intense over the gneiss terrane where vertical tectonic processes led to the development of a number of mantled gneiss domes (Sims, 1976; Morey, 1978b) and fault-bounded anticlinal blocks (Cannon, 1973). A second distinctive feature of the Penokean orogen is the presence of a number of metamorphic nodes characterized by the progressive appearance of biotite, garnet, staurolite and sillimanite (James, 1955; Morey, 1979) on the flanks of the reactivated blocks or domes of gneissic basement rock.

Toward the end of the Penokean orogeny and mainly after its termination, plutons of generally calcalkaline affinity, ranging in composition from tonalite or diorite to granite, intruded the stratified and volcanic rocks, particularly in east-central Minnesota and northeastern Wisconsin. In general, the older and more mafic plutons are somewhat deformed or cataclasized, whereas the younger granitic plutons are generally undeformed. Younger, post-Penokean rhyolite and coeval epizonal granite (Smith, 1978; Van Schmus, 1978) and platform quartzite (Dott and Dalziel, 1972) were deposited generally south of the erosional edge of the lower Proterozoic rocks in the interval 1760 m.y. to 1600 m.y. (Van Schmus, 1980). A large anorogenic intrusion of granite and anorthosite, the Wolf River batholith, was emplaced in central Wisconsin about 1500 m.y. ago (Van Schmus et al., 1975a).

About 1100 m.y. ago, a north-northeast-trending branch of the Midcontinent rift system separated the Animikie basin into two distinct segments. A second, generally north-northwest-trending branch of this rift system also truncated what is now the eastern end of the basin. Lastly, much of the Lake Superior region is covered by generally flat-lying sedimentary rocks of Phanerozoic age.

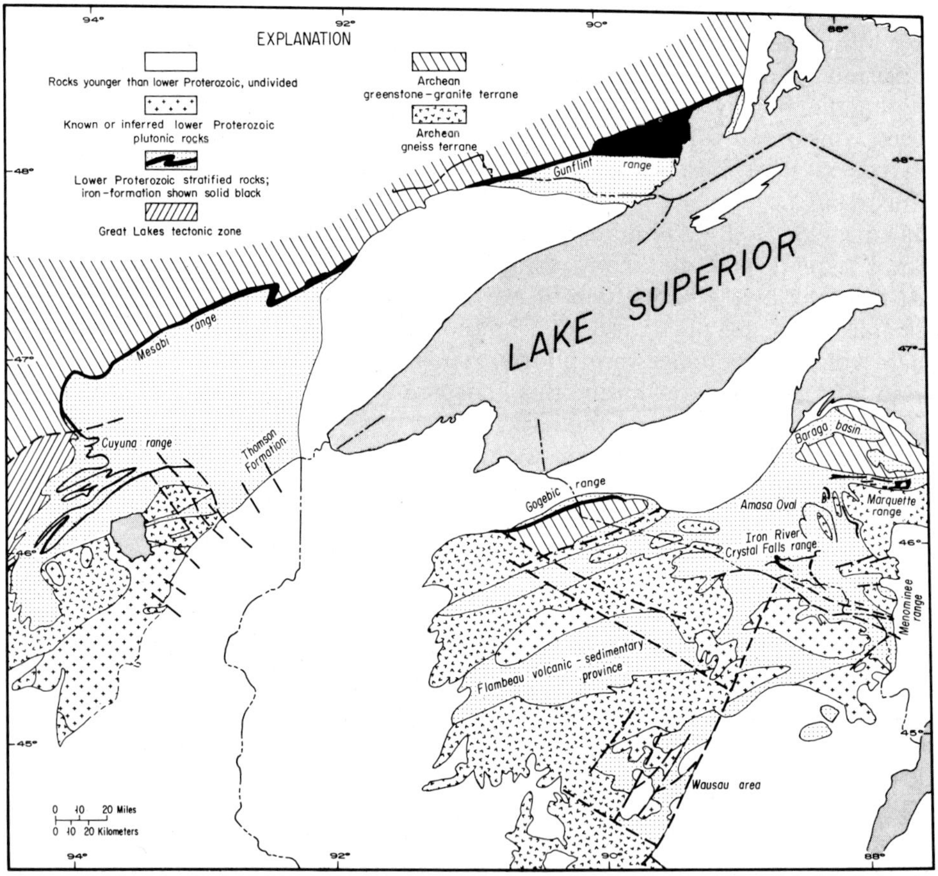

Fig. 2-3. Generalized geologic map showing the locations of the major iron ranges with respect to the principal geologic features of the Animikie basin.

Folding and an extensive cover of Pleistocene drift together restrict outcrops of the Animikie iron-formations to a number of separate "iron ranges." These include the Gunflint, Mesabi and Cuyuna ranges of Minnesota and adjoining parts of Ontario, and the Gogebic, Marquette, Menominee, and Iron River—Crystal Falls ranges of northern Wisconsin and Michigan (Fig. 2-3).

GEOCHRONOLOGY

A major stratigraphic problem concerning the lower Proterozoic stratified rocks of the Animikie basin has been their possible correlation with the Huronian Supergroup on the north shore of Lake Huron, but the Animikie basin strata are now considered to be younger (Van Schmus, 1976). In parts of northern Michigan they unconformably overlie Archean basement rocks that were affected by a metamorphic episode at about 2000 m.y. ago (Banks and Van Schmus, 1971, 1972; Van Schmus and Woolsey, 1975). Elsewhere in northern Michigan the supracrustal rocks unconformably overlie "granitic rocks" (James et al., 1961) yielding a zircon U-Pb concordia intercept age of approximately 2060 m.y. (Banks and Van Schmus, 1971). In northern Minnesota, the lower Proterozoic stratified rocks unconformably overlie approximately 2000-m.y.-old dike rocks (Hanson and Malhotra, 1971), and somewhat older stratified rocks in east-central Minnesota unconformably overlie a migmatitic terrane affected by a metamorphic event 2100 to 2000 m.y. ago (Goldich, 1973). These events in the Archean basement rocks correspond to the approximate time of a tectono-thermal event that affected rocks of the Huronian Supergroup (Van Schmus, 1965, 1976; Fairbairn et al., 1969; Gibbins and McNutt, 1975).

Lower Proterozoic rocks of the Animikie basin were metamorphosed during a major tectono-thermal event — the Penokean orogeny of Goldich et al. (1961) — at about 1860 ± 50 m.y. (Aldrich et al., 1965; Peterman, 1966; Van Schmus, 1976). Metamorphosed igneous rocks within the sequence have been dated at about 1860 ± 25 m.y. (Aldrich et al., 1965), and Banks and Van Schmus (1971, 1972) have obtained a U-Pb age of about 1920 ± 25 m.y. for zircon from a rhyolite unit intercalated in the stratified sequence. Thus the lower Proterozoic rocks were deposited between 2100 m.y. and 1850 m.y. ago, and therefore are younger than the Huronian Supergroup.

A second major stratigraphic problem concerns the position of the mafic to felsic volcanic rocks and associated sedimentary rocks that crop out over much of north-central, west-central and southern Wisconsin (Fig. 2-2). This volcanic-sedimentary sequence has geologic attributes very similar to the Archean greenstone belts of northern Minnesota. However, the volcanic rocks in north-central Wisconsin have yielded a zircon U-Pb age of about 1850 m.y. (Banks and Rebello, 1969) and similar volcanic rocks from west-central and southern Wisconsin have yielded similar ages (Van Schmus et al., 1975b;

Van Schmus, 1980). Model lead ages from two massive sulfide deposits in the volcanic-sedimentary belt in north-central Wisconson yielded values of about 1820 ± 50 m.y. (Stacey et al. as cited in Sims, 1976) and U-Pb ages of 1859 ± 20 m.y. (Van Schmus, 1980). The model lead values have been interpreted as primary ages mainly because the sulfide deposits from which they are derived are believed to be synvolcanic in origin (May, 1976; Schmidt et al., 1978).

The volcanic rocks were intruded by a number of granitic rocks presumably during the later stages of the Penokean orogeny. These include tonalitic rocks in south-central Wisconsin at 1842 ± 10 m.y. and 1824 ± 25 m.y. (Van Schmus, 1980), granodioritic and granitic rocks in northeastern Wisconsin at 1840 to 1820 m.y. (Banks and Cain, 1969), and granitic to tonalitic rocks at 1885 ± 65 m.y. in north-central Wisconsin (Sims and Peterman, 1980). Although the stratigraphic position of the volcanic-sedimentary sequence is equivocal relative to the stratigraphic position of the rocks in the Animikie basin proper, all of the geochronometric data indicate that the two sequences are broadly correlative.

DOCUMENTATION OF THE BASIN

Literally hundreds of papers have been written about various aspects of the Animikie basin and its contained iron-formations (Fig. 2-4) during 126 years of work by many geologists of diverse disciplines. It is possible to define four phases in the evolution of our understanding of the geology of the Animikie basin (Table 2-I). The first phase, from the discovery of iron-formation in northern Michigan in 1844 until about 1911 when mining was underway in all major ranges of the region, was characterized by exploration for commercial-size concentrations of high-grade ore within the iron-formations. Because of their economic importance, the high-grade ore deposits were studied in great detail, but mainly as to distribution, structural controls, mineralogy, grade, and tenor.

By around 1904 all major iron-formations in the Animikie basin had been found, mapped, and described in some detail (Table 2-II). The exploration period culminated with the publication of U.S. Geological Survey Monograph 52, "The Geology of the Lake Superior Region" (Van Hise and Leith, 1911), which described the geology and ore deposits of the individual iron-bearing districts in great detail. Van Hise and Leith also established many of the interrange and intrarange correlations that are in use today, and proposed several theories to explain the origin of the iron-formations and their high-grade ore deposits.

The extensive geologic studies of the period from 1844 to 1904 led to the belief that iron-ore reserves were sufficient to support national needs for many years. Therefore the period from 1912 to the end of World War II was characterized by relatively little interest in iron-formation geology on the part of the mining companies and the federal and state surveys. However, during that period the U.S. Geological Survey published a geologic map of the Lake Superior region (Leith et al., 1935) that revised many of the inter-range correlations first proposed by Van Hise and Leith (1911). Also during that period, John W. Gruner published several papers on various aspects of iron-formation stratigraphy and petrology (Gruner, 1924, 1933, 1946), particularly on the Mesabi range. Gruner's work also included discussions as to the origin of iron-formations by weathering processes (1924), the origin of the iron ores by hydrothermal (1926, 1930) or mixed hydrothermal/meteoric processes (1937b), the paragenetic relationships between magnetite, martite and hematite in natural ore bodies (1926), and the structures and compositions of greenalite (1936), stilpnomelane (1937a) and minnesotaite (1944).

TABLE 2-I

Exploration phases in the study of the Animikie basin

I. *Natural ore exploration period 1840—1911*
(Rapidly increasing natural ore reserves)
1. Identification of major iron ranges
2. Geological mapping of each range
   a. Distribution of natural ore bodies
      1. Structural/stratigraphic controls
      2. Mineralogy, grade, tenor
3. Origin of natural ore bodies

II. *Natural ore mining period 1911—1945*
(Stable to gradually declining natural ore reserves; abrupt decline in reserves during World War II)
1. Origin of natural ore bodies
2. Origin of iron-formations
3. Mineralogy and crystal chemistry of iron-formations and ore minerals

III. *Iron-formation exploration period 1945—1970*
(Rapidly declining natural ore reserves; rapidly increasing "taconite" reserves)
1. Remapping of major iron ranges
   a. Stratigraphy of the iron-formations
   b. Mineralogy and chemistry of iron-formations
   c. Diagenetic and metamorphic modification
2. Sedimentary facies of iron-formations
   a. Interrelationships between textural attributes and facies
3. Origin of iron-formations

IV. *Present phase 1970—*
(Abundant taconite reserves)
1. Regional geological syntheses
2. Geochronology and isotopic chemistry
3. Implications as to the evolution of the biosphere, hydrosphere, and atmosphere

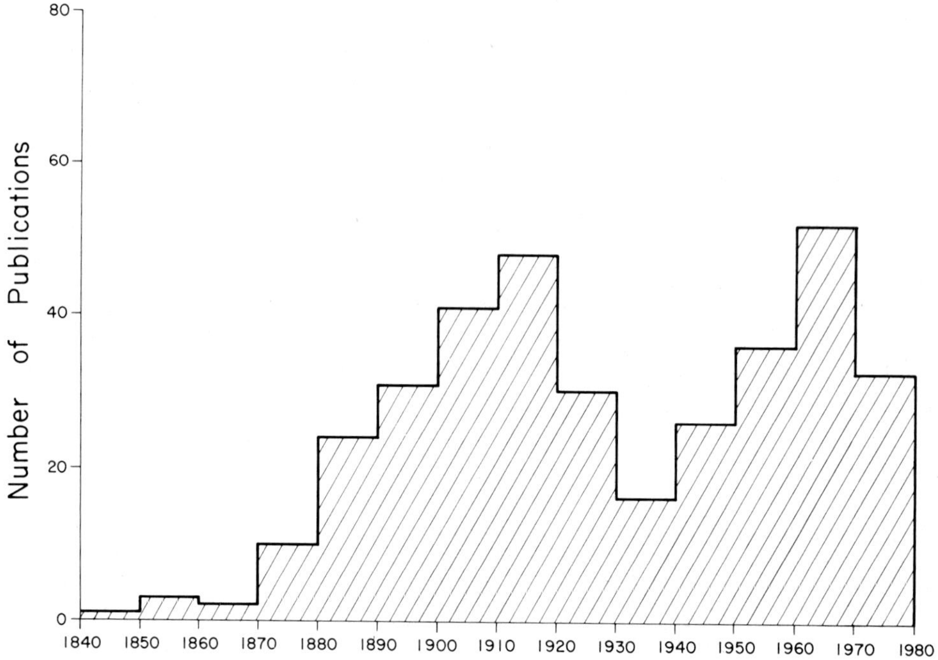

Fig. 2-4. Publications regarding iron-formations of the Lake Superior region by 10-year intervals from 1840 to 1980.

TABLE 2-II

Exploration history of the iron-mining ranges in the Animikie basin

Range	Discovery		Mining started
	Iron-formation	Ore	
Marquette	1844	1845	1848/1856[1]
Menominee	1848	1874	1877
Gogebic	1849	1880	1884
Gunflint	1850	—[2]	—
Iron River/Crystal Falls	1855	1880	1882
Mesabi	1886	1890	1892
Cuyuna	1893	1904	1911

[1] Mining from 1848 to 1852 was financially unsuccessful; continuous mining did not occur until 1856.

[2] No secondary enrichment deposits of commercial size have been found, nor is the iron-formation a source of magnetite.

National needs during World War II greatly depleted the high-grade ore reserves of the Lake Superior region. Consequently, new mining and beneficiation techniques were developed to use the iron-formations themselves as a source of iron. These processes were based on the magnetic character of magnetite and thus required a very comprehensive knowledge as to its distribution within any given iron-formation. As a consequence all of the iron ranges of the region were reevaluated during the period from 1946 to about 1970, with emphasis on the sedimentological, diagenetic and metamorphic attributes of the various iron-formations. These studies led directly to the development of facies concepts in iron-formations (James, 1954), to several detailed sedimentological studies of specific iron-formations (White, 1954; Goodwin, 1956; Huber, 1959; James, 1966), to the detailed chemistry of iron-formations (Lepp, 1963, 1966, 1968), and to renewed interest in the origin of the iron-formations (Lepp and Goldich, 1964).

Since about 1970, interest in the iron-formations *per se* has again declined and has been replaced by an interest in the sedimentological regime of the Animikie basin as a tectonic entity. Several regional syntheses have resulted, including those of Morey (1973a, 1979), Bayley and James (1973), Sims (1976). LaRue and Sloss (1980), and Sims et al. (1981). These studies were made possible in part by the radiometric studies of Goldich et al. (1961), Goldich (1968), Van Schmus (1976) and numerous others, whose data provide the chronometric framework within which correlations can be made.

DESCRIPTION OF THE BASIN

Because rocks of the Midcontinent rift system separate the Animikie basin into two physically isolated segments, the strata in the northwestern segment are assigned to the Animikie and Mille Lacs Groups (Morey, 1973a, 1978a), whereas those in the southeastern segment are assigned to the Marquette Range Supergroup (Cannon and Gair, 1970). Although the two sequences are lithologically similar and therefore broadly correlative, the Marquette Range Supergroup is thicker, more diverse, and interrupted by numerous unconformities which divide it into the Chocolay, Menominee, Baraga, and Paint River Groups (James, 1958, p. 30). Although physical continuity between the rocks of the several ranges in both segments has not been firmly established, the stratigraphic successions in each of the iron-mining ranges have been correlated as shown in Fig. 2-5 (James, 1958; Bayley and James, 1973; Sims, 1976).

Northwestern segment

The lower Proterozoic stratified rocks in the northwestern segment of the Animikie basin have been divided into the Animikie and Mille Lacs Groups

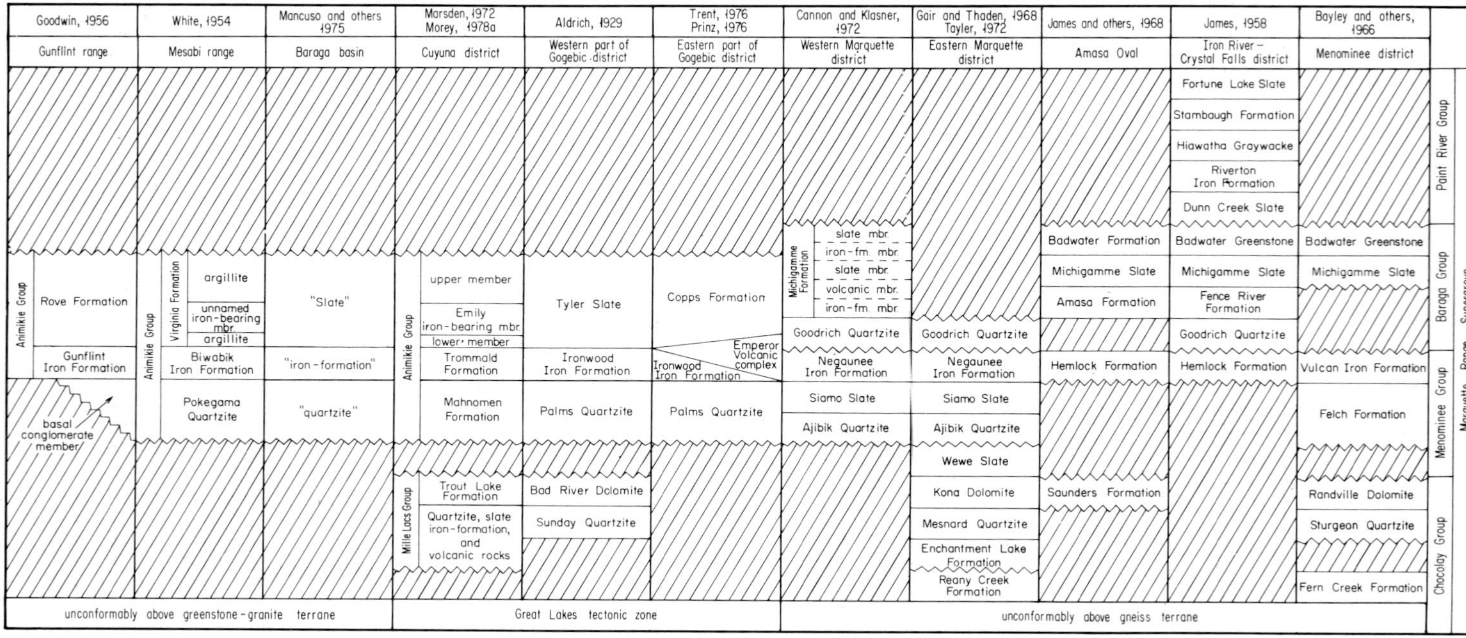

Fig. 2-5. Correlation chart of lower Proterozoic bedded rocks in the Lake Superior region. See Fig. 2-3 for locations of individual ranges.

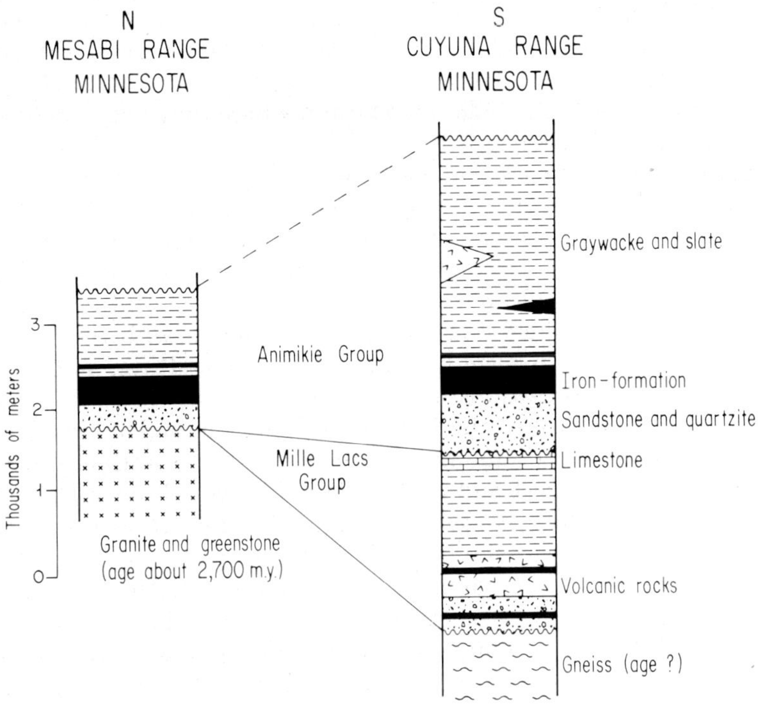

Fig. 2-6. Selected stratigraphic sections in the northwestern segment of the Animikie basin. Note the extensive development of lower Proterozoic rocks in the Cuyuna range where they overlie Archean basement rocks of the Great Lakes tectonic zone.

(Morey, 1978a). Rocks of the Animikie Group are no more than 750—800 m thick where they unconformably overlie Archean greenstone and granite on the Mesabi and Gunflint ranges (Fig. 2-6). However, in east-central Minnesota, and particularly over the Great Lakes tectonic zone, the Animikie Group is at least 1 km thick, and is underlain by a sequence of strata at least 1 km thick assigned to the Mille Lacs Group (Morey, 1978a).

As defined by Morey (1978a), the Denham Formation at the base of the Mille Lacs Group consists dominantly of quartz-rich conglomerate and sandstone of arenitic affinity, dolomite, and lesser amounts of oxide-facies iron-formation and subaqueous volcanogenic rocks of mafic to intermediate composition. In places the Denham Formation passes laterally into large, thick bodies of mafic to intermediate subaqueous volcanic rocks intercalated with appreciable quantities of carbonate-facies iron-formation and pyrite-rich, carbonaceous argillite; these dominantly volcanic sequences have been named the Glen Township and Randall Formations. All of the named units pass gradationally upward into a thick sequence of interbedded quartz-rich wacke, siltstone and shale with lesser amounts of volcanogenic rocks named the Little Falls Formation. Lenses and beds of impure dolomite or limestone also are present throughout the Little Falls Formation, but are particularly abundant in the upper part of the Mille Lacs Group where they compose the Trout Lake Formation of Marsden (1972).

An unconformity separates rocks of the Mille Lacs Group from those of the overlying Animikie Group on the Cuyuna range (Marsden, 1972). The Animikie Group there and on the Mesabi and Gunflint ranges represents a single cycle, starting with well-sorted clastic detritus (Kakabeka, Pokegama and Mahnomen formations), followed by a major phase of iron-formation (Gunflint, Biwabik and Trommald formations), and ending with fine sand and mud characteristic of a deep basin with poor circulation (Rove, Virginia, Rabbit Lake and Thomson Formations) (Morey, 1973a).

The well-sorted clastic detritus of the Kakabeka, Pokegama and Mahnomen formations is no more than several meters thick on the Gunflint range (Goodwin, 1956) and at the east end of the Mesabi range (Gundersen and Schwartz, 1962). However, the Pokegama Quartzite thickens to as much as 100 m at the westernmost end of the Mesabi range and to at least 600 m to the south on the Cuyuna range. Apparently subsidence was relatively greater in the southern part of the basin, particularly over the Great Lakes tectonic zone and the gneiss terrane, and sedimentation more or less kept pace with subsidence because these rocks constitute a southward-thickening wedge of fine-grained detritus fringed by a thin strandline deposit of sandstone and conglomerate (Morey, 1973a).

The overlying Gunflint, Biwabik, and Trommald formations generally are 100 to 200 m thick and are characterized by intercalated lithotopes indicative of shallow and "deeper water deposition" (Morey, 1973a). Volcanic rocks, mainly of pyroclastic origin, are sparingly present on the Mesabi range (French, 1968), but pyroclastic and extrusive volcanic rocks occur in the iron-formation on the Gunflint (Goodwin, 1956) and Cuyuna ranges (Schmidt, 1958).

Crustal instability ultimately led to the cessation of iron-formation deposition and to the accumulation of more than 1000 m of intercalated carbonaceous mudstone and siltstone assigned to the Rove, Virginia, Rabbit Lake and Thomson Formations. The lower 60 m or so of this depositional phase is characterized by black, carbonaceous shale and siltstone, iron-formation (e.g., the Emily Iron-formation Member of the Rabbit Lake Formation on the Cuyuna range), and igneous material including pyroclastic deposits, flows and thin hypabyssal dikes or sills. Much of this sequence, however, contains appreciable quantities of graywacke, deposited by southward-flowing turbidity currents (Morey, 1969; Morey and Ojakangas, 1970), and beds of carbonaceous, sulfide-facies iron-formation, mafic tuff, lava flows and coeval diabasic intrusions (Morey, 1978a).

Southeastern segment

Stratified rocks over the greenstone-granite basement in the northern and western parts of the southeastern segment of the Animikie basin have stratigraphic relationships similar to those in the northwestern segment (Fig. 2-5). For example, Menominee Group strata in the Baraga basin (Fig. 2-3) of northern Michigan are similar to the Animikie Group (Mancuso et al., 1975). In the western part of the Gogebic range in Wisconsin, the rocks of the Menominee Group are remarkably similar to those of the Animikie Group in that they represent a shallow-water (Palms Quartzite), to iron-formation (Ironwood Iron Formation) to deeper-water (Tyler Formation) depositional sequence (Schmidt, 1980). Moreover, these rocks overlie erosional remnants of quartzite (Sunday Quartzite) and dolomite (Bad River Dolomite) assigned to the Chocolay Group. Thus the Animikie and Menominee Groups and the Mille Lacs and Chocolay Groups are lithologically similar, and this similarity provides much of the evidence for correlation of strata between the northwestern and southeastern segments.

Stratigraphic relationships in the eastern part of the southeastern segment are more complex than in the western part of the Gogebic range (Fig. 2-7). Near the Wisconsin—Michigan border for example, the Ironwood Iron Formation is interlayered with a thick sequence of mafic volcanic rocks, the Emperor Volcanic Complex of Trent (1976), which in turn is unconformably overlain by a multifacies unit consisting dominantly of graywacke and slate named the Copps Formation (Allen and Barrett, 1915). The Emperor Volcanic Complex may be correlative with volcanic rocks assigned to the Hemlock Formation in the Iron River—Crystal Falls range (Prinz, 1976). If so, the Ironwood Iron Formation, the Negaunee Iron Formation of the Marquette range, and the Vulcan Iron Formation of the Menominee range occupy similar stratigraphic positions in the Menominee Group (Fig. 2-5).

Although the iron-formations of the Menominee Group can be correlated from the northwest to the southeast, overall stratigraphic relationships within the Marquette Range Supergroup are much more complex, particularly over gneissic basement rocks (Fig. 2-5). In both the Marquette and Menominee ranges, the Chocolay and Baraga Groups are considerably thicker than on the Gogebic range. Moreover, the Baraga Group in the Iron River—Crystal Falls range is overlain by yet another thick succession of strata assigned to the Paint River Group (Fig. 2-5). Sedimentation of the Chocolay Group began in the eastern part of the Marquette range with deposition of the Enchantment Lake Formation, a conglomeratic and texturally immature arenitic unit, in a fault-controlled trough (LaRue and Sloss, 1980). This phase was followed by the widespread accumulation of pure quartz sands (Mesnard, Sunday and Sturgeon Quartzites), dolomite with stromatolitic structures (Kona, Randville, Bad River and Saunders formations), and locally argillaceous material (Wewe Slate). Although dominantly an epiclastic unit, parts of the Wewe

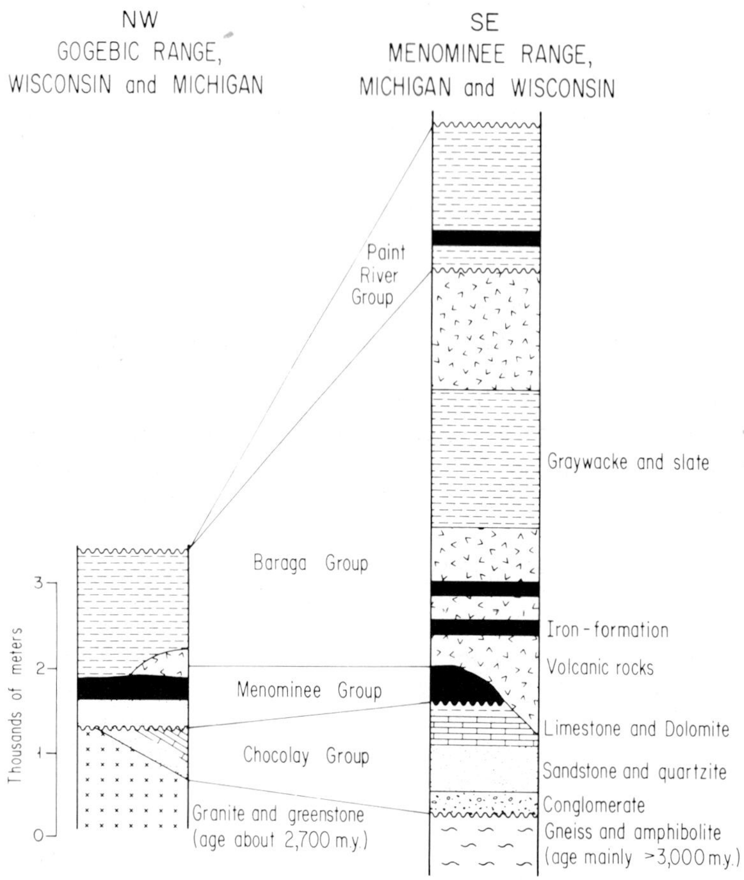

Fig. 2-7. Selected stratigraphic sections in the southeastern segment of the Animikie basin. Note the extensive development of lower Proterozoic rocks in the Menominee range where they overlie Archean gneisses.

may be volcanic in origin (Gair and Thaden, 1968). Because of irregularities on the basement surface and subsequent erosion, the thickness of the Chocolay Group varies considerably, ranging from a feather edge in the western Marquette range to about 1400 m in the Menominee range.

Following a period of mild uplift and erosion, strata assigned to the Menominee Group were deposited directly on Archean basement rocks or on eroded remnants of the Chocolay Group. This cycle of sedimentation started with the deposition of 360 to 750 m of quartzitic material (Ajibik, Felch and Palms formations) that locally, as on the Marquette range, grades upward into at least 600 m of interbedded argillite, quartzite, and detritus-choked iron-formation assigned to the Siamo Slate (LaRue, 1979). These clastic rocks are overlain by iron-rich strata assigned to the Ironwood, Negaunee and Vulcan

Iron Formations. The differing thicknesses, stratigraphic details, and facies types of these iron-formations from range to range suggest that they were deposited in isolated fault-bounded, second-order troughs within the larger basin (Bayley and James, 1973). Iron-formation sedimentation was accompanied by the eruption of 600 to 4500 m of subaqueous basalt and associated volcanogenic rocks (Hemlock Formation of Prinz, 1976) and deposition of lesser amounts of interbedded felsic volcanic rocks, iron-rich strata, and conglomerate of fluvial origin (Johnson, 1975).

After yet another period of uplift, deformation, and erosion leading in places to the nearly total removal of rocks assigned to the Menominee Group, sedimentation of the Baraga Group ushered in a period of pronounced crustal disturbance and sedimentation in a number of grabenlike depositional basins (Cannon, 1973). In places, sedimentation started with deposition of as much as 450 m of conglomeratic quartzite (Goodrich Quartzite) having sedimentary attributes indicative of considerable relief (Tyler and Twenhofel, 1952; Nordeen and Spiroff, 1962), whereas in other places it started with the deposition of at least 550 m of a cherty iron-formation and ferruginous slate assigned to the Amasa Formation (Fig. 2-5). However, the bulk of the Baraga Group consists of more than 3000 m of interbedded graywacke and slate assigned to the Tyler and Michigamme Formations (Cannon and Klasner, 1975). Although the Michigamme Formation is a rather monotonous turbidite deposit, it contains thick sequences of volcanic rocks of mostly mafic composition, iron-formation, and several other kinds of clastic rocks including quartzite and black carbonaceous and pyritic shale (Boyum, 1975). Volcanic lenses in the Baraga Group in the southern part of the basin (Badwater Greenstone) reach local thicknesses of 3000 m or more. Widespread gabbroic diabase dikes and sills and differentiated gabbroic plutons, such as the Peavy Pond Complex, were probably emplaced at this time, mainly as the subvolcanic equivalents of the mafic extrusive rocks.

The deep-water environment initiated during deposition of the Baraga Group continued with the additional accumulation of about 2000 m of strata assigned to the Paint River Group — a thick sequence of graywacke and slate (Dunn Creek Slate, Hiawatha Graywacke and Fortune Lake Slate) and several intercalated iron-rich units including the Stambaugh Formation, and the Riverton Iron Formation (Fig. 2-5). Although the Paint River Group probably had a much greater areal extent and thickness, it is now preserved only in the deeper downfolds of the Iron River—Crystal Falls district near the Michigan—Wisconsin border.

Volcanic rocks of central and northeastern Wisconsin

Volcanic rocks of early Proterozoic age occur as a series of east-trending belts in northeastern and central Wisconsin (Dutton and Linebaugh, 1967; Medaris and Anderson, 1973; LaBerge, 1976; Mudrey, 1978). Although they apparently underlie a large area (Fig. 2-2), the volcanic rocks are exposed or have been studied only in a few places in northeastern (Bayley et al., 1966; Lahr, 1972; Schmidt et al., 1978), west-central (May, 1976), and south-central Wisconsin (LaBerge and Myers, 1972; LaBerge, 1976). In northeastern Wisconsin the volcanic rocks are assigned to the Quinnesec Formation, which can be divided into a pyroclastic and extrusive sequence of generally rhyodacitic to rhyolitic composition, and a pillowed lava and pyroclastic sequence of dominantly mafic composition. Each sequence contains minor amounts of the other, and both contain thin beds of argillaceous material, impure quartzite, arkose, and iron-formation. A similar but apparently more felsic succession (May, 1976) occurs at the western end of an east-trending belt which Myers (1974) termed the "Flambeau volcanic-sedimentary province" (Fig. 2-3). However, the eastern end of this belt is characterized by mafic and intermediate flows, pyroclastic rocks, and a carbonaceous black shale; felsic volcanic rocks are present only locally. The felsic portions at either end of the Flambeau province contain appreciable quantities of massive copper- and zinc-bearing sulfide deposits that have textural and spatial attributes similar to those observed in stratabound, volcanogenic massive sulfide deposits in greenstone belts of Archean age in Canada (May, 1976; Schmidt et al., 1978). Volcanic rocks including subaqueous flows and pyroclastic rocks of intermediate to felsic composition also occur near Wausau in south-central Wisconsin where they are underlain locally by pillowed volcanic rocks of mafic composition (LaBerge, 1969).

Deformation, metamorphism and igneous activity

The stratified rocks of the Animikie basin can be divided into two broad longitudinal zones on the basis of contrasting styles of deformation and grades of metamorphism (Sims, 1976) — a northern stable cratonic zone and a southern deformed zone termed the Penokean foldbelt (Sims et al., 1980). The tectonic front separating the two zones coincides with the inferred northern edge of the Great Lakes tectonic zone (Sims et al., 1980), and is marked in the northwestern segment of the Animikie basin by the northern limit of a penetrative cleavage (Marsden, 1972; Morey, 1978a).

North of the tectonic front, the supracrustal rocks, which unconformably overlie Archean greenstone-granite complexes, were virtually undeformed and unmetamorphosed during the Penokean orogeny. Strata on the Mesabi and Gunflint ranges in northern Minnesota and Ontario dip gently southward, and the surface of the basement rocks appears to be relatively undisturbed (Fig. 2-8). The metamorphic grade of the overlying supracrustal rocks ranges from the zeolite facies to the lower greenschist facies (Hanson and Malhotra, 1971; Perry et al., 1973; Morey, 1973a, 1978b; Floran and Papike, 1975; Lucente, 1978).

In contrast, rocks south of the tectonic front are extensively metamor-

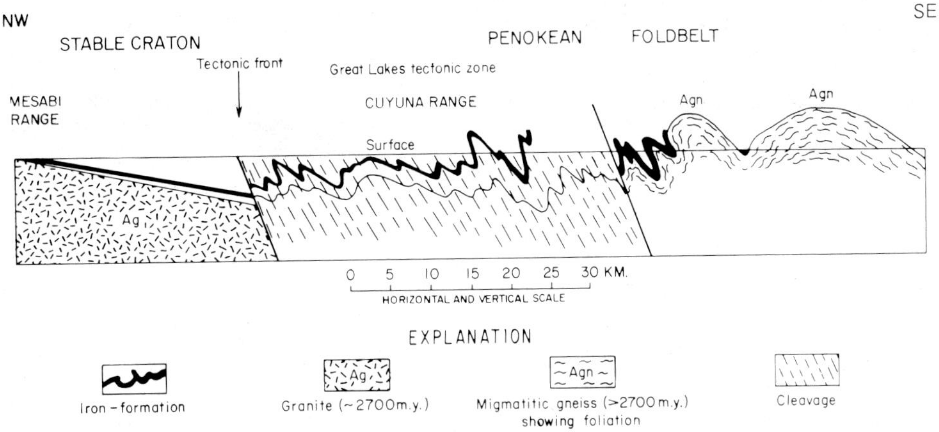

Fig. 2-8. Geologic section across the northwestern segment of the Animikie basin (modified from Morey, 1979).

phosed and both the supracrustal rocks and the underlying basement rocks are complexly infolded. Deformation was manifested principally by vertical tectonic processes leading to the development of fault-bounded blocks (Cannon, 1973) or mantled gneiss domes (Morey and Sims, 1976; Sims, 1976; Morey, 1978b).

The Penokean foldbelt itself is characterized by several contrasting tectonic styles which differ from north to south partly because different structural levels are exposed (Fig. 2-8). Nonetheless, these contrasting tectonic styles serve to divide the foldbelt into three subzones. The northernmost subzone occurs in east-central Minnesota immediately south of the tectonic front. It is some 60 to 70 km wide, and is characterized by a number of large anticlines and synclines with numerous coaxial second- and third-order folds on their limbs. The folds have nearly vertical, straight to broadly curvilinear axial planes that trend in a generally eastward direction (Fig. 2-9). Metamorphism reached the biotite grade in this subzone.

The next subzone to the south is characterized in east-central Minnesota by superposed folds that are steeply overturned to the northwest and have a steep southeast-dipping penetrative cleavage. Furthermore, the axial planes of the superposed folds are subparallel to basement-cover contacts. Much of the southeastern segment lies within the intermediate subzone of the Penokean foldbelt where it is further characterized by a nodal distribution of metamorphic zones (James, 1955; Morey, 1978b). Through numerous structural studies in the several iron-mining ranges, this part of the Penokean foldbelt, particularly in northern Michigan, has become the "type area" for the Penokean orogen (Cannon, 1973; Klasner, 1978). In this area, an early generation of east-trending folds, possibly formed by a regional episode of gravity sliding

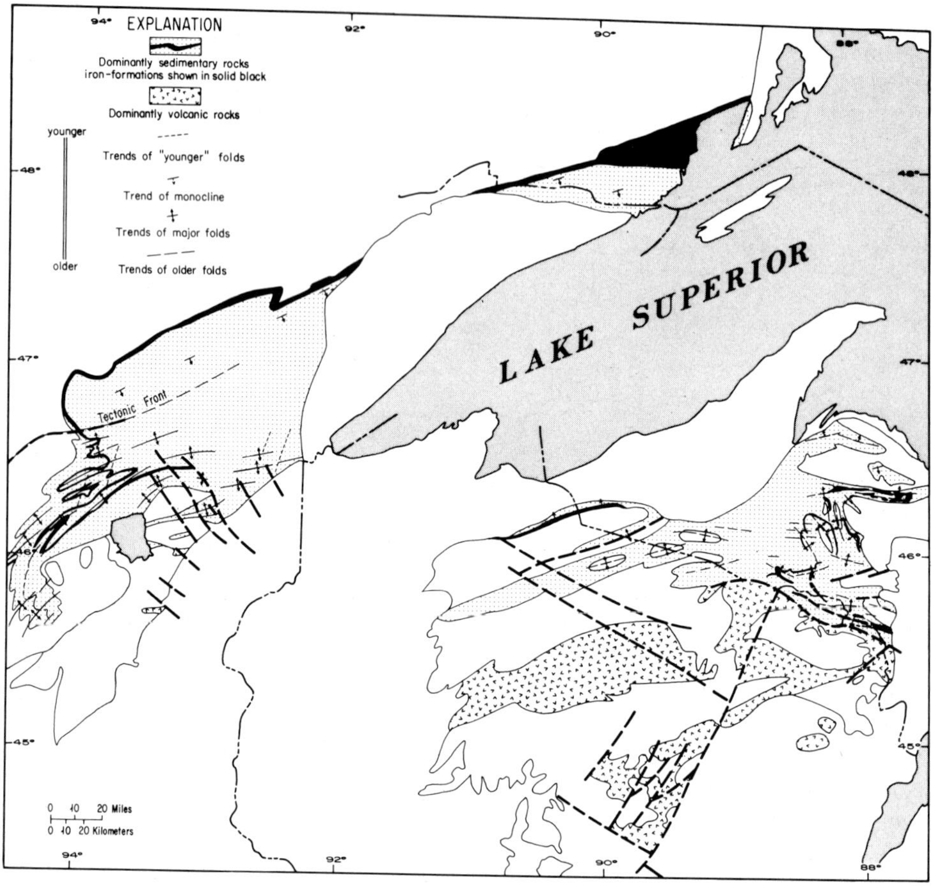

Fig. 2-9. Tectonic map showing major fold axes in the lower Proterozoic rocks of the Animikie basin (Michigan data modified from Cannon, 1973).

(Fig. 2-9), has been modified by a later generation of folds related to the diapiric uplift of gneiss domes and fault-bounded blocks of diverse orientation. The fold axes associated with the second period of deformation also tend to exhibit diverse orientations because they were controlled to a large extent by older structures in the basement rocks. Metamorphism began during deformation and continued after deformation ceased, as diabase dikes which are not themselves folded cut folds, and are metamorphosed to regional grade (Cannon, 1973). Metamorphism of the low-pressure type, in which andalusite and sillimanite are the stable aluminosilicates (James, 1955), is associated spatially in places with uplifted blocks or domes of gneissic rock (Sims, 1976). Metamorphism of the supracrustal rocks was accompanied by internal recrystallization along cataclastic zones and partial anatexis of the basement rocks.

Details of the deformation and metamorphism in the southernmost subzone of the Penokean foldbelt are not well understood. This subzone, which is well-developed only in the volcanic rocks of northern Wisconsin, is characterized by linear, rather than nodal patterns of deformation (Sims and Peterman, 1980). Although the fold trends are linear, they are discontinuous and diversely oriented (Sims et al., 1978), and the rocks themselves have been variably metamorphosed from the greenschist to lower amphibolite facies (Morey, 1978b). In part at least, the deformation pattern is tentatively interpreted as indicating separate regimes within different crustal blocks, implying strong control by Archean basement structures.

Sedimentological implications

The stratified rocks of the Animikie basin have been divided longitudinally into two zones on the basis of pronounced differences in facies and thickness (Sims, 1976). These include: (1) a thin succession (250—2000 m) of predominantly sedimentary rocks in the north; and (2) a much thicker succession (> 19,000 m) of intercalated sedimentary and volcanic rocks in the south. The thicker succession is intensely deformed, metamorphosed, and intruded locally by granitic plutons. The close correspondence between the sedimentological and tectonic patterns implies that both sedimentation and tectonism were part of a tectonic continuum that began with the development of the depositional basin and culminated with the major tectono-thermal pulse of the Penokean orogeny.

Most sedimentological models for the Animikie basin rely on the work of James (1954) who concluded that the Marquette Range Supergroup evolved from a "stable-shelf" sequence to a "geosynclinal" assemblage during early Proterozoic time. This transition is especially evident in the northwestern segment of the basin (Fig. 2-10) where sedimentation can be divided into five depositional phases (Morey, 1979). The first three phases constitute a miogeosynclinal sequence that thickens more or less symmetrically toward the axis of the basin. The fourth phase forms a transitional sequence as the shelf foundered, whereas the last phase forms a eugeosynclinal, southward-facing, clastic wedge (flysch) deposited by southward-flowing turbidity currents. Rocks of the earliest phase include a discontinuous veneer of coarse- to fine-grained, generally well-sorted clastic rocks and a thick complex of pillowed basalt, agglomerate, cherty iron-formation and black carbonaceous slate. The well-sorted rocks were deposited under shallow-water conditions along the fringes of the basin, whereas the volcanic complex was deposited in the axial part of the basin, probably in fault-bounded troughs. All of these rocks are overlain by the second or quartzite phase which forms a basinward-thickening wedge of compositionally mature quartzite, quartz-rich siltstone, mudstone and shale (Fig. 2-11) derived from both the greenstone-granite terrane to the north and the gneiss terrane to the south (Peterman, 1966;

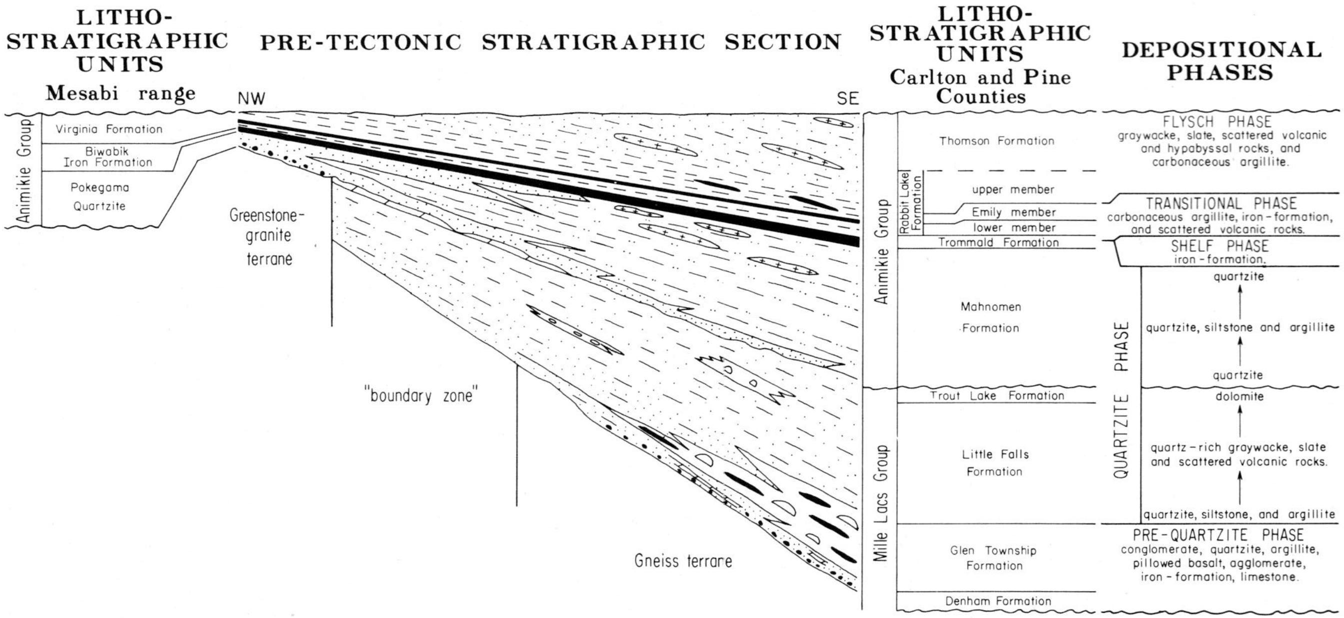

Fig. 2-10. Pretectonic north—south stratigraphic section showing the relationship between lithostratigraphic nomenclature and depositional phases in the evolution of the northwestern segment of the Animikie basin (modified from Morey, 1979).

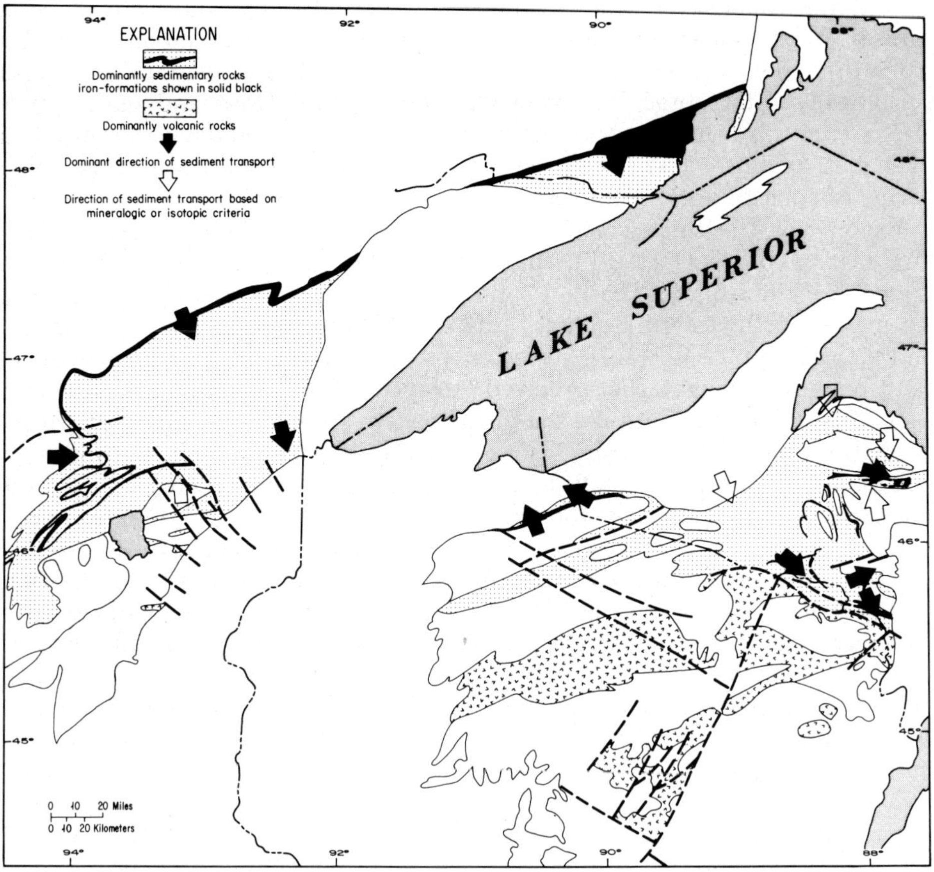

Fig. 2-11. Generalized geologic map showing inferred directions of sediment transport for the lower Proterozoic clastic rocks of the Animikie basin (modified from Nilsen, 1965; Peterman, 1966; Morey, 1969, 1973a; Morey and Ojakangas, 1970; Keighin et al., 1972, Alwin, 1979; LaReu, 1979).

Keighin et al., 1972). The ratio of mud to sand increases stratigraphically upward, and much of the sequence is characterized by rather monotonous interbeds of quartz-rich wacke, subgraywacke, siltstone, and shale punctuated by scattered beds of sandy dolomite and quartz-pebble grit or conglomerate.

Rocks of the quartzite phase provide the foundation for the third phase — the formation of a southward-facing shelf characterized by various kinds of iron-formation having shallow-water attributes to the north and west and deeper water attributes to the south and east (Morey, 1973a). The shelf deposits are gradationally overlain by the fourth phase — a thin succession of dominantly black, laminated mudstone deposited in a starved environment that formed as the shelf began to founder. As the shelf foundered into deeper water, mud deposition was periodically interrupted by southward-flowing turbidity currents which deposited beds of feldspathic graywacke and siltstone (Schmidt, 1963; Morey and Ojakangas, 1970). Sedimentation was also periodically interrupted by the deposition of both felsic and mafic pyroclastic rocks, by the extrusion of microdiabasic flows, and by the injection of diabasic gabbro sills. All of these rocks constitute a southward-thickening wedge of considerable thickness.

Except for the thick volcanic rocks intercalated in the youngest or fifth depositional phase (Fig. 2-7), the depositional history of the southeastern segment of the Animikie basin is similar to that of the northwestern segment. As in the northwestern segment, crustal instability during the pre-quartzite and quartzite phases was manifested principally by the presence of local fault-bounded troughs that received "deeper water" sediments (LaRue and Sloss, 1980). However, unlike the northwestern segment, the succeeding iron-formations were not deposited on a gradually deepening shelf, but rather in a number of preexisting fault-bounded troughs. Crustal instability also was manifested at this time by the extrusion of thick sequences of subaqueous volcanic rocks and by erosion as recorded by many local unconformities.

Crustal instability in the southeastern segment intensified after deposition of the major iron-formations, and led to rapid local subsidence and the accumulation of thick graywacke turbidite sequences. Increasing amounts of clastic detritus were derived from positive areas within the basin (Fig. 2-11), and possibly from its southern margin. Volcanism continued through this phase. The extreme variability in distribution and thickness of the volcanic units implies that they accumulated in local basins having considerable structural relief. Existing deep depressions continued to founder, expecially in the very southern part of the southeastern segment, with continued deposition of clastic detritus, iron-formation, and black pyritic and graphitic shale.

Tectonic implications

Most authors (Bayley and James, 1973; Cannon, 1973; Van Schmus, 1976; Sims, 1976; Morey, 1979) agree that the lower Proterozoic rocks of the Animikie basin were deposited in a rift-like basin that increased in size with time. Unfortunately, however, there is no general consensus as to the driving forces that caused the basin to form, or as to the deformational, metamorphic and igneous events associated with the Penokean orogeny.

The abrupt changes in sedimentologic and tectonic patterns at the boundary between the two Archean basement terranes clearly indicate that the Great Lakes tectonic zone was instrumental in the evolution of the Animikie basin. The tectonic zone appears to have been a zone of weakness in early Proterozoic time where crustal extension, faulting and concurrent subsidence provided the depressions in which the sediments accumulated. During compressional stages the contrasting basement rocks seem to have exhibited vastly different tectonic stabilities; the gneissic basement rocks were reactivated with elevated geothermal gradients that provided diapiric uplift and probably expansion, accompanied by some lateral transport of this crustal segment against the more rigid greenstone-granite crust to the north.

The patterns of early Proterozoic sedimentation, deformation, metamorphism, and volcanic-plutonic igneous activity in the Lake Superior region are notably asymmetrical from north to south, and thus do not coincide entirely with tectonic patterns that characterize the Phanerozoic Era. Consequently, Sims (1976) and Sims et al. (1981) have proposed that the tectonic processes that led to the opening and closing of the Animikie basin were unique to early Proterozoic time. However, the fact that the sedimentological record has a strong resemblance to the stratigraphic history of Phanerozoic geosynclines (Pettijohn, 1957, p. 640) has led to several attempts to explain the evolution of the Animikie basin by various kinds of Phanerozoic plate-tectonic processes. Van Schmus (1976), for example, proposed that a north-dipping subduction zone existed south of the Animikie basin in early Proterozoic time. In this model, the volcanic rocks of Wisconsin represent the island-arc region and the granitic rocks the eroded roots of this arc. The rocks in Michigan would then occur between the island arc to the southeast and the shoreline to the northwest, with those in northern Minnesota and Ontario being closest to the shoreline. The Penokean orogeny in this model was the product of a consuming continental margin with ocean floor to the south subducted toward the north under the foreland basin.

All of the lower Proterozoic rocks, including those of volcanic affinity in northern Wisconsin, appear to have been deposited on continental crust (Cannon, 1973; Sims, 1976; Van Schmus and Anderson, 1977; Morey, 1978b). Furthermore, the sedimentary rocks in the Animikie basin were derived from preexisting silicic rocks like those now exposed both to the north and the south of the basin. These observations preclude the occurrence of oceanic crust to the south of the Animikie basin at the start of early Proterozoic time. They also seem to preclude any analogy with plate-tectonic processes involving an oceanic/continental crustal boundary that could subsequently become the site of a subduction zone (Van Schmus and Anderson, 1977).

The fact that the stratified rocks were deposited on continental crust does not, however, preclude the possibility that the Animikie basin formed by continental rifting processes somewhat akin to those proposed to explain the present Atlantic Ocean margins (Cambray, 1977, 1978a, b). In this model, rifting was followed by subsidence at the margin of an expanding ocean followed by reversal of plate movement, subduction, compression, and metamorphism. The Penokean orogeny would then reflect the ultimate closing of the basin by subduction of newly formed oceanic crust between two colliding continental plates.

An intracontinental rifting model in which the Animikie basin represents the north side of a gradually opening rift zone has many appealing sedimentological aspects. However, it also has some problems. Geologic phenomena associated with the presumed south side of such a rift zone have not been recognized in either northern Wisconsin or east-central Minnesota. Furthermore, there is no evidence either that rifting proceeded to the stage where oceanic crust was developed between disrupted crustal segments, or that the Penokean orogeny reflects the ultimate closing of the basin by subduction of that crust between two colliding continental plates. Geologic evidence, such as large foreland-directed overthrusts and associated melanges, or paired high-pressure/low-temperature and low-pressure/high-temperature metamorphic belt is lacking, as is any evidence for the presence of a suture zone.

In summary, there are arguments both for and against invoking Phanerozoic plate-tectonic processes to explain the evolution of the Animikie basin. The advocates of each point of view have taken what appear to be mutually exclusive positions. Both views, however, ignore the fundamental possibility that while the driving forces for tectonism were the same as in Phanerozoic time, a considerably different geothermal regime might well have led to considerably different near-surface manifestations in early Proterozoic time. Even in the Phanerozoic, cause-and-effect relationships between geothermal regimes, tectonic forces, and the near-surface manifestations of those forces are not well understood. Therefore, until these relationships can be documented, the significance of the presence or absence of Phanerozoic near-surface phenomena in the Proterozoic rocks of the Lake Superior region should be evaluated with caution.

THE IRON-FORMATIONS AND THEIR DEPOSITIONAL ENVIRONMENTS

Iron-formations are widely distributed in the Animikie basin. This section briefly summarizes the kinds of iron-formation in the basin, particularly as they relate to the sedimentary-tectonic evolution of the basin. Although the geologic literature has focused on iron-formations of possible or proven economic importance, the published data suffice to demonstrate that the iron-formations are not restricted to any one sedimentological regime within the basin, a factor that must be recognized in any hypothesis intended to explain their origin.

Iron-formations of the northwestern segment

Major attributes of iron-formations in the northwestern segment of the Animikie basin are summarized in Table 2—III. The major iron-formations of this segment appear to represent a continuous blanket deposit formed on a southward-sloping shelf that had an initial strike length of more than 640 km. However, as Table 2-III shows, numerous other iron-formations occur throughout the sequence. For the most part these iron-formations are thin, restricted units that are laminated to thin bedded and composed of mineral assemblages ranging from the oxide facies to the sulfide facies as defined by James (1954).

The major iron-formations of the northwestern segment include the Biwabik Iron Formation of the Mesabi range, the Gunflint Iron Formation of northeastern Minnesota and Ontario, and the Trommald Formation of the Cuyuna range in east-central Minnesota. The Biwabik Iron Formation ranges in thickness from 30 m to 225 m. Its basal contact is defined by an abrupt change from iron-poor quartzite to iron-bearing, granular, cherty material. The top of the iron-formation is similarly well defined by a thin, but persistent, limestone-bearing unit that contains a few interbeds of chert, but little iron (Fig. 2-12). The limestone-bearing unit pinches out in the western part of the Mesabi range, and the upper part of the iron-formation to the west consists of laminated to thin-bedded chert and siderite. The cherty siderite unit also pinches out farther west, and the upper part of the Biwabik Iron Formation consists of several tens of meters of iron-rich carbonaceous argillite.

Because recognizable lithologic units consisting of various proportions of rock strata having "cherty" (granular) or "slaty" (nongranular) textures occur over long distances, Wolff (1917) subdivided the Biwabik Iron Formation into four units, from bottom to top: lower cherty, lower slaty, upper cherty, and upper slaty. These units, which subsequently were redefined as informal members (Gruner, 1946; White, 1954), can be traced along most of the Mesabi range and throughout the Minnesota part of the Gunflint range (Broderick, 1920). Subsequently, Goodwin (1956) divided the Gunflint Iron Formation in Ontario into four informal members (Fig. 2-13). The boundaries of these members do not coincide with the boundaries of the older four-fold classification scheme, but the two schemes can be correlated with only slight difficulty (Fig. 2-13).

The lower cherty and upper cherty units of the Biwabik Iron Formation are, as the names imply, characterized by discrete layers of chert having granular and oolitic textures and varying proportions of magnetite, siderite, ankerite, iron silicates and rarely hematite. Conglomeratic zones with hematite-bearing oolites and algal structures composed largely of chert and hematite occur at the base of the lower cherty unit and in the middle part of the upper cherty unit.

The lower slaty and upper slaty units also contain cherty beds, but these units consists dominantly of dark-colored, laminated to thin-bedded iron-formation. Individual beds or laminae may be composed entirely of chert, magnetite, iron silicates or siderite, or they may be composed of widely varying proportions of these constituents. In addition, the lower slaty member contains a pronounced ash-fall unit called the intermediate slate.

The Gunflint Iron Formation ranges in thickness from 100 m to 160 m. As on the Mesabi range, the lowermost facies in the Lower Gunflint member consists of algal chert. It lies on the Archean basement or on conglomeratic

TABLE 2-III

Major attributes of iron-formation in the northwestern segment of the Animikie basin

Depositional phase	Formation	Member	Facies[1]	Mineralogy	Textures	Length	Thickness	Geometry	Lithic association	Comments
Flysch	Thomson	—	sulfide	pyrite, carbonaceous slate	laminated	several hundreds of meters	several tens of meters	lenticular	graywacke, slate, volcanic rocks	numerous beds. Morey, 1978a
Flysch	Rabbit Lake	"upper"	?	ferruginous chert	laminated	several hundreds of meters	several tens of meters	lenticular	graywacke, slate	several beds. Schmidt, 1963
Starved basin	Virginia	—	carbonate	chert, siderite	thin bedded	several hundred km	0—60 m	lenticular	carbonaceous slate	White, 1954
Starved basin	Rabbit Lake	Emily	carbonate	siderite, ankerite, chert	laminated	several hundred km	> 60 m	lenticular	pyritic, carbonaceous slate	Marsden, 1972
	Gunflint	—	oxide/silicate	chert, silicates ± carbonate ± magnetite ± hematite	granular, oolitic	> 288 km	> 10—150 m	blanket	quartzite below, carbonaceous argillite above, interlayered volcanic rock	facies interbedded as members. Goodwin, 1956; Floran and Papike, 1975
			silicate/carbonate	silicates, carbonates ± chert ± magnetite	laminated					
Shelf	Biwabik	—	oxide/silicate	chert, magnetite, silicates ± carbonates	granular, oolitic	> 192 km	100—240 m	blanket	quartzite below, carbonaceous argillite above, interlayered volcanic rock	minor oxide facies, algal structures. White, 1954
			silicate/carbonate	silicates, carbonates ± magnetite	laminated					

	Trommald	—	oxide	chert, magnetite, hematite	granular, oolitic	100 km	3—15 m	blanket	quartzite below, carbonaceous argillite above, inter-layered volcanic rock	Mn carbonate facies inter-layered as members. Schmidt, 1963
			carbonate/ silicate	silicates, carbonates ± chert, ± magnetite	laminated					
Quartzite	None	—	—	—	—	—	—	—	—	—
Pre-quartzite	Randall	—	oxide?	chert, hematite	laminated	several km and less	tens of meters or less	lenticular	pillowed basalt ± quartzite	Morey, 1978a
	Glen Township	—	carbonate	chert, ankerite	thin to thick bedded, non-granular	several km and less	tens of meters or less	lenticular	pillowed basalt ± quartzite	facies inter-layered. Morey, 1978a
		—	sulfide	pyrite, graphite	laminated					
	Denham	—	oxide	hematite, magnetite, chert	laminated	several km and less	tens of meters or less	lenticular	quartzite, conglomerate, dolomite ± pillowed basalt	Morey, 1978a

[1] Nomenclature of James (1954).

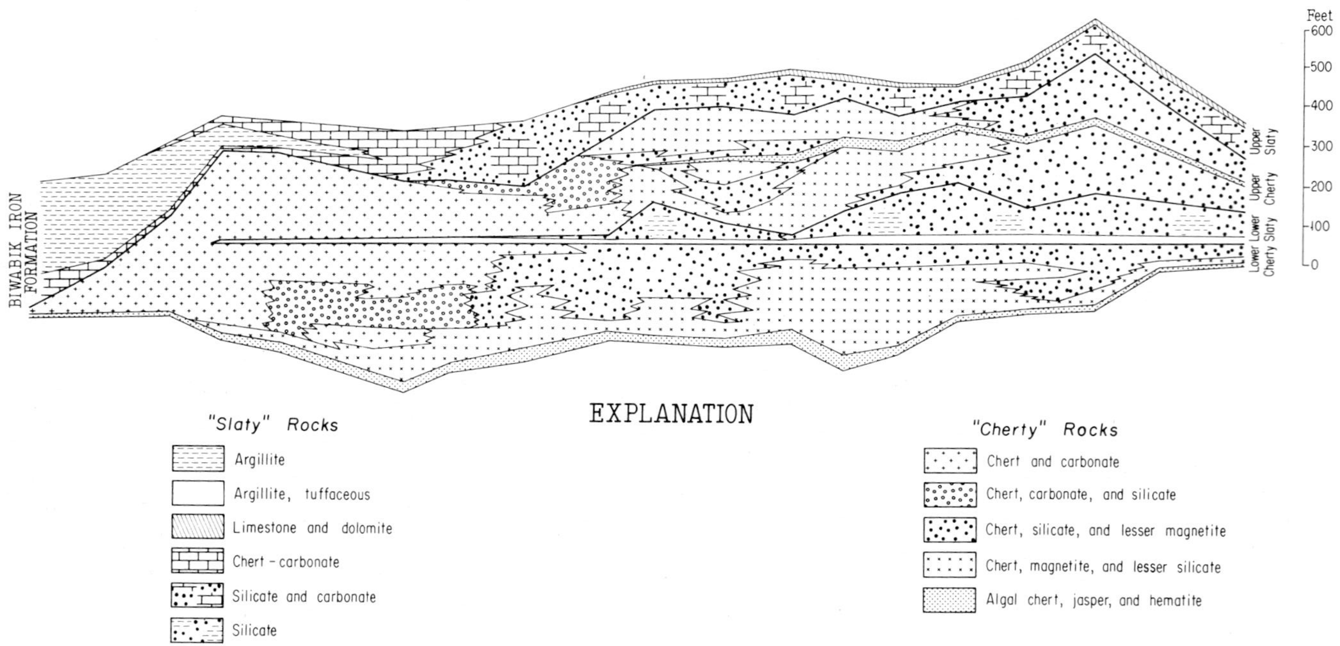

Fig. 2-12. Generalized east—west stratigraphic section showing the distribution of facies within the Biwabik Iron Formation (modified by Morey, 1973b, from the data of White, 1954).

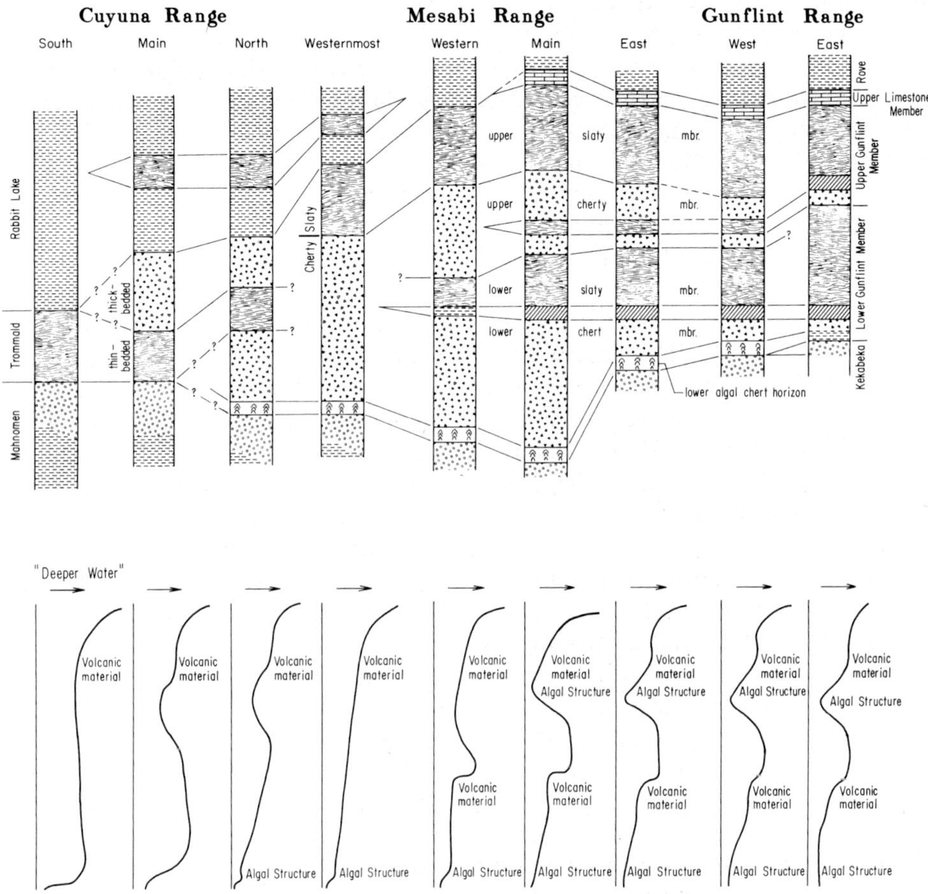

Fig. 2-13. Summary of inferred lithologic correlations of various lithotopes and their inferred sedimentologic settings in rocks of the Gunflint, Biwabik, and Trommald formations (modified from Morey, 1973a).

material that Goodwin (1956) named the basal conglomerate member of the Gunflint Iron Formation. It is overlain by a thin cherty unit consisting dominantly of chert and magnetite, which in turn is overlain by a tuffaceous shale unit that appears to be correlative with the intermediate slate on the Mesabi range. The tuffaceous shale is succeeded by a granule-bearing cherty unit that consists dominantly of chert, greenalite and siderite. This unit passes transitionally to the north-northeast into a unit consisting dominantly of interlayered chert and carbonate, which in turn grades to the north-northeast into a granule-bearing cherty unit. The basal facies of the succeeding Upper Gunflint member in the southwestern part of the range consists of algal chert and conglomerate, and may be equivalent to the algal chert unit in the upper cherty unit of the Biwabik Iron Formation. The algal chert beds are overlain by a second unit of tuffaceous shale that forms a persistent bed of time-stratigraphic significance throughout much of the Gunflint range in Ontario. To the southwest the tuffaceous shale is overlain by a thick unit of laminated silicates with interbeds of granule-bearing chert, whereas to the northeast it is overlain by interbedded chert and carbonate. Both units are overlain by the "upper limestone member" that can be correlated with similar strata in the Biwabik Iron Formation.

The lower and upper tuffaceous shale beds of the Gunflint Iron Formation and the intermediate slate of the Biwabik Iron Formation are the products of explosive volcanism and, together with several lava flows of basaltic composition in the Gunflint Iron Formation, indicate that volcanism and iron-formation deposition were more or less contemporaneous.

The Trommald Formation ranges in thickness from 13 m to more than 150 m. There is little doubt that it is correlative with the Biwabik Iron Formation, but the two units display considerably different sedimentological attributes (Fig. 2-13). In the western part of the Biwabik Iron Formation the lower slaty beds are absent and the lower and upper cherty units are joined as a continuous sequence composed dominantly of interlayered chert and iron carbonates. Both the thickness and the iron content of this cherty unit diminish westward, and at the far western end of the range, the unit is only 6 m thick and contains almost no iron-bearing minerals except for hematite associated with algal units at the base of the formation. However, the overlying slaty beds persist to the west where interlayered beds and laminae of chert and siderite are intercalated with thick beds of argillite that generally contain more iron than the overlying Virginia Formation. In contrast, the Trommald Formation in the northern part of the Cuyuna range (Fig. 2-13) consists of two iron oxide-rich, thick-bedded cherty units separated by a thin-bedded unit composed largely of iron silicates and iron carbonates (Marsden, 1972). As on the Mesabi range, algal structures occur in the basal part of the lower thick-bedded unit. The algal-bearing beds appear to pinch out to the south, and either thick-bedded or thin-bedded iron-formation may make up the entire Trommald Formation at a given locality. However, the Trommald Formation in the main part of the Cuyuna range appears to consist of a thin-bedded unit overlain by the thick-bedded unit (Schmidt, 1963). The thick-bedded unit also pinches out to the south, and the Trommald Formation in the southern part of the Cuyuna range appears to consist entirely of thin-bedded iron-formation (Harder and Johnston, 1918; Marsden, 1972).

The thin-bedded units throughout the Cuyuna range are typically evenly layered and laminated. The layering and lamination reflect varying proportions of chert, siderite, magnetite, stilpnomelane, minnesotaite, and chlorite (Schmidt, 1963). The thick-bedded facies consists partly of evenly bedded iron-formation separated by beds of chert that range in thickness from several centimeters to several meters, and partly of wavy-bedded rock in which chert and iron minerals alternately dominate in layers. The thick-bedded units typically are granule-bearing and locally are oolitic. The granules consist of various mixtures of magnetite, chert, and iron silicates, whereas the oolites consist of hematite and chert; locally the oolites have cores of clastic quartz. In general, the thin-bedded unit can be classed as carbonate facies iron-formation, whereas the thick-bedded units are classed as oxide facies.

The various textural and compositional aspects of the Gunflint, Biwabik and Trommald formations result from deposition under differing environmental conditions, and a close relationship has been documented between the inferred physical and chemical environment, and the composition and textural character of the precipitate. LaBerge (1967) suggested that the slaty, or thin-bedded units are similar in many respects to siltstone or argillite, and that many of the granules in the cherty rocks were derived from material texturally akin to that in the slaty units. The granules commonly occur in cherty strata having graded bedding, cross-bedding, or mixtures of chert and carbonate pebbles, fragments of algal structures, oolites, and detrital quartz. These textural associations imply that the granules behaved as particulate detritus (Mengel, 1965). Because many of the granules were reworked from previously deposited material, the cherty rocks appear to be akin to oolitic and intraclastic limestones. Thus, during iron-formation deposition, granule-bearing sediments were deposited in a shallow-water, agitated environment, whereas slaty or thin-bedded sediments were deposited in deeper, less active water (Fig. 2-13). These textural phenomena are consistent with White's (1954) earlier suggestion that the intercalated cherty (oxide-silicate facies) and slaty (silicate-carbonate facies) units in the Biwabik Iron Formation resulted from deposition near a transgressing and regressing strandline. They also are consistent with Goodwin's (1956) suggestion that the same vertical facies arrangement in the Gunflint Iron Formation resulted from deposition at various water depths during periods of crustal instability, and that subsidence periodically modified the basin configuration and, in turn, the facies distribution.

Iron-formations of the southeastern segment

Major attributes of iron-formations in the southeastern segment of the Animikie basin are summarized in Table 2-IV. As in the northwestern segment, thin and areally restricted units of iron-formation occur throughout the sequence, and the main iron-formations are thought to be correlative. However, the detailed stratigraphy of the iron-formations on the Gogebic, Marquette, and Menominee ranges is so different from range to range that it seems likely that they never were entirely continuous.

The Ironwood Iron Formation of the Gogebic range is the least deformed iron-formation in the southeastern segment. It has a strike length of about 100 km and ranges in thickness from 180 m to 300 m. In general the internal stratigraphy of the Ironwood Iron Formation (Huber, 1959; Schmidt, 1980)

TABLE 2-IV

Major attributes of iron-formations in the southeastern segment of the Animikie basin

Depositional phase	Formation	Member	Facies[1]	Mineralogy	Texture	Length	Thickness	Geometry	Lithic association	Comments
	Stambaugh	—	carbonate	chert, carbonate	laminated	?	12 m	lenticular	argillite, carbonaceous	James et al., 1968
		—	sulfide	pyrite, chert	even bedded	?	30 m	lenticular		
	Riverton	—	carbonate	chert, siderite, stilpnomelane	laminated	> 50 km	45—180 m	lenticular	argillite, carbonaceous	James et al., 1968
	Dunn Creek	—	carbonate	chert, siderite	even bedded	?	30—150 m	lenticular	carbonaceous slate	James et al., 1968
		Wauseca	sulfide	pyrite, siderite, chert, greenalite	laminated	?	?	lenticular		
	Badwater	—	carbonate	chert, carbonate	chert, breccia	?	?	lenticular	volcanic rocks	James et al., 1968
		upper part	sulfide?	pyrite	laminated	?	?	lenticular	carbonaceous slate	James et al., 1968
		upper part	oxide?	chert, magnetite after siderite	laminated to thin bedded	?	?	lenticular	schist, quartzite	James et al., 1961
		upper part	?	hematite, chert, martite ± dolomite ± pyrite	even bedded	?	?	lenticular	graywacke, slate	clastic quartz. Bayley et al., 1966
Flysch	Michigamme	—	—	magnetite, stilpnomelane, chert	indistinctly bedded	?	< 30 m	lenticular	magnetite-rich argillite	Bayley et al., 1966
		—	?	chert, goethite, hematite	even bedded	?	?	lenticular	ferruginous slate	Bayley et l., 1966
		Greenwood	silicate?	silicates ± magnetite, rare chert	laminated to thin bedded	?	330 m	lenticular	argillite	no chert beds. Cannon, 1975
		Bijiki	silicate?	chert, silicates, magnetite	thin bedded	?	30—50 m	lenticular	graywacke	Cannon and Klasner, 1975
		"lower slate unit"	silicate	silicates, rare chert	"banded"	?	?	lenticular	detrital material	Cannon, 1975

	Fence River	—	oxide/ silicate	silicates, chert, magnetite	laminated	?	30 m	lenticular	volcanic rocks	major clastic component. Cannon and Klasner, 1975
	Amasa	—	carbonate/ oxide	chert, siderite, hematite	oolitic, granular	?	544 m	lenticular	dominantly ferruginous slate, pyritic slate	James et al., 1961
	Hemlock	Bird	oxide	chert, magnetite, hematite	oolitic, granular	?	60 m	lenticular	ferruginous quartzite	James et al., 1968
	Vulcan	—	oxide	jasper, magnetite, hematite ± iron silicates	laminated, oolitic and granular	> 50 km	180—230 m	lenticular	ferruginous slate	detrital component. Bayley et al., 1966
		—	oxide	chert, hematite	oolitic, granular					vertical facies, detrital component. Gair, 1975
Shelf	Negaunee	—	silicate	chert, magnetite, silicates	granular	60 km	0—1000 m	tabular	quartzite	
		—	carbonate	chert, siderite	laminated					
			carbonate	siderite, chert, magnetite	laminated				quartzite below, carbonaceous argillite above	interlayered facies, volcanics. Huber, 1959; Schmidt, 1980
	Ironwood	—				100 km	100—300 m	blanket		
			oxide	chert, magnetite, silicates	granular, oolitic					
	Felch	—	?	grunerite, garnet, magnetite	layered	?	1 m	blanket	clastic rocks	James et al., 1961
Quartzite				siderite, chert, magnetite, chlorite, stilpnomelane	laminated granular				clastic rocks	Gair, 1975
	Siamo	Goose Lake	carbonate	chert, jasper, siderite ± minnesotaite	even bedded	?	15—30 m	"marker bed"	clastic rocks	Puffett, 1969

[1] Nomenclature of James (1954).

is very similar to that of the Biwabik and Gunflint Iron Formations of the northwestern segment (Fig. 2-14). The Ironwood has been divided into five members based on the predominance of irregularly to wavy-bedded cherty material or evenly bedded to laminated material. As in the northwestern segment, the irregularly bedded cherty units are characterized by granular and oolitic structures and have mineral assemblages of the oxide facies, whereas the evenly bedded units have textures indicative of deeper water deposition and mineral assemblages of the carbonate facies.

Iron in the irregularly bedded cherty rocks occurs principally as magnetite and the iron silicates minnesotaite and stilpnomelane, probably derived from greenalite (Huber, 1959). Some primary hematite is preserved in the oolitic beds, but most of the oolites have been replaced by magnetite and siderite. As in the Biwabik Iron Formation, hematite-bearing algal structures occur near the base (Huber, 1959) and in the middle part of the Ironwood Iron Formation.

The evenly bedded rocks are mineralogically complex and consist of chert, siderite, iron silicates and magnetite. Each of these minerals may constitute a given bed or lamina or may be accompanied by one or more of the other minerals (Schmidt, 1980).

The Ironwood Iron Formation passes eastward with a strong facies change into a dominantly argillaceous and mafic volcanic sequence (Trent, 1972). The presence of an extensive volcanic sequence in this area implies that the Ironwood and Negaunee depositional basins were separate structural entities.

The Negaunee Iron Formation of the Marquette range (Anderson, 1968; Simmons, 1974; Clark et al., 1975) is confined to a westward-plunging synclinorium about 53 km long and 5—10 km wide, and to a smaller northwest-plunging syncline called the Republic trough (Cannon, 1975). The Negaunee is more than 1000 m thick at the eastern end of the Marquette range, but it thins rapidly to the west, partly because of the nature of the original depositional basin and partly because of post-Negaunee, pre-Baraga Group erosion. The distribution of the iron-formation implies that it was deposited in a narrow, deep, east-trending trough that shallowed abruptly to the west. The trough was bordered to the north and south by positive areas of Archean basement rocks. These positive areas, particularly the southern one, contributed detritus to the trough while iron-formation was being precipitated.

No formal subdivisions of the Negaunee Iron Formation have yet been devised that have more than local application (Anderson, 1968). At the eastern end of the Marquette trough, the Negaunee is divided into three parts — a lower unit consisting of laminated chert and siderite, a middle unit consisting of alternating thin layers in which magnetite, iron silicates or chert are the dominant constituents, and an upper unit consisting of thinly layered chert and hematite. Riebeckite and aegirine-augite occur in a zone about 125 m thick in the top unit. The soda content of this zone, which varies from 0.5% to 6% (Gair, 1973, 1975), and the clastic strata contained in or associ-

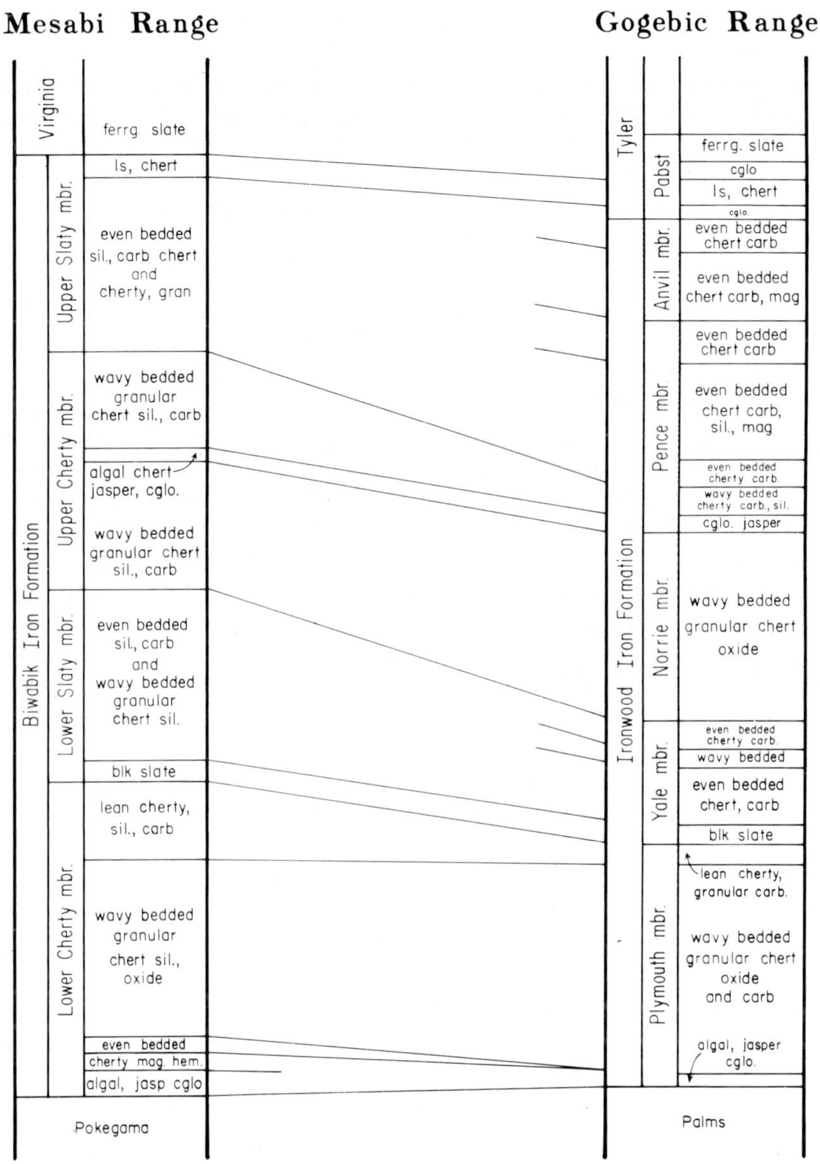

Fig. 2-14. Inferred correlations of specific lithotopes in the Biwabik Iron Formation of the northwestern segment with those in the Ironwood Iron Formation of the southeastern segment of the Animikie basin (modified from Grout and Wolff, 1955, and Huber, 1959). The similarity between the Biwabik and Ironwood Iron Formations is believed to indicate more or less contemporaneous sedimentation near strandlines on opposite sides of the Animikie basin (Bayley and James, 1973). Original physical continuity between the two units is unlikely.

ated with this zone imply sedimentation in shallow water under evaporite conditions (Gair, 1973).

At the western end of the Marquette trough, the uppermost oxide facies makes up most of the Negaunee Iron Formation, and at some localities beds rich in hematite are interlayered with beds rich in magnetite (Cannon and Klasner, 1972). Only the two upper units appear to occur in the Republic trough and in areas to the west of the Marquette trough (Cannon and Klasner, 1972). Thus it seems likely that deposition in early Negaunee time was confined to the eastern end of the Marquette trough and only later as the trough was progressively filled did deposition of the iron-formation spread to the western end and to the Republic trough. The extent to which iron-formation was deposited beyond the limits of the present exposures is unknown.

Several iron-formations of significant dimensions also occur stratigraphically above the Negaunee Iron Formation in the Baraga Group at the western end of the Marquette range and in the area of the Amasa Oval some 25 km to the southeast (Fig. 2-3). In particular the Michigamme Formation in the western Marquette range contains two units of iron-formation, the Greenwood and Bijiki Members, generally separated by bedded metavolcanic rocks of mafic to intermediate composition and lesser amounts of argillaceous material (Fig. 2-15). The lowermost or Greenwood Member is dominantly thin-bedded, silicate- and magnetite-rich iron-formation as much as 330 m thick. Chert forms the groundmass around the iron minerals, but pure chert beds are absent (Cannon and Klasner, 1977). In contrast, the Bijiki Member generally is 30—50 m thick and consists of thin-bedded, cherty, iron-silicate- and magnetite-rich iron-formation (Cannon and Klasner, 1976) with some interbeds of graywacke (Cannon and Klasner, 1977).

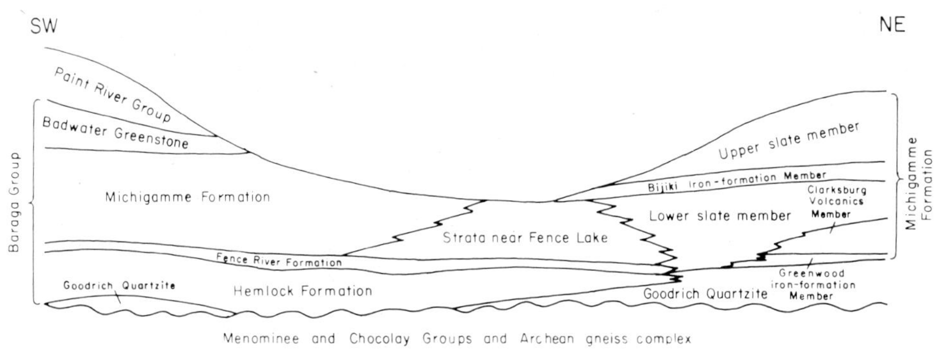

Fig. 2-15. Pretectonic stratigraphic section (modified from Cannon and Klasner, 1975) showing inferred correlations of stratigraphic units in the Baraga Group in Iron and Dickinson Counties to the southwest and the Marquette trough to the northeast (width of section is about 50 km; no vertical scale is implied). See text for discussion, and note that the strata near Fence Lake represent a transitional unit consisting of amphibolitic schist, graywacke, iron-formation, conglomerate, and pyroclastic rocks.

Three iron-bearing units are present in the Baraga Group in the area of the Amasa Oval. Two units occur within the ellipsoidal Hemlock Formation which is mainly metabasalt. The basal Mansfield Member of Bayley (1959) is about 150 m thick and consists dominantly of chert and siderite. The overlying Bird Member of Bayley (1959) is as much as 60 m thick and consists of hematite-rich, oolitic, cherty iron-formation, ferruginous slate and quartzite (James et al., 1968). The third iron-bearing unit is the Amasa Formation which unconformably overlies the Hemlock Formation, and is as much as 545 m thick along the western side of the oval (James et al., 1961). The Amasa consists mostly of ferruginous slate and quartzite, but iron-formation components include layered chert and hematite in which oolitic structures are common. The Amasa Formation is probably correlative with the Fence River Formation, a silicate-chert-magnetite iron-formation about 30 m thick, on the Iron River—Crystal Falls range. All of the various iron-formations of the Baraga Group appear to have been deposited in small basins that existed during brief quiescent periods within an active volcanogenic regime (Bayley and James, 1973).

The Vulcan Iron Formation of the Menominee Group (Fig. 2-5) is the principal iron-formation of the Menominee range (Bayley et al., 1966). The range consists of two separate, steeply dipping belts of Vulcan Iron Formation separated by high-angle faults. The Vulcan also crops out in the Calumet and Felch troughs, 14 km and 19 north of the Menominee range proper. Both trend to the east and open to the west into broad areas of lower Proterozoic strata.

In the Menominee range proper, the Vulcan ranges in thickness from about 180 m to 230 m and can be divided into four members: (1) a basal member, about 15 m thick, of ferruginous slate of clastic origin; (2) an iron-bearing member, 18—60 m thick, of thin-bedded to laminated iron-formation made up of jasper and interlayered magnetite and hematite; (3) a ferruginous slate member, 30—90 m thick; and (4) an uppermost iron-bearing member, 24—60 m thick, of thin-bedded to laminated iron-formation containing jasper, magnetite and hematite. Oolites and granules occur in both iron-bearing members and are particularly common in the uppermost member. The iron-formation has been deeply oxidized, but the original assemblages probably contained chert, hematite and magnetite, as well as minor quantities of silicates and carbonates.

The Vulcan Iron Formation in the Calumet trough has a maximum known thickness of about 60 m, and consists almost entirely of alternating thin, irregular beds and laminae of chert and magnetite (James et al., 1961). Both the chert- and magnetite-rich beds contain scattered grains of iron silicates, mostly the metamorphic mineral grunerite. In contrast, the Vulcan Iron Formation in the Felch trough is about 255 m thick and consists of four members (Cumberlidge and Stone, 1964): (1) a basal unit of magnetite-bearing recrystallized chert, 9—12 m thick, overlain by; (2) as much as 70 m of poorly layered oolitic, hematite-rich iron-formation with minor magnetite and silicates; (3) as much as 91 m of laminae rich in chert, magnetite and silicates, and interlayered thicker beds of admixed hematite and chert that typically are oolitic; and (4) an uppermost member about 91 m thick which consists of uniformly layered chert, magnetite and iron silicates.

The various mineral assemblages and the ubiquitous presence of oolitic structures imply that the Vulcan Iron Formation was deposited as an oxide facies. However, the contrast between the internal stratigraphy of the iron-formation in the main Menominee range and that in the Calumet and Felch troughs implies that sedimentation occurred in separate basins.

The large (> 48 km in maximum dimension) triangular synclinorium of the Iron River—Crystal Falls range (Fig. 2-3) is the northwestern extension of the general Menominee structure (James et al., 1968; Dutton, 1971). The major iron-formations are contained within tightly folded strata of the Paint River Group (Fig. 2-5). As such they are younger than the main iron-formations of the Baraga Group, and represent sedimentation in a deep-water environment dominated by submarine volcanism and sedimentation of turbidites (Fig. 2-16).

The Riverton Iron Formation forms the principal iron-bearing unit of the Paint River Group on the Iron River—Crystal Falls range. It ranges in thick-

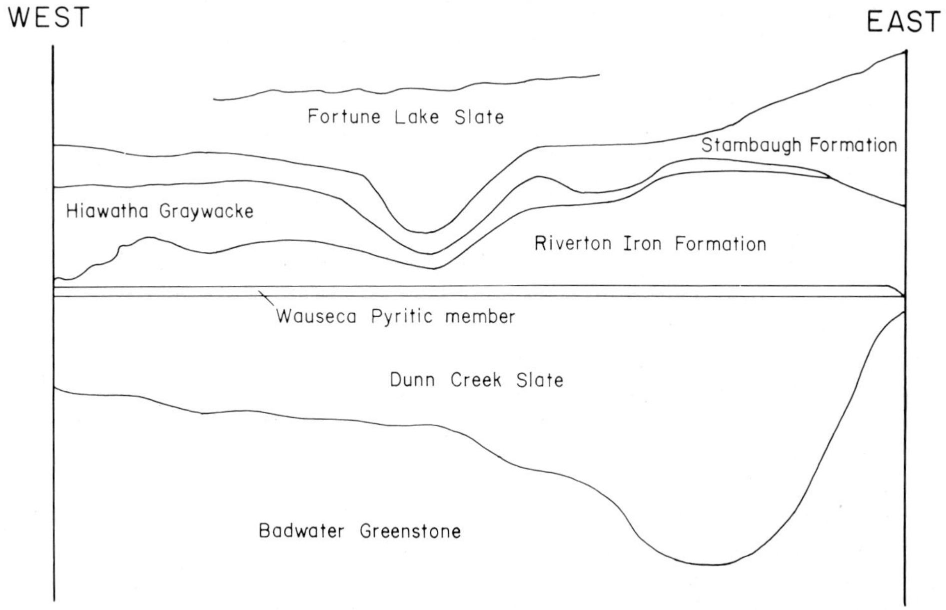

Fig. 2-16. Pretectonic stratigraphic section (modified from James et al., 1968) showing thickness variations of lithostratigraphic units in the Paint River Group of the Iron River—Crystal Falls range (width of section about 30 km; maximum thickness about 5000 m).

ness from about 45 m to 180 m and consists dominantly of evenly bedded layers of siderite and chert. Partings of carbonaceous argillite are common in the upper part of the formation, as are nodules of chert that are rimmed and veined by pyrite.

The Riverton is overlain by the iron-rich Hiawatha Graywacke, which in turn is overlain by a magnetite-bearing unit assigned to the Stambaugh Formation. This iron-bearing unit is about 30 m thick and is partly of clastic origin; locally it contains rhythmic alternations of pyrite and porcelaneous chert (James et al., 1968).

Iron-formations of north-central Wisconsin

Iron-formation occurs in north-central Wisconsin in several isolated areas where the bedrock geologic relationships are poorly exposed (Mudrey, 1978). In general, the iron-formations are evenly bedded and dominantly composed of magnetite and chert, with lesser amounts of hematite or the iron silicate grunerite (Dutton and Bradley, 1970). They occur as lenticular bodies intercalated with pillowed basalt, quartzite, slate, and mica schist (Allen and Barrett, 1915; Sims and Peterman, 1980).

Although the rocks appear to be Proterozoic in age, the association of iron-formation with metavolcanic, metasedimentary and granitic rocks of approximately the same age is remarkably similar to that observed in the greenstone-granite complexes of Archean age in the Superior province of the Canadian Shield (Schmidt et al., 1978; Sims and Peterman, 1980).

Genetic implications

From the foregoing discussion it is obvious that iron-formation sedimentation persisted throughout the entire depositional history of the Animikie basin. Furthermore, the variety of iron-formations in the basin implies that local conditions were responsible for the present array of facies types, and that the chemical requirements necessary to precipitate iron-formation were met in a variety of sedimentological regimes.

There still are many unanswered questions as to how specific iron-formations were formed in specific depositional regimes. Thus understanding the detailed sedimentological history of a particular iron-formation in the basin is still an important objective. However, because the waters of the Animikie basin were a reservoir for high concentrations of iron and silica throughout deposition in the basin, the more fundamental problem involves the ultimate source of these constituents.

A number of theories have been proposed to account for the iron, and to a lesser extent for the silica. For example, Van Hise and Leith (1911, p. 516) concluded that most of the iron and silica were derived from magmatic springs, and some from the reaction of seawater with hot submarine flows. However, the absence of geologic evidence for either mechanism is significant, particularly when the vast quantities of both constituents are considered. Therefore, Gruner (1922) concluded that weathering of a landmass of basaltic rocks under humid or semitropical conditions could have supplied the necessary iron and silica to streams emptying into the Animikie sea. However, the detrital rocks in the Animikie basin were derived from a largely granitoid terrane. These source rocks are incompatible with a model whereby the iron and silica in the iron-formations were derived through the weathering of a basaltic terrane. Therefore, Lepp and Goldich (1964) suggested that the silica and iron were derived by the lateritic weathering of a granitic terrane under atmospheric conditions characterized by an absence or marked deficiency of free oxygen. This is a chemically appealing theory, but it requires some mechanism to transport the iron and silica to the basin of deposition — presumably rivers flowing from the source area. However, even the major iron-formations of the northwestern segment, which for the most part were deposited near a strandline, lack appreciable quantities of clastic detritus. Therefore, the rivers that presumably supplied the iron and silica to the basin must have been sluggish and incapable of carrying a significant bed or suspended load. However, all existing rivers sufficiently large and mature to have low velocities, have deltas built from transported debris. No evidence for such deltas, as far as I know, has been found in any iron-formation in the basin. Therefore, it seems improbable that extensive weathering of the adjoining land and subsequent transport of the weathering products could have provided the necessary iron and silica. Thus the iron and silica must have been derived from sources within the Animikie basin itself.

Postulated intrabasin sources for the iron and silica most commonly involve volcanic rocks that were extruded within the basin (e.g., Goodwin, 1956). Indeed there is a close spatial and temporal relationship between many of the iron-formations and volcanic rocks of one kind or another. However, there is no geologic evidence to indicate that the iron-formations themselves are of volcanic origin, and the large volumes of leached volcanic material that could have provided the needed amounts of silica and iron are apparently lacking (Gruner, 1924; Tyler and Twenhofel, 1952; James, 1954).

A second intrabasin source could have been the water of the basin itself. It is generally accepted that early Proterozoic time was characterized by an oxygen-deficient atmosphere. Thus any iron and silica could have remained in solution in much higher concentrations than would be possible with an oxygen-rich atmosphere. Thus Borchert (1960) suggested that deep water might have contained a reasonably large concentration of ferrous iron, that circulation brought this deep water in contact with a somewhat oxygenated environment, and that the oxidation of ferrous to ferric iron was followed by the precipitation of ferric hydroxide. The oxygen itself was probably supplied by processes related to algal photosynthesis (Cloud, 1973). Furthermore, iron concentrations of a few milligrams per liter are quantitatively reasonable if the water was saturated with respect to siderite and calcite (Holland, 1973). In theory, therefore, iron-rich oxides and carbonates would accumulate in those warm, shallow-water environments that were not receiving any appreciable input of terrigenous or volcanic material. Inasmuch as the water also was probably saturated with respect to silica, and inasmuch as the solubility of amorphous silica decreases with decreasing pressure, deep seawater rising isothermally toward the surface would also tend to become supersaturated with respect to amorphous silica. Thus all of the major components of the iron-formations would have been available in the water of the basin itself. This argument is necessarily hypothetical, and any additional conclusions must await further geologic and geochemical data.

SECONDARY ENRICHMENT DEPOSITS

Much of the iron ore mined from the Lake Superior region was produced from hematite- or goethite-rich deposits resulting from the oxidation and leaching of the primary iron-formations by circulating waters. The proportions of hematite and goethite vary widely from place to place and appear to be related to the mineralogy of the iron-formations from which the deposits developed (Van Hise and Leith, 1911; Gruner, 1946). In general, iron-formations of the oxide facies yield secondarily enriched deposits containing martite (after magnetite), goethite (after silicates), hematite (unaffected), and some kaolinite and quartz, whereas iron-formations of the carbonate facies yield deposits containing hematite (after siderite) and goethite (after silicates), and some martite and quartz.

Descriptions of the secondarily enriched deposits and particularly those of commercial-ore grade are available in many publications including Van Hise and Leith (1911), Gruner (1926, 1946), Hotchkiss (1919), Wolff (1915), Royce (1942), Tyler (1949), Mann (1953), White (1954), Bailey and Tyler (1960), Schmidt (1963), and James et al. (1968). In general, the secondary enrichment deposits occur from near the present bedrock surface to depths ranging from 240 m on the Mesabi range to as much as 2000 m on the Gogebic and Marquette ranges. Although the deposits vary considerably in size and shape from range to range, they all exhibit forms that have a close relationship to the structure and stratigraphy of the iron-formation in which they occur (Royce, 1942). Thus there is little doubt that enrichment occurred by the action of water circulating in porous and permeable zones associated with primary rock types, or in faults, joints, and other fractures associated with upward-opening structural traps, such as synclinal troughs.

There is general agreement that the secondary enrichment deposits formed by two largely concurrent processes. One process involved the oxidation and hydration of primary, diagenetic or low-grade metamorphic minerals to various iron oxides with loss of volume and a concurrent increase in secondary porosity and permeability (Van Hise and Leith, 1911, p. 187; Gruner, 1946). The resulting oxidized iron-formations are in themselves not of ore grade, but their increased porosity and permeability greatly enhanced the ability of solutions to leach silica, phosphorus, magnesium and calcium. Leaching in turn led to even more porosity and permeability and to ore deposits that in some places formed many hundreds of meters away from the original channelways.

Although it is generally agreed that oxidizing and leaching solutions moving along structural channelways led to the formation of the iron ores of the Lake Superior region, there is less agreement as to the source of the solutions. Four theories of origin have current status: (1) action of meteoric waters—essentially deep weathering by surface waters carrying oxygen and carbon dioxide from the atmosphere (Van Hise and Leith, 1911); (2) action of ground water during a postulated period of aridity and extraordinarily deep water table (James et al., 1968); (3) action of hydrothermal solutions derived from magmatic sources at depth (Gruner, 1930); and (4) action of waters dominantly meteoric in origin but augmented and activated by fluid from magmatic sources (Gruner, 1937b).

Much has been written regarding the pros and cons of each of these theories (for an excellent summary see James et al., 1968) because none has proven to be entirely satisfactory. Neither the hydrothermal nor the mixed hydrothermal/meteoric theory has much geologic evidence to support it. The mineral assemblages of the secondarily enriched deposits are not consistent with a hydrothermal origin. Furthermore, there is a general absence of characteristic hydrothermal mineral assemblages in rocks closely associated with the iron-formations, an absence of secondarily enriched bodies in downward-facing structures, and an absence of hydrothermally altered rocks beneath the enriched bodies. One exception to this lack of evidence occurs on the Cuyuna range where near-vertical bodies of largely hematitic ore are accompanied by relatively large amounts of boron which can not have been derived from the original iron-formation. Schmidt (1963) suggested that these ore bodies formed by the action of heated ground water admixed with magmatic emanations. When the hematitic bodies formed is not known, but Peterman (1966) has proposed that rocks associated with them were affected by "hydrothermal" leaching during the interval from 1500 to 1600 m.y. ago, a time well after the end of the Penokean orogeny. A second exception involves the formation of magnetite ore on the Marquette range (Cannon, 1976). There, iron-bearing metamorphic fluids that formed during the Penokean orogeny moved along structurally or stratigraphically controlled channelways in the Negaunee Iron Formation and precipitated iron where they encountered oxidizing conditions.

The meteoric theory of Van Hise and Leith (1911) is supported by geologic observations from most of the other ore deposits of the Animikie basin. The mineral assemblages are those that are stable under highly oxygenated, primarily atmospheric conditions. Furthermore, the atmosphere appears to be the only adequate source for the large volumes of oxygen required in the formation of the secondary minerals.

There are two major theoretical objections to the meteoric theory. These involve: (1) the inability of meteoric water to remove silica on the large scale required; and (2) the apparent inability of any reasonable ground-water system to circulate to the great depths required, and yet be able to continue to oxidize and leach primary iron-formation. Moreover, most of the secondarily enriched bodies occur along the axial zones of downward-facing structures where the ground water would be less free to circulate and thus more likely to become relatively stagnant. In contrast, the steep limbs of many folds do not appear to have been particularly favorable places for oxidation and leaching, yet they are the places where vigorous ground-water circulation would have occurred. Therefore, James et al. (1968) suggested a somewhat modified meteoric theory involving three stages that were repeated many times. The first stage involves the oxidation of iron-formation to great depths in a zone of aeration and vadose water during an epoch of aridity and extraordinarily deep-water

TABLE 2-V

Summary of dominant ore types, postulated processes, and time of formation of secondary enrichment deposits in the Animikie basin

Range	Dominant ore type	Processes	Age
Mesabi	hematite, goethite, martite	weathering	Early Cretaceous
	clastic	reworking and placer concentrates	Late Cretaceous
Cuyuna	hematite-rich, tabular	hydrothermal	1700 m.y.
	goethite-rich, blanket	weathering	Early Cretaceous
Iron River—Crystal Falls	hematite, goethite	ground-water weathering	pre-Late Cambrian
Menominee	hematite, goethite	ground-water leaching	pre-Late Cambrian
Gogebic	hematite, goethite	ground-water leaching	?
Marquette	soft ore: hematite, martite	ground-water leaching	pre-Late Cambrian
	hard ore: specularite, magnetite	weathering, metamorphism, hydrothermal activity	post-Negaunee/ pre-Goodrich 1800 m.y.
	clastic ore	reworking and placer concentrates	Goodrich time
Gunflint	no secondary enrichment deposits		

table. The next stage involves the entrapment of water in upward-facing structures, and the gradual dissolution of chert and consequent supersaturation of silica in the trapped water. The last stage involves the periodic expulsion of the silica-charged waters in artesian systems of brief duration.

In conclusion, it should be emphasized that none of the available field, experimental, or theoretical evidence bearing on the origin of the secondarily enriched deposits can be considered at this time as proof for either deep weathering or cyclic ground-water activity. However, the hypothesis of James et al. (1968) accounts for many of the geologic attributes associated with the secondary enrichment deposits of the Lake Superior region. Although they suggested that the ore-forming processes took place during a unique period of Precambrian time, it now seems more likely that the processes continued during Phanerozoic time in at least two additional episodes (Table 2-V). Some hematite ore on the Marquette range formed by the leaching of silica from the Negaunee Iron Formation during a period of weathering and erosion immediately after deposition of the iron-formation, followed by deformation and metamorphism to its present hard specular form during the Penokean orogeny (Cannon, 1976). Other large bodies of enriched ore in Wisconsin and Michigan appear to have formed between late-middle Proterozoic and Late Cambrian time, inasmuch as boulders of goethite and hematite occur in overlying Upper Cambrian strata. It also seems likely that the secondarily enriched ores in Minnesota formed during Late Mesozoic (Late Jurassic through Early Cretaceous) time (Tyler and Bailey, 1961; Symons, 1966), inasmuch as placer deposits of hematite ore occur in Upper Cretaceous strata (Table 2-V). Both intervals in the Phanerozoic correspond to periods of intense chemical weathering as indicated by the widespread development of thick pre-Upper Cambrian and pre-Upper Cretaceous saprolite in the Lake Superior region (Ostrom, 1967; Parham, 1970; Morey, 1972a). Thus even though questions regarding the specific processes by which oxidation and leaching occurred have not been completely resolved, it seems likely that most of the high-grade ore deposits of the Lake Superior region are the products of near-surface processes that were not unique to a specific period in earth history.

REFERENCES

Aldrich, H.R., 1929. The geology of the Gogebic Iron Range of Wisconsin. Wis., Geol. Nat. Hist. Surv., Bull., 17: 279 pp.

Aldrich, L.T., Davis, G.L. and James, H.L., 1965. Ages of minerals from metamorphic and igneous rocks near Iron Mountain, Michigan. J. Petrol., 6: 445—472.

Allen, R.C. and Barrett, L.P., 1915. Contributions to the pre-Cambrian geology of northern Michigan and Wisconsin. Mich., Geol. Surv. Publ. 18, Geol. Ser., 15: 13—164.

Alwin, B., 1979. Sedimentology of the Tyler Formation. Geol. Soc. Am., Abstr. Programs, 11: 225.

Anderson, G.J., 1968. The Marquette district. In: J.D. Ridge (Editor), Ore Deposits of the United States 1933—1967. Am. Inst. Min., Metall. Pet. Eng., New York, N.Y., pp. 508—517.

Bailey, S.W. and Tyler, S.A., 1960. Clay minerals associated with the Lake Superior iron ores. Econ. Geol., 55: 150—175.

Banks, P.O. and Cain, J.A., 1969. Zircon ages of Precambrian granitic rocks, northeastern Wisconsin. Geol. Soc. Am. Bull., 80: 907—910.

Banks, P.O. and Rebello, D.P., 1969. Zircon ages of Precambrian rhyolite, northeastern Wisconson. Geol. Soc. Am. Bull., 80: 907—910.

Banks, P.O. and Van Schmus, W.R., 1971. Chronology of Precambrian rocks of Iron and Dickinson Counties, Michigan — Part I. Inst. Lake Superior Geol., 17th Annu. Meet., Duluth, Minn., Abstr. Field Guides, pp. 9—10.

Banks, P.O. and Van Schmus, W.R., 1972. Chronology of Precambrian rocks of Iron and Dickinson Counties, Michigan — Part II. Inst. Lake Superior Geol., 18th Annu. Meet., Houghton, Mich., Abstr. Field Guides, paper 23.

Bayley, R.W., 1959. Geology of the Lake Mary quadrangle, Iron County, Michigan. U.S., Geol. Surv., Bull., 1077: 112 pp.

Bayley, R.W. and James, H.L., 1973. Precambrian iron-formations of the United States. Econ. Geol., 68: 934—959.

Bayley, R.W., Dutton, C.E. and Lamey, C.A., 1966. Geology of the Menominee iron-bearing district, Michigan and Wisconsin. U.S., Geol. Surv., Prof. Pap., 513: 96 pp.

Borchert, H., 1960. Genesis of marine sedimentary iron ores. Trans. Inst. Min. Metall., 69: 261—277.

Boyum, B.H., 1975. The Marquette mineral district of Michigan. Inst. Lake Superior Geol., 21st Annu. Meet., Marquette, Mich., Supplement, pp. 109—261.

Broderick, T.M., 1920. Economic geology and stratigraphy of the Gunflint iron district, Minnesota. Econ. Geol., 15: 422—452.

Cambray, F.W., 1977. The geology of the Marquette district — a field guide. Mich. Basin Geol. Soc., 62 pp.

Cambray, F.W., 1978a. Plate tectonics as a model for the environment of sedimentation, the Marquette Range Supergroup and the subsequent deformation and metamorphism associated with the Penokean orogeny. Inst. Lake Superior Geol., 24th Annu. Meet., Milwaukee, Wis., Abstr. Field Guides, p. 6.

Cambray, F.W., 1978b. Plate tectonics as a model for the development of deposition and deformation of the early Proterozoic (Precambrian X) of northern Michigan. Geol. Soc. Am., Abstr. Programs, 10: 376.

Cannon, W.F., 1973. The Penokean Orogeny in northern Michigan. In: G.M. Young (Editor), Huronian Stratigraphy and Sedimentation. Geol. Assoc. Can., Spec. Pap., 12: 251—271.

Cannon, W.F., 1975. Bedrock geology of the Republic quadrangle, Marquette County, Michigan. U.S. Geol. Surv., Misc. Invest. Ser., Map I-862, scale 1 : 24,000.

Cannon, W.F., 1976. Hard iron ore of the Marquette range, Michigan. Econ. Geol., 76: 1012—1028.

Cannon, W.F. and Gair, J.E., 1970. A revision of stratigraphic nomenclature for middle Precambrian rocks in northern Michigan. Geol. Soc. Am. Bull., 81: 2843—2846.

Cannon, W.F. and Klasner, J.S., 1972. Guide to Penokean deformational style and regional metamorphism of the western Marquette Range, Michigan. Inst. Lake Superior Geol., 18th Annu. Meet., Houghton, Mich., Field Guides, pp. B1—B38.

Cannon, W.F. and Klasner, J.S., 1975. Stratigraphic relationships within the Baraga Group of Precambrian age, central upper Peninsula, Michigan. U.S., Geol. Surv., J. Res., 3: 47—51.

Cannon, W.F. and Klasner, J.S., 1976. Geologic map and geophysical interpretation of the Witch Lake quadrangle, Marquette, Iron, and Baraga Counties, Michigan. U.S. Geol. Surv., Misc. Invest. Ser., Map I-987, scale 1 : 62,500.

Cannon, W.F. and Klasner, J.S., 1977. Bedrock geologic map of the southern part of the Diorite and Champion $7\frac{1}{2}$-minute quadrangles, Marquette County, Michigan. U.S., Geol. Surv., Misc. Invest. Ser., Map I-1058, scale 1 : 24,000.

Cannon, W.F., Powers, S.L. and Wright, N.A., 1978. Computer-aided estimates of concentrating-grade iron resources in the Negaunee Iron-formation, Marquette district, Michigan. U.S., Geol. Surv., Prof. Pap., 1045: 21 pp.

Clark, L.D., Cannon, W.F. and Klasner, J.S., 1975. Bedrock geologic map of the Negaunee S.W. quadrangle, Marquette County, Michigan. U.S., Geol. Surv., Geol. Quad. Map GQ-1226, scale 1 : 24,000.

Cloud, P., 1973. Paleoecological significance of the banded iron-formation. Econ. Geol., 68: 1135—1143.

Cumberlidge, J.T. and Stone, J.G., 1964. The Vulcan iron formation at the Groveland mine, Iron Mountain, Michigan. Econ. Geol., 59: 1094—1106.

Dott, R.J., Jr. and Dalziel, I.W.D., 1972. Age and correlation of the Precambrian Baraboo Quartzite of Wisconsin. J. Geol., 80: 552—568.

Dutton, C.E., 1971. Geology of the Florence district, Wisconsin. U.S., Geol. Surv., Prof. Pap., 633: 54 pp.

Dutton, C.E. and Bradley, R.E., 1970. Lithologic, geophysical, and mineral commodity maps of Precambrian rocks in Wisconsin. U.S., Geol. Surv., Misc. Geol. Invest., Map I-631, scale 1 : 1,000,000.

Dutton, C.E. and Linebaugh, R.E., 1967. Map showing Precambrian geology of the Menominee iron-bearing district and vicinity, Michigan and Wisconsin. U.S., Geol. Surv., Misc. Geol. Invest., Map I-466, scale 1 : 25,000.

Fairbairn, H.W., Hurley, P.M., Card, K.D.H. and Knight, C.J., 1969. Correlation of radiometric ages of Nipissing Diabase and Huronian metasediments with Proterozoic orogenic events in Ontario. Can. J. Earth Sci., 6: 489—497.

Floran, R.J. and Papike, J.J., 1975. Petrology of the low-grade rocks of the Gunflint Iron-formation, Ontario-Minnesota. Geol. Soc. Am. Bull., 86: 1169—1190.

French, B.M., 1968. Progressive contact metamorphism of the Biwabik Iron-formation, Mesabi Range, Minnesota. Minn., Geol. Surv., Bull., 45: 103 pp.

Gair, J.E., 1973. Iron deposits of Michigan. In: Genesis of Precambrian Iron and Manganese Deposits. Proc. Kiev Symp., 1970, UNESCO, Earth Sci., 9: 374—375.

Gair, J.E., 1975. Bedrock geology and ore deposits of the Palmer quadrangle, Marquette County, Michigan. U.S., Geol. Surv., Prof. Pap., 769: 159 pp.

Gair, J.E. and Thaden, R.E., 1968. Geology of the Marquette and Sands quadrangles, Marquette County, Michigan. U.S., Geol. Surv., Prof. Pap., 397: 77 pp.

Gibbins, W.A. and McNutt, R.H., 1975. The age of the Sudbury nickel irruptive and the Murray Granite. Can. J. Earth Sci., 12: 1970—1989.

Goldich, S.S., 1968. Geochronology of the Lake Superior region. Can. J. Earth Sci., 5: 715—724.

Goldich, S.S., 1973. Ages of Precambrian banded iron-formations. Econ. Geol., 68: 1126—1134.

Goldich, S.S., Nier, A.O., Baadsgaard, H., Hoffman, J.H. and Krueger, H.W., 1961. The Precambrian geology and geochronology of Minnesota. Minn., Geol. Surv., Bull., 41: 193 pp.

Goodwin, A.M., 1956. Facies relations in the Gunflint iron-formation. Econ. Geol., 51: 565—595.

Grout, F.F. and Wolff, J.F., Sr., 1955. The geology of the Cuyuna district, Minnesota, a progress report. Minn. Geol. Surv., Bull., 36: 144 pp.

Gruner, J.W., 1922. Origin of sedimentary iron formations — the Biwabik Formation of the Mesabi range. Econ. Geol., 17: 407—460.

Gruner, J.W., 1924. Contributions to the geology of the Mesabi range with special reference to the magnetites of the iron-bearing formation west of Mesaba. Minn., Geol. Surv., Bull., 19: 71 pp.

Gruner, J.W., 1926. The Soudan Formation and a new suggestion as to the origin of the Vermilion ores. Econ. Geol., 21: 629—644.

Gruner, J.W., 1930. Hydrothermal oxidation and leaching experiments, their bearing on the origin of Lake Superior hematite-limonite ores. Econ. Geol., 25 (7): 697—719; (8): 837—867.

Gruner, J.W., 1933. The Mesabi range. In: Lake Superior Region. Int. Geol. Congr., 16 Sess., Washington, D.C. 1933, Guideb., 27: 88—101.

Gruner, J.W., 1936. The structure and composition of greenalite. Am. Mineral., 20: 475—483.

Gruner, J.W., 1937a. Composition and structure of stilpnomelane. Am. Mineral., 22: 885—860.

Gruner, J.W., 1937b. Hydrothermal leaching of iron ores of the Lake Superior type — a modified theory. Econ. Geol., 32: 121—130.

Gruner, J.W., 1944. The composition and structure of minnesotaite, a common iron silicate in iron-formations. Am. Mineral., 29: 363—372.

Gruner, J.W., 1946. The Mineralogy and Geology of the Taconites and Iron Ores of the Mesabi Range, Minnesota. Office of Commissioner Iron Range Resources and Rehabilitation, St. Paul, 127 pp.

Gundersen, J.N. and Schwartz, G.M., 1962. The geology of the metamorphosed Biwabik Iron-formation, eastern Mesabi district. Minnesota. Minn., Geol. Surv., Bull., 43: 139 pp.

Hanson, G.N. and Malhotra, R., 1971. K-Ar ages of mafic dikes and evidence for low-grade regional metamorphism in northeastern Minnesota. Geol. Soc. Am. Bull., 82: 1107—1114.

Harder, E.C. and Johnston, A.W., 1918. Preliminary report on the geology of east-central Minnesota including the Cuyuna iron-ore district. Minn., Geol. Surv., Bull., 15: 178 pp.

Holland, H.H., 1973. The oceans: a possible source of iron in iron-formations. Econ. Geol., 68: 1169—1172.

Hotchkiss, W.O., 1919. Geology of the Gogebic range and its relation to recent mining developments. Eng. Min. J., 108: 443—501, 537—577.

Huber, N.K., 1959. Some aspects of the origin of the Ironwood Iron-formation of Michigan and Wisconsin. Econ. Geol., 54: 82—118.

James, H.L., 1954. Sedimentary facies of the Lake Superior iron-formations. Econ. Geol., 49: 253—293.

James, H.L., 1955. Zones of regional metamorphism in the Precambrian of northern Michigan. Geol. Soc. Am. Bull., 66: 1455—1488.

James, H.L., 1958. Stratigraphy of pre-Keweenawan rocks in parts of northern Michigan. U.S., Geol. Surv., Prof. Pap., 314-C: C27—C44.

James, H.L., 1966. Chemistry of the iron-rich sedimentary rocks. U.S., Geol. Surv., Prof. Pap., 440-W: W1—W61.

James, H.L., Clark, L.D., Lamey, C.A. and Pettijohn, F.J., 1961. Geology of central Dickinson County, Michigan, U.S., Geol. Surv., Prof. Pap., 310: 176 pp.

James. H.L., Dutton, C.E., Pettijohn, F.J. and Wier, K.L., 1968. Geology and ore deposits of the Iron River-Crystal Falls district, Iron County, Michigan. U.S., Geol. Surv., Prof. Pap., 570: 134 pp.

Johnson, D.J., 1975. Petrology of a Portion of the Hemlock Formation, Iron County, Michigan. M.S. Thesis, Mich. Technological Univ., Houghton, 26 pp. (unpubl.).

Keighin, C.W., Morey, G.B. and Goldich, S.S., 1972. East-central Minnesota. In: P.K. Sims and G.B. Morey (Editors), Geology of Minnesota: A Centennial Volume. Minn., Geol. Surv., pp. 240—255.

King, P.B., 1969. Tectonic Map of North America. U.S. Geol. Surv., scale 1 : 5,000,000.

Klasner, J.S., 1978. Penokean deformation and associated metamorphism in the western Marquette range, northern Michigan. Geol. Soc. Am. Bull., 89: 711—722.

LaBerge, G.L., 1967. Evidence on the physical environment of iron-formation deposition. Inst. Lake Superior Geol., 13th Annu. Meet., East Lansing, Mich., Abstr., p. 25.

LaBerge, G.L., 1969. Preliminary Report on the Geology of the Northern Part of the Wausau East Quadrangle, Wisconsin. Wis., Geol. Nat. Hist. Surv., Open-file Rep., 13 pp.

LaBerge, G.L., 1976. Major structural lineaments in the Precambrian of central Wisconsin. In: R.A. Hodgson, S.P. Gay Jr. and J.Y. Benjams (Editors), Proceedings 1st International Conference on the New Basement Tectonics. Utah Geol. Assoc., pp. 508—518.

LaBerge, G.L. and Myers, P.E., 1972. Progress Report on Mapping of Precambrian Geology of Marathon County, Wisconsin. Wis., Geol. Nat. Hist. Surv., Open-file Rep., 28 pp.

Lahr, M.M., 1972. Precambrian Geology of a Greenstone Belt in Oconto County, Wisconsin and Geochemistry of the Waupee Metavolcanics. M.S. Thesis, Univ. Wis., Madison, 62 pp. (unpubl.).

LaRue, D.K., 1979. Sedimentary History Prior to Chemical Iron Sedimentation of the Precambrian X Chocolay and Menominee Groups (Lake Superior Region). Ph.D. Dissert., Northwestern Univ., Evanston, Ill., 173 pp. (unpubl.).

LaRue, D.K. and Sloss, L.L., 1980. Early Proterozoic sedimentary basins of the Lake Superior region. Geol. Soc. Am. Bull., pt. 1, 91: 450—452.

Leith, C.K., Lund, R.J. and Leith. A., 1935. Pre-Cambrian rocks of the Lake Superior region. U.S., Geol. Surv., Prof. Pap., 184: 34 pp.

Lepp, H., 1963. The relation of iron and manganese in sedimentary iron-formations. Econ. Geol., 58: 515—526.

Lepp, H., 1966. Chemical composition of the Biwabik Iron Formation, Minnesota. Econ. Geol., 61: 243—250.

Lepp, H., 1968. The distribution of manganese in the Animikian iron formation of Minnesota. Econ. Geol., 63: 61—75.

Lepp, H. and Goldich, S.S., 1964. Origin of Precambrian iron formation. Econ. Geol., 59: 1025—1060.

Lucente, M.E., 1978. Geology of the Middle Precambrian Virginia Formation in Northeastern Minnesota. M.S. Thesis, Univ. Kansas, Lawrence, 116 pp. (unpubl.).

Mancusco, J.J., Lougheed, M.S. and Shaw, R., 1975. Carbonate-apatite in Precambrian cherty iron-formation, Baraga County, Michigan. Econ. Geol., 70: 583—586.

Mann, V.I., 1953. The relation of oxidation to the origin of soft iron ores of Michigan. Econ. Geol., 48: 251—281.

Marsden, R.W., 1972. Cuyuna district. In: P.K. Sims and G.B. Morey (Editors), Geology of Minnesota: A Centennial Volume. Minn., Geol. Surv., pp. 227—239.

Marsden, R.W., 1978a. Iron ore reserves of the Mesabi range, Minnesota — a minerals availability system report. In: L.R. Graven (Editor), Productivity in Lake Superior District Mining. Proc. 39th Annu. Mining Symp. Dept. Conf., Univ. Minn., Minneapolis, pp. 23/1—23/57.

Marsden, R.W., 1978b. Iron ore reserves of Wisconsin — a mineral availability system report. In: L.R. Graven (Editor), Productivity in Lake Superior District Mining. Proc. 39th Annu. Mining Symp. Dept. Conf., Univ. Minn., Minneapolis, pp. 24/1—24/8.

May, E.R., 1976. Flambeau — a Precambrian supergene enriched massive sulfide deposit. Soc. Min. Eng., Preprint 76-I-55, 34 pp.

Medaris, L.G., Jr. and Anderson, J.L., 1973. Preliminary geologic maps of the Green Bay Sheet, Wisconsin and the Iron Mountain Sheet, Wisconsin and Michigan. In: Guidebook to the Precambrian Geology of Northeastern and North-central Wisconsin. Inst. Lake Superior Geol., 19th Annu. Meet., Madison, Wis., scale 1 : 250,000.

Mengel, J.T., Jr., 1965. Precambrian taconite iron-formations — a special type of sandstone. Geol. Soc. Am., 79th Annu. Meet., Kansas City, Mo., Program, p. 106 (abstr.).

Morey, G.B., 1969. The geology of the Middle Precambrian Rove Formation in northeastern Minnesota. Minn., Geol. Surv., Spec. Publ., SP-7: 62 pp.

Morey, G.B., 1972a. Mesabi range. In: P.K. Sims and G.B. Morey (Editors), Geology of Minnesota: A Centennial Volume. Minn. Geol. Surv., pp. 204—217.

Morey, G.B., 1972b. Pre-Mt. Simon regolith. In: P.K. Sims and G.B. Morey (Editors), Geology of Minnesota: A Centennial Volume. Minn. Geol. Surv., pp. 506—508.

Morey, G.B., 1973a. Stratigraphic framework of Middle Precambrian rocks in Minnesota. In: G.M. Young (Editor), Huronian Stratigraphy and Sedimentation. Geol. Assoc. Can., Spec. Pap., 12: 211—249.

Morey, G.B., 1973b. Mesabi, Gunflint and Cuyuna ranges, Minnesota. In: Genesis of Precambrian iron and manganese deposits. Proc. Kiev. Symp., 1970, UNESCO, Earth Sci., 9: 193—208.

Morey, G.B., 1978a. Lower and Middle Precambrian stratigraphic nomenclature for east-central Minnesota. Minn., Geol. Surv., Rep. Invest., 21: 52 pp.

Morey, G.B., 1978b. Metamorphism in the Lake Superior region, U.S.A., and its relation to crustal evolution. In: J.A. Fraser and W.W. Heywood (Editors), Metamorphism in the Canadian Shield. Geol. Surv. Can., Pap., 78—10: 283—314.

Morey, G.B., 1979. Stratigraphic and tectonic history of east-central Minnesota. In: N.H. Balaban (Editor), Field Trip Guidebook for Stratigraphy, Structure and Mineral Resources of East-central Minnesota. Minn., Geol. Surv., Guideb. Ser., 9: 13—28.

Morey, G.B., 1980. A brief review of the geology of the western Vermilion district, northeastern Minnesota. Precambrian Res., 11: 247—265.

Morey, G.B. and Ojakangas, R.W., 1970. Sedimentology of the Middle Precambrian Thomson Formation, east-central Minnesota. Minn., Geol. Surv., Rep. Invest., 13: 32 pp.

Morey, G.B. and Sims, P.K., 1976. Boundary between two Precambrian W terranes in Minnesota and its geologic significance. Geol. Soc. Am. Bull., 87: 141—152.

Mudrey, M.G., Jr., 1978. Zinc-copper resources of Wisconsin. Skillings Min. Rev., 67 (12): 1, 16—19, 28.

Myers, P.E., 1974. Precambrian rocks of the Chippewa region. Tri-state Geol. Field Conference, 38th, Univ. Wis., Eau Claire, 58 pp.

Nilsen, T.E., 1965. Sedimentology of the Middle Precambrian Animikian quartzites, Florence County, Wisconsin. J. Sediment. Petrol., 35: 805—817.

Nordeen, S.C. and Spiroff, K., 1962. Significance of some depositional features in the Siamo Slate. Geol. Soc. Am., Spec. Pap., 68: 240—241 (abstr.).

Ostrom, M.E., 1967. Paleozoic stratigraphic nomenclature for Wisconsin: Wis., Geol. Nat. Hist. Surv., Inf. Circ., 8: 1 sheet.

Parham, W.E., 1970. Clay mineralogy and geology of Minnesota's kaolin clays. Minn., Geol. Surv., Spec. Publ., SP-10: 142 pp.

Perry, E.C., Jr., Tan, F.C. and Morey, G.B., 1973. Geology and stable isotope geochemistry of the Biwabik Iron Formation, northern Minnesota. Econ. Geol., 68: 1110—1125.

Peterman, Z.E., 1966. Rb-Sr dating of Middle Precambrian metasedimentary rocks of Minnesota. Geol. Soc. Am. Bull., 77: 1031—1041.

Peterman, Z.E., 1979. Geochronology and the Archean of the United States. Econ. Geol., 74: 1544—1562.

Pettijohn, F.J., 1957. Sedimentary Rocks. Harper and Row, New York, N.Y., 718 pp.

Prinz, W.C., 1976. Correlative iron-formations and volcanic rocks of Precambrian X age, northern Michigan. Proc. Inst. Lake Superior Geol., 22nd Annu. Meet., St. Paul, Minn., p. 50 (abstr.).

Puffett, W.D., 1969. The Reany Creek Formation, Marquette County, Michigan. U.S., Geol. Surv., Bull., 1274-F: 25 pp.

Royce, S., 1942. Iron ranges of the Lake Superior district. In: W.H. Newhouse (Editor), Ore Deposits as Related to Structural Features. Princeton Univ. Press, Princeton, N.J., pp. 54—63.

Schmidt, P.G., Dolence, J.D., Lluria, M.R. and Parson III, G., 1978. Geology of Crandon massive sulfide deposit in Wisconsin. Skillings Min. Rev., 67 (18): 8—11.

Schmidt, R.G., 1958. Titaniferous sedimentary rocks in Cuyuna district, central Minnesota. Econ. Geol., 53: 708—721.

Schmidt, R.G., 1963. Geology and ore deposits of Cuyana North range, Minnesota, U.S., Geol. Surv., Prof. Pap., 40' : 96 pp.

Schmidt, R.G., 1980. The Marquette Range Supergroup in the Gogebic iron district, Michigan and Wisconsin. U.S., Geol. Surv., Bull., 1460: 96 pp.

Simmons, G.C., 1974. Bedrock geologic map of the Ishpeming quadrangle, Marquette County, Michigan. U.S., Geol. Surv., Geol. Quad. Map, GQ-1130, scale 1 : 24,000.

Sims, P.K., 1976. Precambrian tectonics and mineral deposits, Lake Superior region. Econ. Geol., 71: 1092—1127.

Sims, P.K. and Peterman, Z.E., 1980. Geology and Rb-Sr age of lower Proterozoic granitic rocks, northern Wisconsin. In: G.B. Morey and G.N. Hanson (Editors), Selected Studies of Archean Gneisses and Lower Proterozoic Rocks, Southern Canadian Shield. Geol. Soc. Am., Spec. Pap., 182: 139—146.

Sims, P.K. and Peterman, Z.E., 1981. Archean rocks in the southern part of the Canadian Shield — a review. Spec. Publ., Geol. Soc. Aust., 7: 85—98.

Sims, P.K., Cannon, W.F. and Mudrey, M.G., Jr., 1978. Preliminary geologic map of Precambrian rocks in part of northern Wisconsin. U.S., Geol. Surv., Open-file Rep., 78—318, scale 1 : 1,000,000.

Sims, P.K., Card, K.D., Morey, G.B. and Peterman, Z.E., 1980. The Great Lakes tectonic zone — a major crustal structure in North America. Geol. Soc. Am. Bull., pt. 1, 91: 690—698.

Sims, P.K., Card, K.D. and Lumbers, S.B., 1981. Evolution of early Proterozoic basins of the Great Lakes region. In: F.H.A. Campbell (Editor), Proterozoic Basins of Canada. Geol. Surv. Can., Pap., 81—10: 379—397.

Smith, E.I., 1978. Precambrian rhyolites and granites in south-central Wisconsin, field relations and geochemistry. Geol. Soc. Am. Bull., 89: 875—890.

Stockwell, C.H., McGlynn, J.C., Emslie, R.F., Sanford, B.V., Norris, A.W., Donaldson, J.A., Fahrig, W.F. and Currie, K., 1970. Geology of the Canadian Shield. In: R.J.W. Douglas (Editor), Geology and Economic Minerals of Canada. Geol. Surv. Can., Econ. Geol. Rep., 1: 44—150.

Symons, D.T.A., 1966. A paleomagnetic study of the Gunflint, Mesabi, and Cuyuna iron ranges of the Lake Superior region. Econ. Geol., 61: 1336—1361.

Taylor, G.L., 1972. Stratigraphy, Sedimentology, and Sulfide Mineralization of the Kona Dolomite, Ph.D. Dissert., Mich. Technological Univ., Houghton, 111 pp. (unpubl.).

Trendall, A.F., 1968. Three great basins of Precambrian banded iron depositon. Geol. Soc. Am. Bull., 79: 1527—1544.

Trent, V.A., 1972. Three-phase deformation associated with the Penokean orogeny, east Gogebic range, Michigan. Inst. Lake Superior Geol., 18th Annu. Meet., Ishpeming, Mich., Technical Sess. Abstr., paper 20, 4 pp.

Trent, V.A., 1976. The Emperor Volcanic Complex of the east Gogebic range, Michigan. In: G.V. Cohee and W.B. Wright (Editors), Changes in Stratigraphic Nomenclature by the U.S. Geological Survey, 1975, U.S., Geol. Surv., Bull., 1422-A; A69—A74.

Tyler, S.A., 1949. Development of Lake Superior soft iron ores from metamorphosed iron-formation. Geol. Soc. Am. Bull., 60: 1101—1124.

Tyler, S.A. and Bailey, S.W., 1961. Secondary glauconite in the Biwabic Iron-formation of Minnesota. Econ. Geol., 56: 1033—1044.

Tyler, S.A. and Twenhofel, W.H., 1952. Sedimentation and stratigraphy of the Huronian of upper Michigan, Parts I and II. Am. J. Sci., 250: 1—27, 118—151.

Van Hise, C.R. and Leith, C.K., 1911. The geology of the Lake Superior region. U.S., Geol. Surv., Mon., 52: 641 pp.

Van Schmus, W.R., 1965. The geochronology of the Blind River—Bruce mines area, Ontario, Canada. J. Geol., 73: 755—780.

Van Schmus, W.R., 1976. Early and middle Proterozoic history of the Great Lakes area, North America. In: Global Tectonics in Proterozoic Times. Philos. Trans. R. Soc. London, Ser. A, 280: 605—628.

Van Schmus, W.R., 1978. Geochronology of the southern Wisconsin rhyolites and granites. Wis., Geol. Nat. Hist. Surv., Geoscience Wisconsin, 2: 19—24.

Van Schmus, W.R., 1980. Chronology of igneous rocks associated with the Penokean orogeny in Wisconsin. In: G.B. Morey and G.N. Hanson (Editors), Selected Studies of Archean Gneisses and Lower Proterozoic Rocks, Southern Canadian Shield. Geol. Soc. Am., Spec. Pap., 182: 159—168.

Van Schmus, W.R. and Anderson, J.L., 1977. Gneiss and migmatite of Archean age in the Precambrian basement of central Wisconsin. Geol. Soc. Am., Geology, 5: 45—48.

Van Schmus, W.R. and Woolsey, L.L., 1975. Rb-Sr geochronology of the Republic area, Marquette County, Michigan. Can. J. Earth Sci., 12: 1723—1733.

Van Schmus, W.R., Medaris, L.G., Jr. and Banks, P.O., 1975a. Geology and age of Wolf River batholith, Wisconsin. Geol. Soc. Am. Bull., 86: 907—914.

Van Schmus, W.R., Thurman, E.M. and Peterman, Z.E., 1975b. Geology and Rb-Sr chronology of Middle Precambrian rocks in eastern and central Wisconsin. Geol. Soc. Am. Bull., 86: 1255—1265.

White, D.A., 1954. Stratigraphy and structure of the Mesabi range, Minnesota. Minn., Geol. Surv., Bull., 38: 92 pp.

Wolff, J.F., 1915. Ore bodies of the Mesabi range. Eng. Min. J., 100: 89—94, 135—139, 178—185, 219—224.

Wolff, J.F., 1917. Recent geologic developments on the Mesabi range, Minnesota. Trans. Am. Inst. Min. Metall. Eng., 56: 142—169.

Chapter 3

THE HAMERSLEY BASIN

A.F. TRENDALL

INTRODUCTION

The term Hamersley Basin was first used in a formally defined sense by Trendall (1968a) to refer to the depositional basin of the Precambrian Hamersley Group, of the northwestern part of Western Australia. MacLeod (1966, p. 11) had earlier applied the term Hamersley Iron Province to the area of outcrop of the Hamersley Group; although MacLeod (1966, p. 64) had also referred to the basin in which the Hamersley Group accumulated as the Hamersley basin of sedimentation he neither defined nor consistently used this term. Its scope was later (Trendall and Blockley, 1970, p. 17) extended to apply also to the depositional basin of the underlying Fortescue Group and the overlying Wyloo Group. They regarded the basin as a continuing tectonic entity which was infilled by these three successive groups, which had been previously included by Halligan and Daniels (1964) within the Mount Bruce Supergroup. This concept of the Hamersley Basin followed a programme of regional mapping at a scale of 1 : 250,000 by the Geological Survey of Western Australia, to which a total of 14 geologists contributed.

Trendall and Blockley's (1970, p. 278) proposed reconstruction of the basin envisaged a closed, crudely ovoid intracratonic depression with an area of about 100,000 km^2; a connection with the open ocean, possibly across a volcanic ridge, was envisaged, and an analogy was drawn, in degree of restriction, with the present Okhotsk Sea. Horwitz and Smith (1978), challenging this barred basin concept for the deposition of the Hamersley Group, proposed instead a shelf environment at the margin of a deep ocean; they nevertheless retained the term Hamersley Basin. Horwitz (1978, 1980) has used, but not defined, the term Hamersley Province, apparently in a similar sense to these earlier usages of Hamersley Basin. He (Horwitz, 1981) has subsequently introduced the term Hamersley Shelf. More recently, Morris and Horwitz (1981, 1983) have pointed out that a model for deposition of the Hamersley Group on an oceanic platform comparable with the present Bahama Platform can be argued, and have used the term "Hamersley Platform".

In this paper, following the conventional practice of the Geological Survey of Western Australia (GSWA, 1975, p. 30) the term "basin" is used interchangeably in two senses. Firstly, it denotes the actual present outcrop area of a substantial thickness of sedimentary rocks which possess unifying characteristics of stratigraphy and structure, due to their deposition during a regionally restricted episode of crustal depression, or a related sequence of such episodes. However, it refers also to the actual crustal depression in which those sediments accumulated. The use of the term in these two different senses need cause no confusion, as the context clarifies which sense is intended. The "depression" referred to is a depression relative to sea level, resulting in submersion and not simply a depression relative to the surface of the lithosphere: the vast bulk of sedimentation throughout geological time has probably taken place on relative elevations of the lithosphere — the immediate margins of the continents. The term "basin" used in this way carries no implications of the physical configuration of the depositional area of the sediments concerned, and could include equally well a continental margin shelf or a closed intracratonic sea.

Several descriptive summaries of the Hamersley Basin have already been published, and are noted below the succeeding heading. The main purpose of the present paper is to provide an account comparable with the other basin descriptions in this volume. Apart from the inclusion of results more recent than those presented earlier reviews, this paper differs in two further respects. It uses for the first time a modified nomenclature for banding in the BIFs which has been proposed by Trendall et al. (in prep.); and it includes a detailed analysis of the published geochronological evidence, as a basis for a new interpretation of the tectonic development of the basin.

DOCUMENTATION

The rocks of the Hamersley Basin first received serious geological attention, by A.G. Maitland, in 1903. During the following decade, Maitland (1904, 1905, 1906, 1908, 1909) and H.W.B. Talbot (1920) demonstrated the unconformable relationship of the rocks of the basin over the underlying Pilbara Block, and established them in the literature as the "Nullagine Series". A few notes and descriptions of specific rocks or areas of the Hamersley Basin were published later by Forman (1938), Finucane (1939) and Miles (1942), but the main geological boundaries established by Maitland and Talbot were not modified until the area was systematically mapped at a scale of 1 : 250,000, about 40 years after the completion of their work.

That programme of mapping, by the Geological Survey of Western Australia, has served as a foundation for much subsequent work on particular aspects of basin geology, and is still continuing for the production of second edition sheets. For convenience, the boundaries and names of all the 1 : 250,000 sheets which include parts of the Hamersley Basin are set out in Table 3-I, together with the years of publication.

MacLeod (1966) provided an excellent synthesis of the early results of

TABLE 3-I

1 : 250,000 scale map sheets covering the Hamersley basin

Sheet name	Latitude (°S) of boundaries North	South	Longitude (°E) of boundaries West	East	Reference (* = Second edition)
Dampier and Barrow Island	20°	21°	115° 30′	117°	Kriewaldt, 1964
Roebourne	20°	21°	117°	118° 30′	Ryan, 1966
Port Hedland	20°	21°	118° 30′	120°	Low, 1965; *Hickman, 1977
Yarrie	20°	21°	120°	121° 30′	Wells, 1959; *Hickman and Chin, 1977
Yarraloola	21°	22°	115° 30′	117°	Williams, 1968
Pyramid	21°	22°	117°	118° 30′	Kriewaldt and Ryan, 1967
Marble Bar	21°	22°	118° 30′	120°	Noldart and Wyatt, 1962; *Hickman and Lipple, 1978
Nullagine	21°	22°	120°	121° 30′	Noldart and Wyatt, 1962; *Hickman, 1978
Wyloo	22°	23°	115° 30′	117°	Daniels, 1970
Mount Bruce	22°	23°	117°	118° 30′	de la Hunty, 1965
Roy Hill	22°	23°	118° 30′	120°	MacLeod and De la Hunty, 1966
Balfour Downs	22°	23°	120°	121° 30′	de la Hunty, 1964
Turee Creek	23°	24°	117°	118° 30′	Daniels, 1968
Newman	23°	24°	118° 30′	120°	Daniels and MacLeod, 1965
Robertson	23°	24°	120°	121° 30′	de la Hunty, 1969

this regional mapping and his plate 2, a map at a scale of 1 : 500,000, is still the best geological map covering most of the basin in a single sheet. An earlier paper by MacLeod et al. (1963) was superseded by the 1966 bulletin, but retains documentary significance as being the first formal definition of much of the basin stratigraphy. The later bulletin of Trendall and Blockley (1970) focussed mainly on the iron-formations of the Hamersley Group, but provided also a general summary of the development of the basin.

Detailed contributions to specialised aspects of Hamersley Basin geology since 1970 include work on isotopes of carbonate carbon, oxygen (Becker and Clayton, 1972, 1976), carbon (Oehler et al., 1972) and strontium (Van der Wood, 1977), on the chemical composition of particular stratigraphic units (Trendall and Pepper, 1977; Ewers and Morris, 1980, 1981; Davy, in press), and also on the phosphorus distribution within a single unit (Morris, 1973). Mineralogical work on one of the BIF units (Klein and Gole, 1981) and detailed mineralogy has also been undertaken by Smith, both alone (Smith, 1975) and with others (Smith et al., 1982) in support of a study of metamorphism. Smith (1976) has also contributed local stratigraphic detail, as have Horwitz (1976, 1978), Blockley (1979) and Trendall (1979). A great deal of geochronological work has been carried out on both the Hamersley Basin and the structurally underlying Pilbara Block, and the relevant references in this field appear in Table 3-V. Papers dealing with aspects of the broad development and regional relationships of the basin include those of Horwitz and Smith (1978) and Gee (1979). The most important recent advance in the study of the iron ore deposits has been made by Morris (1980). Other references on iron ore are mentioned under the appropriate heading below.

Three papers which are practical field guides to persons wishing to visit the basin independently are worth mentioning. These are a paper by Trendall (1966) which was intended as a field guide to the outstanding exposures of the Dales Gorge Member at Dales Gorge, a similar guide (Trendall, 1968b) to the superb exposures of the Joffre Member in the gorges south of Wittenoom, and an excursion guide prepared for the 1976 International Geological Congress (Trendall, 1976a), which permits independent repetition of a 6-day excursion across the basin. It may also be useful to note here that several reviews of Hamersley Basin geology (Trendall, 1975a, 1975b, 1979) contain no primary data not included in other papers referred to elsewhere in this paper.

Much stable-isotope and palaeontological work has recently been carried out on some units of the Hamersley Basin by the "Precambrian Paleobiology Research Group", led by J.W. Schopf and based at UCLA. I am indebted to this group for providing me with its early results and approval to use them here. However, it has seemed more appropriate in this paper, which is largely a review of published work, to note the forthcoming appearance of those results without any attempt to summarise them.

LOCATION, AREA, SHAPE, AND OUTCROP LIMITS

The position of the Hamersley Basin within the Australian continent is shown in Fig. 3-1. The present outcrop area of the three stratigraphic groups now regarded as representing the contents of the basin (see following heading) is somewhat over 100,000 km^2. Of this area the Fortescue Group crops out over about 40,000 km^2, while the outcrop of the overlying Hamersley Group occupies some 60,000 km^2. The uppermost Turee Creek Group covers only about 1200 km^2. These three groups together comprise the Mount Bruce Supergroup.

The likely shape and extent of the depositional basin at different stages in its development are discussed under a later heading. Attention here is limited to certain features of the present outcrop, and in particular of its boundaries, as they appear in Fig. 3-1. Virtually all present outcrops of the Mount Bruce Supergroup lie within an ellipse with a minor axis about 400 km long and a major axis of about 600 km, lying approximately west-northwest; the centre of this ellipse is about a third of the way along a straight line from Wittenoom

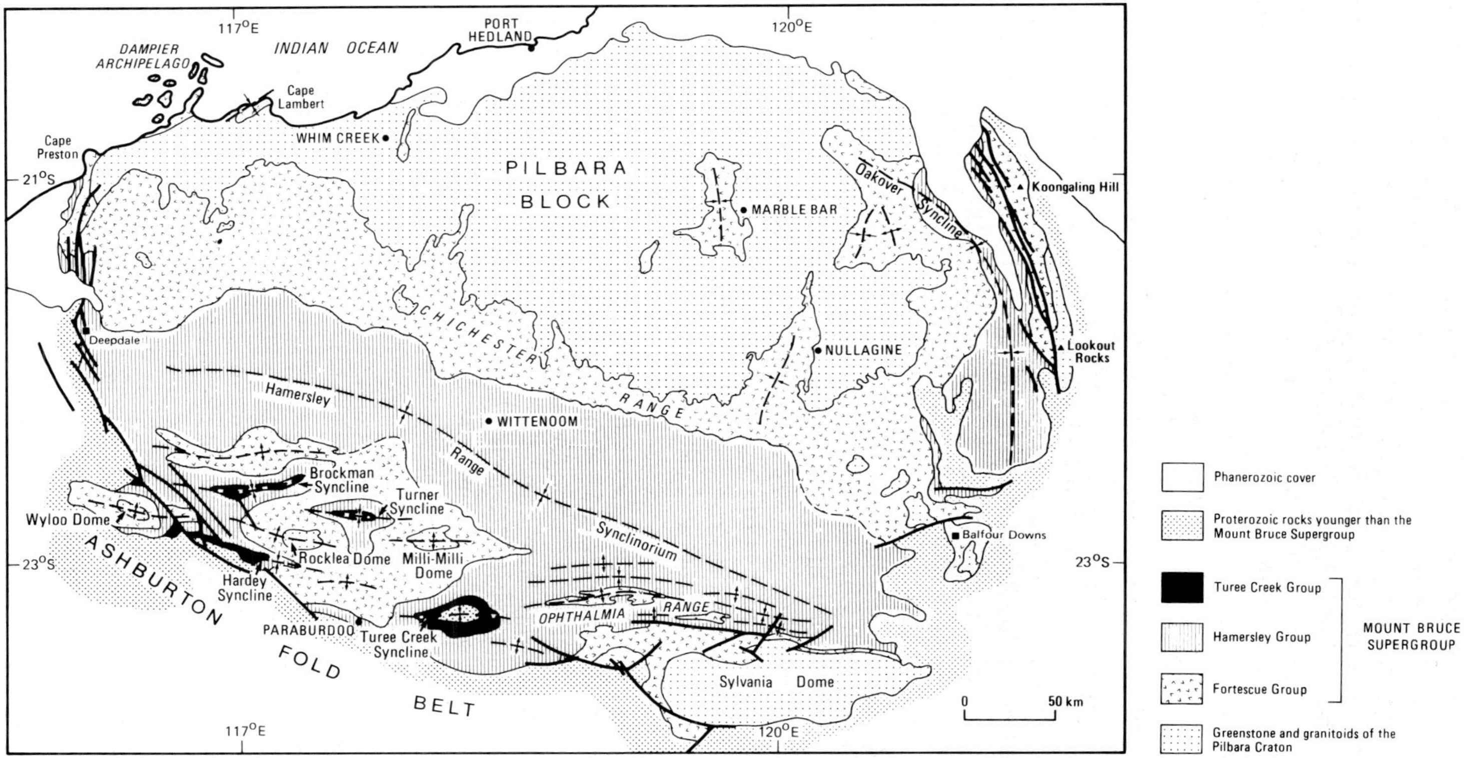

Fig. 3-1. Geological map of the Hamersley Basin, based on GSWA (1979).

to Marble Bar. Many structural features of the basin within and adjoining the onshore part of this ellipse reinforce its geological significance, and this is discussed below a later heading (The "Pilbara egg"); much of the northern part is offshore, and purely conceptual. It is nevertheless useful to use this concept as a basis for the following commentary.

The major axis of the ellipse divides the outcrop of the Mount Bruce Supergroup into two contrasting halves. The southern half, or main outcrop area, is underlain almost entirely by this Supergroup, with rare and scattered inlying domes of Archaean rocks. The situation in the northern half is reversed, with Archaean rocks of the Pilbara Block forming most of the area, with only scattered outliers of the Mount Bruce Supergroup. Only in the east, along the Oakover Syncline, does the main outcrop area have significant northward continuity. The major axis of the ellipse thus follows closely the line of a regional unconformity at which the gently south-dipping Fortescue Group overlies the Pilbara Block. Although it is later questioned whether the tectonic significance of this unconformity is as great as its clarity of expression at first suggests, the fact of that expression, both on the map and on the ground at the foot of the northfacing scarp of the Chichester Range (Fig. 3-1), cannot be doubted.

As far as the outer outcrop limits are concerned the gentle northwesterly dip and southwesterly strike of the Fortescue Group in the Dampier Archipelago, and the south-southwesterly syncline at Cape Lambert, emphasize the conceptual continuity of the bounding ellipse into the Indian Ocean. Anticlockwise along the ellipse from these locations, in the first mainland outcrops of the main outcrop area, the Fortescue and Hamersley Groups, at Cape Preston and nearby James Point, strike north—south and dip westwards; outcrop is limited to the west by overlying Phanerozoic rocks. This situation continues southwards as far as Deepdale. Between Deepdale and the Wyloo Dome a narrow fault zone, swinging steadily to the southeast, sharply demarcates the Hamersley Basin from the younger sediments (Wyloo Group) of the Ashburton Fold Belt (of Gee, 1979) to the west. The junction between the Mount Bruce Supergroup and the Wyloo Group in the Wyloo Dome—Hardey Syncline area is of particular significance and is referred to in more detail later. Between the Hardey Syncline and the Sylvania Dome the edge of the basin is both poorly exposed and incompletely understood. Although much of this section was originally mapped as faulted (Daniels, 1968) it now appears that the line of poor exposure, which here tends to follow the contact between the Mount Bruce Supergroup and the Wyloo Group, may conceal an unfaulted but steeply dipping unconformity (Bourn and Jackson, 1979). The nature of the contact on the north side of the Sylvania Dome is also poorly understood, and is assumed or deduced (Blockley et al., 1980) to be an unconformity, although folding of the Mount Bruce Supergroup immediately north of the dome is intense, and some movement along the contact would not be surprising. Horwitz (1976) has described the basal section at one location.

Between the eastern end of the Sylvania Dome and the southern end of the Gregory Range, south of Lookout Rocks the edge of the Hamersley Basin has an entirely different character; it is limited in the poorly exposed country of the Balfour Downs area by the irregular boundary of the unconformably overlying and virtually undeformed, younger Proterozoic sediments of the Bangemall and Yeneena Groups. Along the Gregory Range, which runs north-northwest between Lookout Rocks and Koongaling Hill, there is another abrupt change. A zone of sub-parallel faults clearly defines the arcuate edge of the ellipse; the unconformable base of the Yeneena Group here follows the same arc a few kilometres to the east. Geological relationships in this section of the basin margin are enigmatic, and are discussed further below. From Koongaling Hill the northern completion of the ellipse anticlockwise back to the Dampier Archipelago, is masked, if it exists at all, by younger sediments or by the Indian Ocean.

STRATIGRAPHY

Major stratigraphic components

Trendall's (1968a) account of the Hamersley Basin, in which that term was introduced, was based largely on the work led by MacLeod (1966). This had established three main stratigraphic subdivisions of the Mount Bruce Supergroup: the basal, largely volcanic or volcaniclastic Fortescue Group; the succeeding, mainly chemical, Hamersley Group; and the final, and principally clastic, Wyloo Group. However, early reservations (Trendall and Blockley, 1970, p. 295; Trendall, 1975a, p. 119) were expressed concerning the validity of the Wyloo Group as a component of the same basin as that in which the Fortescue and Hamersley Groups were laid down. Later stratigraphic revision (Trendall, 1979) formalized this view, by redefining the Mount Bruce Supergroup to include the Fortescue Group, Hamersley Group, and a newly defined Turee Creek Group. The Turee Creek Group, as the former Turee Creek Formation, had been the lowest unit of the Wyloo Group, which was shown to overlie the Turee Creek Group with marked unconformity. The revised Wyloo Group, comprising the main bulk of the group as first understood, is now considered to have been laid down in the later, and separate, Ashburton Trough of Gee (1979), the folded content of which forms the Ashburton Fold Belt (Fig. 3-1). Further consideration of the Wyloo Group is outside the scope of this paper.

Thus the post-1979 view of Hamersley Basin stratigraphy differs significantly from that prevailing during the preceding decade, although the primary threefold volcanic-chemical-clastic subdivision is unchanged. The stratigraphy of the Hamersley Basin as it is now understood is remarkably consistent throughout the preserved basin area. While some local facies variations and hiatuses are present, and are noted below, the general succession described here has basinwide validity.

Fortescue Group

This group crops out (Fig. 3-1) over all of the fifteen 1 : 250,000 scale map sheets listed in Table 3-I. Regional mapping of these sheets was carried out with some independence, with the result that local variations of stratigraphic nomenclature arose. Trendall (1975a) grouped these into a lithologically generalised correlation table. More recently, Hickman (1980) has employed a standardised nomenclature on a map covering those parts of the Fortescue Group within or adjacent to the Pilbara Block. In Table 3-II this nomenclature is used as a basis for summarising the salient stratigraphic features of the group over its entire outcrop area.

Two points concerning Fortescue Group stratigraphy are not included in Table 3-II: firstly, an important local difference in the sequence exposed on the eastern side of the Oakover Syncline (Fig. 3-1), and secondly the distribution of several major stratiform intrusions near the base of the succession. In the Gregory Range area, between Koongaling Hill and Lookout Rocks (Fig. 3-1), Hickman (1980) indicates the identifiable presence only of the Jeerinah Formation, the Tumbiana Formation, and the Kylena Basalt. Between the first two of these he applies the local name Pearana Basalt to basalts believed to represent both the Maddina Basalt and the Nymerina Basalt, the Kuruna Siltstone being absent. In this area also, the Kylena Basalt directly overlies a 1000-m-thick succession of felsic volcanic rocks named the Koongaling Volcanics by Hickman (1975), who regards it as a felsic correlative of the Mount Roe Basalt.

Below the Koongaling Volcanics, Hickman (1980) indicates a major granophyre intrusion (formerly the Isabella Porphyry) believed to separate the Koongaling Volcanics from unconformably underlying granitoids of the Pilbara Block, at and north of Lookout Rocks. The significance of the relationships between these units for basin developments is discussed in more detail later. Other major stratiform bodies which intrude the Fortescue Group include the Gidley Granophyre, on the peninsula immediately south of the Dampier Archipelago (Fig. 3-1), the Cooya Pooya Dolerite, in the westernmost part of the Chichester Range, and the Spinaway and Bamboo Creek Porphyries, around and north of Nullagine. None of these is individually shown on Figs. 3-1 or 3-11.

The major feature of Fortescue Group stratigraphy, evident from Table 3-II, is the broad alternation of extrusive volcanic (lava) units with clastic sedimentary units. The lava units consist of extensive flows of dark, fine-grained, massive, amygdaloidal or vesicular basalt, locally pillowed. Of the intervening clastic units both the Kuruna Siltstone and Tumbiana Formation have a major tuffaceous content, and lapilli tuffs are common. Shallow-water

TABLE 3-II

Stratigraphic summary of the Fortescue Group

Formation	Principal lithologies	Dampier	Roebourne	Port Hedland	Yarrie	Yarraloola	Pyramid	Marble Bar	Nullagine	Wyloo	Mount Bruce	Roy Hill	Balfour Downs	Turee Creek	Newman	Robertson	Main local subdivisons or nomenclatural variations
		Reported maximum thickness (m) from most recent 1 : 250,000 scale geological mapping. For sheet locations and references see Table 3-I															
Jeerinah Formation	Siltstone, shale and chert, sandstone; local felsic lava, tuff and agglomerate; basalt; mafic sills locally abundant	?	0	0	200	210	107	0	500	NM	920	175	60	900	760	305	
Maddina Basalt	Vesicular and amygdaloidal basalt and andesite		0	0	100	610	610	NM	200			60			0		Mount Jope Volcanics (Maddina Basalt to Kylena Basalt); Little De Grey Lava upper unit (Maddina Basalt to Nymerina Basalt)
Kuruna Siltstone	Sandstone and siltstone; local shale, ooidal sediments, pisolitic tuff and banded siliceous limestone		0	0	20			20	20			20	60		0		Pillingi i Tuff (Kuruna Siltstone to Tumbiana Formation)
Nymerina Basalt	Coarse-textured basalt; some amygdaloidal basalt, pillow lava and interbedded pyroclastics and sediments	2100	0	0	300	305	122	50	300	"Thick" NM	2133	150		NM	0	122	
Tumbiana Formation	Siliceous limestone (stromatolitic) and pisolitic tuff; subordinate tuff, basalt, and sediments		0	0	150			60	200			75	45		0		
Kylena Basalt	Massive amygdaloidal and vesicular basalt and andesite; subordinate but widespread interbedded sediments		0	880	500	460	213	300	500			30	0		0		
Hardey Sandstone	Sandstone and conglomerate; local tuff, shale, mudstone and basalt		305	20	1000	230	243	1000	1000	0	1220	0	15	1000	?		
Mount Roe Basalt	Basalt and andesite with thin basal sedimentary rocks	?	NM	500	300	60	305	50	300	0		0	0	0	0		

0 = not exposed or not developed on sheet. NM = exposed but not measured.

stromatolitic carbonate intercalations are locally well developed. The shale component of the uppermost clastic unit, the Jeerinah Formation, has little demonstrable volcanic content, but in many areas of its outcrop pillowed basalts and associated massive basaltic flows or sills are abundant. The lowest clastic unit, the Hardey Sandstone, differs in being composed largely of granitic debris, with only a minor tuffaceous content.

The lateral continuity of the lowest unit, the Mount Roe Basalt, is very restricted; this is discussed in more detail further below and illustrated in Fig. 3-11. Each successive overlying formation not only tends to be more laterally extensive, but also to have a more uniform regional thickness. The Jeerinah Formation represents the culmination of this trend, apparently spreading as a blanket over the entire area of the basin.

Regional variations in the total Fortescue Group thickness and in the individual thicknesses of individual formations, are not well known. Trendall (1975a, pp. 126—127) compiled and displayed the thickness data obtained during systematic mapping programmes of the Geological Survey (Table 3-I). From this it appeared that, with some local irregularity, the thickest development of the Fortescue Group, of some 4.3 km (de la Hunty, 1965) occurred in the Mount Bruce map sheet area (Table 3-I), in the west-central part of the basin, and closely contiguous with the thickest area of the Dales Gorge Member (Fig. 3-3). Thickness data from regional mapping, including that from the second edition maps (Table 3-I) not available to Trendall (1975a), are included in Table 3-II. This thickness variation of the Fortescue Group has been challenged by Horwitz and Smith (1978), Horwitz (1980) and Morris and Horwitz (1983). All these authors (e.g., Morris and Horwitz, 1983, fig. 2B) interpret, on the basis of more recently measured sections in the southernmost part of the Fortescue Group area between the Rocklea Dome and Paraburdoo (Fig. 3-1), a general thickening of the Fortescue Group from northeast to southwest. Blight (in prep.) in a detailed study of the lower part of the Fortescue Group in the Mount Bruce and Wyloo sheet areas, has presented thickness data which indicate rather complex local variations. Miyano (1976, fig. 6) has also presented a generalisation of Fortescue Group regional thickness variation. More and better documented isopach data are needed from the Fortescue Group before a clear regional appreciation is possible.

Whatever the regional thickness distribution of the Fortescue Group may be, there is no indication from it that any of the volcanic material was derived from a locally confined source. It is probable that each of the major stratigraphically distinct and continuous lava units, such as the Kylena Basalt, represents a complex of coalescing flows from many sources. Smith (1979) has described a sequence of separate thin flows in a section of the Maddina Basalt. Abundant mafic dykes within the Pilbara Block, trending north-northeast, have been linked by Lewis et al. (1975) with the extrusion of Fortescue Group basalts; this possibility is discussed later. Regardless of the validity of that link, it seems likely that the extrusive mafic units within the Fortescue

Group represent periods during which widespread crustal fracturing tapped a deep, uniform, and abundant magma source over a wide area.

Hamersley Group

The stratigraphic subdivision of the Hamersley Group, as established by MacLeod et al. (1963), and modified by Trendall and Blockley (1970), is displayed in Fig. 3-2. The formation thicknesses represented there are those of the central part of the basin, except for the expanded section of the Dales Gorge Member, on which the type section thicknesses are shown. Isopachs are available only for the Dales Gorge Member of the Brockman Iron Formation (Fig. 3-3). These show maximum thicknesses along a west-northwest-trending trough in the central part of the basin, with a general outward decrease in all directions. The regional variation in thickness of the other formations, with the exception of the Woongarra Volcanics, appears to follow a similar pattern, although no systematic data are available.

In Table 3-III one important feature of the vertical distribution of the sedimentary components of the Hamersley Group is emphasised; this is a broad alternation between shale (or shale and dolomite) and banded iron-formation (BIF), usually with subordinate shale, which alternation extends down into the Fortescue Group. Each of the (even-numbered) BIF units in this sequence is lithologically different from the others, and in the case of the topmost unit the Weeli Wolli Formation and Boolgeeda Iron Formation also differ from each other to give a total of five major lithologically distinctive BIF units. These are described under a separate heading below.

TABLE 3-III

Vertical distribution of the sedimentary components of the Hamersley Group

	Unit	Approximate thickness (m)
8.	Weeli Wolli Formation and Boolgeeda Iron Formation	380
7.	Yandicoogina Shale Member of Brockman Iron Formation	60
6.	Joffre Member of Brockman Iron Formation	370
5.	Whaleback Shale Member of Brockman Iron Formation	60
4.	Dales Gorge Member of Brockman Iron Formation + top of Mt McRae Shale	180
3.	Wittenoom Dolomite + Mount Sylvia Formation + lower part of Mt McRae Shale	260
2.	Marra Mamba Iron Formation	180
1.	Jeerinah Formation of Fortescue Group	150—300

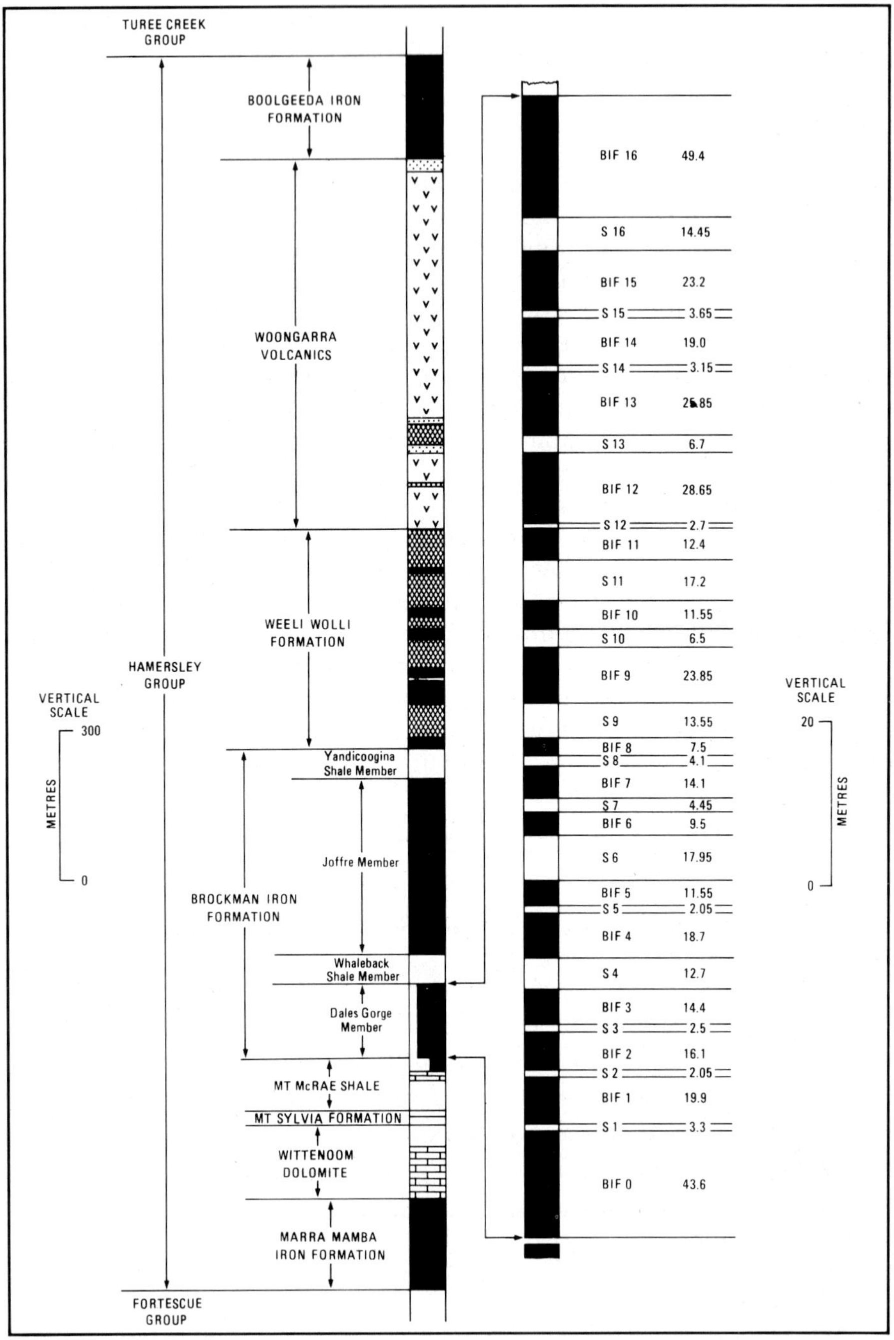
TUREE CREEK GROUP
BOOLGEEDA IRON FORMATION
WOONGARRA VOLCANICS
WEELI WOLLI FORMATION
HAMERSLEY GROUP
VERTICAL SCALE
300
METRES
0
Yandicoogina Shale Member
Joffre Member
BROCKMAN IRON FORMATION
Whaleback Shale Member
Dales Gorge Member
MT McRAE SHALE
MT SYLVIA FORMATION
WITTENOOM DOLOMITE
MARRA MAMBA IRON FORMATION
FORTESCUE GROUP
BIF 16 49.4
S 16 14.45
BIF 15 23.2
S 15 3.65
BIF 14 19.0
S 14 3.15
BIF 13 25.85
S 13 6.7
BIF 12 28.65
S 12 2.7
BIF 11 12.4
S 11 17.2
BIF 10 11.55
S 10 6.5
BIF 9 23.85
S 9 13.55
BIF 8 7.5
S 8 4.1
BIF 7 14.1
S 7 4.45
BIF 6 9.5
S 6 17.95
BIF 5 11.55
S 5 2.05
BIF 4 18.7
S 4 12.7
BIF 3 14.4
S 3 2.5
BIF 2 16.1
S 2 2.05
BIF 1 19.9
S 1 3.3
BIF 0 43.6
VERTICAL SCALE
20
METRES
0

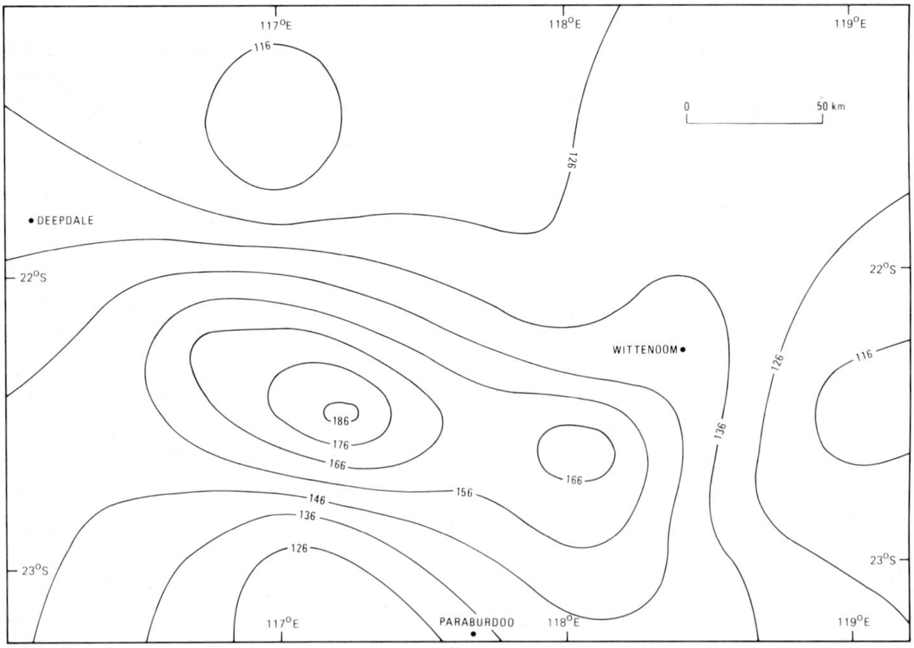

Fig. 3-3. Isopach map of the total thickness (all macrobands) of the Dales Gorge Member of the Brockman Iron Formation. This map is based on the data used for plate 3 of Trendall and Blockley (1970), but with some deletions and additions; the computer-generated lines shown here have the same regional pattern as Trendall and Blockley's "eyed-in" isopachs.

Comparatively little work has been carried out on the sedimentary units other than the iron-formations (that is, the odd-numbered units in Table 3-III). The units which contain shale as the main, or an important, constituent include the Mount McRae Shale, the Wittenoom Dolomite, the Mount Sylvia Formation, and the Whaleback and Yandicoogina Shale Members of the Brockman Iron Formation. Although it is poorly exposed, much of the shale appears to share common characteristics: it is very fine-grained, dark grey or green to black when fresh, and white or buff when weathered, often finely laminated, and usually with a proportion of inter-mixed chert and carbonate, and sometimes a little iron-formation. The upper part of the Mount McRae Shale contains thin bands of shard-bearing volcanic ash, but there is no other direct

Fig. 3-2. Stratigraphic subdivision and lithological summary of the Hamersley Group (left-hand column), and internal detail of the Dales Gorge Member of the Brockman Iron Formation (right-hand column). The thicknesses shown are generally representative of the west-central part of the main outcrop area (Fig. 3-1), except for the Dales Gorge Member, which shows the type section (Trendall and Blockley, 1968). Lithology is represented as follows: BIF — solid black; shale — white; dolerite — diagonal hatching; acid lava — v; tuff — stipple; dolomite — brick pattern.

evidence for a volcanic contribution. In spite of its apparent uniformity, there seem to be at least two varieties of shale: firstly, very fine-grained, soft, poorly laminated black shale rich in free carbon and pyrite, and secondly, dark green, finely laminated shale often with some aftbanded chert and with evidence of volcanic activity. Intermediate types *may*, both lithologically and in depositional significance, represent mixtures of these two end-members, but insufficient work has yet been done to test this suggestion.

The lower part of the Wittenoom Dolomite is composed of massive, medium to thin-bedded dolomite with rare beds of black chert. Where fresh, the dolomite is finely crystalline and brown, pink, or grey with faint colour banding. It contains sedimentary structures such as cross-beds and slumps, and features such as stylolites and chert nodules. Stromatolites are present in the outlying stratigraphic equivalent of the Wittenoom Dolomite, the Carawine Dolomite, in the Oakover Syncline area of the northeastern part of the basin, but are absent over the main outcrop area. Davy (1975) has provided chemical and petrographic details of the lowermost part of the Wittenoom Dolomite, and Button (1976) has made comparative comments on the Wittenoom Dolomite and the extensive dolomites of the South African Transvaal Basin.

The igneous rocks which form over 40% of the Hamersley Group comprise thick dolerite sills largely within the Weeli Wolli Formation, and acid volcanics of the overlying Woongarra Volcanics. Over most of the outcrop area more than half of the thickness of the Weeli Wolli Formation consists of dolerite in several separate sills. Although local variations have been recorded in the number and thickness of the sills, down to their complete absence in the Yarraloola Sheet area, there is insufficient information to relate these to other aspects of regional stratigraphy. A dolerite sill within the Woongarra Volcanics may be assumed to be the uppermost representative of this group of sills. The dolerite is uniformly massive and homogeneous, dark green, and medium to coarse grained (De Laeter et al., 1974). It invariably shows strong deuteric modification, with uralite, albite, epidote and leucoxene replacing the presumably primary pyroxene, labradorite and ilmenite.

The acid Woongarra Volcanics, the thickest formal stratigraphic unit of the Hamersley Group, includes discrete bands of both non-porphyritic and porphyritic rhyolite and dacite as well as thinner horizons of stratified tuff and agglomerate, at least one major intercalation of BIF, and a dolerite sill.

Fig. 3-4. Stratigraphic subdivision and lithological summary of the Turee Creek Group in the Hardey Syncline and Wyloo Dome areas (Fig. 3-1), showing the major angular unconformity at the base of the Wyloo Group. The figure is modified from fig. 42 of Trendall (1979), in which the relationship between the Wyloo Group and Mount Bruce Supergroup was redefined. The "unnamed quartzite unit 3", which lies above the Turee Creek Group in the Hardey Syncline area, is known only from the core of that structure, and has been disregarded in stratigraphic descriptions of the Mount Bruce Supergroup in this paper, as being incidental to its broad subdivision into three components: Fortescue Group, Hamersley Group, and Turee Creek Group.

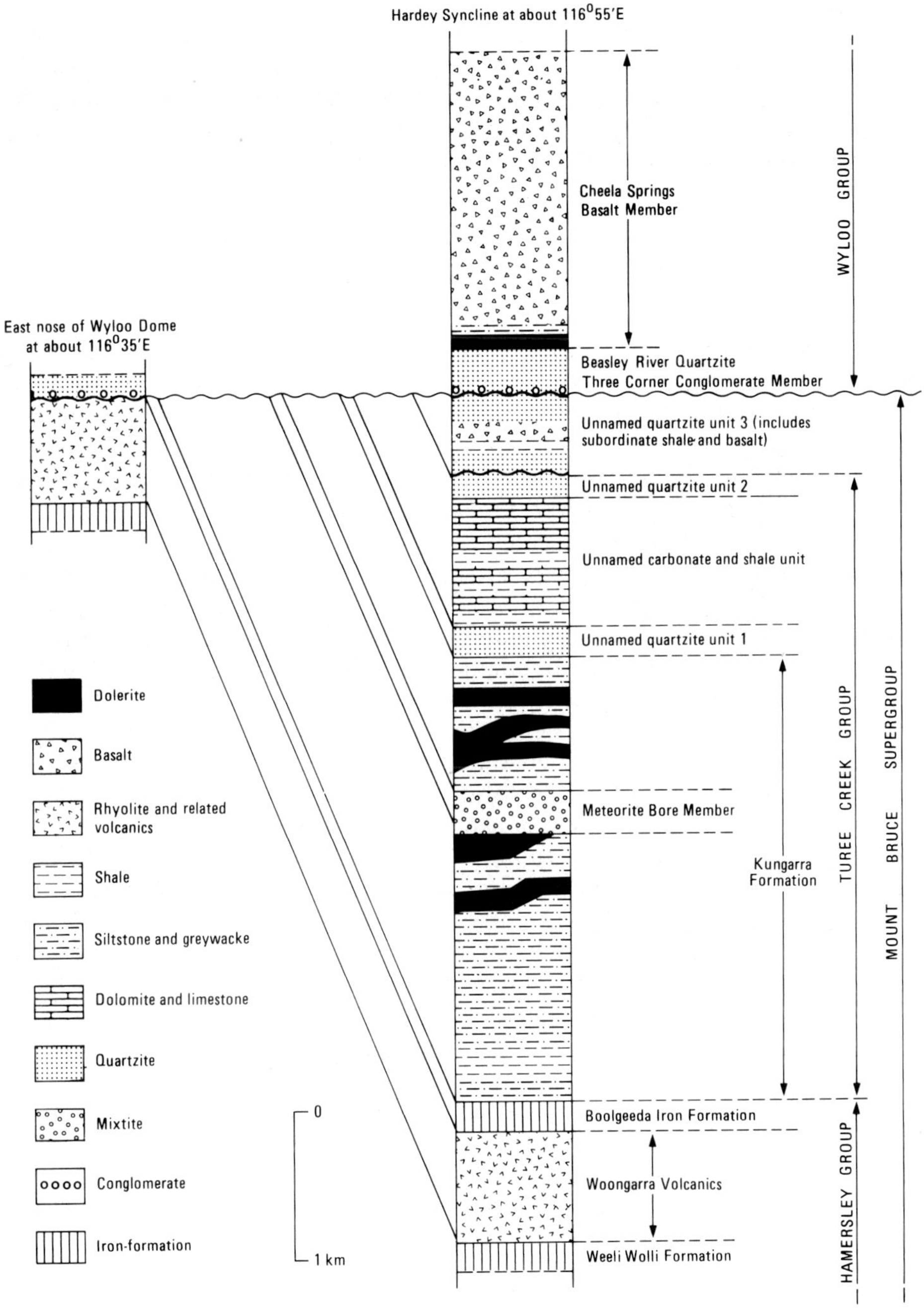
Hardey Syncline at about 116°55'E
East nose of Wyloo Dome at about 116°35'E
Cheela Springs Basalt Member
Beasley River Quartzite
Three Corner Conglomerate Member
Unnamed quartzite unit 3 (includes subordinate shale and basalt)
Unnamed quartzite unit 2
Unnamed carbonate and shale unit
Unnamed quartzite unit 1
Meteorite Bore Member
Kungarra Formation
Boolgeeda Iron Formation
Woongarra Volcanics
Weeli Wolli Formation
WYLOO GROUP
TUREE CREEK GROUP
HAMERSLEY GROUP
MOUNT BRUCE SUPERGROUP
Dolerite
Basalt
Rhyolite and related volcanics
Shale
Siltstone and greywacke
Dolomite and limestone
Quartzite
Mixtite
Conglomerate
Iron-formation
0
1 km

There is a high order of lateral stratigraphic continuity and concordance with the overlying rocks and these features contributed to their original interpretation as extrusive. Recent unpublished re-examination has made an intrusive origin much more likely. A recent paper by Kokelaar (1982) reports many features produced at the margins of undoubted sills which would earlier have been accepted as supporting extrusion.

Turee Creek Group

Trendall (1979) defined and described the Turee Creek Group in the Wyloo Dome and Hardey Syncline areas (Fig. 3-1), and included within it the thick basal Kungarra Formation and a number of overlying units not formally named; this revised stratigraphy, and its earlier equivalent, is shown in Fig. 3-4.

The base of the Kungarra Formation is rarely well exposed, but wherever it is seen it consists of greenish siltstone overlying the Boolgeeda Iron Formation with perfect conformity. The bulk of the formation consists of a monotonous sequence of greyish-green siltstone, fine-grained greywacke, and fine-grained sandstone, in which the normally thin bedding is weakly defined by slight colour changes. Thin carbonate beds occur in the upper part of the formation and dolerite sills are abundant. In the Wyloo Dome and Hardey Syncline areas a strongly developed slaty cleavage hinders sedimentological interpretation. The 300-m-thick Meteorite Bore Member within the Kungarra Formation consists of mixtite, in the sense of Schermerhorn (1966), or diamictite (Trendall, 1981) in which abundant boulders of sandstone and acid volcanic rock up to 0.3 m across lie randomly scattered in a matrix of greenish-grey siltstone; the volcanic boulders are petrographically indistinguishable from the Woongarra Volcanics of the Hamersley Group. A small proportion of both sandstone and volcanic clasts exhibit striation or grooving of a type suggesting glacial origin (Trendall, 1976b).

Above the Kungarra Formation in the Hardey Syncline area the Turee Creek Group contains a substantial thickness of well sorted white or grey quartzites (Fig. 3-4) the topmost of which lies unconformably below the Three Corner Conglomerate Member of the Beasley River Quartzite, the lowest formation of the Wyloo Group. Of the scattered outlying exposures of the Turee Creek Group shown on Fig. 3-1, those in the Brockman and Turner Synclines probably contain only Kungarra Formation. The differentiation into Turee Creek Group and Wyloo Group shown in the Turee Creek Syncline must at present be considered as conceptual only.

Band nomenclature

The nomenclature introduced by Trendall (1965a) to distinguish different forms of banding in the Dales Gorge Member of the Brockman Iron Formation was based on a hierarchy of three scales of band: "macroband", "mesoband" and "microband". Later work (Ewers and Morris, 1981) has revealed conceptual and practical inadequacies in this system, and the band nomenclature used in the following pages is the revised scheme of Trendall et al. (in prep.), which is briefly described in the Introduction to this volume. But to retain continuity with previously published work on Hamersley Group iron-formations this new nomenclature has been used to the minimum necessary extent. In practice this has meant that, except for Fig. 3-9, in the caption to which some features of the new nomenclature are explained, only one systematic change in terminology is employed: the new term "aftband" replaces directly the discarded term "microband".

Lithology and petrography

The stratigraphic position of the iron-formations has already been described, and is summarised in Table 3-III and Fig. 3-2. Because the Dales Gorge Member is the most intensively studied, and therefore the best known, of the five iron-formation units, this description begins with it.

The type section of the Dales Gorge Member of the Brockman Iron Formation was defined by Trendall and Blockley (1968) in drill core from Wittenoom Gorge and Yampire Gorge (Fig. 3-5). The division of the 142.2 m of type section core into two main lithologies is shown in the right-hand column of Fig. 3-2. Each of the 33 numbered subdivisions of the member is designated a macroband. The 17 BIF macrobands are numbered upwards from 0 to 16 and each of these except the lowermost is underlain by a similarly numbered shale, or S, macroband. Trendall and Blockley (1968), in their published description of the type section, included continuous photographic coverage of the defining drill core at a scale of one fifth natural size; one of their figures is reproduced here as Fig. 3-6.

The S macrobands consist mainly of dark green to black, iron-rich, stilpnomelane-bearing shale, often finely laminated, and of interbanded chert and green siderite, which may be very finely laminated, more or less structureless, or thinly bedded, with a ghost clastic structure defined by slight colour variations within the fine-grained siderite. Limestone and breccia bands also occur locally within the S macrobands. Thin bands of stilpnomelane within the shale have textural variations which define the shapes of volcanic shards (La Berge, 1966).

Fig. 3-5. View of the eastern side of Wittenoom Gorge immediately north of the old Wittenoom Mine buildings, showing the macrobands of the Dales Gorge Member of the Brockman Iron Formation (Fig. 3-2). The lack of postdepositional disturbance, other than uplift, and spectacular surface exposure, is typical of much of Hamersley Group exposure of the Hamersley Range Synclinorium (Fig. 3-1). The base of the cliff is at about the midpoint of BIF O, and the S macroband sequence (Fig. 3-2) up the cliff is marked.

The BIF macrobands of the Dales Gorge Member conform in general lithology with "typical" Precambrian BIF of many continents. Thin bands (mesobands) of chert alternate with mesobands of fine-grained iron-rich material, designated chert-matrix by Trendall and Blockley (1970). It is worth emphasizing that this term was introduced as a non-genetic one, based on the purely geometrical status of iron-rich mesobands as a matrix in which the iron-poor (chert) mesobands and pods lie. The name, like its predecessor QIO (Trendall, 1965a), is not a satisfactory one, but no other name exists specifically to denote the material which forms iron-rich mesobands.

Chert mesobands are mostly 5—15 mm thick; the frequency of thicker cherts falls off rapidly up to the measured maximum of 87 mm. Mesobands of chert-matrix have similar thicknesses, and these two main mesoband types, chert and chert-matrix make up about 60 and 20% respectively of the total BIF volume (Fig. 3-5). The remainder are magnetite, carbonate, stilpnomelane, riebeckite, and minor miscellaneous types.

Within many chert mesobands, a distinctive type of small-scale regular lamination defined by layers of some iron-bearing mineral within the general fine mosaic of quartz is present, and is known as aftbanding (Fig. 3-7A). Although aftbands were initially defined (as microbands) by Trendall (1965a) to consist of a simple iron-rich/iron-poor couplet, subsequent detailed examination shows that a good deal of fine structure may be distinguished within this. Ewers and Morris (1981) have found as many as twenty subdivisions in complex aftbands, and report that even the simpler types commonly show three or four zones. The defining mineral of the main iron-rich component of each aftband is normally hematite, siderite, or ankerite, but may be magnetite, stilpnomelane or riebeckite. The aftband interval (between the centres of adjacent iron-rich laminae) in different cherts may be between 0.2 and 1.5 mm but there is, by comparison, negligible variation between successive aftbands in a single mesoband. The total iron content of chert mesobands varies inversely with aftband interval, between about 3 and 30%. Chert-matrix mesobands, by contrast with cherts, consist of a fine-grained mixture of quartz, magnetite, hematite, stilpnomelane, ankerite and siderite and have no regular aftbanding, although there is often a vaguely defined and irregular streakiness or sometimes a very regular fine lamination (Ewers and Morris, 1981, fig. 10). Chert-matrix has an average total iron content of about 40%. Although the ideal geometric situation of perfectly planar mesobands is approached by much of the BIF there is also a great variety of laterally discontinuous cherts (pod cherts or cross-pods), at whose lateral terminations the aftbanding passes without disruption, but with an approximate 7 : 1 reduction in thickness, into the vague streaky lamination of the adjacent chert-matrix (Fig. 3-7B).

In many BIF macrobands of the Dales Gorge Member there exists a cyclic sequence of mesoband types, over a stratigraphic thickness of 10—15 cm which Trendall and Blockley (1970, pp. 55—60) have called the Calamina

57
R
56
55
54
53
61
60
59
58
66
65
64
63
62
R
71
70
69
BIF
2
S 2
68
S2
BIF
1
r
75
r
74
R
73
R
R
72
R
R
80
79
R
78
R
R
77
76
84
83
82
81
89
88
BIF
3
S3
87
86
S3
85
BIF
2
R
93
EC10
47A
92
R
91
R
90
r
98
97
R
96
95
47A
EC10
94
102
S4
BIF
3
101
100
R
99
R
R
107
106
105
104
103

cyclothem. In this sequence a single thick chert mesoband with fine aftbanding, or a group of such cherts separated by thin magnetite mesobands, alternates with a mixed mesoband group of chert-matrix, magnetite, and comparatively thin and coarsely aftbanded cherts; these two alternating components of the Calamina cyclothem are called the chert-magnetite group and the mixed group respectively. The thin aftbanded cherts of the chert-magnetite group are normally pink or red, and have their aftbanding defined by hematite, while the more coarsely aftbanded cherts of the mixed group are white, with carbonate-defined aftbanding.

Riebeckite (Fig. 3-6) occurs within restricted stratigraphic sections (riebeckite zones) of the Dales Gorge Member in some areas, mainly in the form of thick massive mesobands (actually consisting of randomly interlocked fibres) which are stratigraphically equivalent to, and result from the replacement of, the chert-magnetite group of particular Calamina cyclothems. Crocidolite, or blue asbestos, is among the several minor textural types of riebeckite occurrence.

The foregoing descriptive summary of the Dales Gorge Member is largely abbreviated from the more complete description of Trendall and Blockley (1970); since that bulletin was published the work of Ewers and Morris (1981) has added further valuable detailed description to the literature.

The other four major stratigraphic units of BIF, and also the comparatively minor BIF development of the Mount Sylvia Formation (Fig. 3-2) are each characteristically different in some way, although all share the fundamental defining characteristics of high iron content and the presence of quartz (chert) and iron oxides as the dominant constituent minerals. They differ in such subtle characters as the nature of the sheet silicate component, the abundance and distribution of riebeckite, and the amount of shale intercalation, and the thickness and degree of irregularity of the chert mesobands; one or more of these have resultant effects on the field expression. These differences are summarised in Table 3-IV.

None of the other BIF units summarised in Table 3-IV has been studied in as much detail as the Dales Gorge Member, and there would hardly be space in a paper of this length for a fuller account than is provided there. However,

---

Fig. 3-6. Part of the type section of the Dales Gorge Member, reduced from Plate 29 of Trendall and Blockley (1968). The type section was defined in diamond drillcore from the Wittenoom area (Fig. 3-1), and the entire section was measured upwards from the base in feet, and correspondingly marked both on the core and the published photographs of it. This figure includes the section from 52.75 feet, at the base of the left-hand column to 107.5 feet, at the top of the right-hand column; part or all of macroband BIF 1, S2, BIF 2, S3, BIF 3, and S4 (Figs. 3-2 and 3-5) are included, and marked. The dark green or black shale of the S macrobands and the bright blue mesoband of massive riebeckite are not distinguishable on the photograph, and massive riebeckite mesobands are therefore marked by the letter "*R*". The alternation of chert (pale) and iron-rich chert-matrix (dark) mesobands within the BIF macrobands is evident.

TABLE 3-IV

Comparion of characters of six BIF units of the Hamersley Group

BIF unit \ Character	Outcrop expression; colour of cleanest available exposure	Internal BIF shale relationship	General character and continuity	
			Thickness and spacing of chert mesobands	Development of pods and related structures
Boolgeeda Iron Formation	Rounded ridges with abundant scree cover; black, dark grey, or dark greenish-grey	Thick central shaly unit	Mesobands absent from most of thickness	Uncommon even where mesobands are present
Weeli Wolli Formation (excluding dolerite Sills)	Sharp "hogback" parallel strike ridges and valleys following shale and dolerite sills; strong reddish cast	Several thick intercalated shales	"Striped facies" lacks mesobands; chert mesobands elsewhere relatively thin and widely spaced	Cross-podding strongly developed at certain levels
Joffre Member of Brockman Iron Formation	High rounded, partly terraced hills; forms extensive plateaus; Bluish grey reddish when weathered	Shales thin and closely spaced	Cherts thick and widely spaced in comparison with Dales Gorge Member	Random and cross-podding developed in both "red" and "white" cherts, but more strongly in latter
Dales Gorge Member of Brockman Iron Formation	Prominent terraced cliffs where flat; conspicuous strike ridges where steep; dark grey where clean; otherwise reddish	Regularly spaced shales form "S macrobands" and make up one quarter of thickness	Cherts thin and relatively closely spaced	Random and cross-podding common in "white" cherts; macules at some levels

of banding			Special mineral-ogical features	
Expression of Calamina cyclothem or equivalent	Aftband characteristics			
	Main defining minerals	Textural features and field appearance	Riebeckite, crocidolite occurrence; principal sheet silicates within BIF	Relevant papers later than Trendall and Blockley (1970)
Not observed to be present	Hematite	Rarely seen	Rare disseminated riebeckite; no known crocidolite; stilpnomelane and chlorite	
Not easily discernable in most of thickness but in places well expressed as red and white chert alternation	Normally hematite	Four intergradational types defined by Trendall (1973); thick "graded" type also present. Often conspicuous bright red	Neither is known to occur; stilpnomelane and chlorite	
Locally very regular alternation of red and white chert mesobands within chert-matrix; usually thinner than Dales Gorge Member	Hematite in red cherts; weak carbonate in white cherts. Often emphasized by riebeckite	Conspicuous red/blue alternation in some "red" cherts is characteristic	Abundant massive and disseminated riebeckite throughout northern part of area; some rare crocidolite; ferrostilpnomelane	Trendall and De Laeter, 1972
Calamina cyclothem commonly 15 cm thick defined by alternation of "chert magnetite" and "mixed" components	Hematitie and ankerite	Very fine in most red cherts; almost always visible in all chert mesobands	Massive riebeckite and crocidolite common throughout northern part of area; ferrostilpnomelane	Morris, 1973 Trendall and Pepper, 1977 MacDonald and Grubb, 1971 Compston et al., 1981 Ewers and Morris, 1981

(continued)

TABLE 3-IV (continued)

Character / BIF unit	Outcrop expression; colour of cleanest available exposure	Internal BIF shale relationship	General character and continuity: Thickness and spacing of chert mesobands	General character and continuity: Development of pods and related structures
B				
Mount Sylvia Formation	BIF units form near-continuous low cliffs on otherwise smooth hill slopes; dark brown or blue-grey	Shales separate three BIF units	Widely spaced cherts	Podding not common
Marra Mamba Iron Formation	Low sinuous ridges flanking main highland areas; often rather bare and covered with pod chert debris; dark grey-brown with yellowish cast	Large-scale alternation of shaly BIF and massive BIF with thin "shale-carbonate macrobands"	Mixed chert thicknesses, but most thick and widely spaced	Extreme development of thick podded cherts, especially in lower part

one feature of special interest which is present in some chert mesobands of the Weeli Wolli Formation is selected for illustration here, and appears as Fig. 3-8. This remarkably regular cyclicity of aftbanding is relevant for later discussion of the depositional environment of the BIFs.

Lateral stratigraphic continuity

From the foregoing descriptions it is apparent that there exists, in the Hamersley Group, a wide range of regular depositional alternations differing in scale and in their exact degree of repetitive regularity. These are summarised in Fig. 3-9.

A remarkable feature of all these scales of stratification is the very high degree of their lateral continuity throughout the area of the basin. No example of significant non-continuity of the major stratigraphic boundaries (F of Fig. 3-9) has yet been well documented, and anomalies earlier thought to exist in the Wyloo Dome area (Trendall and Blockley, 1970; Daniels, 1970) have not

of banding			Special mineralogical features	
Expression of Calamina cyclothem or equivalent	Aftband characteristics			
	Main defining minerals	Textural features and field appearance	Riebeckite, crocidolite occurrence; principal sheet silicates within BIF	Relevant papers later than Trendall and Blockley (1970)
Not observed to be present	Hematite	Not conspicuous. Expressed as slight colour variations in surface exposures	Neither is known to occur; not known	
Not observed to be present	—	Widely spaced microbands often very faintly expressed	Massive riebeckite and crocidolite locally common; minnesotaite	Blockley, 1979 Ewers and Morris, 1980 Klein and Gole, 1981

been confirmed by later work. On the next smaller scale the lateral continuity of the Dales Gorge Member macrobands (E of Fig. 3-9) is unbroken over virtually the whole 60,000 km^2 of the Hamersley Group outcrop area; only in a few southern marginal localities do some S macrobands appear to be absent. The continuity of the smaller-scale stratification (A—D of Fig. 3-9) is not so clearly demonstrable, but Trendall and Blockley (1970), Trendall (1972), and Ewers and Morris (1981) have shown examples, also from the Dales Gorge Member, of basin-wide mesoband continuity, and in the first two of these three papers extensive aftband continuity is also illustrated. While it now appears that there is more lateral variation in fine-scale stratigraphy in the Hamersley Group BIFs than was believed by Trendall and Blockley (1970) it is still uncertain to what extent primary depositional continuity may have been masked or destroyed diagenetically.

Chemical composition

Trendall and Blockley (1970, Chapter 5) provided 48 complete and 19 partial chemical analyses of various samples from the Hamersley Group iron-formations, including BIF, shale, and individual mesobands. More systematic assessments of the chemical composition of the Dales Gorge Member have been made more recently by Trendall and Pepper (1977) and Ewers and Morris (1981). These authors report bulk compositions (weight %) for the Dales Gorge Member, from fresh core material at Wittenoom and Paraburdoo, respectively, shown at the top of p. 96:

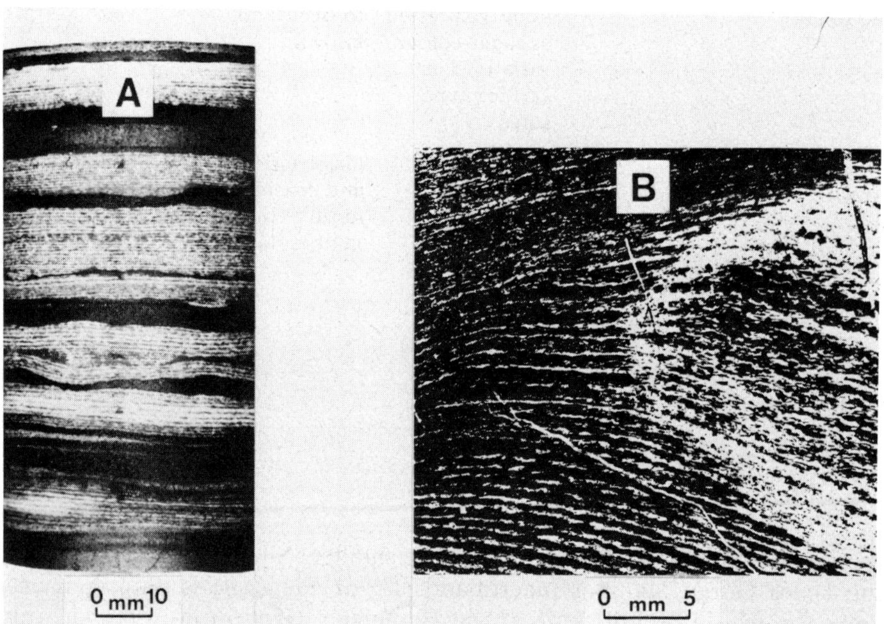

Fig. 3-7. A. Typical drill core of BIF of the BIF macrobands of the Dales Gorge Member. The stratigraphic thickness shown is from 36.2 to 36.5 feet (above the base) in BIF 0 (Fig. 3-2). Pale chert mesobands alternate with dark mesobands of chert- matrix. Each of the chert mesobands shows internal aftbanding. B. Lateral termination of a discontinuous (podded) chert mesoband from the Weeli Wolli Formation, showing the continuity of aft-banding from the chert into the surrounding chert-matrix.

Fig. 3-8. A single chert mesoband, about 10 cm thick, from drillcore through BIF of the Weeli Wolli Formation (Trendall, 1973, fig. 4). Fine regular aftbanding is clearly visible through much of the chert. A cyclic increase and decrease in intensity of definition of the dark (hematite-defined) part of each aftband couplet gives rise to regularly recurring light and dark stripes within the chert. By continuous counting of aftbands, Trendall (1973) showed that centres of dark stripes (marked at right-hand margin of photo) were repeated at an average interval of 23.3 aftbands.

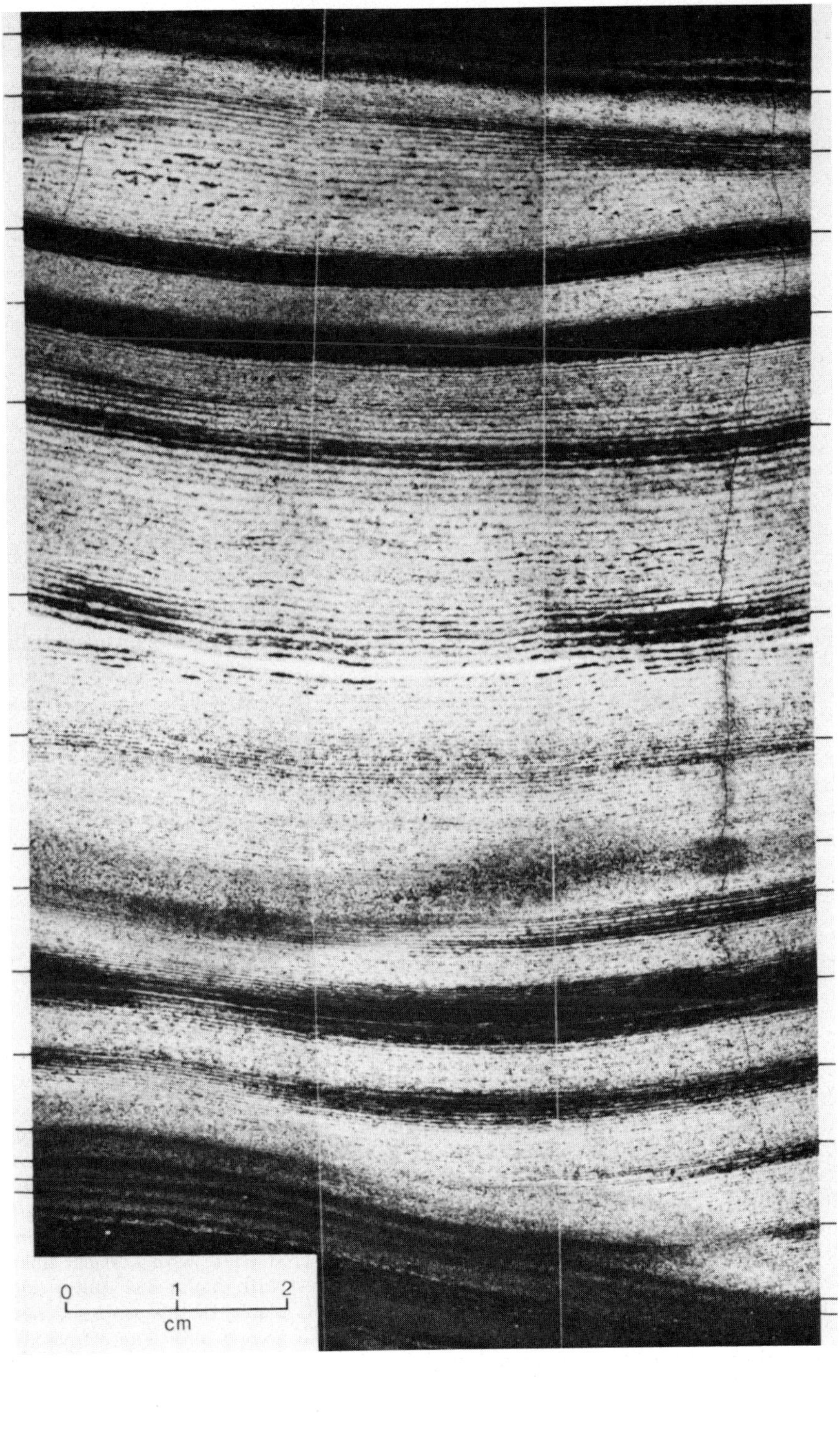
0
1
2
cm

	Trendall and Pepper (1977)	Ewers and Morris (1981)
SiO_2	44.34	43.51
Al_2O_3	0.89	0.36
Fe_2O_3	29.30	43.83[a]
FeO	13.45	
MgO	2.31	3.03
CaO	1.79	1.81
Na_2O	0.53	0.03
K_2O	1.26	0.06
L.O.I.	0.98[b]	5.46[c]
CO_2	4.63	5.81
FeS_2	0.12	—
TiO_2	0.05	0.03
P_2O_5	0.18	0.20
MnO	0.17	0.07
S	—	0.07
Total		98.38

[a] Reported as Fe_2O_3.
[b] Reported as H_2O^+.
[c] Loss on ignition takes into account the loss of CO_2, H_2O, and S, and the gain in oxidising any Fe^{2+} to Fe^{3+}.

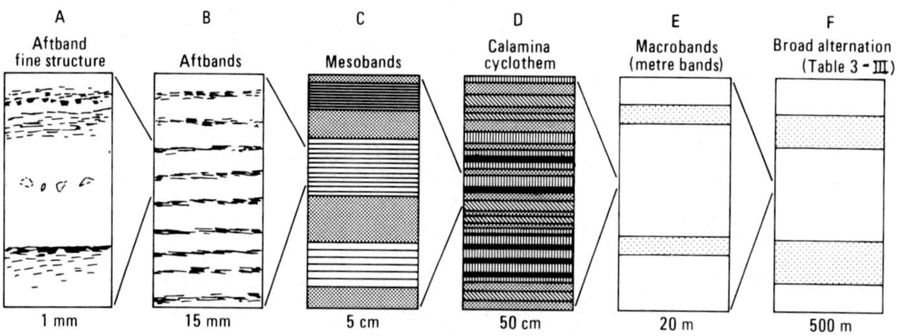

Fig. 3-9. Summary of stratification scales within BIFs of the Hamersley Group. In A, two iron-rich components of successive aftbands are represented, each defined by hematite concentration within chert. In the new nomenclature of Trendall et al., (in prep.) thin recognizable laminae within these concentrations are referred to as micron bands. B shows aftbands (formerly microbands) within a chert mesoband, and emphasises the characteristic regularity of their recurrence (Fig. 3-8). In C, aftbanded chert mesobands of different aftband interval are represented, separated by chert-matrix (dark stipple). In D, the Calamina cyclothem is shown, defined by chert-magnetite groups (red chert with vertical lines, magnetite black) alternating with mixed groups (white chert with diagonal shading). The light stipple in E and F represents shale, while BIF is left blank. In the proposed new nomenclature (Trendall et al., in prep.) macrobands are so named only where formally numbered within a particular stratigraphic unit, such as the Dales Gorge Member; elsewhere, bands of this scale would be referred to as metre bands.

The paper of Ewers and Morris (1981) reported the results of a complete analysis of the Dales Gorge Member, based on the subdivision of a complete drillcore intersection at Paraburdoo into 516 sample units, 254 of which were analysed individually. This permitted an average composition of each of the 33 macrobands to be obtained, as well as separate average compositions of the combined 16 BIF macrobands and the combined 17 S macrobands. They interpreted their chemical data, together with evidence of lateral continuity of banding across the basin, to support an assumption that the primary deposits were chemically similar across the area.

Stable isotope studies

Becker and Clayton (1972, 1976) have carried out carbonate carbon and oxygen isotopic studies on both the iron-formations and some other stratigraphic units of the Hamersley Group, and Oehler et al. (1972) have reported carbon isotope data for kerogen carbon. Becker and Clayton (1972) concluded from their carbon study that the iron-formation was precipitated in a basin isolated from the ocean, but probably in close proximity to it, and that organic activity may have played a significant role in the genesis of the iron-formation. However, they did not exclude a volcanic source for the light carbon. From their (1976) oxygen isotope work the most significant finding was that the minerals of the Dales Gorge Member in the Wittenoom area had undergone isotopic exchange at a temperature estimated to be above 270°C and probably less than 310°C, during burial metamorphism. Oehler et al. (1972) found no anomaly in organic carbon isotopes in the Hamersley Group compared with other Precambrian sediments.

Van der Wood (1977) has reported initial $^{87}Sr/^{86}Sr$ values from the Dales Gorge Member, and has concluded that they are consistent with non-marine deposition. The extensive programme of stable isotope studies undertaken by the Precambrian Paleobiology Research Group, referred to earlier, is still continuing, and is likely to result in important advances in the contribution of this type of evidence to the interpretation of the basin.

STRUCTURE

The most striking structural characteristic of the Mount Bruce Supergroup as a whole, and viewed over the full extent of its outcrop area, is its relatively slight degree of disturbance. Within the conceptual ellipse embracing this outcrop area it is remarkable that although the Pilbara Craton, as a tectonic unit, has ascended an average of perhaps 5 km relative to its position at the time of completion of deposition of the Turee Creek Group, the Hamersley Basin rocks resting on it are widely so undisturbed (Fig. 3-5) as to justify such early intuitive impressions as that of Woodward (1891), who judged them likely to be Devonian in age.

While this regional impression of slight disturbance is valid, two types of folding do affect the Mount Bruce Supergroup: comparatively gentle and open folds which are present over the bulk of the outcrop area, and more intense deformation which is invariably restricted to, and structurally related to, the perimeter zone of the bounding ellipse. The former type, in the northern half of the ellipse mostly trends north—south (Fig. 3-1), and the synclines which mainly define it in this area often run parallel to, and along, synclinorial keels of greenstone belts of the Pilbara Block; this point is discussed in some detail under a later heading. In the southern half of the ellipse the main open folding runs approximately south-southeast, and is well exemplified by the Hamersley Range Synclinorium and the major folds south of it. This open folding of the main southern outcrop area is characterised by axial variations in plunge and en-echelon offsetting of anticlines and synclines to give a striking dome-and-basin pattern on a regional scale. Dips rarely exceed 40°. Both MacLeod (1966, p. 64) and Halligan and Daniels (1964) interpreted this pattern as the result of two interfering fold sets, and the latter gave the separate names "Ophthalmian" and "Rocklean" to these. However, Gee (1979, p. 351) considers that the pattern could have been formed by deflection of folds around basement domes as a response by a sedimentary cover to basement cratonic movements. Whatever the origin or significance of these folds it is evident that they become tighter southwards, and that there is a sharp increase in their intensity within a few kilometres of the bounding ellipse. Along this line, an axial plane cleavage is commonly developed in less competent units, and isoclinal folding with the overturned southern limbs of synclines dipping south is locally present, especially immediately north of the Sylvania Dome. Deformation is also particularly strong in the western part of the Wyloo Dome. A well developed cleavage is also present in the Fortescue Group of the Koongaling Hill—Lookout Rocks area, where it is axial to folding along the eastern edge of the ellipse.

Apart from faulting in the elliptical bounding zone of the Hamersley Basin, already referred to, much of the internal outcrop of the Mount Bruce Supergroup is transected by at least two sets of vertical faults, running roughly southwest and southeast. These are presumed to be later than the folding.

METAMORPHISM

Trendall and Blockley (1970, p. 294) asserted that the "Hamersley Group has nowhere undergone regional metamorphism", and supported that view with unpublished evidence of Hoering, work by Grubb (1967) on riebeckite synthesis, and preliminary oxygen isotope work by Becker and Clayton. A maximum temperature of 160°C was suggested, and it was considered that "the evolution of the Hamersley Basin was not accompanied by heating of the sediments to temperatures above those associated with ordinary geothermal gradients". Since that time a final synthesis of the oxygen isotope data (Becker and Clayton, 1976) indicates temperatures "above 270°C and probably less than 310°C, during burial metamorphism". Ayres (1972), also from a study of material from the Wittenoom area, suggested that there was significant metamorphism of the Dales Gorge Member from a consideration of the mineral paragenesis, and suggested "pressures of 4 to 6 kilobars produced by the load of overlying strata and a temperature of 300°C due to a geothermal gradient of 15°C/km".

The more recent work of Smith et al. (1982; see also Smith, 1976) represents a great advance in understanding of metamorphism in the Hamersley Basin. Selecting metabasic lavas and volcaniclastics for study because of their sensitive response to low-grade metamorphism and general freedom from deuteric alteration, these authors established, almost entirely from Fortescue Group samples, four metamorphic zones: Zone I (ZI) prehnite-pumpellyite zone; ZII, prehnite-pumpellyite-epidote zone, ZIII, prehnite-pumpellyite-epidote-actinolite zone; and ZIV, (prehnite)-epidote-actinolite facies. Their map (Smith et al., 1982, fig. 2) showed the general distribution of these zones in the central-southwest part of the basin, with a broad northern area of prehnite-pumpellyite facies (ZI and ZII) separated by a curved central strip of pumpellyite-actinolite facies (ZIII) from an area of greenschist facies (ZIV) in the south. The isograds separating these zones are shown in Fig. 3-12.

Smith et al. (1982, fig. 3) were able to relate the distribution of these four metamorphic zones to a north—south palinspastic cross-section of the Hamersley Basin at the termination of Turee Creek Group deposition, reconstructed by structural "unfolding", from published stratigraphic sections, and from reasonable stratigraphic assumptions where these were lacking. They showed that with a cross-sectional model assuming a total thickness of the Turee Creek Group at the southern limit of its outcrop of about 3.5 km (cf. Fig. 3-4 of this paper), and thinning northwards, the four metamorphic zones become horizontal, so that metamorphism was controlled by burial depth alone. From detailed chemical analysis of selected minerals they were able to argue that a relatively high geothermal gradient, of 80 to 100°C/km is likely for the shallow part of the sequence, and a gradient of 40°C/km for the deeper part, the change occurring at about 2.5 km.

Although the metamorphic model of Smith et al. (1982) is persuasive, and based on detailed work mainly in the Fortescue Group it is worth noting that a different view is expressed by Ewers and Morris (1981) on the relative grades of metamorphism exhibited by the Dales Gorge Member of the Brockman Iron Formation at Paraburdoo, Wittenoom, and Tom Price, at the eastern end of the Turner Syncline (Fig. 3-1); Ewers and Morris believe from BIF textural data that these places lie in order of increasing metamorphic grade, although the model of Smith et al. (1982) suggests that this order should be Wittenoom, Paraburdoo, Tom Price.

TECTONIC DEVELOPMENT OF THE BASIN

An initial model

It is convenient to conduct this discussion of the evidence for basin development within the framework of a simple initial model, involving the following sequence of events:

(A) There was, before the existence of the Hamersley Basin, an extensive period of formation of Archaean crust over a wide region, including and possibly extending well beyond the area of future basin development.

(B) At some time before the Hamersley Basin was initiated, significant rock-forming events ceased within this Archaean crust: it became "cratonised".

(C) The surface of the resultant craton — the Pilbara Block or Pilbara Craton (Gee, 1979) — was uplifted and eroded.

(D) The Hamersley Basin was initiated by submersion of some part, or parts, of this eroded craton.

(E) The floor of the basin sank more or less continuously during deposition.

(F) Deposition in the basin was terminated when folding, which conspicuously affects some parts, was initiated.

A set of propositions of this kind, based on the general concept that the tectonic development of the "Proterozoic" Hamersley Basin could be considered in isolation from the development of a much older and unconformably underlying "Archaean" crust, is implicit in most previous accounts of Hamersley Basin evolution. Maitland's (1924) inclusion of part of the Fortescue Group (the "Nullagine Formation") as a component of a younger, relatively undeformed and unmetamorphosed Precambrian sequence unconformably overlying an igneous and metamorphic basement, reflected widespread international belief at that time in the worldwide validity of such a twofold Precambrian subdivision; and Trendall and Blockley's (1970, p. 278) account of basin development equally clearly saw the role of the Archaean rocks below the basin as restricted to that of a passive basement which was eroded to a plain before the initiation of the basin.

Relationship between the Pilbara Block and the Hamersley Basin

The evidence reviewed here seems increasingly compatible with a close relationship between the Hamersley Basin and the Pilbara Block. The relevant isotope geochronological evidence is displayed in Fig. 3-10. Each of the 86 "ages" shown in that Figure is listed and appropriately annotated in Table 3-V.

If the sequence of events set out above is accepted, it follows that an oldest possible age for the initiation of the basin (D), ignoring any time for the erosion in C, is set by the time of cratonisation of the Pilbara Block (B), which in turn may be conceived to be the age of the youngest significant rock-forming event within it.

TABLE 3-V

Isotopic age determinations plotted on Figure 3-10

No.	Symbol	Age(m.y.)	Unit and Comments	Reference
1	●	*1684	Boolaloo Granodiorite; cuts Wyloo Group	Leggo et al. (1965)
2	○	*1977±165	Wyloo Group; "acid igneous rock" within group	Compston and Arriens (1968)
3	○	*1811	Wyloo Group; "tuffaceous siltstone" within group	Leggo et al. (1965)
4	○	1950	Woongarra Volcanics; intrude (?) Hamersley Group	Compston et al. (1981)
5	○	2370	Woongarra Volcanics; intrude (?) Hamersley Group	Compston et al. (1981)
6	▲	2470±30	Woongarra Volcanics; intrude (?) Hamersley Group	Compston et al. (1981)
7	○	*2144-100	Nallanarring Volcanic Member; component of Fortescue Group	Compston and Arriens (1968)
8	●	*2150±26	Gidley Granophyre; intrusive along basal uncomformity of Fortescue Group	De Laeter and Trendall (1971)
9	○	2878	Cliff Springs Muscovite; detrital muscovite in sandstone within Fortescue Group	Compston and Arriens (1968)
10	●	2028±168	Mount Roe Basalt; lowest lava of Fortescue Group	De Laeter et al. (in prep.)
11	●	2100-2300	Fortescue Group Lavas; various basaltic units	De Laeter and Trendall (in prep.)
12	●	1800	Weeli Wolli Dolerite; intrudes Hamersley Group	Trendall and De Laeter (in prep.)
13	○	2300	Weeli Wolli Dolerite; intrudes Hamersley Group	Trendall and De Laeter (in prep.)
14	▲	2490±20	Dales Gorge Member, S13, zircon from ash-fall tuff within Hamersley Group	Compston et al. (1981)
15	●	*2760±516	Bamboo Creek Porphyry; intrudes low in Fortescue Group	Trendall (1975d)
16	●	*2079±195	Spinaway Porphyry; intrudes low in Fortescue Group	Trendall (1975d)
17	●	1684±25	Joffre Member; porcelanite (ash-fall tuff) in Hamersley Group	Trendall and De Laeter (1972)
18	▲	2768±24	Spinaway Porphyry; intrudes low in Fortescue Group	R. T. Pidgeon (written commun., 1979)
19	●	2331±42	Mt Brown Rhyolite Member	Sylvester and De Laeter (in prep.)
20	●	2366±60	Mons Cupri Granite	Sylvester and De Laeter (in prep.)
21	●	2684±82	Caines Well Granite	Sylvester and De Laeter (in prep.)
22	●	2610±80	Mons Cupri Porphyry	Sylvester and De Laeter (in prep.)
23	●	*3059±366	Mount Edgar Batholith	De Laeter and Blockley (1972)
24	●	*2889±83	Shaw Batholith	De Laeter et al. (1975)
25	△	2882±60	Newman 70-mile quarry; dark granitoid	Oversby (1976)
26	△	2969±45	Woodstock 1527; gneissic granite	Oversby (1976)
27	△	2961±7	Newman 127-mile quarry; foliated granite	Oversby (1976)
28	△	2936±9	Cooglegong; gneissic granite	Oversby (1976)
29	△	3070±12	Tambourah; migmatite and granite	Oversby (1976)
30	△	2920	Newman 13-mile quarry; K-feldspar	Oversby (1976)

(continued)

TABLE 3-V (continued)

No.	Symbol	Age(m.y.)	Unit and Comments	Reference
31	△	2786±38	Woodstock 1526; gneissic granite	Oversby (1976)
32	△	2769±13	Woodstock 1527; gneissic granite	Oversby (1976)
33	△	2938±33	Tambourah 1512–1513; migmatite and granite	Oversby (1976)
34	▲	3280±20	Mt Edgar Batholith; granite	Pidgeon (1978b)
35	○	2830±30	Pegmatite muscovite (loc. uncertain)	Pidgeon (1978b)
36	▲	3417±40	Shaw Batholith; migmatite	Pidgeon (1978b)
37	●	3280±20	Granodiorite etc. } Corunna Downs	Cooper et al. (1980)
38	●	3227±50	Adamellite } Corunna Downs	Cooper et al. (1980)
39	●	2950±50	Aplite } Corunna Downs	Cooper et al. (1980)
40	●	3087±50	Granite (fine-grained) } Tambourah	Cooper et al. (1980)
41	●	2798±35	Granite (coarse-grained) } Tambourah	Cooper et al. (1980)
42	●	*2819±66	Copper Hills, porphyritic felsite	De Laeter and Trendall (1970)
43	◆	2760	Lynas Find (Pb 403)	Richards et al. (1981)
44	◆	2960±20	Nimerry Creek (Pb 404)	Richards et al. (1981)
45	●	*2280±89	Black Range dolerite	Lewis et al. (1975)
46	●	*2614±95	Moolyella Granite	De Laeter and Blockley (1972)
47	●	*2551±128	Cooglegong Adamellite	De Laeter et al. (1975)
48	●	*2514±37	TR isochron } Cookes Creek Granite;	De Laeter et al. (1977)
49	○	*2552	biotite } Cookes Creek Granite;	De Laeter et al. (1977)
50	◆	2450±40	Hillside (Pb 333)	Richards et al. (1981)
51	□	3490±30	Shaw Batholith	M.McCulloch (pers. commun., 1982)
52	□	3480±20	Shaw Batholith	M.McCulloch (pers. commun., 1982)
53	□	3380±20	Mt Edgar Batholith	M.McCulloch (pers. commun., 1982)
54	□	3810±80	Moolyella Adamellite	M.McCulloch (pers. commun., 1982)
55	□	3300±40	Cooglegong Adamellite	M.McCulloch (pers. commun., 1982)
56	□	3520±30	Duffer Formation	M.McCulloch (pers. commun., 1982)
57	□	3550±30	Duffer Formation	M.McCulloch (pers. commun., 1982)
58	◆	3330	Dooleena Gap (Pb 334)	Richards (1977), Richards et al. (1981)
59	◆	3440	Duffer Formation (Big Stubby)	Sangster and Brook (1977)
60	▲	3452±16	Duffer Formation	Pidgeon (1978a)
61	●	3570±180	North Star Basalt	Jahn et al. (1981)
62	●	3230±280	Duffer Formation	Jahn et al. (1981)
63	■	3556±542	North Star Basalt	Jahn et al. (1981)
64	■	3556±32	Talga Talga Subgroup	Hamilton et al. (1980)
65	◆	3190±10	Lalla Rookh (Pb 401)	Richards et al. (1981)
66	◆	3340	Coppin Gap (Pb 402)	Richards et al. (1981)
67	◆	3480±10	Lennon Find (Pb 406)	Richards et al. (1981)

68	◆	3470±10	Lennon Find (Pb 407)	Richards et al. (1981)
69	◆	3470±10	Lennon Find (Pb 408)	Richards et al. (1981)
70	◆	3470 ±40	North Pole (Pb 412)	Richards et al. (1981)
71	◆	3400±10	North Pole (Pb 413)	Richards et al. (1981)
72	◆	3460±40	North Pole (Pb 414)	Richards et al. (1981)
73	◆	3470±40	North Pole (Pb 415)	Richards et al. (1981)
74	◆	3470±10	Big Stubby; galena from vein in Duffer Formation	Richards et al. (1981)
75	●	2707±151	Hardey Sandstone; low unit of Fortescue Group	Hickman and De Laeter (1977)
76	●	2235±54	Granite; unconformably underlies high Fortescue Group	Blockley et al. (1980)
77	●	1487±305	Pillow lava; uppermost Fortescue Group	Blockley et al. (1980)
78	◆	2720±20	Braeside (Pb 335)	Richards (1977), Richards et al. (1981)
79	◆	2760±20	Braeside (Pb 336)	Richards (1977), Richards et al. (1981)
80	●	*2595±60	Lookout Rocks granite; structurally below Koongaling Volcanics	De Laeter et al. (1977)
81	●	3013±93	Gobbos Granodiorite	Barley and De Laeter (pers. commun., 1982)
82	●	3183±131	Boobina Tonalite Porphyry	Barley and De Laeter (pers. commun., 1982)
83	●	3018±75	McPhee Creek } Duffer Formation	Barley and De Laeter (in press)
84	●	3063±114	Spinaway Creek } Duffer Formation	Barley and De Laeter (in press)
85	●	3457±109	Seven of 10 least altered samples of 83 and 84	Barley and De Laeter (in press)
86	◆	2760±20	Kylena Basalt (Pb 454A); galena within amygdales of basalt	Richards (pers. commun., 1982)

1. Ages marked by an asterisk in column 3 have been reduced by a factor of 1.39/1.42 from the actual values reported by the authors indicated, who used λ^{87} Rb = 1.39×10^{-11} yr^{-1} in their original calculations. Rb-Sr ages not marked by an asterisk are from papers using $\lambda^{87} = 1.42 \times 10^{-11} yr^{-1}$.
2. The only isotopic age determinations known to be omitted from this Table, and from Fig. 3-10, are:
   (a) A number of pre-1965 ages from the Pilbara Block which are consistent with the pattern shown, and which do not affect consideration of the age of the Hamersley Basin (Compston and Arriens, 1968).
   (b) Three "anomalous" galena Pb model ages includes in table 3 of Richards et al. (1981): Pb 331 (Flat Rock), Pb 332 (Andover), and Pb 400 (Green Well). *All* of the 16 other model ages of that table, included above, are from galenas in veins which transect various units of the Pilbara Block or Hamersley Basin. However, these 16 ages are all so consistent with the ages of those units as determined by other methods that it is appropriate to represent them on Fig. 3-10. While the three "anomalous" galenas cannot be ignored, their inclusion on the figure would cause more confusion than enlightenment.
   (c) Some Rb-Sr isochron ages from the layered succession of the Pilbara Block reported by Jahn et al. (1981), which clearly have low reliability.
   (d) Some unpublished work still in progress.
3. All galena model ages except point 59 are shown, at Dr. J.R. Richards' suggestion, as the mean to 3 significant figures of the Cumming-Richards Model III ages "t_{76}, isochron" and "t_6" (Richards et al., 1981), with the error limits as ± half the difference between them; for a full discussion of the reservations to be borne in mind when using ages of this kind, readers are referred to the paper of Richards et al. (1981).

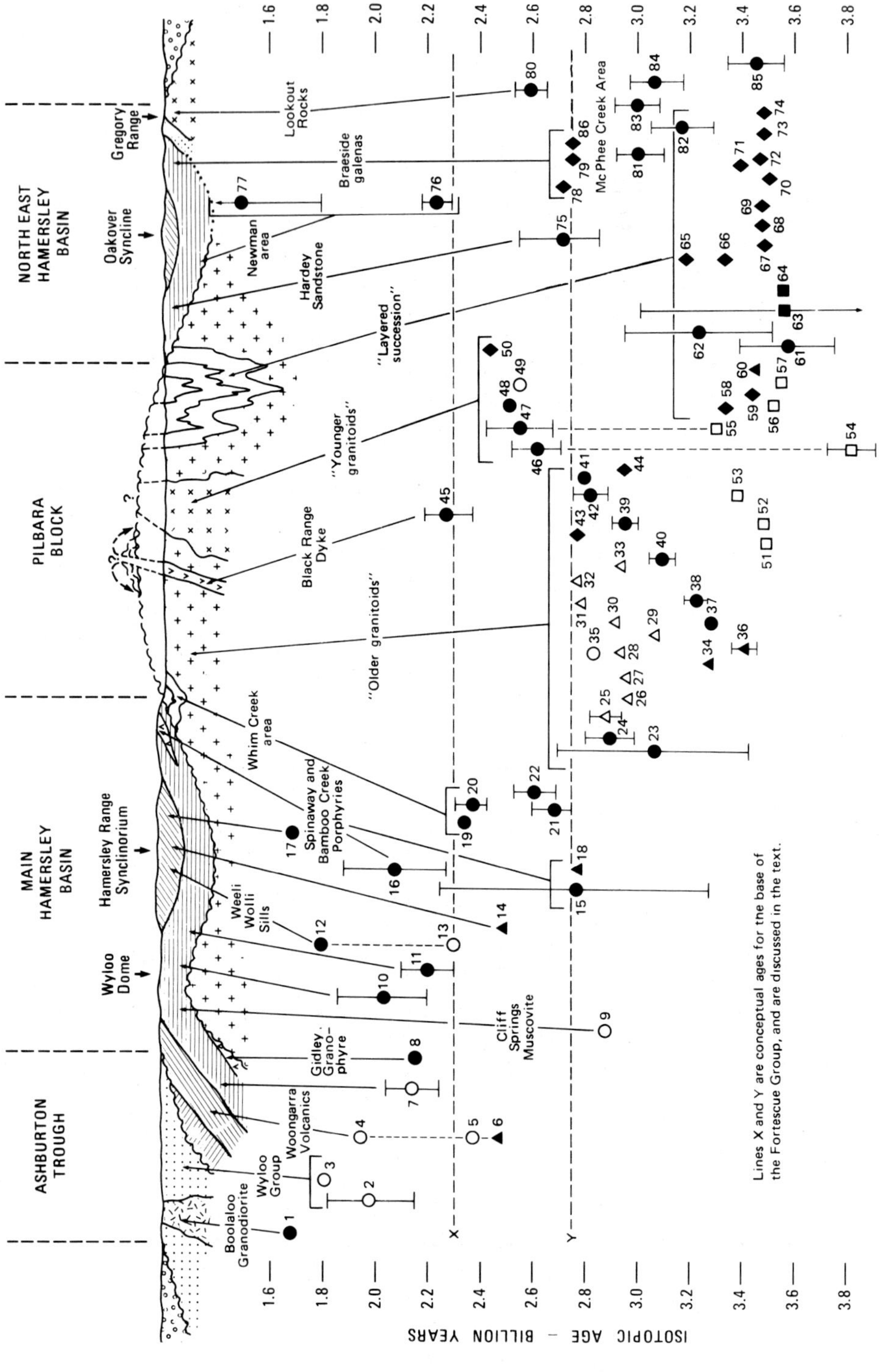
ASHBURTON TROUGH
MAIN HAMERSLEY BASIN
PILBARA BLOCK
NORTH EAST HAMERSLEY BASIN
Wyloo Dome
Hamersley Range Synclinorium
Oakover Syncline
Gregory Range
Boolaloo Granodiorite
Wyloo Group
Woongarra Volcanics
Gidley Granophyre
Cliff Springs Muscovite
Weeli Wolli Sills
Spinaway and Bamboo Creek Porphyries
Whim Creek area
"Older granitoids"
Black Range Dyke
"Younger granitoids"
"Layered succession"
Hardey Sandstone
Newman area
Braeside galenas
Lookout Rocks
McPhee Creek Area
X
Y
ISOTOPIC AGE – BILLION YEARS
1.6
1.8
2.0
2.2
2.4
2.6
2.8
3.0
3.2
3.4
3.6
3.8
Lines X and Y are conceptual ages for the base of the Fortescue Group, and are discussed in the text.

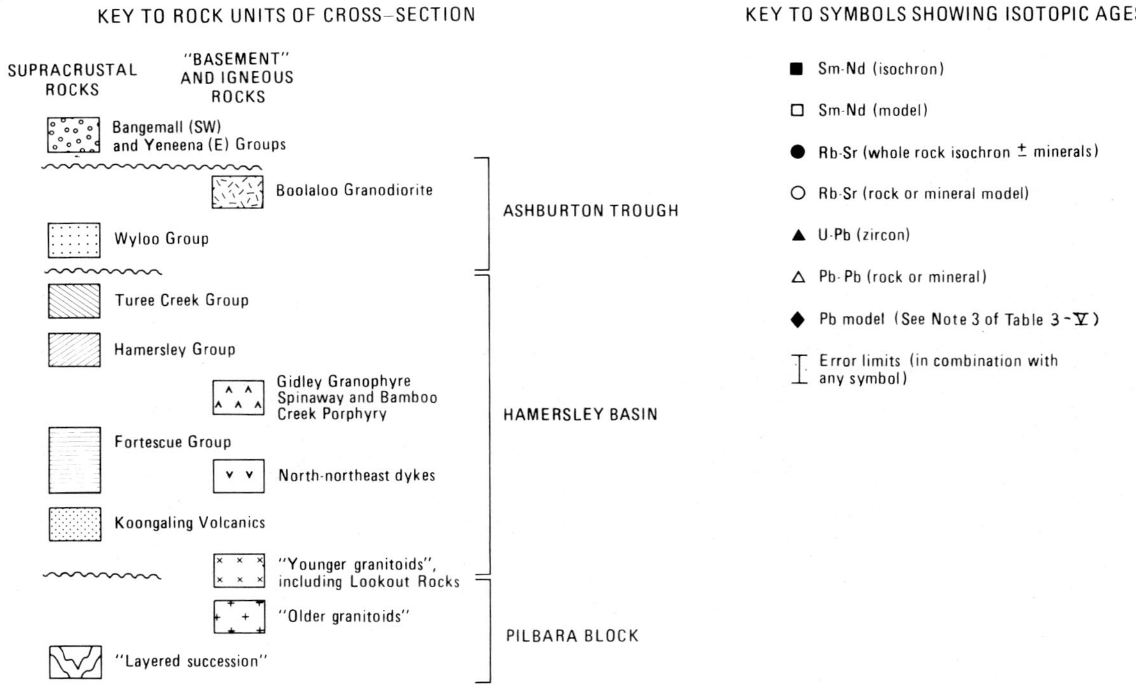

Fig. 3-10. Compilation of most published, and some unpublished, age determinations from rocks of the Hamersley Group, the underlying Pilbara Block, and also units which constrain the upper age limit of the Hamersley Basin. Further details, including references for each numbered symbol, appear in Table 3-V; the notes to that Table explain some omissions. Where no error limits are given these lie within, or trivially outside, the symbol. The cross-section at the top of the figure is a diagrammatic one, not drawn to scale; however, it is a conceptually accurate representation of the relative position and relationships of all the units shown, along a roughly southwest to northeast transect, about 600 km long in total, from the Ashburton Fold Belt to Marble Bar (Fig. 3-1), and thence eastwards to the Gregory Range, which runs between Koongaling Hill and Lookout Rocks. For further discussion and explanation see text.

The Pilbara Block is for the present purpose broadly divisible into three main components (Hickman, 1981): (1) the synclinorial greenstone belts with their complex succession of layered supracrustal rock; (2) the "older granitoids" [migmatites, gneisses, foliated granodiorite, adamellite, and tonalite, including categories (1) and (2) of Hickman, 1981, p. 60], which form the main bulk of the intervening batholithic domes; and (3) the "younger granitoids" [Hickman's, 1981, category (3)] that form relatively small, discrete, post-tectonic plutons of weakly foliated or non-foliated granite and adamellite within the major domes; these latter include the "tin granites" of Blockley (1980). Geological and isotopic evidence are in agreement that this sequence is one of decreasing age, although Hickman (1981) envisages a very wide overlap of granitoid generation and supracrustal deposition. The "layered succession" (Pilbara Supergroup) of the greenstone belts has yielded Sm-Nd, galena model, and U-Pb zircon ages almost all falling between 3450 and 3550 m.y. (points 56—74 of Table 3-V and Fig. 3-10*). Point 85 is the only reported Rb-Sr isochron that registers an age of this order and this was obtained by omission of some analysed samples; there is clearly widespread late Sr isotope equilibration throughout the Pilbara Block.

The "older granitoids" intrude the greenstone belts wherever a primary mutual relationship is discernible, and, consistently with this relationship, give ages mostly within the range 2800 to 3300 m.y. (points 23—44, 51—53, and 81—84). Most of the "younger granitoids" of both the western (point 21) and eastern (points 46—50) Pilbara Block yield ages in the range 2500 to 2700 m.y., but some, such as the Moolyella Adamellite (points 46—54) and Cooglegong Adamellite (points 47 and 55), have yielded much higher Sm-Nd ages, and have very high initial $^{87}Sr/^{86}Sr$ ratios, both features consistent with the view that they are derived largely by remobilisation of older sialic crust. The Mons Cupri Granite, in the west (point 20), gives a Rb-Sr isochron as young as 2366 ± 60 m.y., but as the nearby Mount Brown Rhyolite Member (point 19), regarded by Hickman (in press) as essentially correlative with the upper part of the layered succession of the eastern Pilbara Block, and from his analysis of the data probably about 2900 m.y. old, gives a closely similar age, this may be a thermal up-date unrelated to petrogenesis. All isochrons from the Whim Creek area have relatively high initial $^{87}Sr/^{86}Sr$ ratios.

The cratonisation of the Pilbara Block thus appears to have been completed by about 2500 m.y., or somewhat younger. Line *X* in Fig. 3-10 (2300 m.y.) represents a conceptual age for the lowest Fortescue Group rocks, consistent with data from the Pilbara Block and also with acceptance of assumed stages A and B above. Evidence for its age coming directly from the Fortescue Group itself is represented by points 7, 9, 10, 11, 75, 77, 78 and 79. If the magma source for Fortescue Group lavas is accepted as the major north-northeast

---

* All subsequent citations of point numbers in this discussion refer to Table 3-V and Fig. 3-10.

dykes of the Pilbara Block, then their age (point 45) is also direct evidence for Fortescue Group age. A younger limit for Fortescue Group age is potentially provided by sedimentary rocks overlying it (points 14, 17), and by igneous rocks either intruding it (points 8, 15, 16, 18) or intruding rocks overlying it (points 4—6, 12, 13). Points 7, 8, 11, 16 and 45 are consistent with line *X* as the base of the Fortescue Group, and with the north-northeast dykes as feeders for the volcanics of the group, a possibility indicated by a question mark in Fig. 3-10. However, points 6, 14, 18, 75, 78 and 79, collectively a formidable battery of evidence, are all in direct conflict with line *X*, and are also in approximate agreement with an age for the base of the Fortescue Group of about 2750 m.y. (line *Y*). Of the points noted above as consistent with line *X*, point 8, from an intrusive rock, is equally consistent with line *Y*, while most of the remainder are Rb-Sr isochrons which *could* be interpreted as up-dated; point 7 could also be an up-date.

On the other hand, acceptance of 2750 m.y. as the age of the base of the Fortescue Group then calls into question the concept of a sharply terminated pre-Hamersley Basin cratonisation of the Pilbara Block, by envisaging, as an alternative, that its "younger granitoids" formed concurrently with the sinking of the basin. There is no geological evidence to contradict this, as the Fortescue Group is nowhere seen to overlie proven "younger granitoids" (see question mark in Fig. 3-10), as it is seen to overlie both the greenstone belts and "older granitoids". Neither should there be any difficulty in accepting continuing large-scale plutonic igneous activity in the basement below the Hamersley Basin as it sank. For one thing, small hornblende adamellite diapirs, though not demonstrably consanguineous with the younger granitoids of the Pilbara Block, are known to penetrate the Fortescue Group in the area of the Nullagine 1 : 250,000 Sheet (Hickman, 1978). For another, some 100,000 km^3 of rhyolitic (granitic) magma (the Woongarra Volcanics) were injected into the Hamersley Group not long after deposition, and this material must have been generated within crust underlying the basin. Other massive acid intrusions with a similar origin — the Gidley Granophyre (point 8) and the Spinaway and Bamboo Creek Porphyries (points 15, 16, 18) — were emplaced within the Fortescue Group.

At present, acceptance of about 2750 m.y. for the initiation of the Hamersley Basin (that is, the age of the base of the Fortescue Group, or event D above) must be strongly preferred on the geochronical evidence, and the credibility of this, as well as the credibility of the consequent proposition that assumed stages B and D above are not sequential but have a time overlap of about 200—300 m.y., needs close examination.

The concept of a significant time overlap between Hamersley Basin initiation and cratonisation processes within the Pilbara Block requires a particularly radical revision of earlier interpretations in the Gregory Range area, between Koongaling Hill and Lookout Rocks (Fig. 3-1). De Laeter et al. (1977) accept the 2595 m.y. age of granite at Lookout Rocks (point 80) as that of an outlying part of the Pilbara Block older than the felsic west-dipping Koongaling Volcanics (Fig. 2-10), which in that area form the lowest unit of the Fortescue Group, underlying the mafic Kylena Basalt. If points 78, 79, and 85 do reflect the true age of this basalt, the succession eastwards (stratigraphically downwards) from it, through the Koongaling Volcanics and the Isabella Porphyry into the granite of the Lookout Rocks area must be seen not as a passage into an older and unconformably underlying "basement", but rather as a progression into a younger granite intrusion, shouldering upwards into early lavas (Koongaling Volcanics) which may well have been an earlier manifestation of the same phase of igneous activity, which in this eastern area initiated, and continued through, the early life of the basin. Exact relationships in the Gregory Range are difficult to establish, due to strong north—south (strike) faulting and to extensive desert sand cover, but Dr A.H. Hickman (pers. commun. 1982) who was responsible for mapping much of this area (Hickman, 1981), agrees that this re-interpretation cannot at present be invalidated from field evidence.

The main geological evidence suggesting a wide time gap between cratonisation of the Pilbara Block and initiation of the Hamersley Basin is the clearly unconformable contact along the main northern limit of Fortescue Group outcrop (Fig. 3-1). However, against this two other points must be balanced.

Firstly, there is the tendency for the lowest stratigraphic units of the Fortescue Group, taken over the entire area of the Pilbara Block and including all outliers, to be situated over, and to have their outcrops elongated along, the greenstone belts; the clear resultant impression is that these represent the sites of initial Hamersley Basin deposition. Although this has been challenged by Horwitz (1980; see also Horwitz and Smith, 1978, p. 308) it certainly appears to be generally true from an appraisal of Hickman's (1980) map of the Pilbara Block, from which the essential features have been abstracted for objective comparison in Fig. 3-11. Hickman (in press) affirms the correlation, and Gee (1979, p. 350) independently notes a correspondence of Fortescue Group synclines with Pilbara Block greenstones, and anticlines with granitoid.

Secondly, in many places in these synclines the angular unconformity between the basal Fortescue Group sediments and the underlying Archaean strata of the layered succession is slight, and certainly no more marked, or of more evident tectonic significance, than any of the several earlier unconformities within the 30-km-thick Archaean succession (Hickman, 1981). Almost all of the ages so far determined from this layered succession (points 56—74 of Fig. 3-10) are from its lower stratigraphic units. An age as young as 2800 m.y. for the upper part of the succession implies an average depositional rate of about 40 $m/10^6 yr$. Although this is slow by comparison with average Phanerozoic rates (Hudson, 1964) it is consistent with the only other directly determined Archaean rate (Nunes and Thurston, 1980), and a concept of the Fortescue Group in these synclinal situations as resulting simply from a continuation of the same tectonic event which initiated and controlled Archaean

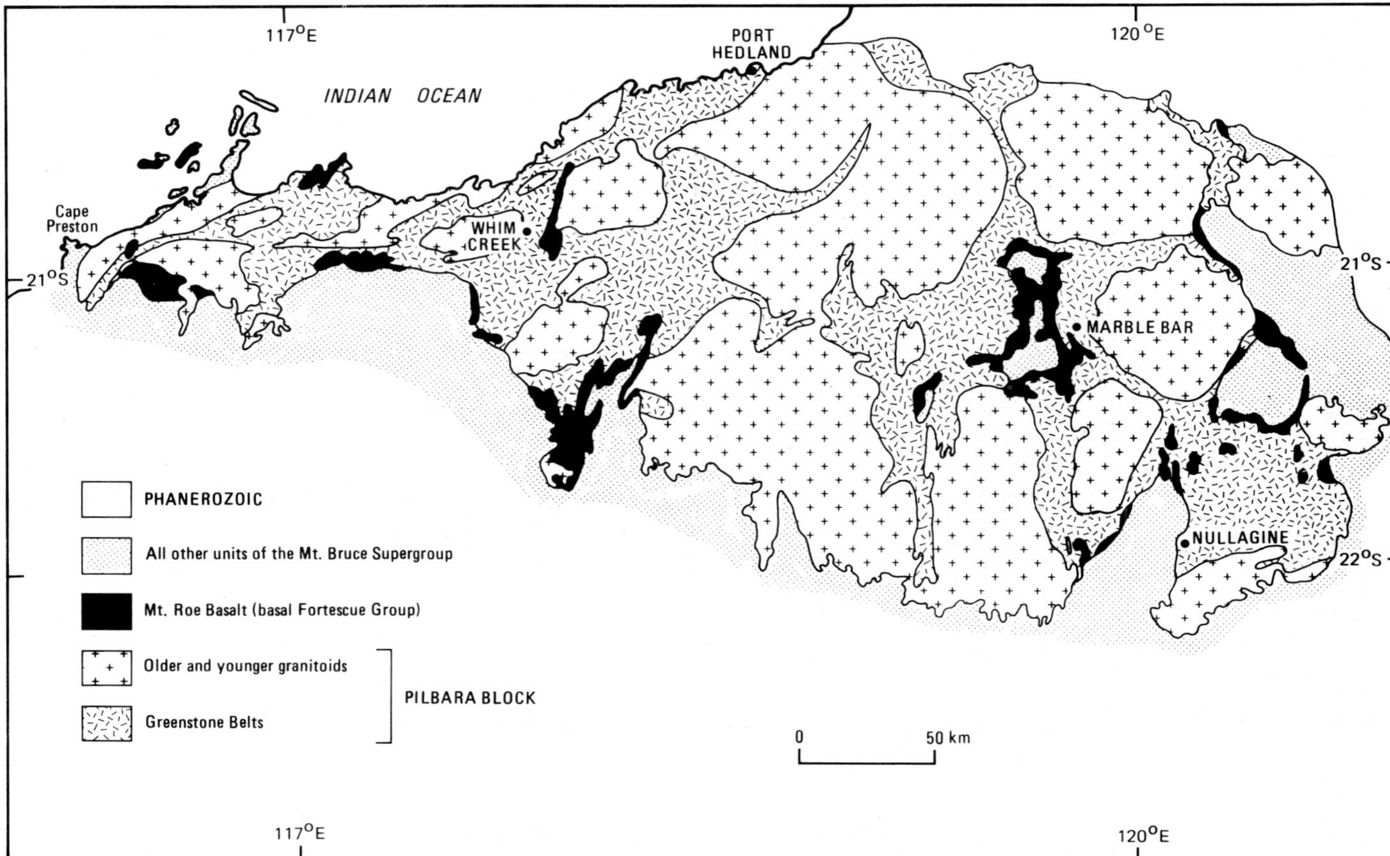

Fig. 3-11. Generalized geological map of the Pilbara Block and immediately overlying parts of the Fortescue Group, after Hickman (1981). While greenstone belts occupy only about 36% of the area of the Pilbara Block, the basal contact of the Mount Roe Basalt lies over greenstone for 80% of its total exposed length, indicating an area-corrected preference of about 7.3:1 for this contact to occur within greenstone belts.

deposition certainly cannot be ruled out.

The question then arises of the reconciliation of this concept with the sharply unconformable base of the Fortescue Group along its main northern outcrop limit. Although the discordance along this contact is evident, three points caution against accepting it as definitive evidence for a long and uniquely significant period of non-deposition — the erosion of event C above. Firstly, even along this boundary north-projecting fingers occur where it crosses greenstone belts (Fig. 3-11), so that it is mainly where the Fortescue Group directly overlies granitoids that the discordance is most evident. And secondly, following from this, at least some of the granitoid domes of the Pilbara Block had already been unroofed well before Fortescue Group deposition, to provide material for the sediments of the Gorge Creek Group, in the higher part of the Archaean layered succession (Eriksson, 1981; Hickman, 1981). Thirdly, the metamorphism of the higher part of the Archaean succession in the central parts of the greenstone belts of the Pilbara Block is of uniformly rather low grade, and gives no support for erosional removal of any significant thickness of overlying material.

In summary, geological and geochronological evidence are consistent with a view that earliest Fortescue Group deposition represents a continuation of the steady accumulation of layered supracrustal rocks which had begun more than 3550 m.y. ago over the entire area of the Pilbara Craton. This accumulation, progressively more confined to the sinking synclinorial greenstone belts, was synchronous with the gradual growth and diapiric ascent of the intervening granitoid plutons, some of which were uncovered before extrusion of the lowest Fortescue Group lavas at about 2750 m.y. These plutons continued to rise actively and more or less concurrently with extrusion to provide debris for the Hardey Sandstone. Emplacement of reactivated granitoid continued well into Fortescue Group time, at least until 2600 m.y. Events A, B, C and D of the simple initial model are thus neither separate nor sequential. The Fortescue Group began as a continuation of the same crust-forming processes that built the Pilbara Craton, and for that reason would not be expected to extend beyond its limits, though it may not have extended so far: A is still accepted. However, it is now supposed that D followed it, and that the greenstone belts were the relatively depressed parts. Event C was a long-lasting and concurrent phase; B took place some time later, and is discussed further below the following heading.

Development of the basin after initiation

The discussion so far has brought the tectonic development to a point where initiation of the basin is postulated to have been marked by volcanicity mainly within the greenstone belts of the underlying Pilbara Craton. This volcanicity was almost exclusively mafic: the acid lavas of the Gregory Range area (Koongaling Volcanics) are apparently unique, and any special significance they may consequently have is not yet understood. The scattered distribution of the outcrops of the mafic lavas — the Mount Roe Basalt (Fig. 3-11) — suggest that volcanism took place at many separate centres. However, the distinctive textural characteristics of these lowest basalts of the Fortescue Group (Hickman, in press; D.F. Blight, pers. commun., 1982) suggest that these eruptions were contemporaneous.

The successively more extensive outcrop areas of the stratigraphic units of the Fortescue Group which immediately succeed the Mount Roe Basalt, noted earlier, have been attributed by several authors to a steady overlap on to topographically high areas formed by the adjacent granitoid plutons (Kriewaldt and Ryan, 1967; Button, 1976; Hickman and Lipple, 1978). However, it is here attributed more to tectonic influence than to the preferential erosion of greenstone belts during a depositional hiatus. It is suggested that the greenstone belts maintained their descent relative to the plutons until midway through Fortescue Group time, and that it was only when this relative movement ceased, and the higher stratigraphic units of the Fortescue Group spread evenly and widely over the whole basin, once more covering the granitoid plutons which had first been unroofed in Gorge Creek Group times, that the Pilbara Craton can be considered to have achieved final "cratonisation". This suggestion is not entirely new: Hickman and Lipple (1978, p. 20) note that in the Marble Bar area "Archaean synclines were slightly tightened during D_4 with the result that the overlying Proterozoic rocks were harmonically folded on broadly Archaean trends. Gravitational down-warping in Lower Proterozoic rocks may have played a minor role in deformation". However, the relative significance of deformation is now emphasised.

A different history of early depression of the Hamersley Basin, during Fortescue Group time, has been suggested by Horwitz and Smith (1978), and emphasised by Horwitz (1980) and Morris and Horwitz (1983). Horwitz and Smith (1978) interpret the pattern of changes in the distribution of successively higher units of the Fortescue Group, and of the overall stratigraphic thickness of the Group, as indicating that "during deposition of the Mount Bruce Supergroup, transgression and progradation resulted in the overlapping on to a broad basement ridge, or geanticline, which was plunging to the northwest and which tilted, or subsided faster, towards the southwest during sedimentation of the Fortescue Group" (Horwitz, 1980, p. 63). They (Horwitz and Smith, 1978, p. 314) also suggested near continuous volcanicity between late Archaean and early Proterozoic in the north.

Lack of detailed information on the lateral extent, thickness variations, and depositional environment of each unit of the Fortescue Group precludes any definite conclusion concerning the tectonic development of the basin during its deposition. Some stratigraphic evidence for the surface configuration and depositional conditions of the basin at that time is available and is discussed later. From a tectonic viewpoint what seems to be generally agreed is that by the close of the period covered by the Fortescue Group the entire present outcrop area had become submerged to form the floor of a single stable depositional basin for the Jeerinah Formation. How far beyond the present outcrop area this basin then extended is a matter which is also discussed later.

The tectonic situation of steady, stable basin-wide crustal depression that was established during deposition of the Jeerinah Formation clearly continued throughout Hamersley Group time. Reliable isopachs are still only available from the Dales Gorge Member of that Group (Fig. 3-3); Trendall and Blockley (1970, pp. 278—281) accepted this pattern as reflecting the ovoid shape of the intracratonic barred basin in which they believed the whole of the Hamersley Group had been deposited. Other possible interpretations are noted later. So far as is known, steady subsidence of this basin continued until completion of the deposition of the Turee Creek Group; however, the presence within the Meteorite Bore Member (Fig. 3-4) of boulders possibly derived from the Hamersley Group, and almost certainly derived from earlier deposited sandstones of the Kungarra Formation, is evidence of greater tectonic instability during Turee Creek Group time than during deposition of the Hamersley Group.

The chronology of this main and essentially continuous period of deposition in the Hamersley Basin is not well established, and the statement in event E of the initial model has no good supporting evidence. However, acceptance of basin initiation at about 2750 m.y. implies acceptance of the points on Fig. 3-10 (6, 14, 18, 75, 78, and 79) which collectively constitute the best evidence for this. Of these, point 14 (2490 m.y.) defines a reliable depositional age for S13 of the Dales Gorge Member. Even with an optimistically high estimate of 8 km for the greatest stratigraphic thickness of any part of the basin between initiation and deposition of the Dales Gorge Member, the consequent sinking rate of 30 $m/10^6yr$ is slower than that suggested above for the layered succession of the Pilbara Block. If this average rate continued to the top of the Turee Creek Group then deposition of this would have been completed by about 2300 m.y.; this is consistent with an age of this order for the intrusive sills of the Weeli Wolli Formation (point 13), whose stratigraphic concordance precludes their emplacement after significant folding of the Hamersley Group. However, basin sinking rates as slow as this are not consistent with Trendall and Blockley's (1970, p. 298) suggested lower limit during Hamersley Group deposition of about 150 $m/10^6yr$, and it is quite likely that the sinking rate varied widely during the life of the basin.

It has been argued by many authors reporting ages less than 2300 m.y. from rocks of the Mount Bruce Supergroup that these correspond to a thermal event associated with folding, and consequently indicate an upper time limit for deposition. Such ages include points 2, 3, 4, 10, 11, 12, 17, and 77, all within the range (disregarding error limits) 2200 and 1500 m.y. The emplacement of the Boolaloo Granodiorite (point 1) at about this time reinforces this argument. However, the 1700 m.y. age of this is itself a poor control for the upper age limit of the basin, because it intrudes the folded Wyloo Group, the earliest deposition of which was accompanied by massive exposure and erosion of the Hamersley Group.

Whenever the main folding of the Hamersley Basin rocks took place, there is general agreement that its style probably reflects block movement and comparatively gentle warping of the lower, "cratonised", crust below them (e.g., Gee, 1979). The Mount Bruce Supergroup, whose depositional history had by then ceased, behaved as an overlying and plastically responsive sheet. The time interval between completion of Mount Bruce Supergroup deposition and this folding remains unknown.

The "Pilbara egg"

The present outcrop limits of the Mount Bruce Supergroup were described above by reference to an imaginary bounding ellipse. This ovoid shape, nearly 200,000 km^2 in area, is a conspicuous feature of the most recently published geological map of Western Australia (GSWA, 1979), and its existence is well known to users of that map. Curiously, however, only one rarely cited paper (Miyano, 1976) has been specifically devoted to its existence and possible significance, although Gee (1979) has noted the ovate shape of the Pilbara Craton, and also the coincidence of this with the Fortescue Regional Gravity Province of Fraser (1976).

This enigmatic ellipse is here informally referred to as the Pilbara egg. Miyano (1976) regards it as a regional dome which began its ascent when the Hamersley Basin was initiated, and which continued to rise during the succeeding 10^9 years, during which it exerted a major tectonic control over peripheral sedimentation. Its extent is shown in Fig. 3-12. The identity of the Pilbara egg as a significant crustal entity is reinforced by at least four independent factors:

(a) The simple definition of its regular shape by the external outcrop limits, often faulted, of the Mount Bruce Supergroup.

(b) An increase in intensity of folding of the group outwards from the centre, and a tendency of the strong peripheral folding to be parallel to the edge.

(c) The tendency for metamorphic isograds of the Fortescue Group to curve in sympathy with its outline.

(d) Its expression as a regional gravity province.

Miyano (1976, fig. 3, with caption of fig. 4) also saw a relationship between the egg and the regional pattern of faults and dykes, but this is less convincing than the features listed above. A real fifth feature, however, is the parallelism of many external regional structures, such as the Ashburton Fold Belt, with the margin of the egg; but this involves consideration of Precambrian geology later than the Hamersley Basin, and is not further referred to here.

In discussion of the deformation of the Hamersley Basin, Gee (1979, p.

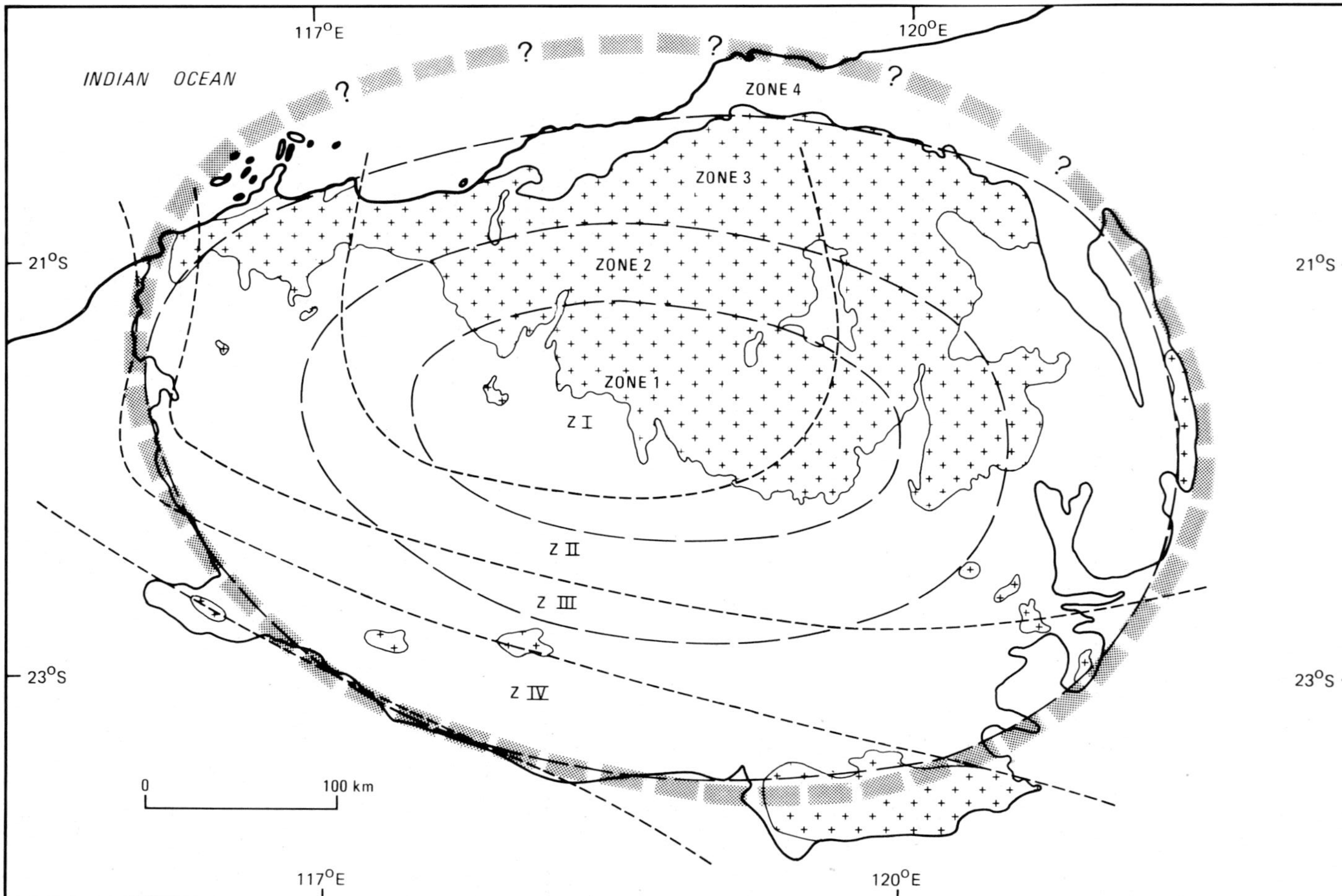

Fig. 3-12. Generalized geologic map emphasising the definition of the Pilbara egg. *Zones 1—4* are zones of varying structural complexity defined by Miyano (1976). *Z I—IV* are the metamorphic zones of Smith et al. (1982). The boundary of the egg itself is

352) noted that "The last deformation affecting the Pilbara Craton is normal faulting in a huge arc which precisely matches the edge of the craton. These faults have a craton-side-up movement, and therefore seem to result from late-stage epeirogenic uplift of the stabilised craton". Gee's "huge arc" is the visible, onshore, boundary of the Pilbara egg, and he correctly describes its marginal structural relationship. But the critical question concerning the egg is whether the Pilbara Craton *never* extended beyond it, its shape having been established by very early Archaean events, so that the late faulting was controlled by the already defined edge; or, alternatively, whether the late faulting was imposed, by some unknown mechanism, upon a regionally extensive "basement" of Archaean crust within which no egg previously existed.

The concept of an identifiable eastern edge of the Pilbara Craton was suggested by De Laeter et al. (1977) on Sr isotopic evidence; however, they did not discuss in detail what the nature of this edge might be. Gee (1979, p. 327) also implies a distinct margin in his reference to the Pilbara Craton as one of two major granitoid-greenstone terrains representing "huge, discrete, somewhat rounded volcanic basins that developed within extensive and perhaps continuous crust". Horwitz and Smith (1978, fig. 7) and Miyano (1976, fig. 7), on the other hand, appear to see an Archaean basement of the Hamersley Basin extending continuously, at least to the south, beyond the edge of the Pilbara egg, and possessing no distinctive characters at its boundary.

Whatever the origin and history of the Pilbara egg, there is at present insufficient evidence for a complete understanding. Such an understanding is believed to be essential for a full reconstruction of both the tectonic development, and, as will appear below, the surface configuration of the basin.

Synthesis

Using a simple initial model as a framework, the foregoing discussion of the tectonic development of the Hamersley Basin has ranged over a wide field, and has taken note of both detailed evidence and speculative possibilities. It is appropriate to present in conclusion a new set of proposals to replace the initial model:

(A) The tectonic and petrogenetic evolution of the Hamersley Basin and the structurally underlying Pilbara Craton should be analysed in terms of continuous processes rather than as a sequence of separate events.

(B) Deposition of the lowest unit of the Mount Bruce Supergroup continued on from that of the highest units of the layered succession (Pilbara Supergroup) of the greenstone belts of the Pilbara Craton.

(C) In the sequence of supracrustal deposition from the lowest part of the Pilbara Supergroup through to the highest part of the Mount Bruce Supergroup there was a corresponding progression in the regional extent of each stratigraphic unit — first a contraction from craton-wide extent to increasing restriction to the evolving synclinorial greenstone belts, then an expansion back to a basin-wide extent of each unit as the greenstone belts stabilised.

(D) Generation of sialic ("granitoid") material began at an early stage of Pilbara Supergroup deposition, and continued during deposition of the Mount Bruce Supergroup. Granitoid plutons ascended steadily as the intervening greenstone belts formed; irruptive bodies periodically rose to higher levels as shallow intrusions, or broke surface as acid volcanics. Lower parts of the Pilbara Supergroup were themselves continuously migmatised as the plutons evolved.

(E) The granitoid plutons were first unroofed during late Pilbara Supergroup time, and continued to supply material for sedimentation until they were completely covered before the end of Fortescue Group deposition.

(F) The whole complex cycle of crustal evolution proceeded as a continuous sequence of inter-related plutonic, volcanic, and supracrustal events from before 3.5 b.y. to about 2.3 b.y. ago, during the latter part of which period the Hamersley Basin was formed.

The inter-related tectonic development of the Pilbara Craton and Hamersley Basin implicit in these statements is summarised graphically in Fig. 3-13. The Pilbara Craton evolution is fully consistent with the interpretation of Hickman (1981), but markedly different from that of Marston and Groves (1981; cf. Fig. 3-13 of this paper with their fig. 5) and Gee (1979, p. 362) who also believes that the granite-greenstone terrain evolved "over relatively

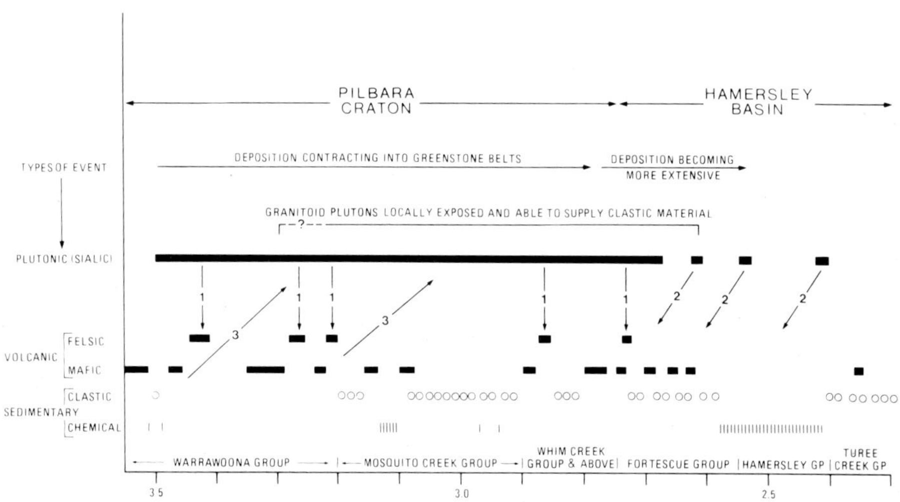

Fig. 3-13. Graphical synopsis of the tectonic development of the Pilbara Craton and Hamersley Basin, emphasising the status of these two entities as products of a continuing crust-forming process. Arrows numbered *1* indicate surface extrusion of felsic magma; arrows marked *2* indicate shallow intrusion of felsic magma; arrows marked *3* indicate migmatitic recycling of the layered succession into granitoid plutons.

short periods of 100—200 m.y.". It is an important general conclusion from the interpretation presented that the "Archaean"—"Proterozoic" distinction has no clear meaning or significance.

SURFACE CONFIGURATION AND DEPOSITIONAL CONDITIONS

Fortescue Group

If the tectonic concepts summarised in Fig. 3-13 are valid the earliest condition of the basin may be envisaged as a series of probably interconnected curvilinear troughs, located on the greenstone belts and separated by granitoid uplands. The troughs contained only shallow water, either lakes, rivers or oceanic inlets (Hickman and De Laeter, 1977). Concurrent active erosion ensured that much of the Mount Roe Basalt was erupted over dissected topography, with a relief of up to 100 m. It seems likely that most of the volcanicity took place within the troughs, rather than from fissures represented by the dyke swarms of the Pilbara Block granitoid domes, although there is no direct evidence of their source. The granitoid uplands provided abundant clastic debris, which spread over the initial lavas and was probably reworked and distributed extensively by braided river systems.

Successive later episodes of volcanicity gradually filled these early valleys and overstepped the granitoid plateaus, while the depth and extent of the water-filled areas increased: the Tumbiana Formation, with its shallow-water stromatolitic carbonates, is of wide regional extent. Following the virtual blanketing of the entire basin area by the Maddina Basalt the extensive regional sea of the Jeerinah Formation marked the initiation of a period of stable chemical sedimentation that prevailed throughout the Hamersley Group.

The supposed extent of the area over which this system of laterally expanding troughs extended is entirely dependent, in the model under discussion, on assumptions made concerning the significance of the Pilbara egg. If it is supposed that the outline of the egg was established early, then the troughs would have been confined within it. The nature of the crust, and its surface configuration, outside the egg are not known.

While the generally shallow-water environment of much of the Fortescue Group is not in dispute there are divergent views concerning the regional configuration of its depositional basin. Horwitz and Smith (1978) envisage the group as having been laid down in a paralic environment on an extensive shelf along the western margin of a land mass. In the area of the Hamersley Basin itself (Horwitz and Smith, fig. 5) they believe the regional distribution of Fortescue Group units immediately overlying the Pilbara Block to be consistent with gradual transgression over a local basement ridge projecting westnorthwest across the shelf. Hall and Goode (1978), Horwitz and Smith (1978, fig. 7) and Horwitz (1976, fig. 1) saw this depositional environment as extending southwards over the area of the Nabberu Basin (Goode et al., 1983, this volume). But it is now generally accepted that with a probable interval of about 500 m.y. between the end of deposition in the Hamersley Basin and its initiation in the Nabberu Basin, these two areas of deposition are unrelated.

Hamersley Group

By the close of Fortescue Group deposition the final form which the basin possessed during deposition of the Hamersley Group appears to have become established. The best evidence for the reconstruction of the basin environment at this stage comes from a consideration of the most characteristic lithology — banded iron-formation (BIF). The arguments used by Trendall and Blockley (1970) to establish a self-consistent hypothesis for BIF deposition were a refinement of Trendall's (1966) earlier suggestions; later revisions (Trendall, 1972; 1973a, b) have not altered its main features so far as direct basin interpretations are concerned, but viewpoints on more speculative problems, such as the source of iron, have since been modified. The immediately following summary more or less reproduces that of Trendall (1976a), in which most of the detail of the evidence was omitted for clarity. Later discussion indicates alternative hypotheses that have been proposed in later publications (Ewers and Morris, 1981; Morris and Horwitz, 1983).

Aftbands are regarded as the only primary depositional features within the BIF, and the regularity of aftband spacing within any one chert mesoband is taken to indicate that the iron-rich/iron-poor cyclicity by which they are defined is related to some equally regular natural environmental event controlling deposition. Only two natural events involving repeated systematic environmental changes have great regularity; the day and the year, controlled respectively by Earth's rotation and revolution. The general order of thickness of aftbands precludes the day as a control, and it is concluded that they are seasonally controlled annual layers: they are varves. The origin of the fine structure within aftbands is not understood.

The regional continuity of aftbands then precludes any depositional mechanism for them other than chemical precipitation. It is difficult to envisage that an undisturbed layer of an even thickness no greater than a few millimetres could be spread over an area of at least 50,000 km^2 by any mechanical means, which left no other obvious evidence of its operation.

The sheer bulk of BIF laid down in the basin makes it unlikely that gross systematic chemical alteration has taken place, and it is therefore assumed that both silica and iron were almost the only stable constituents of the deposited material. If this is so, then its origin as a precipitate from the basin water seems more acceptable than an origin as either volcanic or terrigenous dust (Carey, 1976).

The lack of contemporaneous disturbance in the BIFs indicates an exceptionally still and quiet subaqueous environment. There is complete absence of evidence of any supply of material to the basin by incoming drainage. If the isopachs of the Dales Gorge Member (Fig. 3-3) reflect basin shape, evaporative concentration of iron and silica within a closed basin becomes an attractive hypothesis. A completely arid circum-basinal terrain is consistent with both this, the preceding point, and a close relationship between climate and deposition. Nevertheless, absence of any major stratigraphic disturbance within the Hamersley Group that could be attributed to dessication suggests a limited connection with the open ocean, and a barred basin acting as a circulating cell has potential for replenishment by incoming seawater. If the edges of the basin lay not far beyond the remaining outcrop area of the Hamersley Group a basin area of about 150,000 km^2 is indicated.

If aftbands are chemical varves then an origin needs to be suggested for mesobands which is consistent with this hypothesis. Two questions are relevant: firstly, why does the aftband interval vary between different chert mesobands although it remains virtually constant with any one mesoband, and secondly, what does chert-matrix represent? The question of the status of chert-matrix seems to be resolved by the evidence provided by the lateral terminations of podded cherts; chert-matrix appears to represent grossly compacted material that originally was the chert. If compaction can have such a radical effect then it may possibly also explain the variation of aftband interval between different chert mesobands, in the following hypothesis.

Suppose that, each year, about 5 mm of extremely hydrous precipitate were laid down evenly over the entire basin. As successive layers accumulated, the initially gelatinous colloidal material became compressed and dehydrated, with some silica departing in solution in the water, but the iron remaining stable. An inverse relationship would be produced between aftband thickness and Fe content of chert mesobands. This is the actual situation (Trendall and Blockley, 1970), and the hypothesis is currently accepted by the writer as most consistent with the available evidence. However, it is appropriate to point out that in principle the same textural result could be produced by varying silica and constant iron precipitation (Trendall, 1965b, p. 1066, discusses a logically similar situation).

Within the Weeli Wolli Formation there is evidence for a systematic 23.3-year cyclicity of aftbanding (Fig. 3-8). An increase in the amplitude of variation of whatever depositional conditions were actually responsible for the expression of this rhythm in the Weeli Wolli Formation is believed to have been responsible for mesobanding in the Dales Gorge Member and other BIFs. A longer-term cyclicity of depositional environment is believed to have caused the Calamina cyclothem, and related alternations of red and white chert mesobands, but the cause of this is still doubtful.

Volcanoclastic structures in the S macrobands of the Dales Gorge Member and in some of the major shale units indicate that these represent intervals when "normal" quiet BIF accumulation was prevented by steady influx of volcanic debris. For the Dales Gorge Member, these intervals may be related to spasmodic sinking of the basin.

The immediate cause, or causes, of precipitation in the basin are uncertain, but the writer's preferred hypothesis is that the precipitation of iron was caused by the oxidation of ferrous iron in the basin water by oxygen evolved during photosynthesis by algae, floating either at or below the water surface However, early reports of microfossils in the Brockman Iron Formation (La Berge, 1967; Karkhanis, 1976) are now challenged (Walter and Hoffmann, 1983, this volume), and the isotopically light carbon in carbonates (Becker and Clayton, 1972) could have had a volcanic origin, so that there is little hard evidence in its favour. Morris' (1973) chemical evidence is probably the best yet available for significant biological activity during BIF deposition. The precipitation of silica, on the other hand, may have been caused purely by evaporative concentration. In this hypothesis, both processes would be independently related to annual insolation, so that there would be scope for textural variation due to minor differences each year in such parameters as basin temperature or turbidity. Neither of the main constituents of the precipitate need have been present in large concentrations in the basin water.

Alternative hypotheses very different to several of those embodied in Trendall and Blockley's (1970) reconstruction of the environment and configuration of the Hamersley Basin during BIF deposition, summarised in the foregoing paragraphs, have come from the later work of Ewers and Morris (1981). The main points on which these authors agree with, or differ from, Trendall and Blockley's (1970) view in respect to the depositional conditions of the Dales Gorge Member, are as follows:

(1) The identity of the BIF as chemical sediments which accumulated under conditions of exceptional stability, and without contemporaneous disturbance, is accepted, as also is the view that the chemical composition of the primary precipitate was close to that of the present BIF, except for loss of water.

(2) The identity of S macrobands as marking volcanic episodes during which airborne ash fell into the basin is accepted, although Ewers and Morris (1981) emphasize that the onset and termination of these episodes were gradational, and that during them precipitation of iron and silica continued.

(3) The identity of aftbands as annual layers is accepted, but Ewers and Morris interpret both the mesobanding and the smaller-scale banding within aftbands as the direct result of variations in the primary depositional environment of the basin; this is in sharp contrast to Trendall and Blockley's belief that mesobanding is largely the result of diagenetic differentiation of a rather uniform primary precipitate.

(4) As a corollary of (3), Ewers and Morris believe that most chert pods result from lateral flow of silica during compression.

(5) Ewers and Morris prefer to envisage deposition of the Dales Gorge Member on a shelf rather than in a barred basin, from arguments based on

the logistics of supplying sufficient iron to the area of deposition in dilute solution.

This last point is consistent with the Fortescue Group shelf model of Horwitz and Smith (1978); Morris and Horwitz (1983) have taken the idea a stage further by suggesting that, as the Hamersley Group has a negligible terrigenous content, it may have been deposited on a submarine, essentially volcanogenic platform or bank built on an older Archaean, sialic, northwest-trending shelf protruding into or marginal to, an ocean. They draw attention to the modern Bahamas Platform as the closest modern analogue for the configuration of the area of deposition of the Dales Gorge Member.

It is certainly true that the isopach pattern of the Dales Gorge Member (Fig. 3-3) can be fitted into either a restricted basin or a bank model and there is at present no good evidence for a near-shore facies within the Hamersley Group. A choice between these models must await the availability of more evidence. Foremost amongst the additional data required are regional thickness and facies variation for each individual formation of the group.

Turee Creek Group

The interpretation in terms of basin configuration of the concordant stratigraphic transition from the Hamersley Group to the Turee Creek Group is problematical. Although there is no stratigraphic evidence for tectonic disturbance at the time of transition, two simultaneous and sharp changes in the depositional environment of the basin took place at this time: firstly, the cessation of the stable and chemically distinctive conditions required for BIF deposition, and secondly the sudden start of a continuing supply of terrigenous material.

Trendall (1980) supposes that both changes are related to the establishment of a major connection of the previously restricted basin to the open ocean by subsidence along its southern margin. However, he points out that the possible glacial origin of the Meteorite Bore Member may indicate that a climatic change was also associated with this essentially tectonic event. The work of Smith et al. (1982) has provided good evidence that the Turee Creek Group thins rapidly northwards from its area of thick development in the Hardey Syncline area (Fig. 3-1). Sedimentological studies of this group are needed to establish more closely its environment of deposition.

MINERAL DEPOSITS

The rather sudden rise of the Hamersley Basin to a position of significance in the international literature on iron-formation is closely paralleled by the rapid development of its resources of iron ore. In response to the lifting of an Australian Government embargo on the export of iron ore in 1961, export of high-grade hematite-goethite ore began in 1966 after intensive exploration and development. In 1976, ten years later, over 85 million tonnes of iron ore were mined in Western Australia, of which over 77 million tonnes were exported, representing about 18% of all iron ore transported across international boundaries throughout the world in that year. The bulk of this came from the Hamersley Basin.

Although it is hardly possible to summarise the geology of the Hamersley Basin without reference to this economic aspect it will be sufficient here to indicate where details of the occurrence, distribution, and development of the iron ore associated directly or indirectly with the iron-formations of the Hamersley Group can be found. MacLeod's (1966) bulletin provides a good summary of the geology of iron ore in the Hamersley Basin which has still not been superseded by later work.

Trendall (1975c) reviewed general aspects of ore genesis, and in the same volume, papers by Gilhome, Baldwin, Evans and Clint, Kneeshaw, Ward and others, Neale, and Adair (all 1975) give more detail of individual ore bodies. Lord and Trendall (1976) have provided a review of mining development while a review of the iron ore resources of the Hamersley Basin, totalling over 36 billion (36×10^9) tonnes, is given by Morrison (1978). Both the early mineragraphic studies of Ayres (1971) and the views on genesis of Trendall (1975c) have been superseded by the more recent detailed study of Morris (1980). This latter work has been linked into a model for genesis of iron ore by Morris et al. (1980) which represents the most satisfying general theory yet proposed.

Iron ore is overwhelmingly the most important exploitable mineral of the Hamersley Basin. Blue (riebeckite) asbestos, or crocidolite, was mined up to 1966 (Trendall and Blockley, 1970), while small deposits of lead, zinc, silver, copper, and gold have been worked (Blockley, 1975), all of them within the Fortescue Group.

ACKNOWLEDGEMENTS

The list of persons and organisations to whom I owe a deep debt of gratitude for help and advice of a variety of kinds during my 20 years of association with the geology of the Hamersley Basin is far too long for inclusion here, and many must be covered in a general expression of thanks. In the preparation of this paper I have received assistance from many colleagues on the Geological Survey of Western Australia, including Dr D.F. Blight, Mr J.G. Blockley, Mr R.R. Connolly, Dr R. Davy, Dr R.D. Gee, Dr A.H. Hickman, Dr J.R. Myers, and Dr W.G. Libby; Miss B.M. Nash was particularly helpful in translation of the paper of Miyano (1976). Outside the Geological Survey I am indebted to Mr R.C. Morris, of the CSIRO Division of Mineralogy, co-editor of this volume, for supplying the computer-drawn isopachs of Fig.

3-3, for much constructive comment, and for his patient encouragement. Final thanks go to Drs. M.R. Barley, J.R. de Laeter, M.J. McCulloch, J.R. Richards, and G.C. Sylvester for permission to include unpublished geochronological results on Table 3-V and Fig. 3-10.

REFERENCES

Adair, D.L., 1975. Middle Robe River iron ore deposits. In: C.L. Knight (Editor), Economic Geology of Australia and Papua-New Guinea. 1 Metals. Australas. Inst. Min. Metall., Monogr., 5: 943—945.

Ayres, D.E., 1971. The hematite enrichment ores of Mount Tom Price and Mount Whaleback, Hamersley Iron Province. Australas. Inst. Min. Metall., Proc., 238: 47—58.

Ayres, D.E., 1972. Genesis of iron-bearing minerals in banded iron formation mesobands in the Dales Gorge Member, Hamersley Group, Western Australia. Econ. Geol., 67: 1214—1233.

Baldwin, J.T., 1975. Paraburdoo and Koodaiderie iron ore deposits, and comparisons with Tom Price iron ore deposits, Hamersley Iron Province. In: C.L. Knight (Editor), Economic Geology of Australia and Papua-New Guinea. 1 Metals. Australas. Inst. Min. Metall., Monogr., 5: 898—905.

Barley, M.E. and De Laeter, J.R., in press. Disturbed Rb-Sr systems of the Archaean Duffer Formation, Pilbara Block, Western Australia. J. R. Soc. West. Aust.

Becker, R.H. and Clayton, R.N., 1972. Carbon isotopic evidence for the origin of a banded iron-formation in Western Australia. Geochim. Cosmochim. Acta, 36: 577—595.

Becker, R.H. and Clayton, R.N., 1976. Oxygen isotope study of a Precambrian banded iron-formation, Hamersley Range, Western Australia. Geochim. Cosmochim. Acta, 40: 1153—1165.

Blight, D.F., in prep. The lower Fortescue Group in the southern Hamersley Basin. Geology and economic potential.

Blockley, J.G., 1975. Hamersley Basin — mineralization. In: C.L. Knight (Editor), Economic geology of Australia and Papua-New Guinea. 1. Metals. Australas. Inst. Min. Metall., Monogr., 5: 413—415.

Blockley, J.G., 1979. A contribution to the stratigraphy of the Marra Mamba Iron Formation. West. Aust., Geol. Surv., Annu. Rep., 1978: 71—73.

Blockley, J.G., 1980. The tin deposits of Western Australia. West. Aust., Geol. Surv., Miner. Resour. Bull., 12: 184.

Blockley, J.G., Trendall, A.F., De Laeter, J.R. and Libby, W.G., 1980. Two "anomalous" isochrons from the vicinity of Newman. West. Aust., Geol. Surv., Annu. Rep., 1979: 93—96.

Bourn, R. and Jackson, D.G., 1979. A generalized account of the Paraburdoo iron ore bodies. Australas. Inst. Min. Metall., Annu. Conf., Perth, pp. 187—201.

Button, A., 1976. Transvaal and Hamersley Basins — review of basin development and mineral deposits. Mineral. Sci. Eng., 8: 262—293.

Carey, S.W., 1976. The Expanding Earth. Developments in Geotectonics, 10. Elsevier, Amsterdam, 488 pp.

Compston, W. and Arriens, P.A., 1968. The Precambrian geochronology of Australia. Can. J. Earth Sci., 5: 561—583.

Compston, W., Williams, I.S., McCulloch, M.T., Foster, J.J., Arriens, P.A. and Trendall, A.F., 1981. A revised age for the Hamersley Group. Geol. Soc. Aust., 5th Annu. Conv., Perth, Abstracts, 3: 40.

Cooper, J.A., James, P.R. and Rutland, R.W.R., 1980. Rb/Sr dating of granitic intrusions in relation to the stratigraphic and deformational history of the Pilbara region. Second Int. Archaean Symp., Perth, Abstracts, p. 14.

Daniels, J.L., 1968. Turee Creek, W.A. West. Aust., Geol. Surv., 1 : 250 000 Geol. Ser., Explan. Notes.

Daniels, J.L., 1970. Wyloo, W.A. West. Austr., Geol. Surv., 1 : 250 000 Geol. Ser., Explan. Notes.

Daniels, J.L. and MacLeod, W.N., 1965. Newman, W.A. West. Austr., Geol. Surv., 1 : 250 000 Geol. Ser., Explan. Notes.

Davy, R., 1975. A geochemical study of a dolomite-BIF transition in the lower part of the Hamersley Group. West. Aust., Geol. Surv., Annu. Rep., 1974: 88—100.

Davy, R., in press. The Mount McRae Shale and the upper part of the Mount Sylvia Formation in core RD1, Rhodes Ridge, W.A.

De Laeter, J.R. and Blockley, J.G., 1972. Granite ages in the Pilbara Block. J. Geol. Soc. Aust., 19 (3): 363—370.

De Laeter, J.R. and Trendall, A.F., 1970. The age of the Copper Hills Porphyry. West. Aust., Geol. Surv., Annu. Rep., 1969: 54—59.

De Laeter, J.R. and Trendall, A.F., 1971. The age of the Gidley Granophyre. West Aust., Geol. Surv., Annu. Rep., 1970: 62—67.

De Laeter, J.R. and Trendall, A.F., in prep. Rb-Sr isotopic data from lavas of the Fortescue Group, Western Australia.

De Laeter, J.R., Peers, R. and Trendall, A.F., 1974. The petrography, chemical composition and geochronology of two dolerite sills from the Precambrian Weeli Wolli Formation, Hamersley Group. West. Aust., Geol. Surv., Annu. Rep., 1973: 82—91.

De Laeter, J.R., Lewis, J.D. and Blockley, J.G., 1975. Granite ages within the Shaw Batholith of the Pilbara Block. West. Aust., Geol. Surv., Annu. Rep., 1974: 73—79.

De Laeter, J.R., Hickman, A.H., Trendall, A.F. and Lewis, J.D., 1977. Geochronological data concerning the eastern extent of the Pilbara Block. West. Aust., Geol. Surv., Annu. Rep., 1976: 56—62.

De Laeter, J.R., Seymour, D. and Libby, W.G., in prep. A Rb-Sr date from felsic volcanics within the Mount Roe Basalt of the Wyloo Dome.

De la Hunty, L.E., 1964. Balfour Downs, W.A. West. Aust., Geol. Surv., 1 : 250 000 Geol. Ser., Explan. Notes.

De la Hunty, L.E., 1965. Mount Bruce, W.A. West. Aust., Geol. Surv., 1 : 250 000 Geol. Ser., Explan. Notes.

De la Hunty, L.E., 1969. Robertson, W.A. West. Aust., Geol. Surv., 1 : 250 000 Geol. Ser., Explan. Notes.

Eriksson, K.A., 1981. Archaean platform-to-trough sedimentation, east Pilbara Block, Australia. Spec. Publ., Geol Soc. Aust., 7: 235—244.

Evans, W.J. and Clint, P., 1975. Iron deposits of the Brockman Syncline, Hamersley Iron Province. In: C.L. Knight (Editor), Economic Geology of Australia and Papua-New Guinea. 1. Metals. Australas. Inst. Min. Metall., Monogr., 5: 906—910.

Ewers, W.E. and Morris, R.C., 1980. Chemical and mineralogical data from the uppermost section of the upper BIF member of the Marra Mamba Iron Formation. CSIRO, Perth, Rep. FP23.

Ewers, W.E. and Morris, R.C., 1981. Studies on the Dales Gorge Member of the Brockman Iron Formation. Econ. Geol., 76 (7): 1929—1953.

Forman, F.G., 1938. Notes on a reconnaissance of the Gascoyne, Ashburton, and Fortescue Districts. West. Aust., Geol. Surv., Annu. Rep., 1937: 6—9.

Finucane, K.J., 1939. The blue asbestos deposits of the Hamersley Ranges, Western Australia. Aerial Geol. Geophys. Surv. Northern Australia, Rep. West. Aust., No. 49.

Fraser, A.R., 1976. Gravity provinces and their nomenclature. BMR J. Aust. Geol. Geophys., 1: 350—352.

GSWA, 1975. Geology of Western Australia. West. Aust., Geol. Surv., Mem., 2: 541.

GSWA, 1979. Geological map of Western Australia at 1 : 2 500 000 scale. West. Aust., Geol. Survey.

Gee, R.D., 1979. Structure and tectonic style of the Western Australian Shield. Tectonophysics, 58: 329—369.

Gilhome, W.R., 1975. Mount Tom Price Iron Orebody. In: C.L. Knight (Editor), Economic Geology of Australia and Papua-New Guinea. 1. Metals. Australas. Inst. Min. Metall., Monogr., 5: 892—898.

Grubb., P.L.C., 1967. Asbestos. CSIRO Div. Applied Mineral., Annu. Rep., 1966—67: 6.

Hall, W.D.M. and Goode, A.D.T., 1978. The early Proterozoic Nabberu Basin and associated iron formations of Western Australia. Precambrian Res., 7: 129—184.

Halligan, R. and Daniels, J.L., 1964. Precambrian geology of the Ashburton Valley region, North-West Division. West Aust., Geol. Surv., Annu. Rep., 1963: 38—46.

Hamilton, P.J., Evensen, N.M., O'Nions, R.K., Glikson, A.Y., and Hickman, A.H. 1980. Sm-Nd dating of the Talga Talga Subgroup, Warrawoona Group, Pilbara Block, Western Australia. Second Int. Archaean Conf., Perth, Abstracts, pp. 11—12.

Hickman, A.H., 1975. Explanatory Notes on the Nullagine 1 : 250 000 Geological Sheet, W.A. West. Aust., Geol. Surv., Rec. 1975/5.

Hickman, A.H., 1977. Explanatory notes on the Precambrian part of the Port Hedland — Bedout Island 1 : 250 000 Geological Sheets. West. Aust., Geol. Surv., Rec. 1977/2.

Hickman, A.H., 1978. Nullagine, W.A. West. Aust., Geol. Surv., 1 : 250 000 Geol. Ser., Explan. Notes.

Hickman, A.H., 1980. Lithological map and stratigraphic interpretation of the Pilbara Block (map at 1 : 1 000 000 scale). West. Aust., Geol. Surv. (= Plates 1A and 1B of Hickman, in press; issued in advance of full publication).

Hickman, A.H., 1981. Crustal evolution of the Pilbara Block, Western Australia. Spec. Publ. Geol. Soc. Aust., 7: 57—69.

Hickman, A.H., in press. Geology of the Pilbara Block and its environs. West. Aust., Geol. Surv., Bull. 127.

Hickman, A.H. and Chin, R.J., 1977. Explanatory notes on the Precambrian parts of the Yarrie 1 : 250 000 Geological Sheet. West. Aust., Geol. Surv., Rec. 1976/16.

Hickman, A.H. and De Laeter, J.R., 1977. The depositional environment and age of a shale within the Hardey Sandstone of the Fortescue Group. West. Aust., Geol. Surv., Annu. Rep., 1976: 62—68.

Hickman, A.H. and Lipple, S.L., 1978. Marble Bar, W.A. West. Aust., Geol. Surv., 1 : 250 000 Geol. Ser., Explan. Notes.

Horwitz, R.C., 1976. Two unrecorded basal sections in older Proterozoic rocks of Western Australia. CSIRO, Div. Mineral., Rep. FP17.

Horwitz, R.C., 1978. The Lower Proterozoic of the Wyloo Anticline. SCIRO Div. Mineral., Rep. FP20.

Horwitz, R.C., 1980. Discussion on: A progress review of the Hamersley Basin of Western Australia, by A.F. Trendall. Finl. Geol. Soc. Bull., 53 (1): 63—66.

Horwitz, R.C., 1981. The Ashburton Trough. Geol. Soc. Aust., 5th Annu. Conv., Perth, Abstracts, (3): 33.

Horwitz, R.C. and Smith, R.E., 1978. Bridging the Yilgarn and Pilbara Blocks, Western Australia. Precambrian Res., 6: 293—322.

Hudson, J.D., 1964. Sedimentation rates in relation to the Phanerozoic time-scale. Q. J. Geol. Soc. London, 120S: 37—42.

Jahn, Bor-Ming, Glikson, A.Y., Peucat, J.J. and Hickman, A.H., 1981. REE geochemistry and isotopic data of Archaean silicic volcanics and granitoids from the Pilbara Block, Western Australia: implications for the early crustal evolution. Geochim. Cosmochim. Acta, 45: 1633—1652.

Karkhanis, S.N., 1976. Fossil iron bacteria may be preserved in Precambrian ferroan carbonate. Nature, 261: 406—407.

Klein, C. and Gole, M.J., 1981. Mineralogy and petrology of parts of the Marra Mamba Iron Formation. Am. Mineral., 66: 507—525.

Kneeshaw, M., 1975. Mt Whaleback iron orebody, Hamersley Iron Province. In: C. L. Knight (Editor), Economic Geology of Australia and Papua-New Guinea. 1. Metals. Australas. Inst. Min. Metall., Monogr., 5: 910—916.

Kokelaar, B.P., 1982. Fluidization of wet sediments during the emplacement and cooling of various igneous bodies. J. Geol. Soc. London, 139: 21—33.

Kriewaldt, M.J.B., 1964. Dampier and Barrow Island, W.A. West. Aust., Geol. Surv., 1 : 250 000 Geol. Ser., Explan. Notes.

Kriewaldt, M.J.B. and Ryan, G.R., 1967. Pyramid, W.A. West. Aust., Geol. Surv., 1 : 250 000 Geol. Ser., Explan. Notes.

La Berge, G.L., 1966. Altered pyroclastic rocks in iron-formation in the Hamersley Range, Western Australia. Econ. Geol., 61: 147—161.

La Berge, G.L., 1967. Microfossils and Precambrian iron-formations. Geol. Soc. Am. Bull., 78: 331—342.

Leggo, P.J., Compston, W. and Trendall, A.F., 1965. Radiometric ages of some Precambrian rocks from the Northwest Division of Western Australia. J. Geol. Soc. Aust., 12 (1): 53—65.

Lewis, J.D., Rosman, K.R.J., and De Laeter, J.R., 1975. The age and metamorphic effects of the Black Range dolerite dyke. West. Aust., Geol. Surv., Annu. Rep., 1974: 80—88.

Lord, J.H. and Trendall, A.F., 1976. Iron ore deposits of Western Australia, geology and development. Mem., Am. Assoc. Pet. Geol., 25: 410—417.

Low, G.H., 1965. Port Hedland, W.A. West. Aust., Geol. Surv., 1 : 250 000 Geol. Ser., Explan. Notes.

MacDonald, J.A., and Grubb, P.L.C., 1971. Genetic implications of shales in the Brockman Iron Formation from Mount Tom Price and Wittenoom Gorge, Western Australia. J. Geol. Soc. Aust., 18: 81—86.

MacLeod, W.N., 1966. The geology and iron deposits of the Hamersley Range area. West. Aust., Geol. Surv., Bull. 117.

MacLeod W.N. and de la Hunty, L.E., 1966. Roy Hill, W.A. West. Aust., Geol. Surv., 1 : 250 000 Geol. Ser., Explan. Notes.

MacLeod, W.N., and de la Hunty, L.E., Jones, W.R. and Halligan, R., 1963. Preliminary report on the Hamersley Iron Province, North West Division. West. Aust., Geol. Surv., Annu. Rep., 1962: 44—54.

Maitland, A.G., 1904. Preliminary report on the geological features and mineral resources of the Pilbara Goldfield. West. Aust. Geol. Surv., Bull. 15.

Maitland, A.G., 1905. Further report on the geological features and mineral resources of the Pilbara Goldfield. West. Aust., Geol. Surv., Bull. 20.

Maitland, A.G., 1906. Third report on the geological features and mineral resources of the Pilbara Goldfield. West. Aust., Geol. Surv., Bull. 23.

Maitland, A.G., 1908. The geological features and mineral resources of the Pilbara Goldfield. West. Aust., Geol. Surv., Bull. 40.

Maitland, A.G., 1909. Geological investigations in the country lying between 21°30′ and 25°30′S lat. and 113°30′ and 118°30′E long, embracing parts of the Gascoyne, Ashburton and West Pilbara Goldfields. West. Aust. Geol. Surv., Bull. 33.

Maitland, A.G., 1924. The geology and mineral industry of Western Australia. West. Aust., Geol. Surv., Bull. 89.

Marston, R.J. and Groves, D.I., 1981. The metallogenesis of Archaean base-metal deposits in Western Australia. Spec. Publ., Geol. Soc. Aust., 7: 409—420.

Miles, K.R., 1942. The blue asbestos bearing banded iron formations of the Hamersley Ranges. West. Aust., Geol. Surv., Bull. 100.

Miyano, T., 1976. Geotectonic history of the Proterozoic Erathem in the Hamersley area, Western Australia. Min. Geol., 26: 207—220 (Tokyo, in Japanese).

Morris, R.C., 1973. A pilot study of the phosphorus distribution in parts of the Brockman Iron Formation. West. Aust., Geol. Surv., Annu. Rep., 1972: 75—81.

Morris, R.C., 1980. A textural and mineralogical study of the relationship of iron ore to banded iron-formation in the Hamersley Iron Province of Western Australia. Econ. Geol. 75, 184—209.

Morris, R.C. and Horwitz, R.C., 1981. A conjectural but geologically constrained model for the Hamersley Group sediments. Geol. Soc. Aust., 5th Annu. Conv., Perth, Abstracts, (3): 41.

Morris, R.C. and Horwitz, R.C., 1983. The origin of the BIF-rich Hamersley Group of Western Australia — deposition on a platform. Precambrian Res., 21.

Morris, R.C., Thornber, M.R. and Ewers, W.E., 1980. Deep-seated iron ores from banded iron-formation. Nature, 288: 250—252.

Morrison, R.J., 1978. Iron ore resources of Western Australia as at 31 December 1977. West. Aust., Geol. Surv., Rec. 1978/11.

Neale, J., 1975. Iron ore deposits in the Marra Mamba Formation at Mining Area "C:, Hamersley Iron Province. In: C.L. Knight (Editor) Economic Geology of Australia and Papua-New Guinea. 1. Metals. Australas. Inst. Min. Metall., Monogr., 5: 924—932.

Noldart, A.J. and Wyatt, J.D., 1962. The geology of the portion of the Pilbara Goldfield covering the Marble Bar and Nullagine 4-mile map sheets. West. Aust., Geol. Surv., Bull. 115.

Nunes, P.D. and Thurston, P.C., 1980. An absolute Age Framework for the stratigraphy of the Uchi — Confederation Lakes Metavolcanic — Metasedimentary Belt, District of Kenora. Ont. Geol. Surv. Misc. Pap., 92: 15--23.

Oehler, D.Z., Schopf, J.W. and Kvenvolden, K.A., 1972. Carbon isotopic studies of organic matter in Precambrian rocks. Science, 175: 1246—1248.

Oversby, V.M., 1976. Isotopic ages and geochemistry of Archaean acid igneous rocks from the Pilbara, Western Australia. Geochim. Cosmochin. Acta, 40: 817—829.

Pidgeon, R.T., 1978a. 3450-m.y.-old volcanics in the Archaean layered greenstone succession of the Pilbara Block, Western Australia. Earth Planet. Sci. Lett., 37: 421—428.

Pidgeon, R.T., 1978b. Geochronological investigation of granite batholiths of Archaean granite-greenstone terrain of the Pilbara Block. Western Australia. In: I.E.M. Smith and J.G. Williams (Editors), Proc. 1978 Archaean Geochemistry Conf., pp. 360—362.

Richards, J.R., 1977. Lead isotopes and ages of galenas from the Pilbara region, Western Australia. J. Geol. Soc. Aust., 24 (8): 465—473.

Richards, J.R., Fletcher, I.R. and Blockley, J.G., 1981. Pilbara galenas: Precise isotopic assay of the oldest Australian leads; model ages and growth-curve implications. Miner. Deposita, 16: 7—30.

Ryan, G.R., 1966. Roebourne, W.A. West. Aust., Geol. Surv., 1 : 250 000 Geol. Ser., Explan. Notes.

Sangster, D.F. and Brook, W.A., 1977. Primitive lead in an Australian Zn-Pb-Ba deposit. Nature, 270: 423.

Schermerhorn, L.J.G., 1966. Terminology of mixed coarse-fine sediments. J. Sediment Petrol., 36: 831—835.

Smith, R.E., 1975. Metamorphism of the Proterozoic Fortescue Group, Western Australia — a reconnaissance study. CSIRO Div. Mineral., Rep. FP9.

Smith, R.E., 1976. Exploration significance of a large hydrothermal alteration system in the Hamersley Basin. CSIRO Div. Mineral., Rep. FP14.

Smith, R.E., 1979. Interpretation of volcanic relations and low-grade metamorphic alteration, Maddina Volcanics, Western Australia. CSIRO Div. Mineral., Rep. FP21.

Smith, R.E., Perdrix, J.L. and Parks, T.C., 1982. Burial Metamorphism in the Hamersley Basin, Western Australia. J. Petrol., 23 (1): 75—102.

Sylvester, M.G. and De Laeter, J.R., in prep. Geochronology of the Mons Cupri volcanic centre, Pilbara Block, Western Aust.

Talbot, H.W.B., 1920. The geology and mineral resources of the North-West, Central and Eastern Divisions, between Long. 119° and 122°E, and Lat. 22° and 28°S. West. Aust., Geol. Surv., Bull. 83.

Trendall, A.F., 1965a. Progress report on the Brockman Iron Formation in the Wittenoom—Yampire area. West. Aust., Geol. Surv., Annu. Rep., 1964: 55—65.

Trendall, A.F., 1965b. Origin of Precambrian banded iron formations (Discussion). Econ. Geol., 60: 1065—1070.

Trendall, A.F., 1966. Second progress report on the Brockman Iron Formation in the Wittenoom—Yampire area. West. Aust., Geol. Surv., Annu. Rep., 1965: 75—87.

Trendall, A.F., 1968a. Three great basins of Precambrian banded iron formation deposition. A systematic comparison. Geol. Soc. Am. Bull., 79 (11): 1527—1544.

Trendall, A.F., 1968b. The Joffre Member in the gorges south of Wittenoom. West. Aust., Geol. Surv., Annu. Rep., 1968: 53—57.

Trendall, A.F., 1972. Revolution in earth history. J. Geol. Soc. Aust., 19 (3): 287—311.

Trendall, A.F., 1973a. Iron-formations of the Hamersley Group of Western Australia: type examples of varved Precambrian evaporites. In: Genesis of Precambrian Iron and Manganese Deposits. Unesco, Paris, Earth Sciences, 9: 377—380.

Trendall, A.F., 1973b. Varve cycles in the Weeli Wolli Formation of the Precambrian Hamersley Group, Western Australia. Econ. Geol., 68 (7): 1089—1097.

Trendall, A.F., 1975a. Hamersley Basin. In: Geology of Western Australia. West. Aust., Geol. Surv., Mem., 2: 118—141.

Trendall, A.F., 1975b. The Hamersley Basin — regional geology. In: C.L. Knight (Editor), Economic Geology of Australia and Papua-New Guinea. 1. Metals. Australas. Inst. Min. Metall., Monogr., 5: 411—413.

Trendall, A.F., 1975c. Geology of Western Australian iron ore. In: C.L. Knight (Editor), Economic Geology of Australia and Papua-New Guinea. 1. Metals. Australas. Inst. Min. Metall., Monogr., 5: 883—892.

Trendall, A.F., 1975d. Preliminary geochronological results from two Pilbara porphyry bodies. West. Aust., Geol. Surv., Annu. Rep., 1974: 103—106.

Trendall, A.F., 1976a. Geology of the Hamersley Basin. 25th Int. Geol. Congr., Sydney, Australia, Excursion Guide, No. 43A.

Trendall, A.F., 1976b. Striated and faceted boulders from the Turee Creek Formation — evidence for a possible Huronian glaciation on the Australian continent. West. Aust., Geol. Surv., Annu. Rep., 1975: 88—92.

Trendall, A.F., 1979. A revision of the Mount Bruce Supergroup. West. Aust., Geol. Surv. Annu. Rep., 1978: 63—71.

Trendall, A.F., 1980. A progress review of the Hamersley Basin of Western Australia. Geol. Surv. Finl., Bull., 307: 113—131.

Trendall, A.F., 1981. The Lower Proterozoic Meteorite Bore Member, Hamersley Basin, Western Australia. In: M.J. Hambrey and W.B. Harland (Editors), Earth Pre-Pleistocene Glacial Record. Cambridge University Press, Cambridge, p. 555.

Trendall, A.F. and Blockley, J.G., 1968. Stratigraphy of the Dales Gorge Member of the Brockman Iron Formation, in the Precambrian Hamersley Group of Western Australia. West. Aust., Geol. Surv., Annu. Rep., 1967: 48—53.

Trendall, A.F. and Blockley, J.G., 1970. The iron formations of the Precambrian Hamersley Group, Western Australia, with special reference to the associated crocidolite. West. Aust., Geol. Surv., Bull. 119.

Trendall, A.F. and De Laeter, J.R., 1972. Apparent age and origin of black porcelanite of the Joffre Member. West. Aust., Geol. Surv., Annu. Rep., 1967: 48—53.

Trendall, A.F. and De Laeter, J.R., in prep. A revised age for the dolerite sills of the Weeli Wolli Formation, Hamersley Group, Western Australia.

Trendall, A.F. and Pepper, R.S., 1977. Chemical composition of the Brockman Iron Formation. West. Aust., Geol. Surv., Rec. No. 1976/25.

Trendall, A.F., Morris, R.C. and McConchie, D., in prep., Classification of "bands" and "banding" in banded iron-formation (BIF).

Van der Wood, T.B., 1977. Strontium Isotope Systematics in the Hamersley Range. M. Sc. Thesis, Florida State University.

Ward, D.F., Coles, I.G. and Carr, W.M.B., 1975. Jimblebar and Western Ridge iron ore deposits, Hamersley Iron Province. In: C.L. Knight (Editor), Economic Geology of Australia and Papua-New Guinea. 1. Metals. Australas. Inst. Min. Metall., Monogr., 5: 916—924.

Wells, A.T., 1959. Yarrie — 4 mile Geological Series. Bureau Mineral Resources Aust., Explan. Notes 16.

Williams, I.R., 1968. Yarraloola W.A. West. Aust., Geol. Surv., 1 : 250 000 Geol. Ser., Explan. Notes.

Woodward, H.P., 1891. Annual Report of the Government Geologist for the Year 1890. Perth.

Chapter 4

PALAEOENVIRONMENTAL SETTING OF IRON-FORMATIONS IN THE DEPOSITIONAL BASIN OF THE TRANSVAAL SUPERGROUP, SOUTH AFRICA

N.J. BEUKES

INTRODUCTION

The Proterophytic (Cloud, 1976) Transvaal Supergroup crops out in the Griqualand West and Transvaal structural basins on the Archean Kaapvaal craton (Fig. 4-1). Subtle facies changes between, and physical separation of the two basins are responsible for the different rock-stratigraphic names applied in them (Fig. 4-1). However, the gross stratigraphic subdivision is very similar, consisting of a basal mixed siliciclastic and volcanic unit, followed comformably by a chemical sedimentary unit which is in turn uncomformably overlain by a mixed chemical-volcanic-siliciclastic rock unit. Facies changes are most obvious in the latter unit with siliciclastics dominant in the Pretoria Group of the Transvaal area and chemical and volcanic rocks characterizing the Postmasburg Group in Griqualand West.

Iron-formations are better developed and more abundant in the Griqualand West area than in the Transvaal area. Seven iron-formation units with formation status and more than ten beds or members are present in the Griqualand West sequence, from near the base of the chemical sedimentary Ghaap Group to near the top of the Postmasburg Group. The Kuruman and Griquatown Iron Formations comprising the Asbesheuwels Subgroup (up to 1000 m thick) are the best known and thickest of these. The only iron-formations known in the Transvaal structural basin are the Penge Iron Formation (up to 650 m thick) which is a correlative of the Asbesheuwels Subgroup (Fig. 4-1), and a very thin unit of restricted lateral extent near the base of the Pretoria Group in the western Transvaal.

This chapter will describe the palaeoenvironmental settings of the iron-formations and their relationship with other facies such as ironstones, sedimentary manganese deposits, carbonates, volcanics and siliciclastics. A recent basinal analysis of the Ghaap Group in Griqualand West by Beukes (1978) and one of the Transvaal Supergroup in the eastern Transvaal by Button (1973a) will be the basis of much of the discussion. Recent review papers by Beukes (1973) on the iron-formations, Button (1976a) on the stratigraphy and economic deposits and Tankard et al. (1982) on the genetic stratigraphy could be read in conjunction with this chapter.

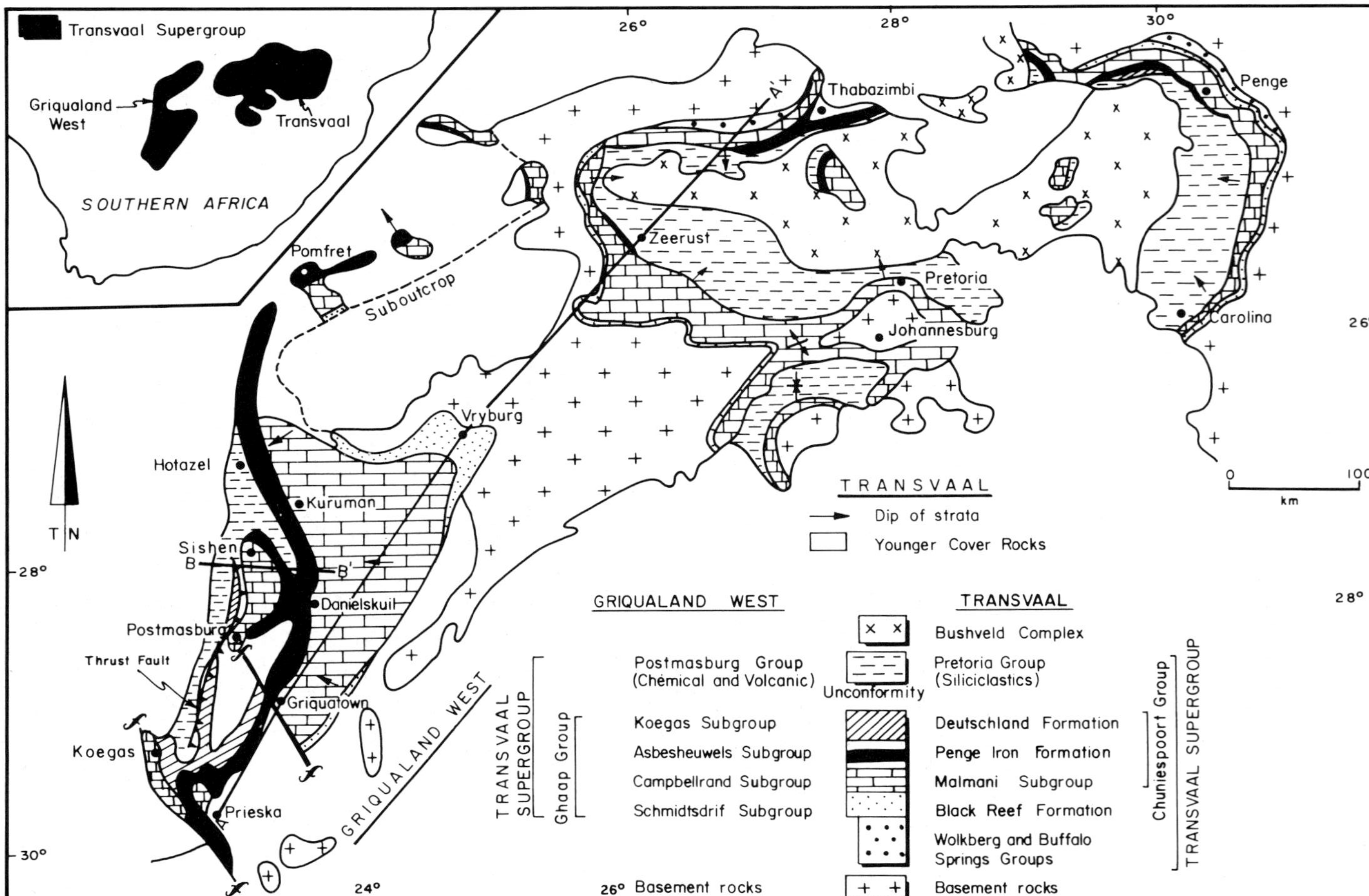

Fig. 4-1. Distribution and gross stratigraphic subdivision of the Transvaal Supergroup in the structural basins of Griqualand West and Transvaal (Section lines AA' and BB' refer to Fig. 4-4B and 4-4C, respectively).

DOCUMENTATION OF THE DEPOSITORY

The history of the documentation of the depository and its iron-formations can be divided into a number of periods each with its own outstanding character (Table 4-I). The first period (Table 4-I, Period I, 1812—1873) was one in which early explorers into the interior of southern Africa first mentioned rocks belonging to the Transvaal Supergroup and also noted the crocidolite asbestos which is associated with the iron-formations.

The first truly geological articles on Transvaal strata were published after the discovery of diamonds near Kimberley in Griqualand West and gold in the eastern Transvaal in circa 1870 (Table 4-I, Period II, 1874—1904). These were reports on reconnaissance stratigraphic traverses.

First mapping of Transvaal strata and the iron-formations took place after the establishment of the Geological Commission of the Cape of Good Hope in 1895 and the Geological Survey of Transvaal in 1903. Reports on these projects were published between 1905 and 1911 (Table 4-I, Period III). The major stratigraphic subdivisions which were to be followed for the next six decades became established during this period.

The Geological Survey of South Africa was formed in 1912 and from then on until 1966 most of the Transvaal Supergroup was mapped and its economic deposits and stratigraphy described. The first detailed descriptions of the stratigraphy and mineralogy of the iron-formations and their economic potential also appeared in this period (Table 4-I, Period IV).

The period 1967—1972 (Table 4-I, Period V) was one of great importance in the sense that it represents a transitional stage from purely descriptive geology to one of stratigraphic and sedimentary modelling of the Transvaal depository. An initiative was also started to change the outdated stratigraphic nomenclature to conform to international standards. Two theses were completed on the crocidolite deposits associated with the Kuruman and Griquatown Iron Formations (Table 4-I, Period V).

The period from 1973 to the present (1981) was one of rapid development in rock-stratigraphic subdivision, basin analyses and sedimentological studies. Also, for the first time, attention was given to the sedimentology of the iron-formation (Table 4-1, Period VI). The next decade will hopefully be one of refined sedimentological, mineralogical and geochemical studies of the various rock units, including the iron-formations and the vast sedimentary manganese deposits associated with them.

STRUCTURE AND METAMORPHISM OF THE STRATA

Geometrically the original Transvaal depository must have represented a gentle warp on the crust of the earth with a preserved plan view dimension of about 100 times more than the maximum known thickness of the strata,

TABLE 4-I

Documentation of the depository of the Transvaal Supergroup*

PERIOD I: 1812—1873. MENTION IS MADE OF TRANSVAAL STRATA BY EARLY PIONEERS	
Griqualand West	Lichtenstein (1812), Burchell (1822), Stromeyer and Hausmann (1831), Moffat (1858)
Transvaal	Mauch (1870), reviews of early work by Harger (1934) and Rogers (1937)
PERIOD II: 1874—1904. RECONNAISSANCE STRATIGRAPHIC TRAVERSING AND NAMING OF ROCK UNITS	
Griqualand West	Stow (1874), Homes (1904a)
Transvaal	Eastern: Cohen (1875), Penning (1885), Bousquet (1896), Wilson-Moore (1896) Central: Penning (1891), Gibson (1892) Western: Francke (1897), Hatch (1904), Holmes (1904b) Review: Molengraaf (1898, 1904)
General review	Hatch and Corstorphine (1905)
PERIOD III: 1905—1911. FIRST MAPPING AND GEOLOGICAL REPORTS BY GEOLOGICAL COMMISSION OF THE CAPE OF GOOD HOPE AND TRANSVAAL GEOLOGICAL SURVEY. STRATIGRAPHIC NOMENCLATURE ESTABLISHED	
Stratigraphy and general geology	
Griqualand West	Rogers (1906 a, b, c, 1907, 1908), Rogers and Du Toit (1909a) Review: Rogers and Du Toit (1909b)
Transvaal	Eastern: Thord-Gray (1905), Hall (1910, 1911) Central: Hall (1905), Kynaston (1907), Kynaston et al. (1911) Western: Holmes (1905), Humphrey (1908), Hall and Humphrey (1910), Humphrey and Kynaston (1911)
Carbonate rocks	Young (1906), von Dessauer (1909)
Volcanic rocks	Hall (1908b)
Metamorphism	Hall (1908a, 1909)
Geochemistry	Horwood (1910)
PERIOD IV: 1912—1966. MAPPING AND DESCRIPTION OF STRATA AND ECONOMIC DEPOSITS. STRATIGRAPHIC DEVELOPMENT ALMOST STAGNANT	
Stratigraphy and general geology	
Griqualand West	Nel (1929), Truter et al. (1938), Visser (1944), Visser (1958), Cilliers (1961), Van Eeden et al. (1963), Hanekom (1966) Review: De Villiers (1961)

TABLE 4-I (continued)

Transvaal	Eastern: Hall (1912, 1913, 1914, 1918), Venter (1934), Brandt and Le Roux (1944), Van Rooyen (1947), Hiemstra and Van Biljon (1959), Visser and Verwoerd (1960), Schwellnus et al. (1962) Central: Kynaston and Mellor (1912), Mellor (1921), Wagner (1927), Kynaston (1929), Lombaard (1931), Boardman (1946), Nel and Jansen (1957), Visser et al. (1961), Cousins (1962), Papenfus (1964) Western: Humphrey and Kynaston (1914), Kynaston and Humphrey (1920), Nel (1935), Nel et al. (1935, 1939), Swiegers (1938), Du Preez (1944), Von Backström (1960), Verwoerd (1964) Review: Coertze (1961) Botswana Review: Boocock (1961)
General reviews	Du Toit (1939), Rogers (1925)
Economic geology	Limestone: Wybergh and Du Toit (1918), Wybergh (1920), Van Biljon (1936), Toens (1966) Gold (Placer-Black Reef Formation): Swiegers (1938, 1939), Frankel (1940), Papenfus (1964) Gold (hydrothermal): Wybergh (1925), Reinecke and Stein (1929), Swiegers (1948), Visser and Verwoerd (1960), Zietsman (1964) Fluorite and Lead/Zinc: Anderson (1915), Kupferburger (1927), Kent et al. (1943), Visser (1958) Chrysotile Asbestos: Van Biljon (1964) Amphibole Asbestos in iron-formation: Wagner (1917), Peacock (1928), Kirkman (1930), Hall (1930), Reinecke and McClure (1934), Du Toit (1945), Vermaas (1952), Genis (1961), Cilliers (1961, 1964), Cilliers and Genis (1964), Hanekom (1966) Iron Ore in iron-formation: Wagner (1920, 1921, 1928), De Villiers (1944), Du Preez (1944), Strauss (1964) Ironstone: Schweigart (1965) Manganese: Hall (1926), Nel (1929), Du Toit (1933), Boardman (1940, 1941, 1961, 1964), De Villiers (1944, 1945, 1956, 1960), Visser (1954), Kupferburger et al. (1956), Frankel (1958), Ortlepp (1964) Tin (hydrothermal): Boardman (1946) Botswana (Review): Boocock (1964)
Iron-formation	Du Toit (1945), Cilliers (1961), Genis (1961), Engelbrecht (1962), Harington (1962), Harington and Cilliers (1963), Cullen (1963), Von Backström (1963), LaBerge (1966), Hanekom (1966)

TABLE 4-I (continued)

Sedimentary structures	Pretoria Group: Rust (1961) Carbonate Rocks: Young (1932, 1933, 1934a,b, 1940a,b, 1943, 1945), Young and Mendelsohn (1948), Toens (1966)
Structure	Visser (1944)
Metamorphism	Van Biljon (1936), Willemse and Bensch (1964)
Volcanism	Truter (1949)
PERIOD V: 1967—1972. CHANGE FROM TIMESTRATIGRAPHIC TO LITHO-STRATIGRAPHIC NOMENCLATURE. FIRST BASIN ANALYSES AND SEDIMENTOLOGICAL STUDIES	
Stratigraphy	Truswell (1967), Wessels (1967), De Villiers (1967), Crockett (1972), Visser and Grobler (1972) Reviews: Haughton (1969), Truswell (1970)
Economic geology	Amphibole asbestos in iron-formation: Fockema (1967), Welch (1969) Manganese: De Villiers (1970) Fluorspar and Lead/Zinc: Hammerbeck (1970) Gold (placer): De Waal and Herzberg (1969)
Iron-formation	Fockema (1967), Welch (1969), Malherbe (1970)
Basin analyses	Southeastern Transvaal: Button (1968) Southwestern Transvaal: Eriksson (1971) Pretoria Group in Transvaal: Visser (1969)
Sedimentology	Carbonate rocks: Eriksson (1972), Truswell and Eriksson (1972), Visser and Grobler (1972) Pretoria Group: Visser (1971, 1972), Rhodes (1971)
Geotectonic setting	Anhaeusser et al. (1969)
PERIOD VI: 1973—1981. RAPID DEVELOPMENT IN LITHOSTRATIGRAPHIC SUBDIVISIONS, BASIN ANALYSES AND DEPOSITIONAL MODELS	
Lithostratigraphic subdivision	Button (1973a), Beukes (1978, 1979, 1980a,b), Eriksson and Truswell (1974), Stear (1977), Tyler (1978), S.A.C.S. (1980), Van Wyk (1980)
Basin analyses	Eastern Transvaal: Button (1973a) Northern Cape Province (Ghaap Group): Beukes (1978) Wolkberg Group: Button (1973b), Tyler (1978, 1979a,b)
General geology	Northern Cape: Potgieter (1973), Smit (1973), Van der Merwe (1973), McLaren (1974), Vajner (1974a), Potgieter and Visser (1978), Van Wyk (1980)
Economic geology	Review: Anhaeusser and Button (1976), Button (1976a), Coetzee (1976), Button and Tyler (1981)

TABLE 4-I (continued)

	Gold: Minnitt et al. (1973) Fluorite: Martini (1976), Crocker (1979) Lead/Zinc: Beukes (1977b) Manganese: Litherland and Malan (1973), Söhnge (1977), Taljaardt (1979), Beukes (1977b)
Iron-formation	Beukes (1973, 1978, 1980a,c), Button (1976b), Klemm (1979)
Sedimentology	Iron-formation: Button (1976b), Beukes (1978) Carbonate rocks: Button (1973c,d), Eriksson (1977), Eriksson and Truswell (1973, 1974), Eriksson et al. (1975, 1976), Truswell and Eriksson (1973, 1975), Beukes (1977a, 1978), MacGregor et al. (1974) Pretoria Group: Eriksson (1973), Button (1975), Button and Vos (1977), Stear (1977) Postmasburg Group: De Villiers and Visser (1977)
Volcanic rocks	Button (1974), Grobler and Botha (1976)
Geotectonic setting	Jansen (1975)
Metamorphism	Button (1976c), Klopp (1978), Hatingh et al. (1979)
General reviews	Button (1976a), Truswell (1977), Tankard et al. (1982)
Papers related to age	Van Niekerk and Burger (1964, 1978), Van Niekerk (1968), Davies et al. (1969), Harding et al. (1974), Crampton (1974), Key (1976), Hamilton (1977), Coertze et al. (1978)

* It is not a complete list of references but intended to give reader some idea of literature available.

which is 12 km (Button, 1976a). Later tectonism resulted in a range of deformation styles, most of which are very gentle. Some of them are related to tectonism associated with the metamorphic complexes that surrround the Kaapvaal craton, some are related to fundamental structures of the craton itself, and some to the intrusion of the Bushveld Complex.

A craton-marginal fold belt with a north—south grain corresponding to that of the Groblershoop metamorphic complex is present along the western side of the Kaapvaal craton (Fig. 4-2). The Transvaal strata become progressively more strongly folded towards the craton margin and a major thrust fault zone is present immediately to the east of it (Fig. 4-2). Deformation along the craton margin took place after deposition of the Paleophytic (Cloud, 1976) Olifantshoek Group (Visser, 1944) which overlies the Transvaal Supergroup along the craton margin (Fig. 4-2). Away from the fold belt the strata dip gently at 1°—6° towards the west.

Along the southwestern edge of the Kaapvaal craton the trend of the fold axis in the craton-marginal fold belt swings towards the southeast to parallel the Doringberg fault zone (Vajner, 1974b) which has parallels in the 1000 m.y.-old Namaqua—Natal metamorphic complex and against which outcrops of the Transvaal Supergroup terminate (Fig. 4-2). The Griquatown fault zone which parallels the Doringberg fault zone and is situated to the northeast of it, close to the southwestern edge of the Kaapvaal craton (Fig. 4-2), represented an active growth fault during deposition of the Transvaal Supergroup (Beukes, 1978).

The age relationships between the Transvaal Supergroup and the metasediments and volcanics in the Namaqua-Natal and Groblershoop metamorphic provinces are uncertain. Vajner (1974a) was of the opinion that the metasediments and metavolcanics are of Archean age and in fault contact with the Transvaal and Olifantshoek sequences, whereas Botha et al. (1976) suggested that they are in part lateral equivalents of the latter.

On the Kaapvaal craton the Transvaal strata are, as a rule, only gently folded. The folding is related to large open dome and basin structures in the Archean granite-greenstone basement. These structures are apparently the result of two directions of folding, one in a northwest—southeasterly sense, the other in a northeast—southwesterly direction. Major faults tend to parallel these two fundamental fold directions on the Kaapvaal craton (Fig. 4-2). Some of the domes and basins affected deposition of the Transvaal strata, especially during early stages of its development. As deposition continued, the basement topography and structural highs and lows tended to become smoothed out (Button, 1973a, b; Beukes, 1978, 1979; Tyler, 1979b).

Apart from the domelike structures, linear greenstone belts (which parallel the grain of the Limpopo metamorphic complex) have undergone tectonic rejuvenation. This resulted in a linear fold belt in the Transvaal strata on the Kaapvaal craton. The Mhlapitsi fold belt overlies the Murchison greenstone belt, and reflects the tectonic grain of the latter (Button, 1973a, 1976a). This fold belt also conforms to the Selati trough (Fig. 4-2) in which a thick pile (2000 m) of sedimentary and volcanic rocks accumulated early in the depositional period of the Transvaal Supergroup (Button, 1973a).

In the Transvaal area, the above-mentioned structural configuration was modified by the intrusion of the Bushveld Complex. Sagging of the Transvaal strata took place below the Bushveld Complex resulting in a saucer-shaped structural basin (Fig. 4-2). The floor of the Bushveld Complex cuts across the Transvaal sequence (Button, 1973a. 1976c) and stoping of the roof led to the incorporation of large xenoliths of Transvaal strata into the complex. High-grade thermal metamorphism took place in the xenoliths and along the floor of the Bushveld Complex. Further away from the floor, the degree of metamorphism decreases gradually, and eventually no influence can be detected (Hall, 1908a, 1908; Willemse and Bensch, 1964; Button, 1973a, 1976c; Hattingh et al., 1979). The Penge Iron Formation outcrops within the limits of the metamorphic aureole of the Bushveld Complex and has been altered to a grunerite-bearing iron-formation (Cilliers, 1964; Beukes, 1973).

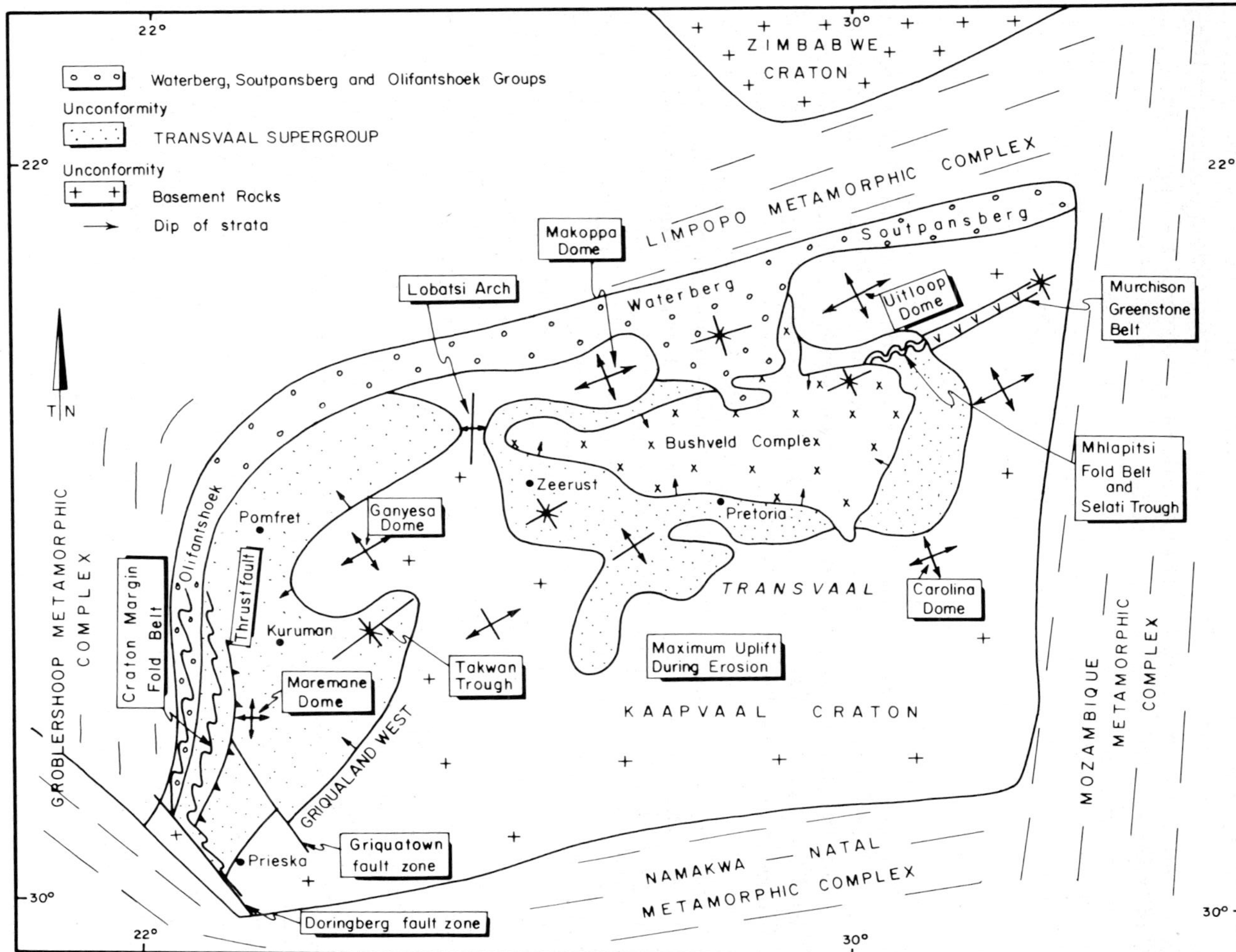

Fig. 4-2. Gross structural features on the Kaapvaal craton related to Transvaal strata.

Outside the Bushveld metamorphic aureole the grade of metamorphism is of the lower greenschist facies and is related to burial metamorphism (Button, 1973a). Along the western marginal fold belt there is an increase in the grade of metamorphism to upper greenschist and lower amphibolite facies (Vajner, 1974a). However, most of the iron-formations of the Griqualand West area have been very little affected by metamorphism or structural deformation (Cilliers, 1961; Hanekom, 1966; Fockema, 1967; Beukes, 1978).

AGE

In many areas the Transvaal Supergroup unconformably overlies acidic and mafic lavas of the Ventersdorp Supergroup. The age of the lavas is uncertain because radiometric determinations display a bimodal distribution. Rb-Sr analysis tend to cluster around an age of 2300 m.y. (Van Niekerk and Burger, 1964; Harding et al., 1974; Key, 1976) whereas U-Pb determinations cluster in the period between 2500 and 2700 m.y. ago (Van Niekerk, 1968; Van Niekerk and Burger, 1978). Acid volcanics of uncertain correlation underlying Transvaal strata in the eastern part of the Bushveld Complex were dated at 2460 ± 120 m.y. by the U-Pb method (Coertze et al., 1978).

The Hekpoort Basalt of the Pretoria Group of the Transvaal Supergroup has yielded a Rb-Sr whole-rock age of 2224 ± 21 m.y. (D. Crampton, quoted in Button, 1976a) while shales underlying it are dated by the same method at 2263 ± 85 m.y. (Hamilton, 1977). These figures suggest that the lower units of the Transvaal Supergroup may well be older than the 2300 m.y. Ventersdorp age and that the lower U-Pb age limit of between 2460 and 2700 m.y. is perhaps more reliable.

The Transvaal Supergroup is almost certainly older than 2100 m.y. since the age of mafic igneous rocks of the Bushveld Complex according to Rb-Sr determinations is 2095 ± 24 m.y. (Hamilton, 1977). Granitic rocks of the Bushveld Complex have been dated at 1954 ± 30 m.y. (Davies et al., 1969) and 1870 ± 190 m.y. (Hamilton, 1977). The upper age limit of 2100 m.y. for the Transvaal Supergroup is further manifested by a 2070 ± 90 m.y. Rb-Sr age obtained for a lava of the overlying Olifantshoek Group (Crampton, 1974). Summarizing, it can thus be stated that deposition of the Transvaal Supergroup and its iron-formations probably began around 2460—2500 m.y. ago and ended prior to about 2100 m.y. ago.

NOMENCLATURE

Iron-formation is defined as an iron-rich sedimentary rock consisting of microcrystalline iron minerals (femicrite) and chert. The chert occurs either as chert mesobands and microbands or as microcrystalline interstitial grains between femicrite particles. The iron-formations display textural and structural features resembling those of carbonate rocks (Dimroth and Chauvel, 1973) and the nomenclature and classification of iron-formation based on carbonate terminology by Beukes (1980c) will be used in this chapter. According to this classification allochemical iron-formation is composed of allchems such as pebbles and grains whereas orthochemical iron-formation consists of felutite (Fig. 4-3). Microbanded iron-formations are referred to as ferhythmites which for the most part represent autochthonous iron-formations. An adjective is used to describe the mineralogical composition of the iron-formation and the dominant mineral is placed first (Fig. 4-3).

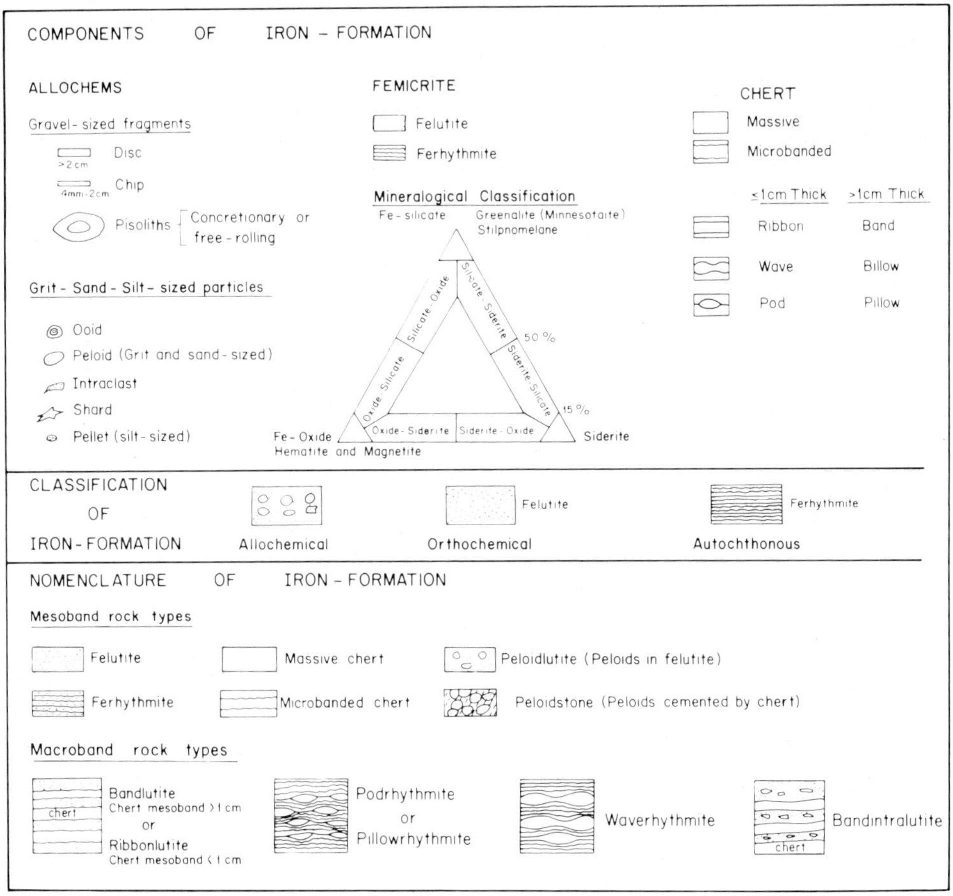

Fig. 4-3. The components, classification and nomenclature of iron-formation.

It should be noted that all the iron-formations of the Transvaal Supergroup are weathered to depths of several tens of meters below surface. However, mineralogical data were obtained from fresh rocks from diamond drill cores and mines as well as from the characteristic weathered appearance of some mineralogical facies.

Chert mesobands which are thinner than 10 mm are called ribbons when they are of even thickness along strike, waves when they pinch and swell, and pods when they are lenticular. Thicker chert mesobands displaying the same series of structures are respectively referred to as bands, billows and pillows (Fig. 4-3). Mesobanded units of iron-formation consisting of evenly alternating femicrite and chert mesobands are named through a combination of chert and femicrite mesoband names, such as for example pillowlutite and bandlutite (Fig. 4-3).

Different orders of sedimentary cycles are present in the iron-formations and the rocks associated with them. Microcycles are defined by microbanding and mesocycles by mesobanding. Mesobanded rock units are called macrobands, and they in turn build macrocycles i.e. first-order cycles according to a convention by Beukes (1978). Higher-order rock units and cycles such as second (megacycles), third and fourth order are defined in the same manner.

The rock-stratigraphic subdivision of the South African Committee for Stratigraphy (S.A.C.S., 1980) will be used for the Transvaal structural basin. However, this committee's subdivision for the Griqualand West basin is not sufficiently detailed and that of Beukes (1978, 1979, 1980a, b) and Van Wyk (1980) will be used with minor modifications.

TECTONO-SEDIMENTARY AND STRATIGRAPHIC SETTING OF THE IRON-FORMATIONS

Initially, deposition of Transvaal strata was restricted to the proto-basin of the Wolkberg and Buffalo Springs Group (Fig. 4-1) in the northern Transvaal. These sequences consist of basaltic and acidic volcanics, alluvial, deltaic and lacustrine siliciclastics, and minor carbonates and chert (Button, 1973b; Tyler, 1979b). No iron-formations are known to be present in them.

This proto-basinal phase was followed by a major transgression, and resulted in the deposition of a thin veneer of shallow-marine quartzite and offshore argillites of the Vryburg and Black Reef Formations over the entire Kaapvaal craton, smoothing out many basement irregularities. At this early stage of deposition three major tectono-sedimentary elements became established which, according to Beukes (1978), probably controlled the lateral and vertical distribution of sedimentary lithofacies throughout the further depositonal history of the Transvaal Supergroup. These elements are a shallow-water platform on the Kaapvaal craton and a hinge-zone and deeper-water basin along the western and southwestern margin of the craton (Fig. 4-4A). The basinal

and hinge-zone facies are largely obscured by the overlying Olifantshoek Group.

A fourth tectono-sedimentary element situated along the Limpopo metamorphic complex (Fig. 4-4A) served as a positive source area during periods of siliciclastic deposition on the Kaapvaal craton (Button, 1973a, 1976a). However, during deposition of the chemical sedimentary Ghaap and Chuniespoort Groups it may have represented a negative basinal element (Beukes, 1978).

Relative to the above-mentioned tectono-sedimentary elements, iron-formations and other chemical sedimentary rocks are best developed in the basin along the craton margin in Griqualand West. Iron-formations are less abundant in the sequence on the platform in the Transvaal area (Fig. 4-4B).

The lowermost known iron-formation units in the Transvaal Supergroup are interbedded with tuffaceous siltstone and shale of the Lokammona Formation of the Schmidtsdrif Subgroup (Fig. 4-4B). They are represented by two thin beds of banded siderite lutite.

Higher in the sequence, the next iron-formation units are represented by a number of ankerite-banded cherts (proto iron-formation of Button, 1976b) interbedded with chert, carbonaceous shale and clastic-textured limestone and ferruginous dolomite of the basinal Prieska facies of the Campbellrand—Malmani carbonate sequence (Fig. 4-4B). Two tongues of the Prieska facies envelope the stromatolitic Ghaap Plateau facies. Ankerite-banded cherts are also present in these tongues and are especially well developed immediately below the Kuruman and Penge Iron Formations.

The Kuruman and Penge Iron Formations follow conformably on the Campbellrand and Malmani Subgroups, respectively. The Kuruman Iron Formation is between 135 m and 750 m thick and consists essentially of microbanded autochthonous iron-formation (banded ferhythmite). It is conformably overlain by the clastic-textured, orthochemical and allochemical Griquatown Iron Formation which is on average about 250 m thick. Together they constitute the Asbesheuwels Subgroup (Fig. 4-4B). The stratigraphy of the Penge Iron Formation is similar to that of the Asbesheuwels Subgroup in that autochthonous ferhythmites are overlain by clastic-textured iron-formations (Beukes, 1978).

An inverse thickness relationship exists between the Asbesheuwels—Penge iron-formation sequence and the underlying Campbellrand—Malmani carbonate unit. Autochthonous iron-formation reaches a maximum known thickness of 750 m in the basin to the south of the Griquatown fault zone and a minimum known thickness of 135 m on the platform. At the same time, clastic-textured carbonates reach a minimum known thickness of approximately 650 m in the basin relative to an autochthonous stromatolitic carbonate sequence of 1750 m thick on the platform (Beukes, 1978, 1980b; and Fig. 4-4B). In the Griquatown Iron Formation, allochemical iron-formation units are best developed on the platform, and orthochemical iron-formation in the basin (Beukes, 1980a).

South of the Griquatown fault zone, the Griquatown Iron Formation grades upwards into chloritic mudstones of the Pannetjie Formation of the Koegas Subgroup. The Koegas Subgroup thickens basinward from approximately 250 m to over 650 m, and contains the Doradale (< 10 m thick), Kwakwas (0—80 m thick), Rooinekke (approximately 100 m thick) and Nelani (> 200 m thick) Iron Formations (Fig. 4-4B). These iron-formations are interbedded with ferruginous dolomite, ferruginous and manganiferous mudstone, siltstone and quartz wacke (Beukes, 1978; Van Wyk, 1980). The Deutschland Formation (Button, 1973a) which overlies the Penge Iron Formation in the northeastern Transvaal consists of diamictite, carbonaceous shale and dolomite but no iron-formations are known to be present.

A period of uplift and erosion preceded the deposition of the Rooihoogte and Timeball Hill Formations of the Pretoria Group in the Transvaal area. Most uplift took place in the central part of the Kaapvaal craton, with the result that the Penge Iron Formation and part of the Malmani carbonate sequence were eroded before deposition of the Pretoria Group (Button, 1973c, 1976a). For this reason the Penge Iron Formation only crops out along the northern limb of the present day structural basin in the Transvaal area (Fig. 4-1). Towards the edge of the Kaapvaal platform in Griqualand West no erosion apparently took place at this time because the first uncomformity here only occurs at the base of the Makganyene Diamictite — the correlative of which overlies the Timeball Hill Formation in the Transvaal area (Fig. 4-4B). This could mean that the Rooihoogte and Timeball Hill Formations are time-equivalents of part of the Koegas Subgroup. The thin iron-formation near the base of the Rooihoogte Formation in the western Transvaal (Martini, 1976; Klopp, 1978) may thus well be a correlative of one of the uppermost iron-formation units of the Koegas Subgroup (Fig. 4-4B).

This correlation stands in strong contrast to the popular belief that the Rooihoogte and Timeball Hill Formations of the Transvaal area are correlatable with the Gamagara Formation (Fig. 4-4C) of the Sishen-Postmasburg area in Griqualand West as were originally suggested by Wessels (1967) and De Villiers (1967). Van Wyk (1980), however, illustrates that the Nelani Iron Formation of the Koegas Subgroup overlies the Gamagara Formation through a tectonic unconformity (thrust fault) near Postmasburg. The iron-formation is in turn overlain by the Makganyene Diamictite. These relationships reconfirmed the ideas of Rogers (1906a), Nel (1929), Truter et al. (1938) and Visser (1944) that the Gamagara Formation is a correlative of the Paleophytic Olifantshoek Group and that strata of the Proterophytic Transvaal Supergroup have been thrust over it (Fig. 4-4C).

The Manganore Iron Formation of the Sishen-Postmasburg area (Fig. 4-4C) most probably represent relics of the Asbesheuwels Subgroup which have slumped down into sinkhole structures in the Campbellrand carbonate sequence during the erosional period that preceded the deposition of the Gamagara Formation (Rogers, 1906a; Nel, 1929; Van Wyk, 1980). Ideas expressed by Button (1976a) and Beukes (1977b) that the Manganore Iron Formation is part of the Gamagara Formation were disqualified by Beukes (1978) on ground that the Gamagara Formation overlies the iron-formation unconformably. In addition Van Wyk (1980) shows the Wolhaarkop Breccia to be a solution collapse breccia that developed below the Asbesheuwels iron-formation sequence in the Maremane dome (Fig. 4-4C) before deposition of the Gamagara Formation.

Various iron-formation units are present in the Postmasburg Group. Felutite layers are present in the Makganyene Diamictite and jasper beds are interbedded with pillow lava and hyaloclastic breccias of the Ongeluk Lava of the Voëlwater Subgroup (Fig. 4-4B). The Ongeluk Lava is in turn conformably overlain by jasper and jaspilite of the Hotazel Formation. Sedimentary manganese beds of up to 40 m thick are interbedded with it. The Hotazel Formation interfingers laterally with jaspilite, jasper, dolomite, tuff and lava of the Beaumont Formation, and is overlain by clastic-textured light-grey sparatic dolomite of the Mooidraai Formation (Fig. 4-4B). With the Hekpoort Basalt and the Ongeluk Lava as a time-stratigraphic datum, it is evident that the chemical sedimentary rocks of the Hotazel and Mooidraai Formations in Griqualand West most probably represent a distal facies of ironstone and siliciclastics of the Pretoria Group in the Transvaal area (Fig. 4-4B).

A major angular unconformity is present at the base of the Paleophytic Olifantshoek and Gamagara sequences in Griqualand West (Fig. 4-4B and C) and no strata belonging to the Transvaal Supergroup are known to be present above the Mooidraai Dolomite. However, in the Transvaal structural basin several thousand meters of siliciclastics and volcanics are known to be developed above the Hekpoort Basalt and below the Bushveld Complex (Button, 1973a, 1976a).

SCHMIDTSDRIF SUBGROUP

The banded siderite lutites of the Lokammona Formation of the Schmidtsdrif Subgroup (Fig. 4-4B) overlie tuffaceous siltstone which was deposited in a lagoonal setting shoreward of platform edge, carbonate oolite shoal and stromatolite reef deposits (Fig. 4-5). They are overlain by pyritic carbonaceous shale which possibly represents offshore marine muds (Fig. 4-5).

The two felutite beds are of limited lateral extent and less than one meter thick. In outcrop they consist of poorly defined massive chert mesobands interbedded with partly silicified iron oxides and hydroxides representing weathered iron carbonates. Chert nodules, representing partly devitrified lapilli, are present in the tuffaceous siltstone which is associated with the felutites (Beukes, 1978, 1979).

In fresh drill-core samples the banded felutites were seen to consist of microcrystalline chert bands alternating with siderite micrite. Shard-like frag-

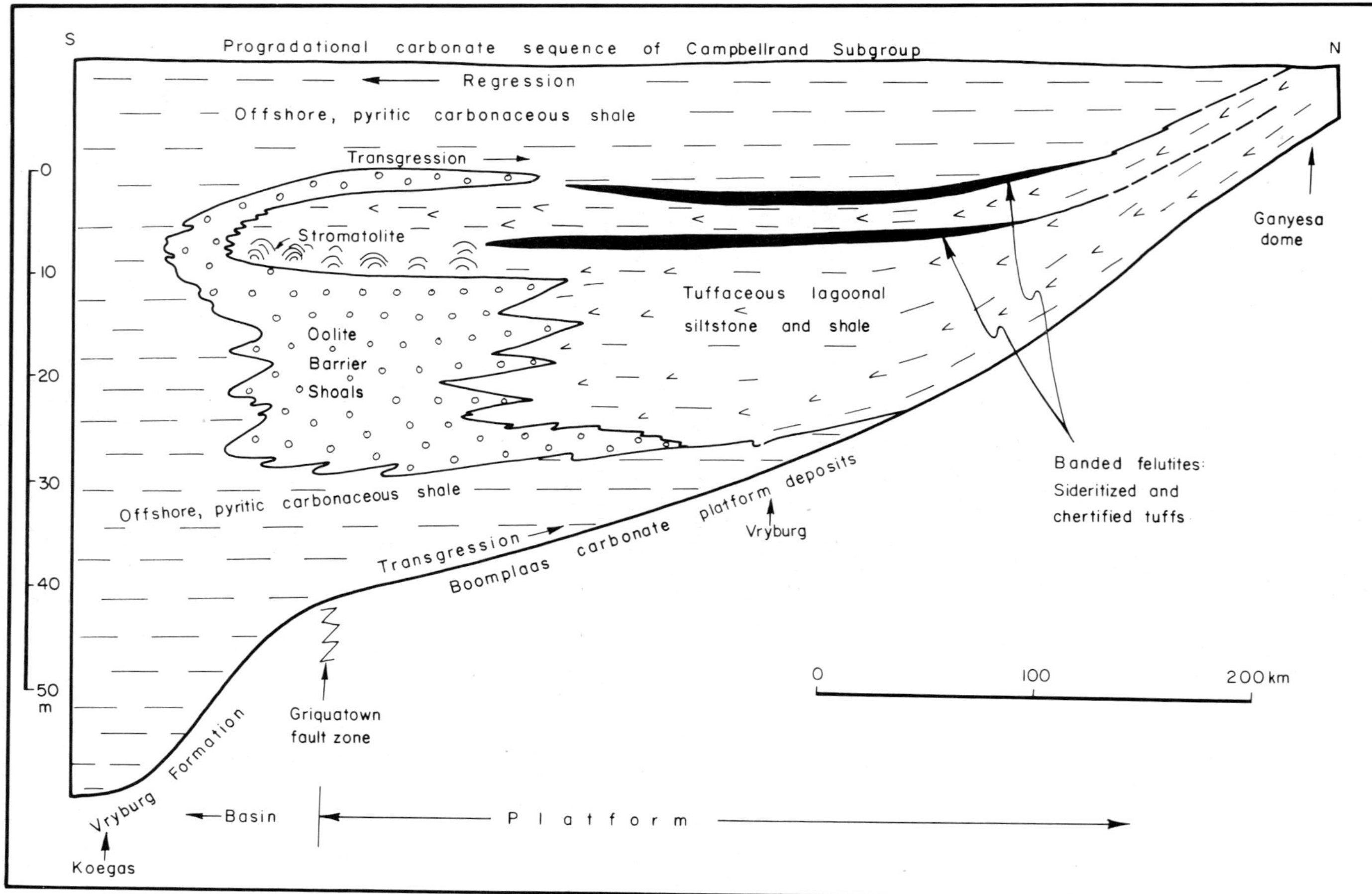

Fig. 4-5. Stratigraphic and palaeoenvironmental setting of iron-formations within the Lokammona Formation of the Schmidtsdrif Subgroup.

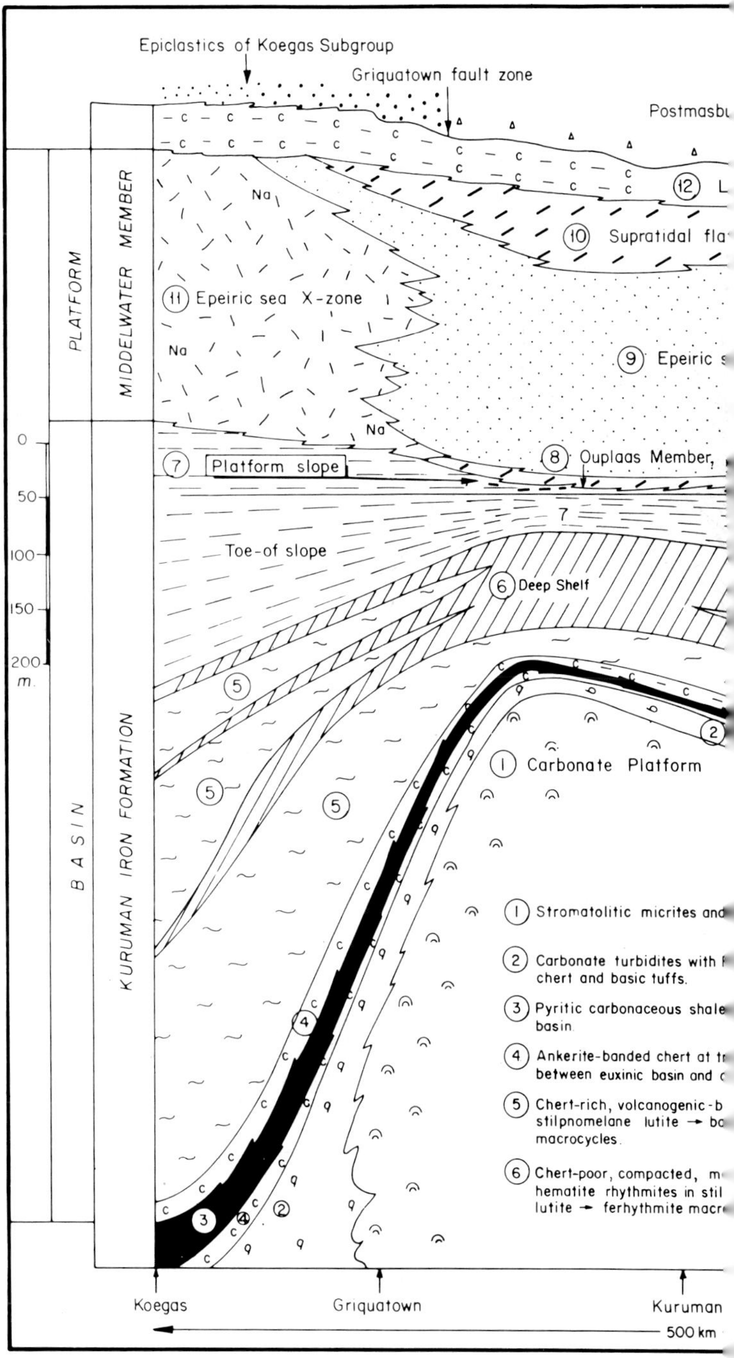

Fig. 4-7. South—north section illustrating stratigraphic relationship
heuwels Subgroup in Griqualand West (after Beukes, 1978).

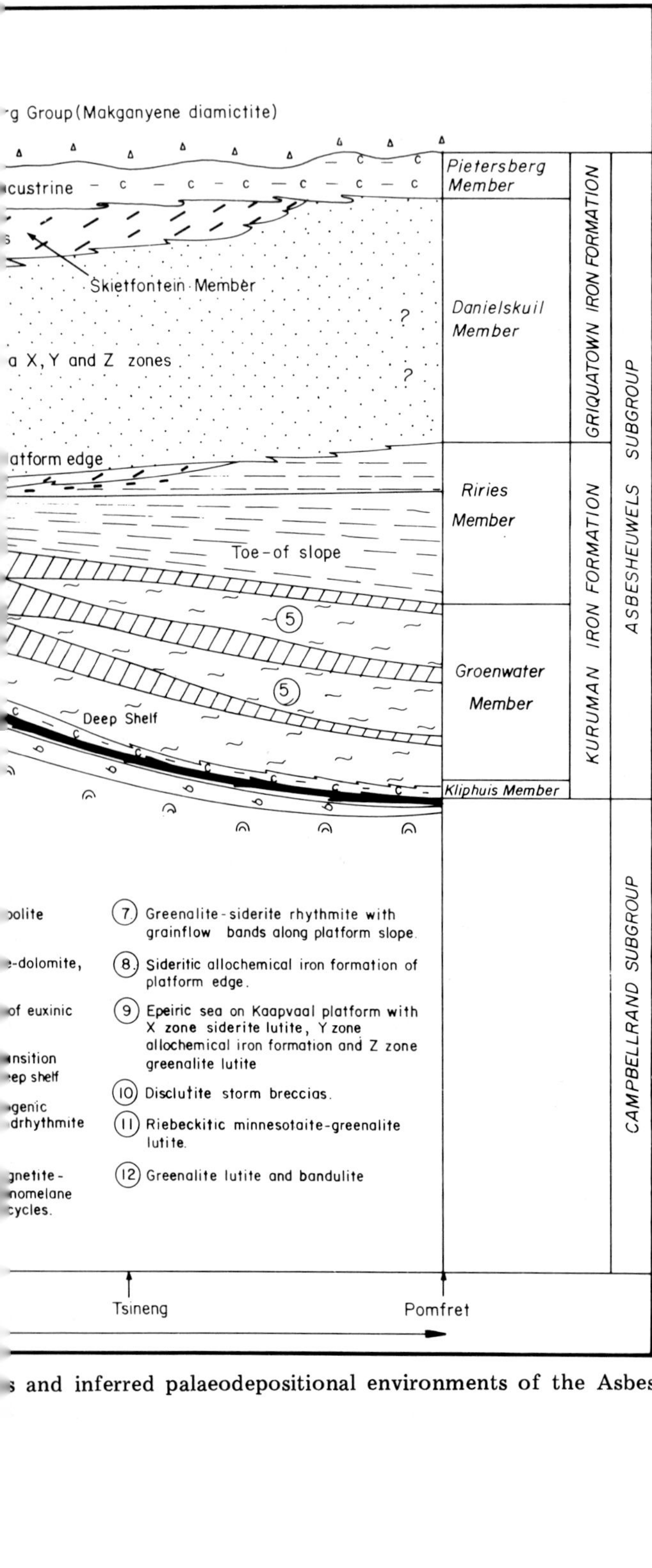

s and inferred palaeodepositional environments of the Asbesments and volcanic particles are present in the siderite and chert mesobands indicating that these two iron-formations represent devitrified, chertified and sideritized volcanic ash beds. Chert bands and nodules, representing devitrified volcanic ash bands and lapilli respectively, are also present in other units of the Schmidtsdrif Subgroup so that the setting of the Lokammona banded felutites in a lagoonal environment may be entirely coincidental.

CAMPBELLRAND—MALMANI CARBONATE SEQUENCE

The clastic-textured, carbonaceous and ferruginous carbonates of the Prieska facies of the Campbellrand—Malmani carbonate sequence, with which ankerite-banded cherts are associated, represent a deeper water carbonate turbidite facies which interfingers with a shallow water stromatolitic carbonate platform sequence onto the Kaapvaal craton (Fig. 4-6)*. The latter unit is referred to as the Ghaap Plateau facies (Beukes, 1978, 1980b) and consists of intertidal to supratidal light-grey cherty dolsparite, platform lagoonal cryptalgal laminated dolmicrites, and subtidal stromatolitic dolmicrite mounds. The platform edge facies (Fig. 4-6) is represented by oolite shoal deposits and elongated stromatolitic columns i.e. stromatolite reefs.

A iron-formation unit only a few meters thick extends from the basinal Prieska facies into the middle of the shallow-water Ghaap Plateau facies (Fig. 4-6). It was most probably deposited during a transgression and is best developed in the basin and near the edge of the carbonate platform. Further onto the carbonate platform it pinches out and is replaced by a chert-in-shale breccia (Fig. 4-6) which according to Beukes (1978) probably represents a lag deposit on the transgression surface. The iron-formation is known as the Kamden Member and contains jasper bands which are significant because they indicate the presence of oxygen fugacity levels high enough for the oxidation of ferrous ions to the ferric state. Chert nodules present in the iron-formation have been disorientated by soft-sediment slumping near to the edge of the platform in the vicinity of the Griquatown growth fault. Chert breccias are also present in the iron-formation in this area, and are considered further evidence for slumping, probably along the slope of the carbonate platform.

Dolomite slump breccias are also present in the slope facies of the Campbellrand—Malmani carbonate sequence (Beukes, 1978). The basinal Prieska facies of the Campbellrand Subgroup consists of pyritic carbonaceous shale and thinly bedded and laminated intramicsparites and intramicrites. The intraclasts consist mainly of algal mat debris. The intramicsparite and intramicrite beds and laminae display sharp erosional basal contacts with micro flame-structures, grade upwards into argillaceous micrite or shale, and display cross-

---

* Figure 4-6 is shown on pp. 151—152.

lamination and horizontal lamination which may represent the upper part of Bouma cycles. Beukes (1978) thus interpreted these clastic-textured carbonates as carbonate turbidites, consisting of debris that was washed down from the adjacent shallow-water Ghaap Plateau carbonate platform onto a deep shelf, defined as an area below storm wave base. In turn the carbonate turbidites interfinger with pyritic carbonaceous shale considered to have been deposited in a still deeper-water euxinic basin (Fig. 4-6).

The ankerite-banded cherts of the Prieska facies (Fig. 4-4B) are preferentially developed along the contact between euxinic shale and deep shelf carbonate turbidites. They consist of chert mesobands alternating with ankeritic or ferruginous dolomitic intramicrite mesobands which are similar in character to adjoining carbonate turbidite beds. These iron-bearing mesobands thus represent ankeritized or ferruginized limestone turbidite beds.

Two types of chert mesobands are present, namely intraclastic mesobands representing chertified limestone turbidite layers and cryptocrystalline mesobands representing primary chert deposits on the sediment—water interface (Beukes, 1978). The ankeritization and chertification of the limestone turbidites to form ankerite-banded cherts, together with the presence of primary chert bands and the carbonaceous and pyritic character of associated rocks, all indicate a deep-water acidic and reducing environment of deposition. Iron carbonates are more stable than calcium carbonates below pH 7.8 explaining the ankeritization and ferruginization of the limestone turbidites. Replacement of limestone turbidites by chert and the precipitation of primary chert bands would also be favoured by acidic deep- and cold-water conditions, if the stability fields of carbonates and silica, as discussed by Berner (1971) and Siever (1971), are taken into consideration.

Mafic tuff beds are interbedded with the carbonate turbidites of the Prieska facies indicating contemporaneous volcanic activity. The ankerite-banded chert units are, however, not associated with them, suggesting that there was no volcanogenic contribution to the deposition of the iron and silica.

Deposition of shallow-water carbonates came to an end on the Kaapvaal craton through a major transgression. This resulted in a veneer of the Prieska facies that underlie the Kuruman and Penge Iron Formations on the craton (Fig. 4-6).

ASBESHEUWELS SUBGROUP

The stratigraphy, mineralogy and asbestos deposits of the Asbesheuwels iron-formations are described by Du Toit (1945), Cilliers (1961, 1964), Genis (1961), Engelbrecht (1962), Hanekom (1966), Fockema (1967), Welch (1969), Beukes (1973, 1978, 1980a) and Dreyer (1974). This section will concentrate on the sedimentology of the iron-formations and its implication for a reconstruction of the Transvaal depository through a correlation with the Penge Iron Formation.

Kuruman Iron Formation

The carbonaceous shale which caps the Campbellrand Subgroup is directly overlain by ankerite-banded chert constituting the Kliphuis Member of the Kuruman Iron Formation (Fig. 4-7)*. These cherts are similar to those of the Prieska facies of the Campbellrand Subgroup and are therefore also thought to represent chertified and ankeritized limestone turbidites.

The ankerite-banded cherts are overlain by stacked stilpnomelane lutite → ferhythmite macrocycles (Beukes, 1980a) constituting the Groenwater Member of the Kuruman Iron Formation (Fig. 4-7). The ferhythmites are usually chert mesobanded. The macrocycles consist in their most simple and complete form of stilpnomelane lutite → siderite-microbanded chert → siderite-magnetite bandrhythmite → magnetite-hematite ribbonrhythmite → siderite-microbanded chert (Fig. 4-8). The arrows point up in the sequence. Individual cycles are of the order of 1—10 m thick and correspond to the alternation of S-macrobands and BIF-macrobands described by Trendall and Blockley (1970) in the Dales Gorge Member of the Brockman Iron Formation in Australia. The stilpnomelane lutite beds, which are only of the order of a few centimeters thick, determine the composition and completeness of the cycles. The chert content and thickness of chert mesobands decrease upwards away from stilpnomelane lutite beds and sharply increase again immediately below such beds (Fig. 4-8). Zones in the sequence which contain a high frequency of stilpnomelane lutite beds are characterized by incomplete cycles, extending only up to siderite-microbanded chert or to siderite-magnetite bandrhythmite (Beukes, 1980a). Sodium (in the form of riebeckite and/or crocidolite) and pyrite are preferentially concentrated immediately below and above stilpnomelane lutite beds (Fig. 4-8).

The stilpnomelane lutite beds most probably represent altered acidic volcanic ash beds as are indicated by the presence of lapilli and tricuspate shards in them, chemical differences between them and adjoining banded ferhythmite, and their correlatability over tens and even hundreds of kilometers (Hanekom, 1966; La Berge, 1966; Fockema, 1967; Beukes, 1978, 1980a). The silica which was deposited in close association with the volcanic ash bands may well have been derived from fumarolic activity. Sharp contacts between chert bands and ash bands, rhythmic alternation between chert and iron mineral microbands, and steep-sided solution hollows on chert bedding planes, indicate that the chert mesobands were deposited at the sediment-water inferface (Beukes, 1978). The water in which the stilpnomelane lutite → ferhythmite macrocycles were deposited i.e. their macro-environment, was thus most probably acidic, because under neutral to alkaline conditions ferrous ions would probably have reacted with the silica to form greenalite after the reaction:

* Figure 4-7 is shown on pp. 153—154.

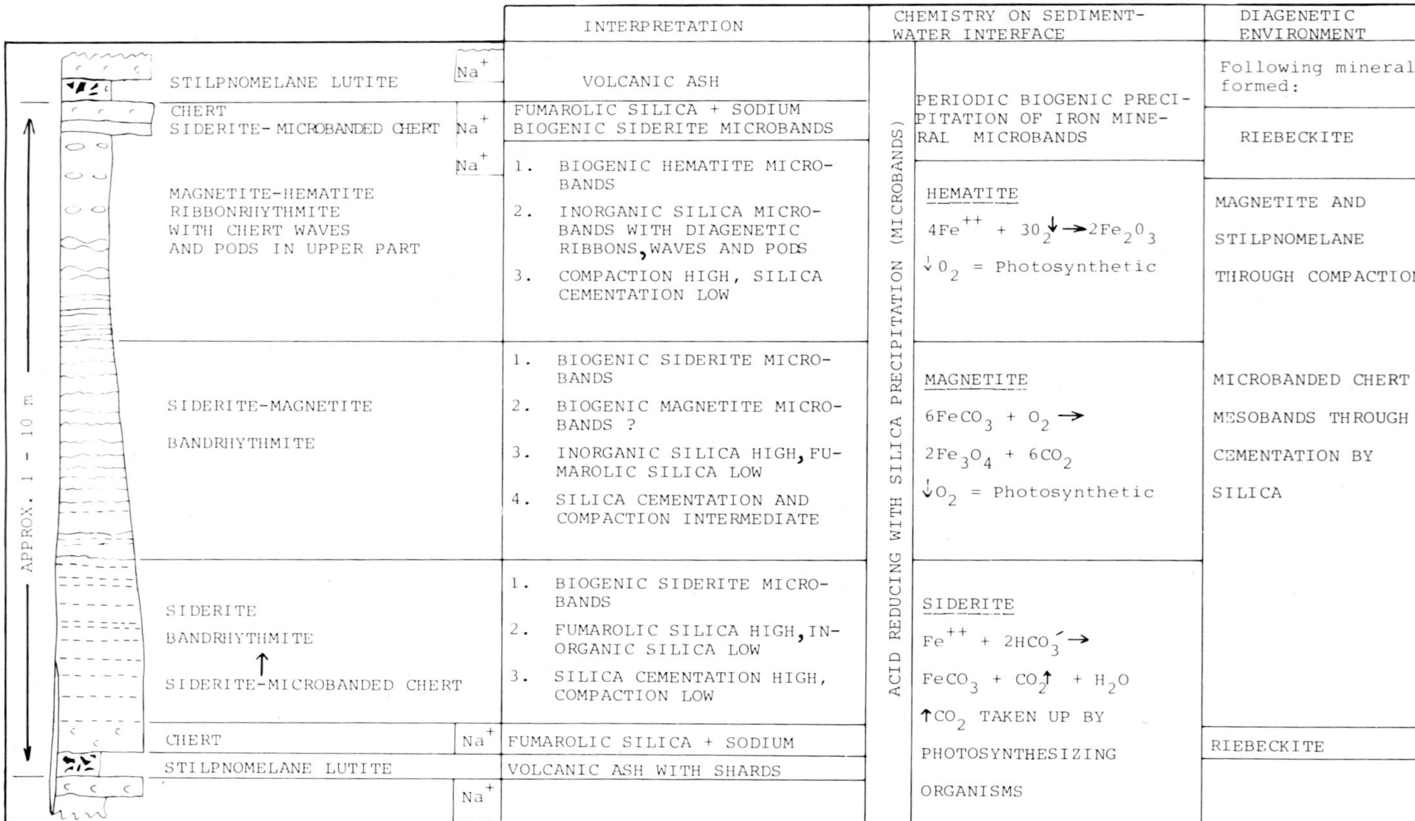

Fig. 4-8. Example of a complete stilpnomelane lutite → ferhythmite macrocycle and the interpretation of its origin (after Beukes, 1978, 1980a).

$$3\,Fe^{2+} + 2H_4SiO_4 + H_2O \rightleftarrows Fe_3Si_2O_5\,(OH)_4 + 6H^+ \qquad (1)$$

given by Eugster and I-Ming Chou (1973).

The above mentioned macro-environment was thus not favourable for femicrite deposition and Beukes (1978) suggested that the iron-mineral microbands in the ferhythmites were deposited through the action of photosynthesizing organisms which flourished seasonally. The chemistry of the basin water was locally changed in the micro-environment of the micro-organisms, and siderite precipitated through the extraction of CO_2 (Fig. 4-8) from the reaction:

$$Fe^{2+} + 2HCO_3 \rightarrow FeCO_3 + CO_2 + H_2O \qquad (2)$$

As volcanism subsided and the water became less turbid and less toxic for the micro-organisms, excess O_2 was produced and dissolved ferrous ions were precipitated as ferric hydroxyoxide which were transformed into hematite micrite (Fig. 4-8). Magnetite may have formed at an intermediate oxygen fugacity level in between the siderite and hematite facies (Fig. 4-8).

The rapid facies changes over relatively small intervals in the stilpnomelane lutite → ferhythmite macrocycles can also be explained by this interaction between volcanic and biochemical activity. However, this is conjectural since it cannot be proved that micro-organisms made a contribution to the deposition of the femicrites. Support for the presence of micro-organisms during deposition comes from Phlug (quoted in Fockema, 1967) and Klemm (1979) who found possible microfossils in some of the cherts, and from Harington (1962) and Harington and Cilliers (1963) who described primitive oils and amino acids which they considered to have been derived from micro-organisms. On the other hand, very little to no positively identifiable organic material is present in the ferhythmites. Some of the carbon could, however, have been used in the reduction of hematite to magnetite. The reaction suggested by Perry et al. (1973) by which this takes place is:

$$6Fe_2O_3 + C \rightarrow 4Fe_3O_4 + CO_2 \qquad (3)$$

Carbon may also have been transported to a deeper euxinic basin — a contemporaneous environment represented by the carbonaceous shale at the base of the Kuruman Iron Formation (Fig. 4-7).

The stilpnomelane lutite → ferhythmite macrocyclicity i.e. macrobanding, is thus considered to be of a mixed volcanic-biological origin, whereas the ferhythmite microbands may represent seasonal geochemical varves produced by photosynthesizing organisms. In contrast, the mesobanding in the banded ferhythmite may be a result of diagenesis. This interpretation is based on microbanded chert pods and pillows which grade laterally into compacted ferhythmite mesobands consisting of alternating microbands of femicrite and chert (Trendall and Blockley, 1970; Beukes, 1978). Apparently early cementation of the sediment by chert prevented compaction of the chert pods and pillows whereas more than 80% compaction took place in the sediment around it. Silica cement may have preferentially crystallized at a specific depth below the sediment—water interface and as more sediment accumulated new crystallization levels could have developed leading to mesobanding. Jenkyns (1974) envisages a similar process for the formation of carbonate nodules (pods or pillows) in pelagic limestones.

Compaction of ferhythmite mesobands between chert mesobands or alongside chert pods and pillows, led to the development of diagenetic magnetite, stilpnomelane and riebeckite. The silica responsible for the cementation of the chert mesobands and the sodium in the riebeckite may in part have been derived from the devitrification of the stilpnomelane volcanic ash bands (Beukes, 1978).

The relatively chert-rich stilpnomelane lutite → ferhythmite macrocycles of the Groenwater Member is overlain by chert-poor greenalite-siderite rhythmite which constitutes the Riries Member of the Kuruman Iron Formation (Fig. 4-7). The greenalite-siderite rhythmites lack well defined chert mesobands and consist of greenalite-siderite lutite laminae alternating with siderite microbands. Neutral to weakly alkaline conditions now existed in the basin so that greenalite precipitated in the place of silica according to the reverse of eq. 1 mentioned above.

The greenalite-siderite rhythmite grades upwards over a small thickness interval into siderite lutite and sideritic allochemical iron-formation (grainstone and discstone) which constitute the Ouplaas Member of the Kuruman Iron Formation on the Kaapvaal platform (Fig. 4-7). This transition probably represents an upward-shallowing sequence towards the edge of a shallow-water platform (Fig. 4-7). That the platform slope must have been steep is suggested by the rapid transition from rhythmites to lutites and grainstones, and by the presence of grainflow (Stauffer, 1967) mesobands and graded bedded lutite-flow felutite mesobands in the greenalite-siderite rhythmites. Chemical conditions also changed upwards from the neutral to weakly alkaline stability field of greenalite (Eugster and I-Ming Chou, 1973) in deeper water, to the alkaline stability field of siderite (Krumbein and Garrels, 1952) in the shallower water (Beukes, 1978).

The Kuruman Iron Formation may thus represent a third order, upward-shallowing progradational sedimentary cycle consisting of pyritic carbonaceous shale at the base, followed in turn upwards by volcanogenic-biogenic stilpnomelane lutite → ferhythmite macrocycles deposited on a deep shelf, greenalite-siderite ferhythmite deposited below wave base at the toe of a slope, ferhythmite with grainflow bands deposited along a slope, and sideritic orthochemical and allochemical iron-formation deposited on a shallow-water platform above wave base (Fig. 4-7).

The thickness variation in the Riries Member (Fig. 4-7) of the Kuruman Iron Formation is probably due to depositional differences between platform and basin. However, in the Groenwater Member, synchronous volcanic ash (stilpnomelane lutite) → ferhythmite macrocycles (first-order cycles) and megacycles (second-order cycles) can be correlated over many kilometers, suggesting that sympathetic thinning of all layers involved are responsible for the decrease in thickness from basin to platform (Fig. 4-7), rather than that certain units pinch out. For example, highly compacted magnetite-hematite ferhythmites are better developed on the platform (Facies 6 of Fig. 4-7) and interfinger with less compacted chert-rich ferhythmites in the basin (Facies 5 of Fig. 4-7). The thickness variation may therefore be related to diagenesis with more silica cementation and less compaction taking place in the basin relative to the platform.

Griquatown Iron Formation

On the Kaapvaal platform the basal part of the Griquatown Iron Formation, i.e. the Danielskuil Member (Fig. 4-7), consists of four upward-coarsening orthochemical → allochemical iron-formation megacycles (second-order cycles). The fourth cycle ends in greenalitic disclutites (edgewise conglomerates) of the Skietfontein Member (Fig. 4-7). Most typically the megacycles consist of a zone rich in banded siderite lutite followed by sideritic grainstones and disclutites which fine upwards into greenalite lutite (Beukes, 1980a).

The megacycles, and also the grainstone and disclutite units within them, form sheetlike sedimentary bodies which are correlatable over hundreds of square kilometers (Engelbrecht, 1962; Hanekom, 1966; Fockema, 1967; Welch, 1969). Beukes (1978) therefore used Irwin's (1965) epeiric sea model to interpret the megacycles as progradational sedimentary increments, consisting of low-energy subtidal X-zone siderite lutite, overlain by sideritic allochemical iron-formation of the high-energy Y-zone, and capped by lagoonal Z-zone greenalite lutite. The megacycles of the Danielskuil Member are usually separated from each other by disclutite beds which may represent lag deposits along transgression surfaces. Sharp-based upward-fining grainstone → felutite mesobands which are present in the X-zone siderite lutite are thought to represent storm wave bands similar in origin to those described by Reineck and Singh (1972) and Smith and Hopkins (1972) from siliciclastic shallow sea environments.

The siderite lutite of the X and Y-zones of the megacycles is most probably of an abiogenic origin indicating that the epeiric sea was weakly alkaline and reducing (Beukes, 1978). This conclusion must, however, be considered tentative because hematite dust is associated with the siderite and the extent of possible post-depositional sideritization has not been evaluated yet. The precipitation of greenalite lutite in the lagoonal Z-zones of the epeiric sea was probably brought about by the inflow of acidic freshwater from a nearby low-lying land area.

Chert mesobands in the orthochemical → allochemical iron-formation cycles are of an early diagenetic origin representing the product of silica cementation and/or silica replacement of felutite. This is indicated by gradational contacts between chert mesobands and felutite mesobands, compaction of felutite and allochemical mesobands around chert pods and pillows, and chert-cement supported fabrics in some grainstones (Beukes, 1978).

The orthochemical → allochemical iron-formation megacycles of the Danielskuil Member interfinger basinwards with riebeckitic minnesotaite-greenalite lutites of the Middelwater Member (Fig. 4-7). The minnesotaite is a metamorphic product of the greenalite which was most probably deposited in a shallow basin below wave base (Fig. 4-7). The basin could have been evaporitic and partly or totally enclosed as is indicated by the presence of a relatively large amount of sodium (now in the form of riebeckite) in it (Cilliers, 1961; Hanekom, 1966; Beukes, 1973, 1978). On the platform in the Danielskuil and Ouplaas Members riebeckite only occurs in grainstone beds. Sodium derived from pore water could have been responsible for the crystallization of the riebeckite during diagenesis and low-grade metamorphism. Magnetite in the Griquatown Iron Formation, as in the case of the Kuruman Iron Formation, always replaces earlier iron mineral assemblages and is, therefore, of a diagenetic or metamorphic origin (Fockema, 1967; Beukes, 1978, 1980a).

The disclutites (edgewise conglomerates) of the Skietfontein Member of the Griquatown Iron Formation (Fig. 4-7) consist of polygonal chert discs set in a matrix of greenalitic felutite or grainstone. The discs come from chert hardgrounds which were broken up by either slumping or wave action (Beukes, 1978). The latter origin is favoured by Malherbe (1970) and Beukes (1978) because of the presence of imbricate structures and pockets of radially arranged, vertically standing chert discs. These pockets probably originated in wave surge eddies. Shrinkage cracks are present in some of the chert mesobands and pillows, and these contributed to the breaking up of the hardgrounds by wave action. Most of the cracks are thought to be due to syneresis but some may represent subaerial desiccation features. The latter possibility together with the poor sorting and angularity of the chert discs led Beukes (1978) to the belief that the disclutites represent storm wave deposits on supratidal flats (Fig. 4-7).

The transition between the Griquatown Iron Formation and the Koegas Subgroup is represented by an upward-coarsening sedimentary cycle, which consists of banded greenalite lutite of the Pietersberg Member overlain by chloritic mudstone, siltstone, and quartz wacke of the Pannetjie Formation. The latter unit forms the basal part of the Koegas Subgroup. The cycle is only preserved off the platform to the south of the Griquatown fault zone. Northwards of the fault zone the Pannetjie Formation and part of the Pietersberg Member were removed by erosion before deposition of the Makganyene diamictite (Fig. 4-7). The upward-coarsening Pietersberg—Pannetjie sedimentary cycle probably represents the infill of a freshwater lake by deltaic sedimentation, with the greenalite lutite as lake bottom deposits, the chloritic mudstone and siltstone as prodelta deposits, and the quartz wackes as delta front or nearshore lake deposits (Beukes, 1978). The freshwater character of

pp. 165—166

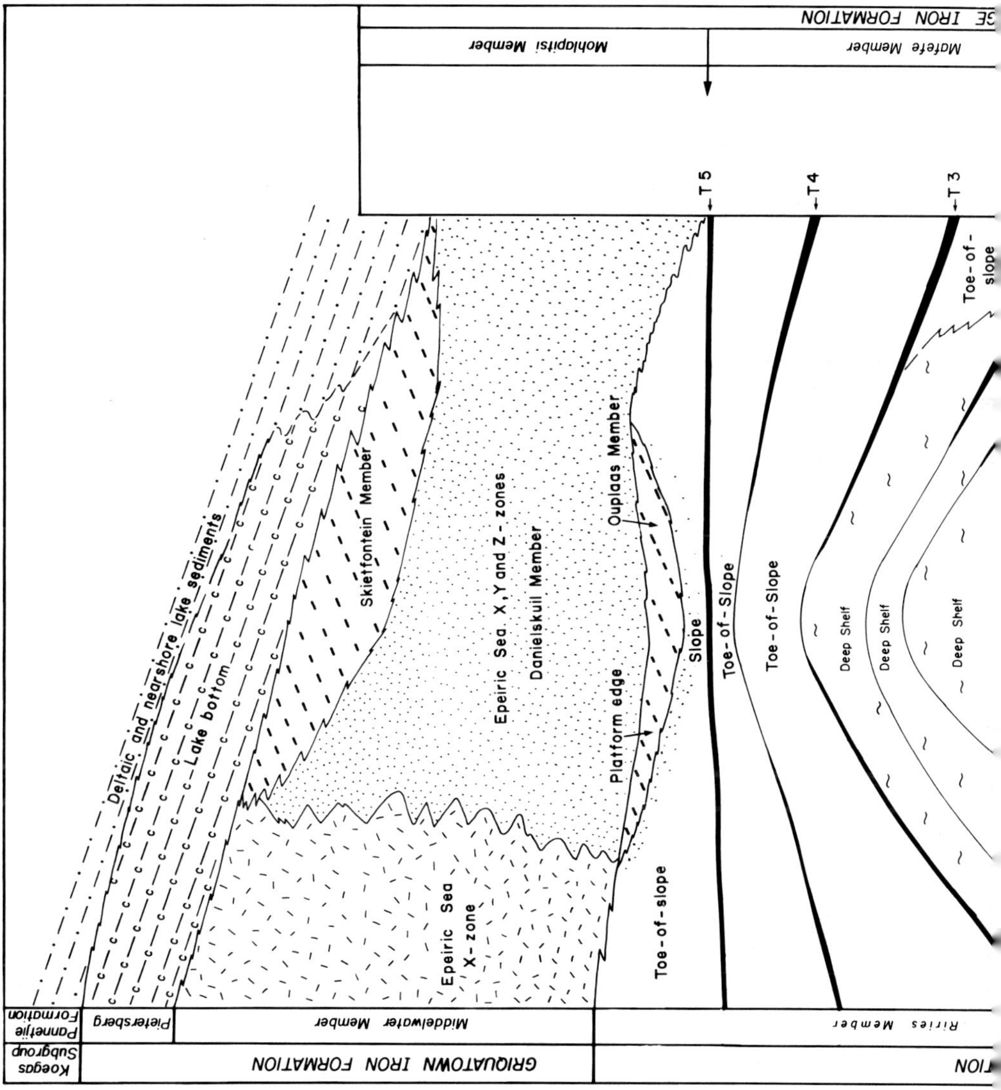

Koegas Subgroup
Pannetjie Formation
GRIQUATOWN IRON FORMATION
Pietersberg
Middelwater Member
Riries Member
TION
IRON FORMATION
Mohlapitsi Member
Mafefe Member
T 5
T 4
T 3
Deltaic and nearshore lake sediments
Lake bottom
Skietfontein Member
Epeiric Sea X, Y and Z - zones
Danielskuil Member
Epeiric Sea X - zone
Ouplaas Member
Platform edge
Slope
Toe-of-slope
Toe-of-Slope
Toe-of-Slope
Toe-of-slope
Deep Shelf
Deep Shelf
Deep Shelf

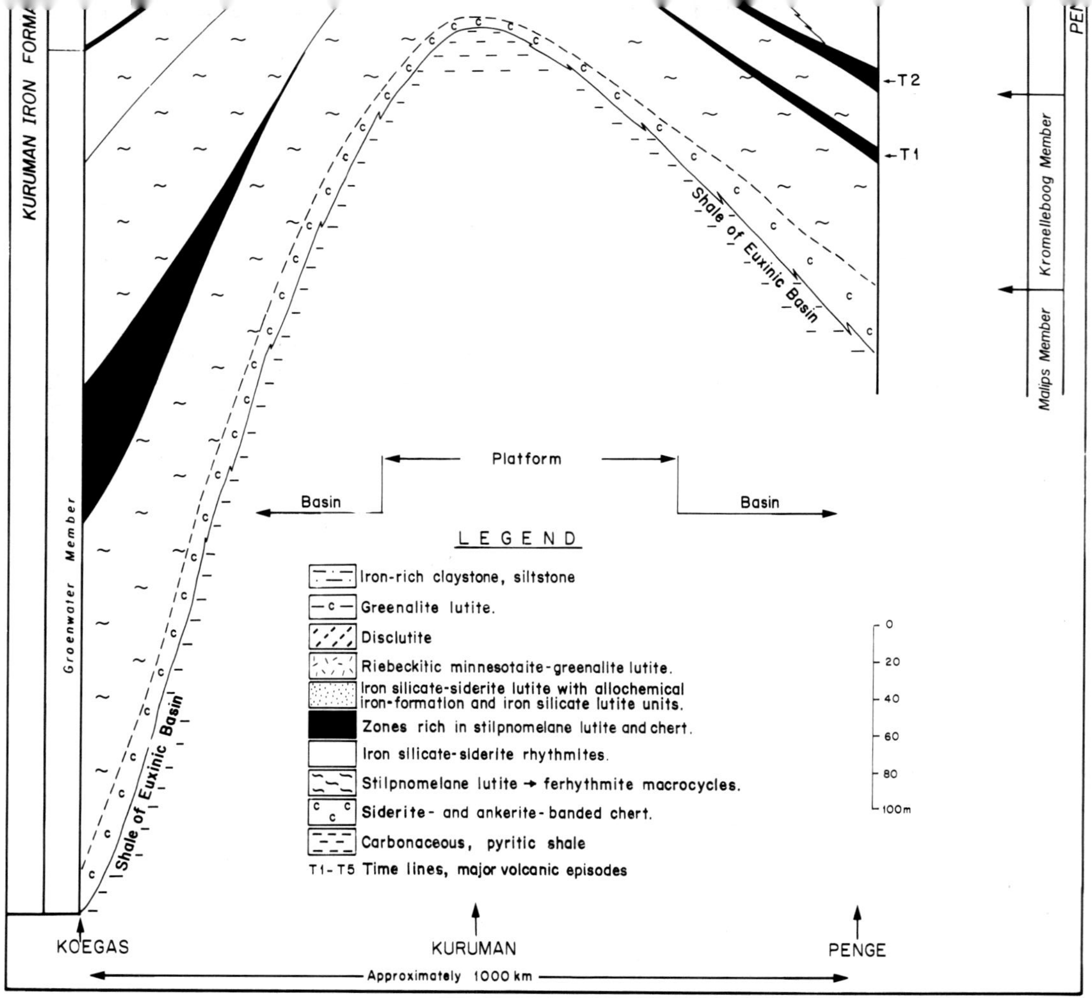

Fig. 4-10. Correlation between the Asbesheuwels and Penge iron-formation sequences with an interpretation of palaeodepositional environments (after Beukes, 1978).

the lake is inferred from the presence of greenalite which, as has been mentioned earlier, would precipitate under neutral to weakly alkaline conditions. Chert mesobands interbedded with the greenalite lutite may have precipitated during periods of abnormally high inflow of freshwater causing the water of the lake to become acidic.

The chlorite present in the Pannetjie Formation of the Koegas Subgroup is iron-rich and represented by thuringite. This may be an indication that some of the iron present in the Griquatown Iron Formation came from the weathering of a nearby source area and that it was transported into the basin adsorbed on to clay minerals. Post depositional recrystallization then led to the formation of thuringite.

Summarizing, it can thus be stated that the Griquatown Iron Formation represents a third-order depositional cycle consisting of hematitic, sideritic and greenalitic orthochemical and allochemical epeiric sea deposits, supratidal storm breccias, and freshwater greenalitic lake bottom felutites.

Asbesheuwels depositional cycle

In order to summarize the depositional history of the Asbesheuwels Subgroup the third-order Kuruman and Griquatown Iron Formation cycles can be combined into a fourth-order progradational cycle of which the major features are given in Fig. 4-9. The base of this cycle is formed by pyritic carbonaceous shale and microcrystalline chert. These were deposited as a distal facies of the Kuruman Iron Formation in a deep-water acidic and euxinic basin. Ankerite-banded cherts which in part represent ankeritized and chertified distal carbonate turbidites, constitute the lowermost part of the Kuruman Iron Formation. They formed along the transition zone between a euxinic basin and a deep-water shelf (Fig. 4-9).

Silica deposition took place on the floor of the deep shelf. The deposition of silica was, however, periodically interrupted by the deposition of volcanic ash beds preserved as shard-bearing stilpnomelane lutite. In addition, femicrite microbands were deposited seasonally by the action of photosynthesizing organisms. The interaction of silica deposition with volcanic and biogenic activity led to the development of stilpnomelane lutite → ferhythmite macrocycles (first-order cycles) on the deep shelf. Many such cycles are present but five zones rich in stilpnomelane lutite mark five major periods of volcanic activity (Fig. 4-9).

The deep-shelf deposits are overlain by greenalite-siderite rhythmite which was deposited along the toe of a platform slope. Chemical conditions were neutral to weakly alkaline so that greenalite precipitated instead of chert (Fig. 4-9). The slope facies itself is characterized by greenalite-siderite rhythmites interbedded with grainflow mesobands. It grades upwards into sideritic orthochemical and allochemical iron-formations which represent a platform edge facies. Following on that are orthochemical → allochemical iron-formation megacycles deposited in an alkaline reducing epeiric sea. Each of these cycles are capped by greenalite lutite deposited in a platform interior, lagoonal environment (Fig. 4-9). Supratidal flats, on which storm-wave chert breccias accumulated, may have bordered the epeiric sea. The most advanced stage of progradation of the fourth order Asbesheuwels depositional cycle is represented by greenalitic and siliciclastic freshwater lake deposits (Fig. 4-9).

The shallow-water depositional environment of iron-formation described above is similar to that in the Campbellrand—Malmani carbonate sequence. The change from carbonate to iron-formation deposition on the Kaapvaal craton thus cannot be ascribed to a change in the type of depositional environments. Rather an extrabasinal reason must be looked for which may, for example, have been a change in climatic conditions i.e. a change from warm (carbonate deposition) to cold (iron-formation deposition) water conditions (Beukes, 1973).

Correlation with Penge Iron Formation and regional depositional model

Subdivisions of the Asbesheuwels Subgroup in Griqualand West are correlatable with those of the Penge Iron Formation in the Transvaal, making possible a very large scale reconstruction of depositional environments (Fig. 4-10). Contact metamorphism related to the intrusion of the Bushveld Complex has caused extensive recrystallization and gruneritization of the Penge Iron Formation (Beukes, 1973, 1978) but the equivalent unmetamorphosed lithofacies are still recognizable. The third and fourth order cyclicity of the Penge Iron Formation is similar to that of the Asbesheuwels Subgroup (Fig. 4-10). Also, stilpnomelane lutite → banded ferhythmite megacycles and even some macrocycles are correlatable from Koegas in the southwestern corner of the Kaapvaal craton to Mafefe, near Penge, in the eastern Transvaal (Beukes, 1978).

The Penge area must have been located along the same depositional strike as the Koegas area because correlations between Koegas and Penge are better than between either of these two areas and the Kuruman area. Coarse-grained allochemical iron-formation is better developed and autochthonous ferhythmites more poorly developed in the Kuruman area, than at Penge and Koegas (Fig. 4-10). This would suggest that both the Koegas and Penge localities were situated in a basinal environment relative to the dominantly shallow-water platform environment of the Kuruman locality (Fig. 4-10). This observation led to a hypothesis by Beukes (1978) that chemical sedimentation developed on the Kaapvaal craton at a stage when the Limpopo metamorphic complex (Fig. 4-4A) represented a negative element, i.e. a basin. Siliciclastic and volcanoclastic material coming from external source areas such as the Zimbabwe craton became trapped in this basin which extended around the northern and western side of the drowned Kaapvaal craton (Fig. 4-11). In turn this led to the development of clear water conditions and the deposition of carbonates and iron-formation on the Kaapvaal craton (Fig. 4-11). The presence of an external siliciclastic and volcanoclastic source area is indicated

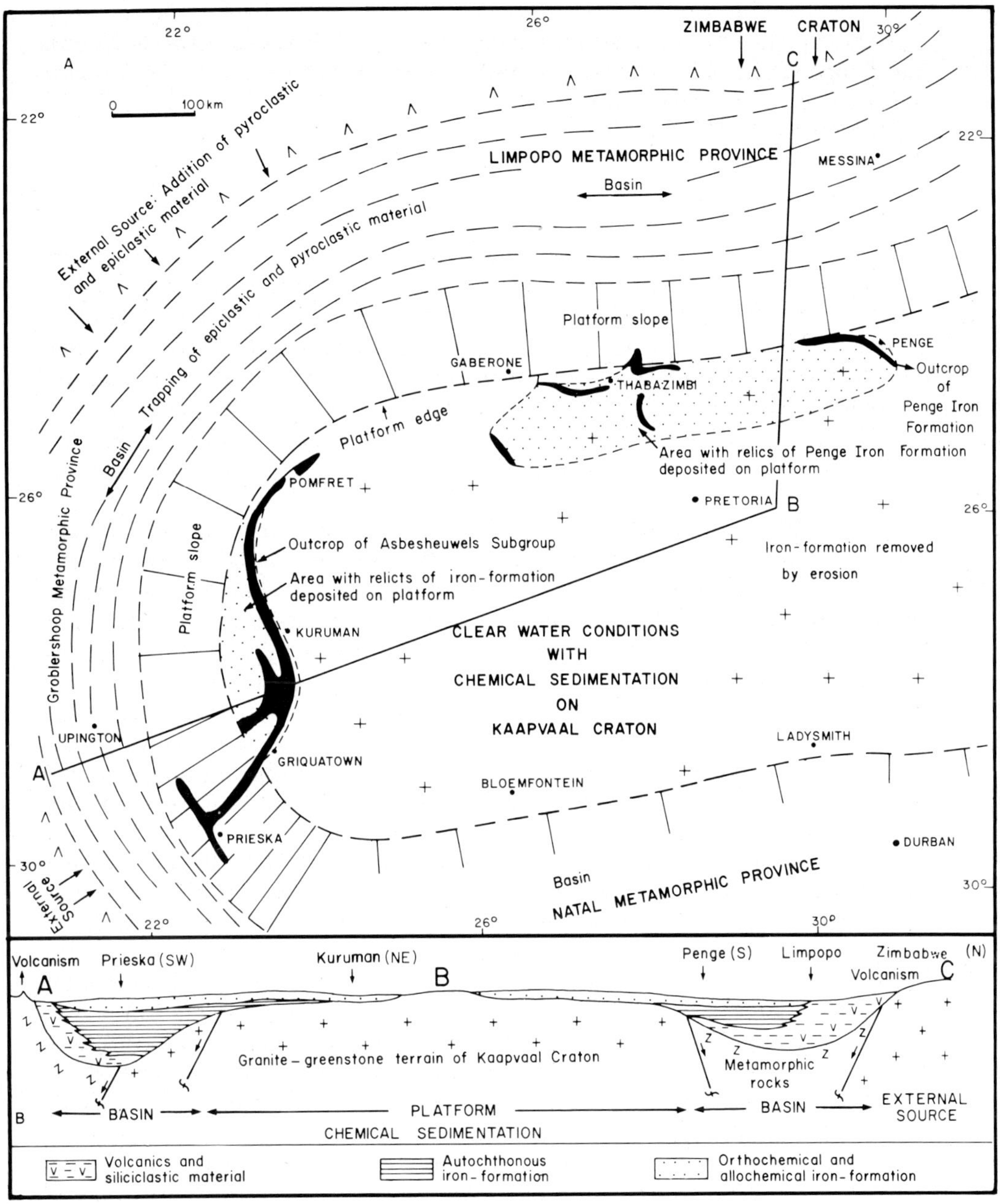

Fig. 4-11. Nature of the depository of the Transvaal Supergroup during the deposition of the Asbesheuwels—Penge iron-formation sequence. Plan view (A) and cross-section (B) (after Beukes, 1978).

by the thickening of both the carbonaceous shale at the base of the iron-formation sequence, and the stilpnomelane volcanic ash zones towards Koegas and Penge (Fig. 4-10). The basin could have been graben-like (Fig. 4-11) as is suggested by the known fault control (Button, 1973a; Beukes, 1977a, 1978, 1980a) on sedimentation.

KOEGAS SUBGROUP

The Doradale, Rooinekke and Nelani Iron Formations of the Koegas Subgroup each forms the base of an upward-coarsening iron-formation → siliciclastic sedimentary cycle (Fig. 4-12). These cycles are described in detail by Beukes (1978) and Van Wyk (1980). Each cycle represents a progradational sedimentary increment consisting of distal iron-formation and proximal ferruginous green-coloured chloritic mudstone, siltstone and quartz wacke. However, in detail the composition and depositional environments of the cycles change upwards in the sequence and facies changes take place along strike.

The lowermost cycle comprising the Doradale, Kwakwas and Naragas Formations is similar in composition and palaeo-environmental setting to the Kuruman → Griquatown → Pannetjie cycle immediately underlying it. However, in the latter cycle iron-formation deposition predominated, whereas in the former, iron-formation deposition was suppressed by the influx of siliciclastics. The base of the Doradale → Naragas cycle is sharp and represented by a transgressive ravinement chert pebble lag conglomerate (Fig. 4-12). The Naragas Formation thins northwards in the direction of the Kaapvaal platform. Thinning is accompanied by an increase in the amount of quartz wacke present, indicating that the siliciclastics came from a source area on the Kaapvaal craton (Fig. 4-13A)*.

Although the Doradale Iron Formation is only a few meters thick it is similar in character to the Kuruman Iron Formation consisting of autochthonous magnetite-siderite ribbonrhythmite in the basin near Koegas. Northwards towards the platform it changes into a siderite bandlutite with interbeds of disclutite which could either represent gravity flow deposits or shallower-water storm wave breccias (Beukes, 1978). The Doradale Iron Formation grades upwards into the Kwakwas Iron Formation which is similar in composition to the Middelwater Member of the Griquatown Iron Formation and consists of riebeckitic minnesotaite-greenalite lutite (Fig. 4-12). It pinches out towards the north (Fig. 4-13A) and, as in the case of the Griquatown Iron Formation, its lutitic character and sodium content probably indicate deposition below wave base in an enclosed or partly enclosed evaporitic basin. The Kwakwas Iron Formation is overlain by an upward-coarsening siliciclastic cycle which, could either represent a progradational deltaic sequence, or the infill of the

* Figure 4-13A is shown on p. 176.

pp. 173—174

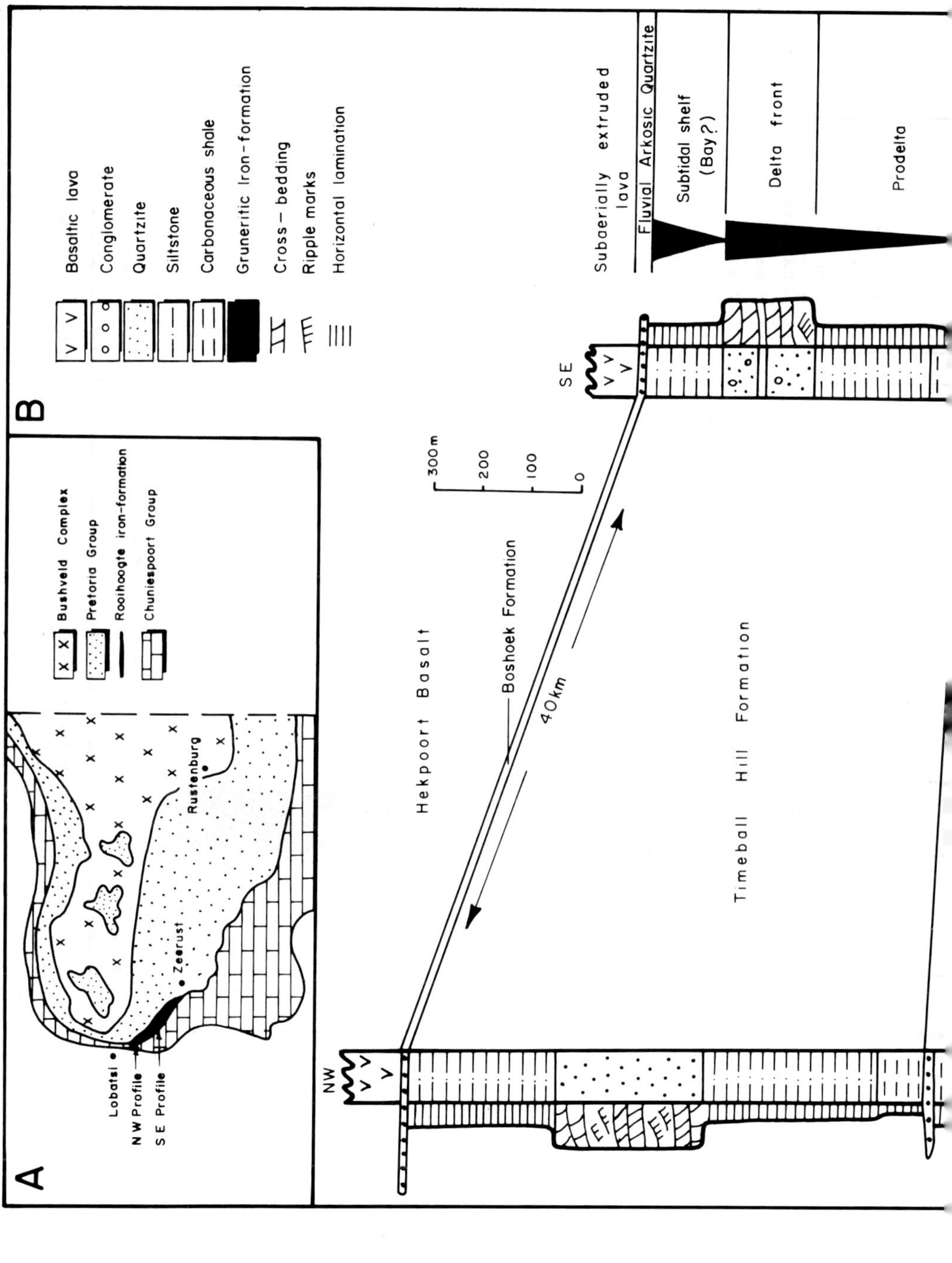
A
B
Lobatsi
NW Profile
SE Profile
Zeerust
Rustenburg
Bushveld Complex
Pretoria Group
Rooihoogte iron-formation
Chuniespoort Group
Basaltic lava
Conglomerate
Quartzite
Siltstone
Carbonaceous shale
Gruneritic Iron-formation
Cross-bedding
Ripple marks
Horizontal lamination
NW
SE
300m
200
100
0
Hekpoort Basalt
Boshoek Formation
40km
Timeball Hill Formation
Subaerially extruded lava
Fluvial Arkosic Quartzite
Subtidal shelf (Bay?)
Delta front
Prodelta

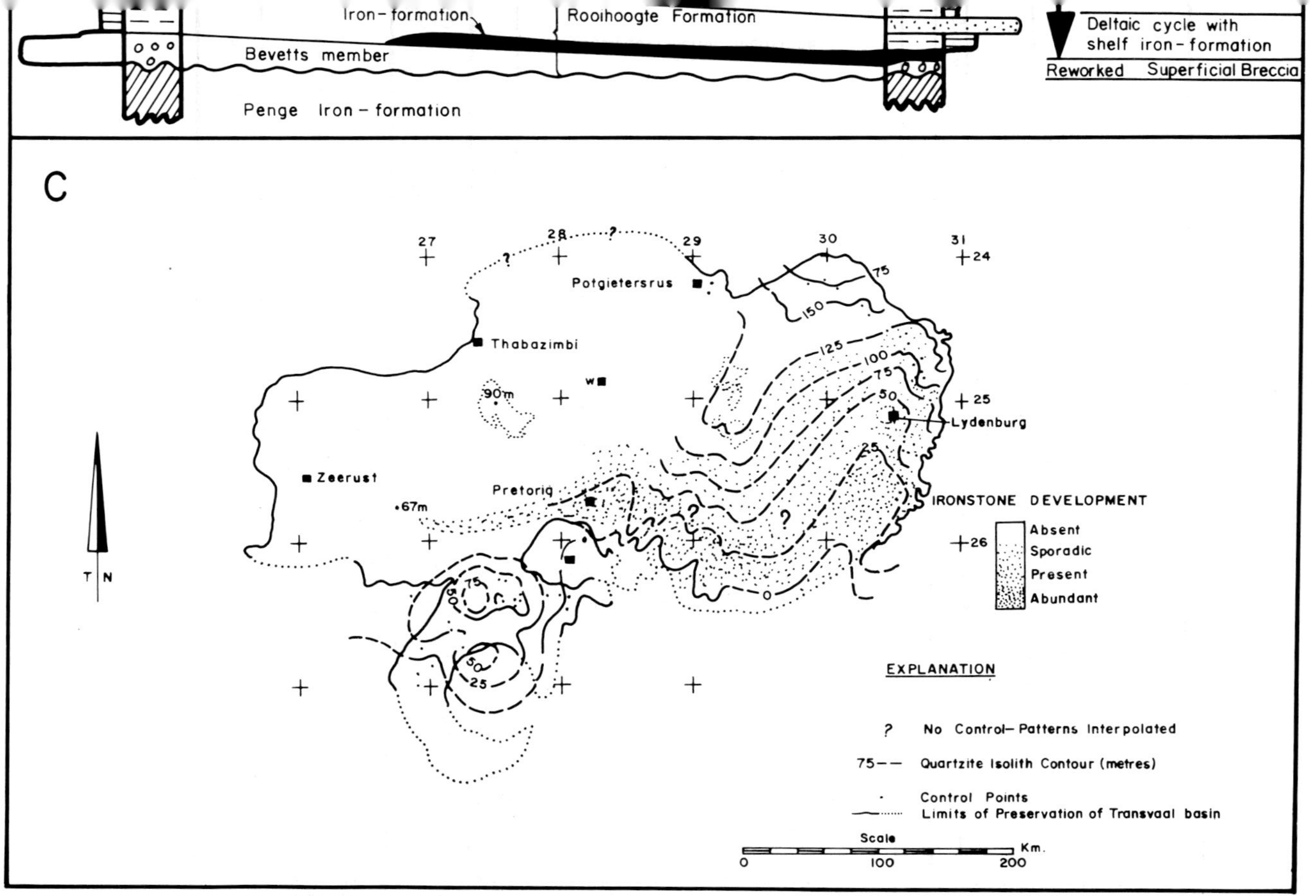

Fig. 4-14. Geographic setting (A) and stratigraphic setting (B) of the iron-formation of the Rooihoogte Formation, Pretoria Group (modified after Klop, 1978). C. Arenite isolith and oolitic ironstone distribution map for the Timeball Hill Formation (after Button, 1973a).

basin by offshore chloritic mudstones and nearshore quartz wackes (Fig. 4-13A).

Manganiferous siderite lutites make their appearance at the top of the Naragas Formation (Fig. 4-12). These units thicken towards the basin and form the base of manganiferous siderite lutite → ferruginous chloritic mudstone cycles heralding a period of transgression below the Rooinekke Iron Formation. Chemically the manganiferous siderite lutites consist of between 31 and 35 wt. % SiO_2, 22—30 wt. % FeO and 4—8 wt. % MnO. Some units contain up to 12 wt. % CaO and 2—4 wt. % MgO but mostly these and other elements are very minor components (Beukes, 1978; Van Wyk, 1980). Most of the manganese is probably taken up in the structure of the siderite whose composition tends to approach that of manganosiderite.

The base of the Rooinekke Iron Formation consists of a manganiferous siderite lutite unit which is overlain by siderite bandlutite and pillowlutite constituting the major part of the formation in the north. Syneresis cracks are abundant in the chert bands in the upper part of the iron-formation. Southward into the basin the Rooinekke Iron Formation becomes more manganiferous and at Koegas it consists mostly of manganiferous siderite lutite (Fig. 4-13A). Steep-sided, stromatolitic, ferruginous dolomite bioherms of up to 40 m high and 100 m wide are present in the Rooinekke Iron Formation (Beukes, 1978). They are concentrated along the contact between manganiferous siderite lutite and siderite bandlutite and probably represent stromatolite reefs that grew along the edge of a platform separating a platform lagoonal environment from a deeper-water basinal environment. More soluble manganiferous siderite lutites were thus deposited in the basin, whereas chert and less soluble siderite lutite were deposited in the more brackish and acidic lagoonal environment (Fig. 4-13B). Locally the pH of the water was sufficiently increased by photosynthesizing blue-green algae (cyanobacteria) to allow the deposition of limestone bioherms. Relics of limestone are present in the bioherms indicating that the dolomite is diagenetic (Beukes, 1978).

The Rooinekke Iron Formation is conformably overlain by interbedded ferruginous chloritic mudstone and chert constituting the Klipputs Member of the Nelani Formation. The chert layers must have constituted early diagenetic hardgrounds in the siliciclastic muds because many of them have been broken up by either slumping or storm wave action to form edgewise conglomerates. This facies is thought to represent nearshore lagoonal deposits of the Rooinekke sedimentary increment (Fig. 4-13B).

Stromatolitic dolomite bioherms mark a transgression surface at the top of the Klipputs Member (Fig. 4-12). Submergence of siliciclastic source areas and the resultant development of clear water conditions could have been responsible for the development of the bioherms during the transgression (Van Wyk, 1980). The Nelani Iron Formation proper consists mainly of allochemical, sideritic and cherty grainstones underlain by a riebeckitic felutite and overlain by an interbedded ferruginous mudstone, chert, gritstone and edgewise

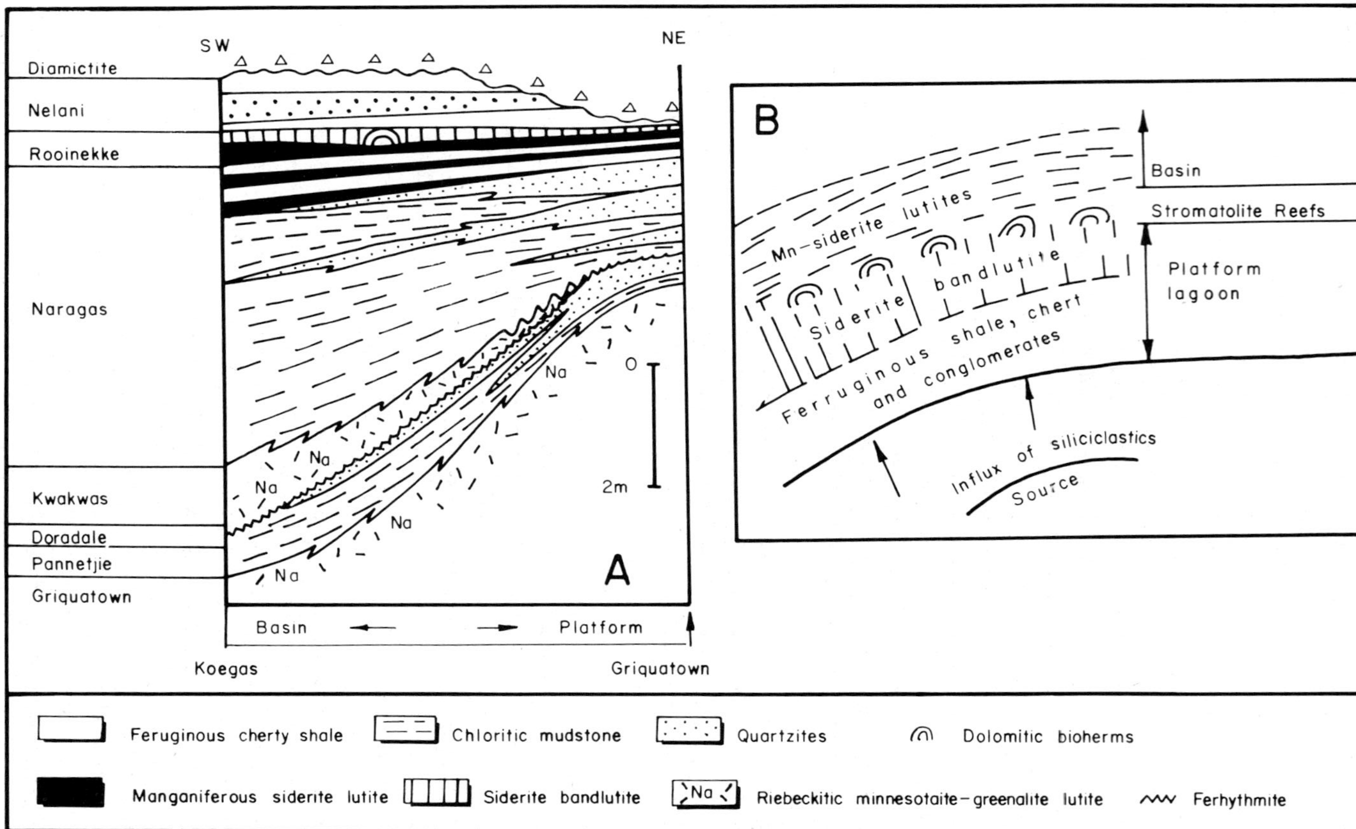

Fig. 4-13. A. Regional stratigraphic relationships in the Koegas Subgroup along a southwest-northeast section. B. Inferred palaeo-depositional environments for the Rooinekke Iron Formation and the Klipputs shale member of the Nelani Iron Formation.

conglomerate facies. This facies is similar to the Klipputs Member and could have been deposited under similar conditions. However, the iron-rich grainstones below it must have been deposited under relatively higher-energy conditions than were the banded felutites of the Rooinekke Iron Formation. The presence of cross-bedded peloidstones and ooidstones led Van Wyk (1980) to believe that the Nelani grainstones were deposited in the high-energy Y-zone of an epeiric sea, with the riebeckitic felutite representing a partly enclosed subtidal X-zone, and the interbedded chert and chloritic mudstones the platform lagoonal Z-zone (Fig. 12). From the Kuruman Iron Formation up to the Nelani Iron Formation there is thus an increase in the grain size of the iron-formations probably reflecting filling of the basin with resultant upward-shallowing and an increase in energy conditions.

ROOIHOOGTE AND TIMEBALL HILL FORMATIONS

Superficial chert rubble derived from the Penge Iron Formation and Malmani dolomite sequence accumulated on the erosion surface that developed before the deposition of the Pretoria Group in the Transvaal area. A subsequent transgression led to reworking of the rubble into chert-pebble conglomerates and sandstone which constitutes the Bevets conglomerate member at the base of the Rooihoogte Formation (Visser, 1969, 1972; Button, 1973a). Following the transgression, progradational siliciclastic sedimentation set in and, to the northwest of Zeerust in the western Transvaal (Fig. 4-14A), an upward-coarsening tide-dominated deltaic cycle developed. It consists of prodelta shale and siltstone, overlain by intercalated quartzite, carbonaceous shale and stromatolitic dolomite representing delta front and tide-dominated delta plain deposits of the Rooihoogte Formation (Fig. 4-14B)*.

A gruneritic banded iron-formation with a maximum known thickness of 10 m (Klopp, 1978) is developed at the base of the deltaic cycle. The iron-formation interfingers to the northwest with prodelta shales (Fig. 4-14B) indicating that it most probably represents a distal offshore shelf facies of the delta sequence. It has a limited lateral extent because it pinches out next to a palaeoerosional high at the base of the Pretoria Group near Zeerust (Martini, 1976). The original extent of the iron-formation to the southwest may, however, have been more extensive (Fig. 4-4B).

The Timeball Hill Formation which overlies the Rooihoogte Formation represents a progradational tide-dominated deltaic increment of sedimentation (Visser 1972; Eriksson 1973; and Fig. 4-4B). Non-cherty, chamositic and hematitic ironstone beds with a maximum known thickness of 8.5 m are associated with the delta front facies of the Timeball Hill Formation (Button, 1973a). These ironstones stand in contrast to the banded iron-formation of

* Figure 4-14 is shown on pp. 173—174.

the Rooihoogte Formation which represents an offshore shelf facies of deltaic sedimentation.

The ironstones vary from a mixture of iron-rich ooids with either mature quartz grains or clay to ferruginous quartzite (Visser, 1972). Ooids consists of concentric shells of chamosite and hematite around mature quartz grains (Schweigart, 1965; Visser, 1969, 1972). Sedimentary structures such as ripple cross-lamination, planebedding with current lineation, trough and planar cross-bedding, runzel marks and clay pellets are associated with the ironstones indicating shallow-water environments of deposition (Button, 1973a). Visser (1972) and Tankard et al. (1982) consider them to have been deposited on shallow subtidal shoals and intertidal flats along the more distal parts of tide-dominated delta plains. The concentration of most ironstones along a linear zone conforming to the pinch-out region of the Timeball Hill arenites (Fig. 4-14C), further confirms this conclusion.

Both Visser (1969) and Eriksson (1973) indicate that the siliciclastics of the Timeball Hill Formation start to wedge out along the extreme southern limit of the Transvaal structural basin. It is thus most probable that in Griqualand West only an offshore shelf iron-formation facies of the deltas was deposited (Fig. 4-4B). A correlation with the upper iron-formations of the Koegas Subgroup thus seems not to be too unreasonable an assumption.

MAKGANYENE DIAMICTITE

Two facies are present in the Makganyene Diamictite of the Postmasburg Group. The first facies consists of interbedded massive diamictite, poorly bedded diamictite, conglomerate, gritstone, coarse-grained quartzite, siltstone and shale. It is confined to the Kaapvaal platform area and Visser (1971) interpreted it as a piedmont glacial and glaciofluvial assemblage.

Along the western margin of the Kaapvaal craton this platform facies interfingers with a second facies consisting of stacked cycles of graded bedded diamictite → greywacke → siderite bandlutite (Fig. 4-15B). The siderite bandlutites (cherty iron carbonates of De Villiers and Visser, 1977) consist of chert mesobands alternating with siderite lutite mesobands. Minor amounts of stilpnomelane are present (De Villiers and Visser, 1977).

De Villiers and Visser (1977) and Van Wyk (1980) believe that the graded diamictite units represent periodic subaqueous debris flow deposits. Presumably the debris was derived from piedmont glaciers (Visser, 1971) that moved into a marine environment where iron-rich chemical sediments (siderite bandlutites) normally accumulated. Post-depositional slumping took place in some of the siderite bandlutites resulting in iron-rich mud-supported chert breccias (Van Wyk, 1980).

The interpretation that the components of the diamictite were glacially derived rests mainly on the abundance of striated chert fragments and pebbles (Visser, 1971; De Villiers and Visser, 1977; Van Wyk, 1980). The fact that

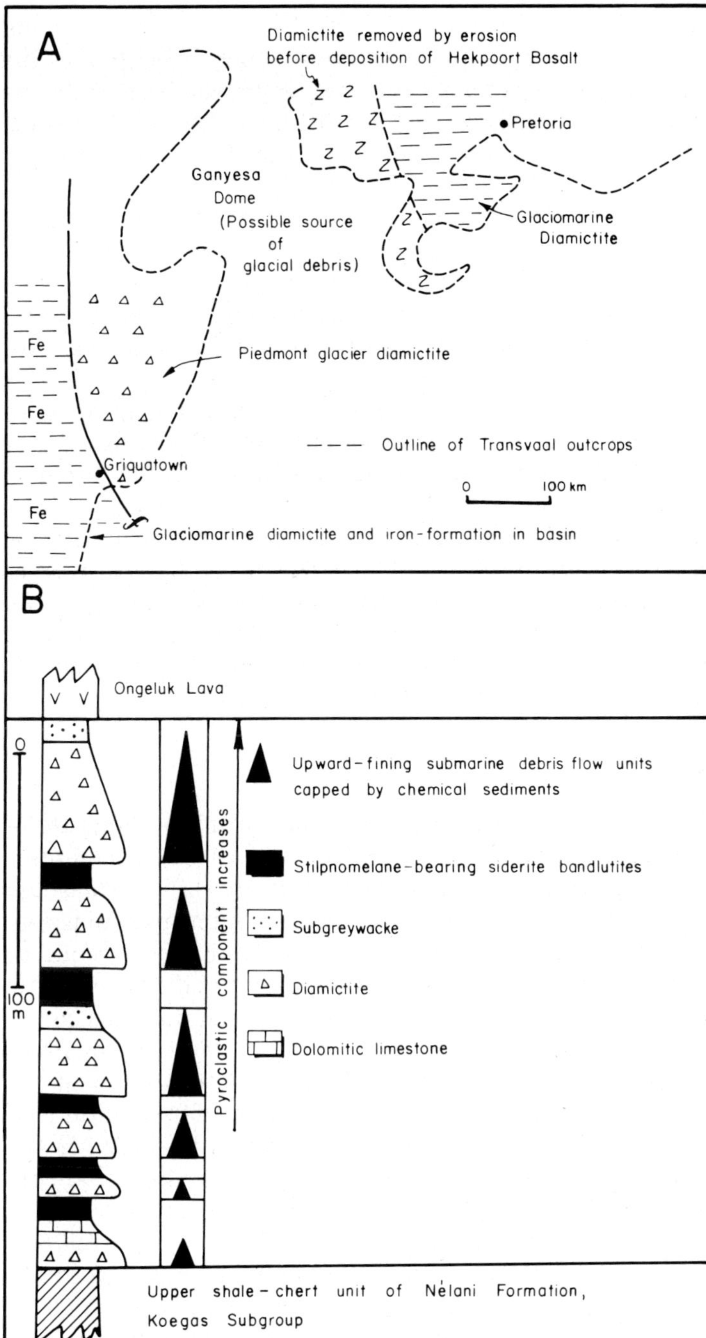

Fig. 4-15. A. Inferred distribution of piedmont glacier and glaciomarine deposits in the Makganyene Diamictite (modified after Visser, 1971). B. Stratigraphic setting of iron-formations in the Makganyene Diamictite near between Sishen and Postmasburg (modified after De Villiers and Visser, 1977).

most of the fragments and pebbles in the diamictite consist of chert is explained by assuming a rather homogeneous source area which consisted of the underlying iron-formation and carbonate sequences. From the basin onto the Kaapvaal platform the unconformity at the base of the Makganyene Diamictite cuts progressively down through the sequence from the Nelani Iron Formation to the Griquatown Iron Formation (Fig. 4-4B).

No glacial pavements are known from the Makganyene Diamictite assemblage and in the eastern Transvaal the correlative of this sequence was interpreted by Button (1973a) as representing density flow deposits which were not necessarily derived from glacial debris. The attribution of a glacial origin to the Makganyene Diamictite thus most probably needs critical re-examination.

Part of the diamictite sequence was removed by erosion in the western Transvaal prior to the deposition of the Hekpoort Basalt (Fig. 4-4B). However, in the basinal area of Griqualand West no unconformity is present at the base of the Ongeluk Lava (Fig. 4-4B). Instead pyroclastic material is included in the upper part of the diamictite sequence (Fig. 4-15B) indicating that the initial phases of volcanism were contemporaneous with the last stages of glacial and/or debris flow sedimentation (De Villiers and Visser, 1977). Some of the silica and iron in the iron-formations could have been derived from volcanic ash or from fumarolic activity. Tectonic instability associated with volcanism could have triggered some of the debris flows.

VOËLWATER SUBGROUP

The Voëlwater Subgroup, comprising the Ongeluk Lava, Mooidraai Dolomite, and Hotazel and Beaumont Formations (Beukes, 1978), represents a mixed volcanogenic-chemical sedimentary rock unit in the Postmasburg Group of Griqualand West (Fig. 4-4B). The Ongeluk Lava outcrops extensively but the other three formations are largely covered by the Cretaceous to Tertiary Kalahari Formation, Paleozoic Karoo Supergroup and the Paleophytic Olifantshoek Group (Fig. 4-16). Knowledge of these formations thus comes almost exclusively from exploration and mining of the vast sedimentary manganese deposits which are interbedded with the iron-formation of the Hotazel Formation. Taljaardt (1979) estimates that about 7500 million tons of 30%-plus manganese ore is present in the deposit which is known as the Kalahari manganese field. However, geological information is considered confidential by mining companies and only a few references are available on the manganese field, i.e. De Villiers (1970), Beukes (1973), Button (1976a), Coetzee (1976), Söhnge (1977), Roy (1981) and Jennings (in press). Recently a sedimentological-mineralogical research project was initiated on the deposits at the Rand Afrikaans University in cooperation with mining and exploration companies. A few preliminary results will be discussed here.

Erosional relics of the Hotazel Formation are preserved in a number of structural (fault-bounded?) basins. The Mamatwan, Middelplaats and Wessels Manganese Mines (Fig. 4-16) are situated in the largest known basin. A bright-red jasper unit forms the base of the Hotazel Formation in this area. It overlies hyaloclastic breccias and pillow lava flows of the mafic Ongeluk Lava (Fig. 4-17). Similar jasper beds also cap pillow lava flows and hyaloclastite units in the upper part of the Ongeluk Lava (Grobler and Botha, 1976) sug-

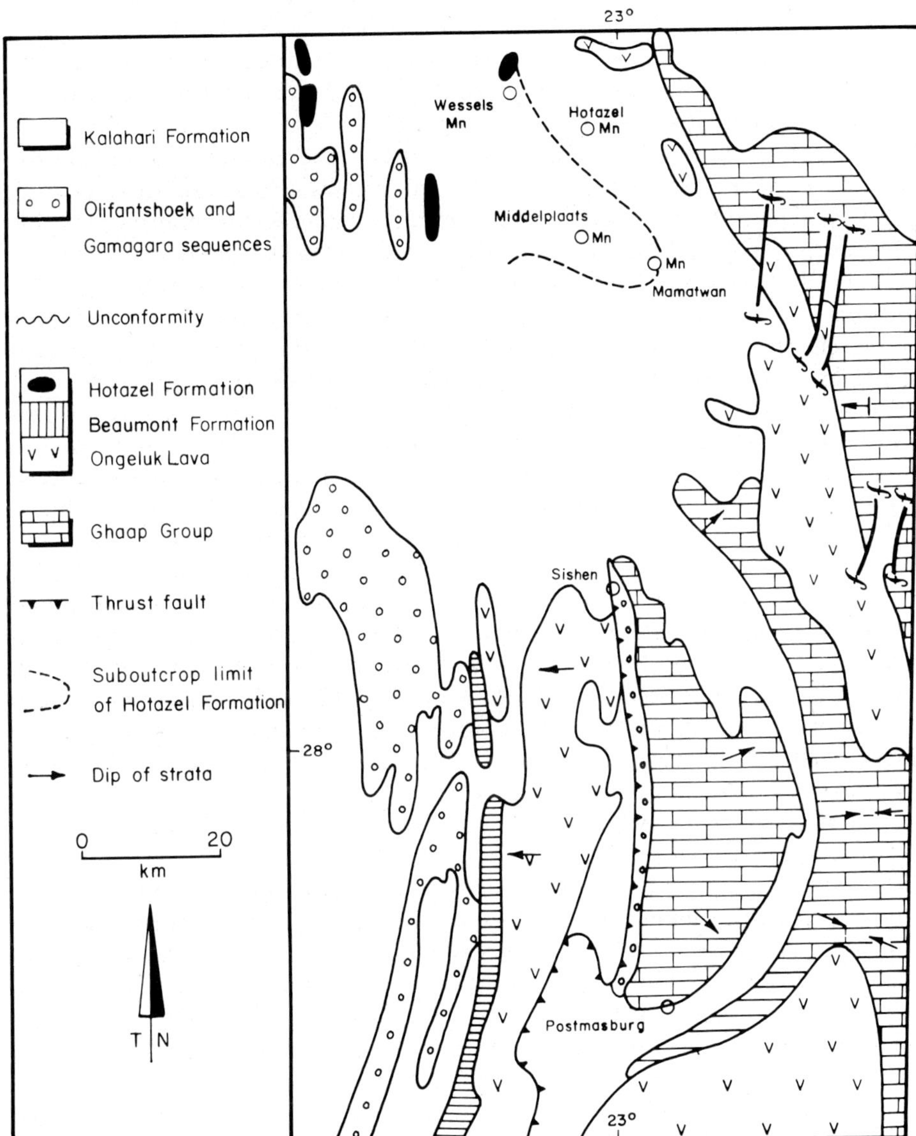

Fig. 4-16. Location of the Beaumont and Hotazel Formations in Griqualand West.

gesting an interfingering relationship between the lava and the Hotazel Formation. Red hematite dust also forms part of the matrix of some of the hyaloclastic breccias.

The jasper beds are either massive, representing cherty hematite lutite beds, or very finely laminated, representing cherty hematite rhythmite units. The jasper consists of submicroscopic hematite dust in a matrix/cement of cryptocrystalline chert. Some of the hematite dust has been transformed to fine-grained euhedral specularite and/or magnetite crystals. In some units jasper ribbons alternate with hematite-magnetite mesobands constituting jaspilites. Two types of jaspilite are present namely hematite-magnetite ribbonrhythmite and hematite-magnetite ribbonlutite.

The chert content of the basal jasper unit of the Hotazel Formation decreases upward in the sequence and the unit grades into kutnahorite-bearing pisolitic hematite lutite, which in turn gradually grades into microcrystalline kutnahoritic pisolitic braunite lutite (Fig. 4-17). The kutnahorite is concentrated in the pisoliths which represent partly compacted, early diagenetic concretions in hematite and braunite lutite. Although pisoliths are most conspicuous in the lutites, ovoids of less than 2 mm in diameter are also present. They fall in the oolith size range but are of a concretionary origin.

The sequence described above represents the lower limb of the lowermost of three symmetrical jasper $\leftrightarrow$ manganolutite sedimentary cycles present in the Hotazel Formation (the double pointed arrow indicates symmetry). Each of the cycles consists of jasper and/or jaspilite $\leftrightarrow$ kutnahoritic hematite lutite $\leftrightarrow$ kutnahoritic braunite lutite (Fig. 4-17). The braunite lutite bed of the lower cycle is between 5 m and 40 m thick and with a manganese content of between 20 and 48 wt. % it represents the major ore unit in the Kalahari manganese field. Ore mined from it in areas of little or no metamorphic recrystallization or superficial enrichment contains 35—38% Mn, 4—7% Fe, 4—7% SiO_2, 12—16% CaO, 3—5% MgO and 15—17% CO_2. Higher-grade ores in areas of secondary enrichment contain less carbonate, manganese oxide minerals like bixbyite, jacobsite and hausmannite, and have the following approximate chemical composition: 44—48% Mn, 11—15% Fe, 6—9% SiO_2, 4—6% CaO, 0.4—1.5% MgO and 1—3% CO_2 (Information brochure, South African Manganese Amcor Ltd., 1979).

The middle manganolutite unit is of the order of 2 m thick, consists essentially of braunitic hematite lutite, and is of no economic significance. The upper manganese ore body rarely exceeds 5 m in thickness and was mined in earlier years. Grey-coloured hematitic and magnetitic minnesotaite ribbon- and bandlutite are present between the lower and middle manganolutite units (Fig. 4-17). Thin chlorite-bearing mesobands interbedded with the iron-formations may represent altered volcanic ash bands.

The above mentioned association of sedimentary manganese deposits with jaspers, hyaloclastites and pillow lavas is similar to the well known greenstone-jasperoid association of volcanogenic-sedimentary manganese deposits review-

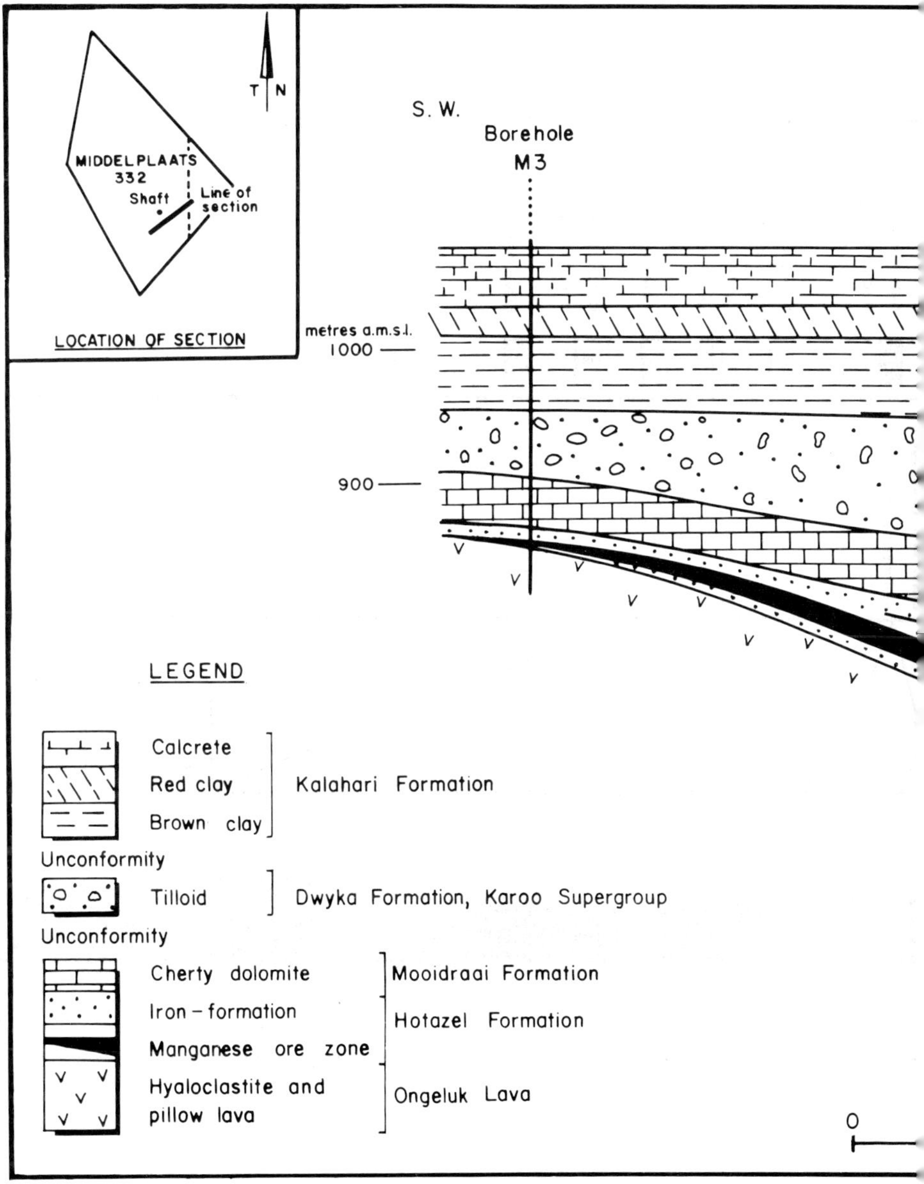

Fig 4-18. Southwest—northeast stratigraphic section at the Middelplaats Manganese Min

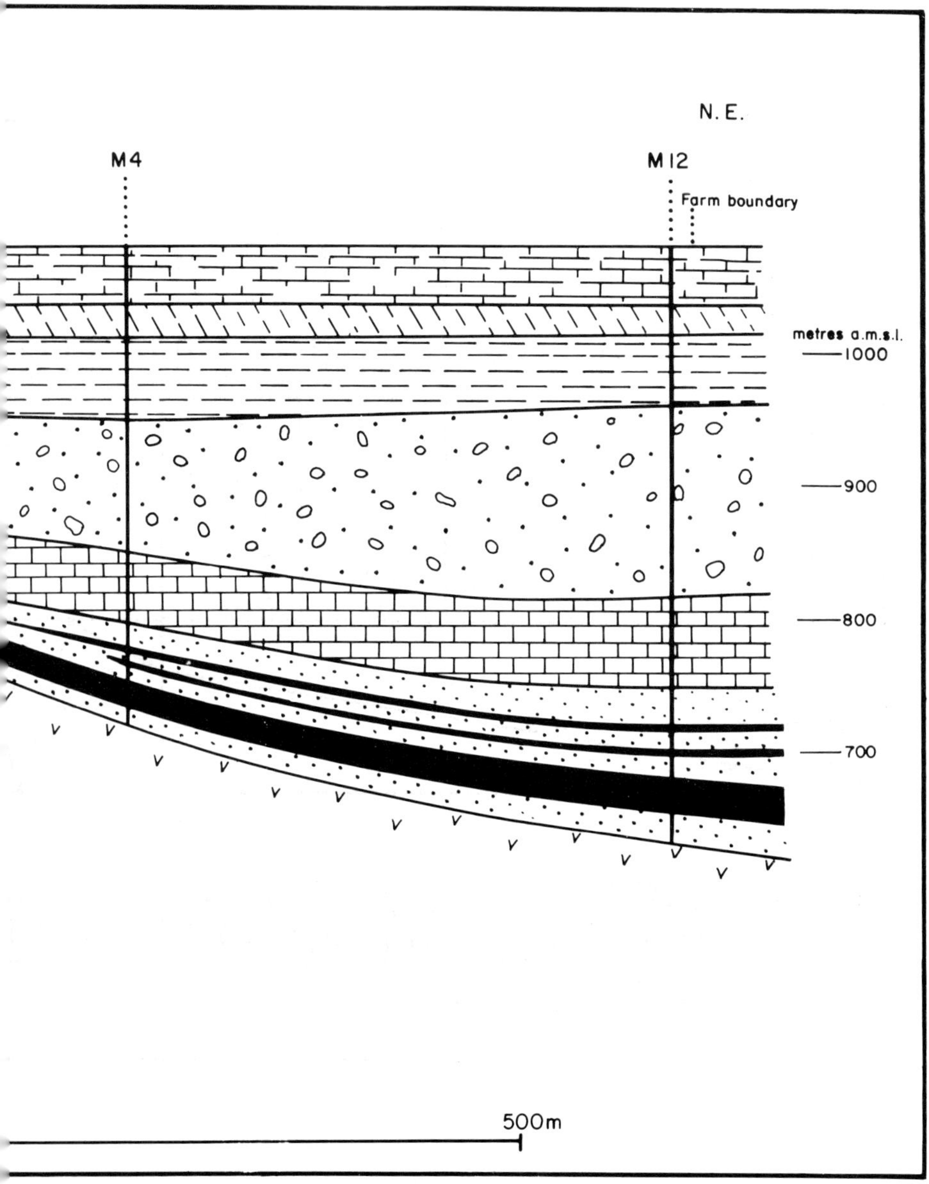

of the Kalahari manganese field (after Jennings, in press).

ed by Shatskiy (1964), Stanton (1972) and Roy (1981). In fact the Kalahari manganese field may represent one of the best developed deposits of this type known, in contrast to ideas expressed by De Villiers (1970), Söhnge (1977) and Roy (1981) that it is a nonvolcanogenic sedimentary manganese deposit.

The jasper which is interbedded with the lavas and also the hematite in the matrix of the hyaloclastics suggest that volcanogenically derived ferrous iron was rapidly oxidized and precipitated in close proximity to the volcanic source. According to models given by Bonatti et al. (1972) divalent manganese ions take longer to be oxidized than ferrous ions, and are, therefore, deposited further away from the volcanogenic source. A similar process could have been responsible for the transition from jasper to manganese oxides in the sedimentary cycles overlying the Ongeluk Lava. Transgressions and regressions relative to a volcanic source would explain the symmetry of the cycles (Fig. 4-17).

The exact geographical relationship of the volcanic source to the manganese deposits has as yet not been fully established. Jennings (in press) shows that the manganolutites pinch out in a southwesterly direction at the Middelplaats Manganese Mine (Fig. 4-18). Still further to the southwest near Postmasburg, the Beaumont Formation outcrops (Fig. 4-16) and consists of jasper and jaspilite interbedded with mafic lava, tuffs and ferruginous dolomite. This sequence may well represent a proximal volcanogenic-sedimentary facies equivalent to the distal, interbedded iron-formations and manganolutite of the Hotazel Formation.

Relative to the basin-platform configuration of tectono-sedimentary elements during Transvaal times (Fig. 4-4A) it can be stated that the jasper ↔ manganolutite cycles are developed in the basin along the western margin of the Kaapvaal craton. Subaqueous pillow lavas and hyaloclastic breccias are also more or less confined to this area. Further onto the platform into the Transvaal structural basin a facies change takes place in the volcanic sequence because, according to Button (1973a) and Klopp (1978), the Hekpoort Basalt was subaerially extruded. The lava sequence also thickens towards the basin and starts to pinch out in the north on the platform (Fig. 4-4B).

The kutnahoritic hematite lutite which overlies the uppermost manganolutite unit is overlain by hematite pillowlutite. In turn this unit starts to become sideritic and a gradation takes place to laminated and cross-laminated siderite ribbonlutites and bandlutites (Fig. 4-17) displaying soft-sediment slump structures. This siderite facies banded iron-formation forms the top of the Hotazel Formation and is overlain by brecciated and laminated, clastic-textured pinkish to light-grey cherty dolomite of the Mooidraai Formation. The breccias are massive and consist of disorientated angular slabs of chert and various types of dolomite. The clastic-textured dolomites which are interbedded with the breccias, display slump folds and consist of very fine-grained algal mat intraclasts set in sparitic micrite. Higher up in the sequence some graded bedded and massive, fine-grained dolarenite units are interbedded

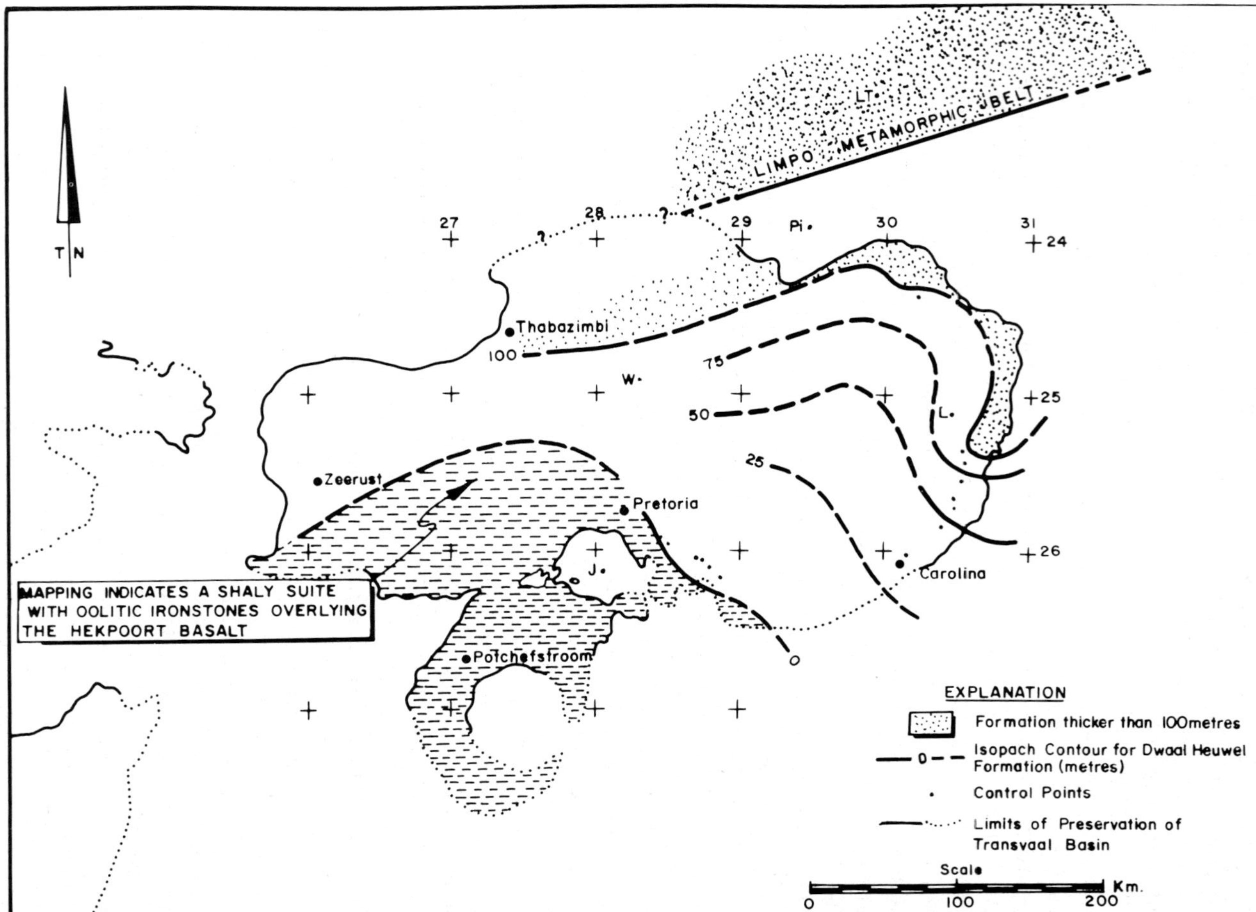

Fig. 4-19. Isopach contour of the Dwaal Heuvel Formation and the distribution of oolitic ironstones relative to it (after Button, 1975).

with the laminated dolomite. They are overlain by thinly bedded dolarenite in which cryptalgal laminations are present (Fig. 4-17).

The slump folds, massive breccias, lamination, graded bedding and clastic texture of the dolomites may indicate that they represent gravity flow deposits in front of a shallow-water carbonate platform similar to the Prieska facies of the Campbellrand—Malmani carbonate sequence. The massive dolomite breccias may represent toe-of-slope deposits with the laminated and graded bedded units as platform slope carbonate turbidites (Fig. 4-17). The upper part of the sequence may have been deposited in shallower water as is indicated by the presence of cryptalgal laminations. The truly shallow platform facies was probably eroded in post-depositional times (Fig. 4-17).

The siderite ribbonlutites and bandlutites may thus very well represent distal carbonate turbidites that became chertified and sideritized. Conditions and processes may have been similar to those that were responsible for the formation of ankerite-banded cherts in the distal basinal facies of the Campbellrand—Malmani carbonate sequence. The overlying laminated carbonates were also partly chertified and ferruginized as is indicated by cross-cutting replacement chert bands in them and the presence of ankerite in the slump breccias. Manganization took place as well because kutnahorite is present in the lower, deeper-water part of the Mooidraai Dolomite (Fig. 4-17).

A carbonate platform sequence thus prograded into the basin after volcanic activity. With the Ongeluk Lava and Hekpoort Basalt as a time-stratigraphic datum line Button (1976a) suggested that the manganese and iron-formation deposits of the Hotazel Formation represent a distal facies of the tide-dominated, shallow-marine and deltaic quartzites of the Dwaal Heuvel Formation in the Transvaal area. However, a period of erosion separates the Hekpoort Basalt from the overlying Dwaal Heuvel Formation (Button, 1973a; Button and Tyler, 1981) so that this distal facies may be represented by the Mooidraai Dolomite with the iron and manganese deposits an integral part of the Ongeluk—Hekpoort volcanic episode. Oolitic ironstones interbedded with shale along the distal parts of the shallow-marine Dwaal Heuvel quartzite (Fig. 4-19) appear to have been deposited under conditions similar to those that controlled the deposition of the Timeball Hill ironstones.

MANGANORE IRON FORMATION

The Manganore Iron Formation is preserved in a number of palaeo-sinkhole structures within the Campbellrand dolomites on the Maremane dome between Postmasburg and Sishen (Fig. 4-20). As has already been mentioned it most probably represents part of the Asbesheuwels iron-formation sequence that slumped into sinkhole structures during the period of erosion that preceded the deposition of the overlying Gamagara Formation. Slumping resulted in folding, brecciation and structural thickening of the unit (Nel, 1929). The Manganore Iron Formation grades down into the Wolhaarkop Breccia which

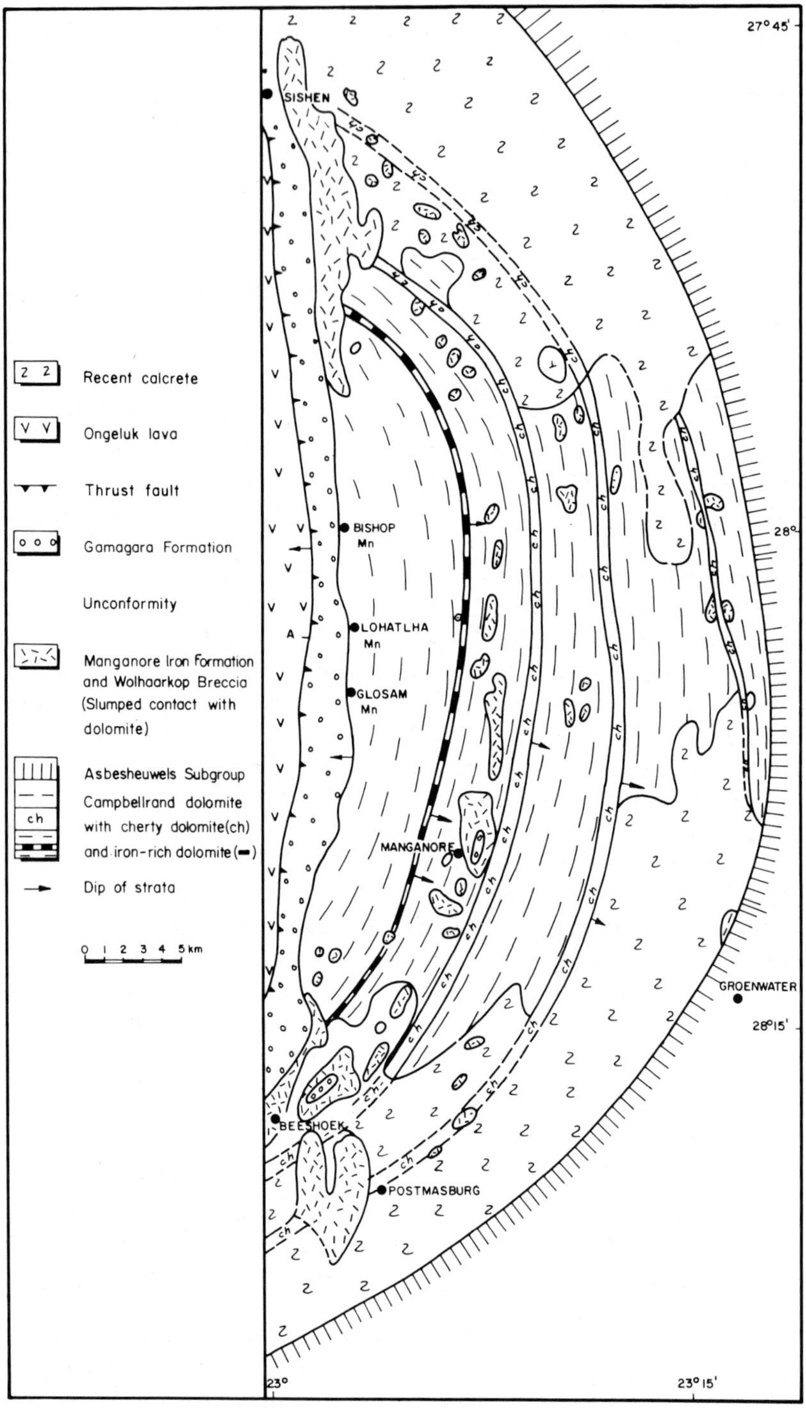

Fig. 4-20. Simplified geological map of the Maremane dome indicating the distribution of the Manganore Iron Formation and Wolhaarkop Breccia.

consist of chert fragments set in a ferruginous and manganiferous matrix of fine-grained silica and drusy quartz (Nel, 1929; Van Wyk, 1980). This breccia is thought to represent accumulations of the insoluble residue of the dolomite in sinkhole structures (Fig. 4-4C). Both the Manganore Iron Formation and Wolhaarkop Breccia are of very variable thickness, ranging from a few meters to several tens of meters thick.

During the period of erosion and slumping most of the Manganore Iron Formation became oxidized and hematitized (Beukes, 1977b) resulting in the high-grade iron ore deposits of Sishen and Postmasburg (Strauss, 1964). Other authors consider these deposits to represent either hematitized shale of the Gamagara Formation (Strauss, 1964) or primary accumulations of ferruginous shale at the base of the Gamagara Formation (Button, 1976a; Tankard et al., 1982). However, I do not agree with these views (Beukes. 1977b, 1978).

Hematitization of the Manganore Iron Formation took place before deposition of the Gamagara Formation because hematite pebbles are present in a conglomerate at the base of the latter formation (Strauss, 1964). The Gamagara Formation can thus be classified as a redbed sequence because epiclastic hematite particles are also present in its shales, siltstones and quartzites.

The correlation of the Gamagara Formation with the Olifantshoek Group through thrust faulting (Fig. 4-4C) implies that only one major unconformity separates these siliciclastic redbed sequences from the Transvaal Supergroup. This unconformity conforms to Cloud's (1976) transition from the Proterophytic to the Paleophytic period, i.e. a transition from a reducing to an oxidizing atmosphere. Data from the Transvaal Supergroup would tend to support such a hypothesis because below this unconformity strong oxidation of pre-existing divalent iron and manganese minerals took place. Sideritic and magnetitic facies of the Asbesheuwels, Manganore and Rooinekke Iron Formations, for example, became hematitized where this unconformity cuts across them (Beukes, 1978; Van Wyk, 1980) whilst at Lohatlha in the core of the Maremane dome, manganese oxides accumulated in sinkhole structures within manganiferous dolomite below the unconformity (Beukes, 1977b; Grobbelaar and Beukes, in press).

VERTICAL DISTRIBUTION OF IRON AND MANGANESE IN THE TRANSVAAL SUPERGROUP

The amount of manganese associated with iron-formation increases upwards in the Transvaal Supergroup. The Asbesheuwels and Penge Iron Formations contain less than 0.5% MnO (Beukes, 1973) and the dominant "primary" iron minerals are siderite, hematite and greenalite. Higher in the sequence manganese-bearing siderite makes its appearance in the Rooinekke Iron Formation of the Koegas Subgroup. Then follow the interbedded jaspers and braunite lutites of the Hotazel Formation.

Iron carbonates and oxides are less soluble under lower pH and Eh conditions than manganese carbonates and oxides so that the upward enrichment in manganese may have been caused by an overall increase with time in the Eh and pH of the basinal water of the Transvaal depository (Fig. 4-21). Siderite and hematite facies iron-formations of the Kuruman Iron Formation have been interpreted as deeper shelf deposits and so have the manganiferous siderite lutites of the Rooinekke Formation and the jaspers and manganolutites of the Hotazel Formation. The upward variation in the valent-state of iron and manganese and the increase in manganese content thus could not be ascribed to different environments of deposition. Rather it is tempting to think of a general increase in the amount of oxygen present in the water of the basin with siderite and hematite being deposited at relatively low oxygen fugacity levels, manganese-bearing siderite at intermediate levels and manganese oxide at a relatively high level (Fig. 4-21).

According to Cloud (1973) the Proterophytic oceans became depleted in dissolved ferrous ions after their oxygenation and therefore iron-formations

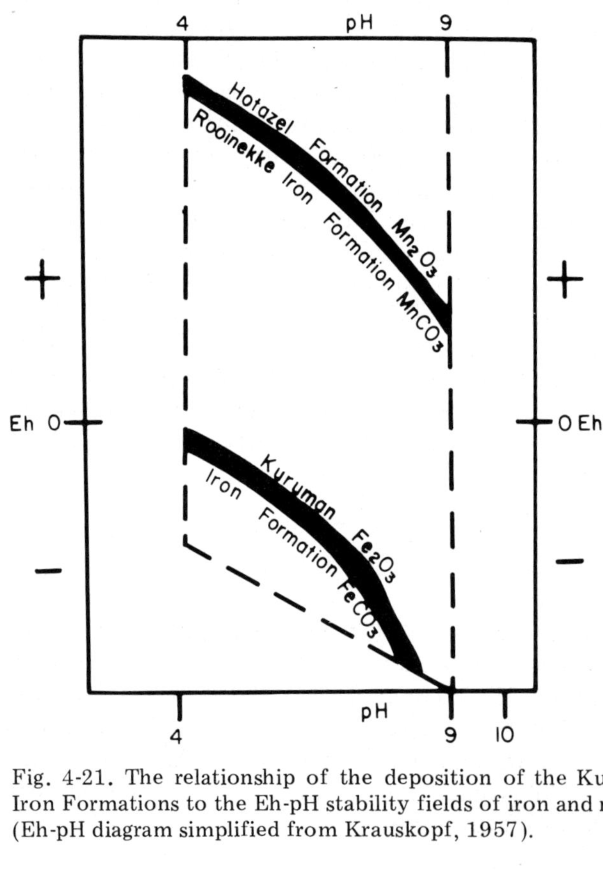

Fig. 4-21. The relationship of the deposition of the Kuruman, Rooinekke and Hotazel Iron Formations to the Eh-pH stability fields of iron and manganese carbonates and oxides (Eh-pH diagram simplified from Krauskopf, 1957).

are much less abundant in the later stratigraphic record of the world. Similarly it can be argued that major precipitation of dissolved divalent manganese ions from Proterophytic oceans may have taken place immediately before the development of the Paleophytic period. The vast volcanogenic manganese deposits of the Hotazel near the top of the Transvaal Supergroup could in part be explained by such a hypothesis. Without suitable Eh and pH conditions the volcanogenically derived manganese may have remained in solution.

The development of a clear-water platform on the Kaapvaal craton during the deposition of the Asbesheuwels—Penge iron-formation sequence may explain the abundance of iron-formations in the Transvaal Supergroup relative to the older Pongola, Witwatersrand and Ventersdorp Supergroups. As in the case of the Pretoria Group the development of iron-formations on the Kaapvaal craton proper may also have been inhibited by the influx of siliciclastics during the deposition of these older platform sequences. Iron-formations may well have been abundant in their individual basinal facies but these have in the meantime either been eroded away or taken up into metamorphic provinces (Beukes, 1978). The overall concentration of iron-formation close to the craton margin in the Griqualand West sequence of the Transvaal Supergroup may be an indication that the upwelling of iron-enriched, cold, deep ocean water onto the shallow-water platform played a part in their deposition.

CONCLUSION

Iron-formation constituted an integral facies of the Transvaal Supergroup throughout its depositional history which lasted from approximately 2460—2500 m.y. ago to about 2100 m.y. ago. The iron-formations represent a variable group of chert-bearing iron-rich rocks about which very few general statements can be made. They are most abundant in the Transvaal sequence along the southwestern margin of the Kaapvaal craton in Griqualand West. Here some of them were deposited as distal facies of either carbonate platform, shallow-marine siliciclastic or volcanic sequences which are, as a rule, better developed on the Kaapvaal craton proper. Such iron-formation units are, however, usually rather thin, ranging from less than a meter thick to a maximum of a few tens of meters thick. They include the Doradale, Kwakwas, Rooinekke and Nelani Iron Formations and iron-formation beds associated with the Campbellrand—Malmani carbonate sequence, the Makganyene Diamictite, Rooihoogte shallow-marine siliciclastics and Hotazel and Beaumont Formations.

Only once did a very thick unit of iron-formation develop on the Kaapvaal craton proper. This unit is between 400 m and 1000 m thick and constitutes the Asbesheuwels—Penge iron-formation sequence. Its deposition was brought about by the development of a clear-water epeiric sea on the Kaapvaal craton at a time when the Limpopo metamorphic complex was a negative tectonic element (basin) in which epicratonic siliciclastics and volcanics were trapped. Under these clear water conditions deposition of iron-formation could go on unhindered and a complete facies range from deep-water shelf deposits to shallow epeiric sea and even supratidal deposits developed. A similar range of depositional environments are present in the Campbellrand—Malmani carbonate sequence which underlies the Asbesheuwels—Penge iron-formation sequence. The change from carbonate to iron-formation deposition thus cannot be attributed to a change in environment of deposition but may be the result of a climatic change from warm to cold water conditions.

The area along the Limpopo metamorphic province normally acted as a positive tectonic element from which siliciclastics were shed into the Transvaal epeiric sea with the result that iron-formation deposition was pushed back to a more distal basinal area along the southwestern margin of the Kaapvaal craton. Along this craton margin the upwelling of deeper, cold, iron-rich ocean water onto a shallow platform could have aided the deposition of the iron-formations in general. However, in detail a number of different genetic types of iron-formations were deposited in a variety of environments. The major groups are *diagenetic iron-formations* representing altered volcanic ash and carbonate turbidite units and *primary iron-formations* of euxinic basin, deep shelf, platform slope and toe-of-slope, epeiric sea (shallow platform), supratidal, and lacustrine environments.

Iron-formation which represents chertified and sideritized volcanic ash units is present in a lagoonal setting in the Lokammona Formation of the Schmidtsdrif Subgroup. The units are less than two meters thick and consist of banded siderite lutite. Iron-formations which represent chertified and ankeritized or sideritized carbonate turbidites are interbedded with pyritic carbonaceous shales and laminated clastic-textured limestones or ferruginous dolomites representing the basinal facies of a shallow-water stromatolitic carbonate platform deposit. They consist of intraclastic ankerite and/or siderite mesobands alternating with microcrystalline or intraclastic chert mesobands. Graded bedded units with sharp basal contacts and horizontal laminations and cross-laminations in intraclastic units are related to Bouma turbidite cycles, whereas microcrystalline chert mesobands represent primary chert beds which were deposited at the sediment—water interface. In this deeper-water turbidite environment, limestone particles were unstable and became replaced by iron carbonates or chert. These iron-formations are represented by the ankerite-banded cherts of the Prieska facies of the Campbellrand—Malmani carbonate sequence and of the basal part of the Kuruman Iron Formation. Clastic-textured laminated and cross-laminated siderite bandlutites at the top of the Hotazel Formation at the base of the Mooidraai dolomite sequence may be of a similar origin.

Microcrystalline siderite, greenalite, hematite dust and chert are the only minerals known that may reflect chemical conditions in the depositional environments of the primary group of iron-formations. All other iron-minerals like specularite, magnetite, riebeckite, stilpnomelane, minnesotaite and grunerite are of a diagenetic and/or metamorphic origin. Care should, however, be taken because all the minerals may have more than one genesis and some siderite and chert are also of a diagenetic origin. The different depositional environments of the primary iron-formations are each characterized by specific sedimentary facies which include mineralogical, textural and structural parameters.

The euxinic basin facies is represented by pyritic carbonaceous shale with some chert bands, and is present at the base of the Kuruman and Penge Iron Formations. The relatively deep-water, open-shelf iron-formations have a distinct volcanogenic component and consist of stacked cycles of altered volcanic ash (stilpnomelane lutite) beds overlain by autochthonous microbanded and mesobanded ferhythmite units. The deep-water environment of deposition of these so called stilpnomelane lutite $\rightarrow$ ferhythmite macrocycles was acidic and favoured the deposition of silica (chert). This deposition was, however, interrupted by the periodic influx of volcanic ash and the seasonal precipitation of iron-mineral microbands, possibly through the action of photosynthesizing micro-organisms. Characteristic features of the deep-shelf iron-formations are well-defined chert meso- and microbanding, extreme lateral continuity of micro-, meso- and macrobands, and rapid transitions over small thickness intervals from siderite facies to hematite facies ferhythmites. This rapid transition is ascribed to biological activity. Some of the silica appear to be associated with the volcanic episodes and may be of a fumarolic origin.

Chert microbands are primary features but chert mesobands formed through early diagenetic silica cementation below the sediment—water interface. Thickness variations in the deep-shelf iron-formations are the result of different compaction ratios and not due to non-deposition of certain units. Sodium was introduced to the system by fumarolic activity or during the devitrification process that led to the formation of stilpnomelane from the volcanic ash bands. Diagenetic riebeckite is thus concentrated in close proximity to stilpnomelane lutite beds. The Groenwater Member of the Kuruman Iron-Formation, the lower part of the Penge Iron Formation, the more basinal facies of the Doradale Iron Formation, and the jaspilite and hematite ferhythmites of the Hotazel and Beaumont Formations belong to the deep-shelf facies. Manganolutites, pillow lavas and hyaloclastites are associated with the hematitic autochthonous iron-formations of the latter two formations.

The platform slope and toe-of-slope facies of iron-formations are represented by greenalite-siderite rhythmites. This rock type probably represents a transition stage between microbanded autochthonous iron-formations and orthochemical felutites because it consists of greenalite-siderite lutite mesobands alternating with siderite microbands. The felutite mesobands represent lutite-flow deposits and may display graded bedding on a microscopic scale. Greenalite was deposited through a reaction of dissolved ferrous ions with silica under neutral to weakly alkaline conditions and that is the reason why well-defined chert mesobands are absent. The platform slope facies differs from the toe-of-slope facies in that massive allochemical grain-flow mesobands with sharp basal and top contacts are present in the former. Iron-formations of this facies type could possibly also be described as laminites and are represented by the Riries Member of the Kuruman Iron Formation and by the middle part of the Penge Iron Formation. In the latter the greenalite has, however, been transformed to grunerite and the siderite to magnetite by metamorphism (Beukes, 1978).

The epeiric sea facies type consists of clastic-textured orthochemical and allochemical iron-formations which, like siliciclastics, display interfingering relationships and thickness changes that can usually be ascribed to non-deposition of units in certain areas. Asymmetrical upward-coarsening orthochemical → allochemical iron-formation megacycles, which may again fine upwards to be capped by orthochemical felutite, is another characteristic feature of this facies type. The three iron-formation units constituting the megacyles were respectively deposited in the subtidal low-energy X-zone, high-energy Y-zone, and platform lagoonal Z-zone of Irwin's (1965) epeiric sea model.

The subtidal low-energy X-zone is represented by felutite in which upward-fining erosively based, storm-wave beds are present. The mineralogical composition of the felutite depends upon chemical conditions in the subtidal shallow-sea area. Alkaline reducing conditions result in the deposition of siderite lutite as in the Danielskuil Member of the Griquatown Iron Formation and the more proximal facies of the Doradale Iron Formation of the Koegas Subgroup. In the Rooinekke Iron Formation somewhat higher Eh and pH conditions led to the deposition of manganiferous siderite lutites in the X-zone. Some hematite femicrite is, however, also present in the Danielskuil Member indicating that under oxidizing conditions the X-zone iron-formations may consist of hematite lutite. In restricted to partly restricted X-zone basins the felutites are riebeckitic and greenalitic like those of the Middelwater Member of the Griquatown Iron Formation, the Kwakwas Iron Formation, and the basal part of the Nelani Iron Formation.

The high-energy Y-zone deposits consist of grainstones and edgewise conglomerates including ooidstone, peloidstone, disclutite and splintlutite. The latter two facies formed through storm-wave action. Depending on oxygen fugacity levels this facies could either be sideritic or hematitic. Siderite facies is well represented in the Nelani Iron Formation, the Ouplaas Member of the Kuruman Iron Formation and the Danielskuil Member of the Griquatown Iron Formation. No truly hematite-rich facies is known from the Transvaal Supergroup but from the Sokoman Iron Formation of Canada (Dimroth and Chauvel, 1973; Chauvel and Dimroth, 1974) it is known that both X- and Y-zone iron-formations are hematitic and jaspery. It should also be remembered that the possible effect of sideritization has not been fully evaluated in the shallow-water iron-formations of the Transvaal Supergroup.

The platform lagoonal Z-zone facies is characterized by greenalite lutite in the Danielskuil Member of the Griquatown Iron Formation. The exact composition of lagoonal felutites is, however, probably also dependant on pH because in the Rooinekke Iron Formation inferred lagoonal sediments consist of siderite lutite. Siliciclastic influx also controls the composition of this facies and where this is relatively high the lagoonal facies may consist of iron-rich shales as in the Rooinekke and Nelani Iron Formations. With major influx of siliciclastics only the distal subtidal X-zone facies of the iron-formations may be developed as in the case of the Doradale and Kwakwas Iron Formations and also the iron-formations interbedded with the Makganyene Diamictite and the offshore shelf iron-formation below the deltaic siliciclastics of the Rooihoogte Formation in the western Transvaal. In the latter association ironstones may be interbedded with delta front and delta plain deposits as in the Timeball Hill and Dwaal Heuvel Formations of the Pretoria Group.

The greenalitic lagoonal iron-formations facies is similar in composition to the inferred lacustrine greenalite lutites at the top of the Griquatown Iron Formation. In both these facies the greenalite probably reflects the inflow of fresh terrestrial water. Lagoonal and lacustrine felutites could thus probably only be distinguished from each other by virtue of the interpretation of facies relationships in sedimentary increments.

Chert mesobands present in the epeiric sea deposits are mostly of a diagenetic origin being characterized by gradational contacts with adjacent felutites, and cement-supported grain fabrics suggesting replacement of some felutite by silica. Some chert mesobands in the greenalitic lagoonal and lacustrine felutites may, however, have been deposited at the sediment—water interface. Such mesobands display dewatering cracks and sharp contacts with adjacent felutite mesobands and were probably deposited under acidic conditions during periods of abnormally high inflow of freshwater.

Iron-rich edgewise conglomerates which are of a possible supratidal origin are present in the Skietfontein Member of the Griquatown Iron Formation. These conglomerates display imbricate structures and radial orientation of vertically standing chert discs. The latter structure may have formed under the influence of storm-wave surge action. Shrinkage cracks which either represent subaerial dessication features or syneresis cracks are present in the chert mesobands that were broken up by wave action to produce the chert discs of the edgewise conglomerates. Care should, however, be taken in the interpretation of such conglomerates because some of them may have formed in high-energy Y-zones of epeiric seas, others may have formed by storm wave action in either subtidal X-zone or lagoonal Z-zone environments and still others may represent lag deposits on transgression surfaces or be of a slump origin. Massive chert breccias in the Kamden iron formation member of the Campbellrand Subgroup are thought to be of the latter origin.

There is an upward increase in the amount of manganese associated with the Transvaal iron-formations. This is ascribed to a gradual increase in the oxygen fugacity levels of the water of the depository through time. Pyroclastic material is associated with most of the iron-formation sequences and volcanism may have contributed to the deposition of the iron-formations. Contemporaneous volcanic activity was, however, not an absolute necessity for iron-formation deposition as is illustrated by the negative correlation between mafic volcanic ash bands and iron-formations in the Prieska facies of the Campbellrand—Malmani carbonate sequence. The primitive Proterophytic oceans probably served as a source for most of the iron and silica present in the Transvaal iron-formations. Apart from volcanogenic addition a small amount of the iron may also have been introduced from the weathering of nearby siliciclastic source areas. The latter type of iron is, however, deposited relatively close to shore as in the lacustrine and lagoonal greenalitic/chloritic lutites of the Griquatown, Rooinekke and Nelani Iron Formations.

ACKNOWLEDGEMENTS

Investigations on the iron-formations and manganese deposits of the Transvaal Supergroup are funded by grants from the South African Council for Scientific and Industrial Research and the Jim and Gladys Taylor Educational Trust. I would also like to thank Eddie Venter for drafting the figures and Charté Niemand for typing the manuscript.

REFERENCES

Anderson, C., 1915. Description of a carbonaceous mineral occuring in the Witkop Mine near Zeerust, Transvaal. Trans. Geol. Soc. S. Afr., 18: 129—131.

Anhaeusser, C.R., Mason, R., Viljoen, M.J. and Viljoen, R.P., 1969. A reappraisal of some aspects of Precambrian shield geology. Geol. Soc. Am. Bull., 80: 2175—2200.

Anhaeusser, C.R. and Button, A., 1976. A review of southern African stratiform ore deposits — Their position in time and space. In: K.H. Wolf (Editor), Handbook of Strata-bound and Stratiform Ore Deposits. Elsevier, Amsterdam, pp. 257—319.

Berner, R.A., 1971. Principles of Chemical Sedimentology. McGraw-Hill, New York, N.Y., 240 pp.

Beukes, N.J., 1973. Precambrian iron-formations of southern Africa. Econ. Geol., 68: 960—1004.

Beukes, N.J., 1977a. Transition from siliciclastic to carbonate sedimentation near the base of the Transvaal Supergroup, northern Cape Province, South Africa. Sediment. Geol., 18: 201—221.

Beukes, N.J., 1977b. Stratigraphy and structure of the Transvaal dolomites in the northern Cape Province — control on lead mineralization in the Griquatown area and iron-manganese mineralization in the Sishen-Postmasburg area. Abstracts Geokongres 77, Geol. Soc. S. Afr., pp. 14—19.

Beukes, N.J., 1978. Die Karbonaatgesteentes en Ysterformasies van die Ghaap-Groep van die Transvaal-Supergroep in Noord-Kaapland. Ph.D. Thesis, Rand Afrikaans University, Johannesburg, 580 pp. (unpubl.).

Beukes, N.J., 1979. Litostratigrafiese onderverdeling van die Schmidtsdrif-Subgroep van die Ghaap-Groep in Noord-Kaapland. Trans. Geol. Soc. S. Afr., 82: 313—327.

Beukes, N.J., 1980a. Lithofacies and stratigraphy of the Kuruman and Griquatown Iron-formations, northern Cape Province, South Africa. Trans. Geol. Soc. S. Afr., 83: 69—86.

Beukes, N.J., 1980b. Stratigrafie en litofasies van die Campbellrand-Subgroep van die Proterofitiese Ghaap-Groep, Noord-Kaapland. Trans. Geol. Soc. S. Afr., 83: 141—170.

Beukes, N.J., 1980c. Suggestions towards a classification of and nomenclature for iron-formation. Trans. Geol. Soc. S. Afr., 83: 285—290.

Boardman, L.G., 1940. The geology of the manganese deposits on Aucampsrust, Potsmasburg. Trans. Geol. Soc. S. Afr., 43: 27—36.

Boardman, L.G., 1941. The Black Rock manganese deposit in the south-eastern Kalahari. Trans. Geol. Soc. S. Afr., 44: 51—60.

Boardman, L.G., 1946. The geology of a portion of the Rooiberg tinfields. Trans. Geol. Soc. S. Afr., 49: 103—132.

Boardman, L.G., 1961. Manganese in the Union of South Africa. Papers 7th Commonw. Min. Metall. Congr., Vol. 1, pp. 201—216.

Boardman, L.G., 1964. Further geological data on the Postmasburg and Kuruman manganese ore deposits, northern Cape Province. In: S.H. Haughton (Editor), The Geology of some Ore Deposits in Southern Africa, Vol. II. Geol. Soc. S. Afr., Johannesburg, pp. 415—440.

Bonatti, E., Kraemer, T. and Rydell, H.S., 1972. Classification and genesis of submarine iron-manganese deposits. In: D.R. Horn (Editor), Ferromanganese Deposits on the Ocean Floor. National Science Foundation, Washington, D.C., pp. 159—166.

Boocock, C., 1961. Notes on the distribution of rocks of Transvaal age in the Bechuanaland Protectorate. Spec. Publ. Geol. Comm. Tech. Cooperation in Africa south of the Sahara, 80: 54—62.

Boocock, C., 1965. Mineral resources of the Bechuanaland Protectorate. Overseas Geol. Miner. Resour., 9: 369.

Botha, B.J.V., Grobler, N.J., Lindström, W. and Smit, C.A., 1976. Stratigraphic correlation between the Kheis and Matsap formations and their relation to the Namaqua metamorphic complex. Trans. Geol. Soc. S. Afr., 79: 304—311.

Bousquet, J.G., 1896. De tegenwoordige toestand van de goudmijnindustrie in de Kaapformatie van de Lijdenburg goudvelden. Jaarrapport van den Staatsmijningenieur over het jaar 1896. Aanhangsel D., Government Printer, Pretoria.

Brandt, J.W. and Le Roux, H.D., 1944. The basal beds of the Transvaal System at Olifants River Poort, N.E. Transvaal. Trans. Geol. Soc. S. Afr., 47: 79—91.

Burchell, W.J., 1822. Travels in South Africa. Vol. I. 359 pp.

Button, A., 1968. Subsurface Stratigraphic Analysis of the Witwatersrand and Transvaal Sequences in the Irene-Delmas-Devon Area, Transvaal. M.Sc. Thesis, Univ. Witwatersrand, Johannesburg, 120 pp. (unpubl.).

Button, A., 1973a. A Regional Study of the Stratigraphy and Development of the Transvaal Basin in the Eastern and Northeastern Transvaal. Ph.D. Thesis, Univ. Witwatersrand, Johannesburg, 351 pp. (unpubl.).

Button, A., 1973b. The depositional history of the Wolkberg proto-basin, Transvaal. Trans. Geol. Soc. S. Afr., 76: 15—25.

Button, A., 1973c. The stratigraphic history of the Malmani Dolomite in the eastern and north-eastern Transvaal. Trans. Geol. Soc. S. Afr., 76: 229—247.

Button, A., 1973d. Algal stromatolites of the early Proterozoic Wolkberg Group, Transvaal sequence. J. Sediment. Petrol., 43: 160—167.

Button, A., 1974. Low-potash pillow basalts in the Pretoria Group, Transvaal Supergroup. Trans. Geol. Soc. S. Afr., 77: 99—104.

Button, A., 1975. A palaeocurrent study of the Dwaal Heuvel Formation, Transvaal Supergroup. Trans. Geol. Soc. S. Afr., 78: 173—183.

Button, A., 1976a. Transvaal and Hamersley basins — review of basin development and mineral deposits. Miner. Sci. Eng., 8: 262—293.

Button, A., 1976b. Iron-formation as an end member in carbonate sedimentary cycles in the Transvaal Supergroup, South Africa. Econ. Geol., 71: 193—201.

Button, A., 1976c. Stratigraphy and relations of the Bushveld floor in the eastern Transvaal. Trans. Geol. Soc. S. Afr., 79: 3—12.

Button, A. and Tyler, N., 1981. The character and economic significance of Precambrian paleoweathering and erosion surfaces in southern Africa. Econ. Geol., 75th Anniv. Vol.: 676—699.

Button, A. and Vos, R.G., 1977. Subtidal and intertidal clastic and carbonate sedimentation in a macrotidal environment: an example from the lower Proterozoic of South Africa. Sediment. Geol., 18: 175—200.

Chauvel, J.-J., and Dimroth, E., 1974. Facies types and depositional environment of the Sokoman Iron-formation, Central Labrador trough, Quebec, Canada. J. Sediment Petrol., 44: 299—327.

Cilliers, J.J. le R., 1961. The Nature and Origin of the Rocks of the Lower Griquatown Stage and the Associated Deposits of Amphibole Asbestos in the Northern Cape, with special reference to the Koegas-Prieska Area. D.Sc. Thesis, Univ. Pretoria, South Africa, 168 pp. (unpubl.).

Cilliers, J.J. le R., 1964. Amosite at the Penge Asbestos Mine. In: S.H. Haughton (Editor), The Geology of some Ore Deposits of Southern Africa, Vol. II. Geol. Soc. S. Afr., Johannesburg, pp. 579—591.

Cilliers, J.J. le R. and Genis, J.H., 1964. Crocidolite asbestos in the Cape Province. In: S.H. Haughton (Editor) The Geology of some Ore Deposits of Southern Africa, Vol. II. Geol. Soc. S. Afr., Johannesburg, pp. 544—570.

Cloud, P., 1973. Paleoecological significance of the banded iron-formation. Econ. Geol., 68: 1135—1143.

Cloud, P., 1976. Major features of crustal evolution. Trans. Geol. Soc. S. Afr., Annexure to vol. 79, 33 pp.

Coertze, F.J., 1961. The Transvaal System in Transvaal. Spec. Publ. Geol. Comm. Tech. Cooperation in Africa south of the Sahara, 80: 54—62.

Coertze, F.J., Burger, A.J., Walraven, F., Marlow, A.G. and Maccaskie, D.R., 1978. Field relations and age determinations in the Bushveld Complex. Trans. Geol. Soc. S. Afr., 81: 1—11.

Coetzee, C.B. (Editor), 1976. Mineral Resources of the Republic of South Africa. Handb. Geol. Surv. S. Afr., 7: 462 pp.

Cohen, E., 1875. Erläuternde Bemerkungen zu der Routenkarte einer Reise von Lijdenburg nach den Goldfeldern und von Lijdenburg nach der Delagoa Bai im östlichen Süd-Afrika. Jahresber. Geograph. Ges. Hamburg., II.

Cousins, C.A., 1962. The stratigraphy, structure and igneous rocks of the Transvaal System at the Western Areas Gold Mine. Trans. Geol. Soc. S. Afr., 65: 13—40.

Crampton, D., 1974. A note on the age of the Matsap Formation in the northern Cape Province. Trans. Geol. Soc. S. Afr., 77: 71—72.

Crocker, I.T., 1979. Fluorite mineralisation in the dolomite of the Transvaal Supergroup. In: A.M. Anderson and W.J. Van Biljon (Editors), Some Sedimentary Basins and Associated Ore Deposits of South Africa. Spec. Publ. Geol. Soc. S. Afr., 6: 73—82.

Crockett, R.N., 1972. The Transvaal System in Botswana: Its geotectonic and depositional environment and special problems. Trans. Geol. Soc. S. Afr., 75: 275—289.

Cullen, D.J., 1963. Tectonic implications of banded ironstone formations. J. Sediment. Petrol., 33: 387—392.

Davies, R.D., Allsopp, H.L., Erlank, A.J. and Manton, W.I., 1969. Sr-isotopic studies on various layered mafic intrusions in southern Africa. Geol. Soc. S. Afr., Spec. Publ., 1: pp. 576—593.

De Villiers, J.E., 1944. The origin of the iron and manganese deposits in the Postmasburg and Thabazimbi areas. Trans. Geol. Soc. S. Afr., 47: 123—135.

De Villiers, J.E., 1945. Some minerals occurring in South African manganese deposits. Trans. Geol. Soc. S. Afr., 48: 17—25.

De Villiers, J.E., 1956. Manganese, Union of South Africa. Papers Int. Geol. Congr. Mexico., Vol. II, Africa, pp. 34—72.

De Villiers, J.E., 1960. The Manganese Deposits of the Union of South Africa. Handb. Geol. Surv. S. Afr., 2, 280 pp.

De Villiers, P.R., 1967. New stratigraphic correlation and interpretation of the geological structure of the Postmasburg-Sishen area. Ann. Geol. Surv. S. Afr., 6: 39—42.

De Villiers, P.R., 1970. The geology and mineralogy of the Kalahari manganese field north of Sishen, Cape Province, Mem. Geol. Surv. S. Afr., 59: 84 pp.

De Villiers, P.R. and Visser, J.N.J., 1977. The glacial beds of the Griqualand West Supergroup as revealed by four deep boreholes between Postmasburg and Sishen. Trans. Geol. Soc. S. Afr., 80: 1—8.

De Villiers, S.B., 1961. The Transvaal System in Griqualand West. Spec. Publ. Geol. Comm. Tech. Cooperation in Africa south of the Sahara, 80: 63—72.

De Waal, S.A. and Herzberg, W., 1969. Uranium and gold mineralisation of the Black Reef Series in the Kaapsehoop area, Nelspruit district, Eastern Transvaal. Ann. Geol. Surv., S. Afr., 7: 111—119.

Dimroth, E. and Chauvel, J.-J., 1973. Petrography of the Sokoman Iron Formation in part of the Central Labrador trough, Quebec, Canada. Geol. Soc. Am. Bull., 84: 111—134.

Dreyer, C.B., 1974. The Geology of the Bewaarkloof Area Pietersburg Asbestos Field. M.Sc. Thesis, Univ. Witwatersrand, Johannesburg, 87 pp. (unpubl.).

Du Preez, J.W., 1944. The structural geology of the area east of Thabazimbi and the genesis of the associated iron ores. Ann. Univ. Stellenbosch, 12, Sect. A: 263—360.

Du Toit, A.L. 1933. The manganese deposits of Postmasburg, South Africa. Econ. Geol., 28: 95—122.

Du Toit, A.L., 1939. The Geology of South Africa. Oliver and Boyd, London, 539 p.

Du Toit, A.L., 1945. The origin of amphibole asbestos deposits of South Africa. Trans. Geol. Soc. S. Afr., 58: 161—206.

Engelbrecht, L.N.J., 1962. Markers in the Lower Griquatown Stage near Kuruman. Ann. Geol. Surv. S. Afr., 1: 71—75.

Eriksson, K.A., 1971. A Basin Analysis of the Transvaal Sequence in the Potchefstroom Synclinorium, Transvaal and Orange Free State. M.Sc. Thesis, Univ. Witwatersrand, Johannesburg, 79 pp. (unpubl.).

Eriksson, K.A., 1972. Cyclic sedimentation in the Malmani Dolomite, Potchefstroom synclinorium. Trans. Geol. Soc. S. Afr., 75: 85—95.

Eriksson, K.A., 1973. The Timeball Hill Formation — A fossil delta. J. Sediment. Petrol., 43: 1046—1053.

Eriksson, K.A., 1977. Tidal flat and subtidal sedimentation in the 2250 m.y. Malmani Dolomite, Transvaal, South Africa. Sediment. Geol. 18: 223—244.

Eriksson, K.A. and Truswell, J.F., 1973. High inheritance elongate stromatolitic mounds from the Transvaal dolomite. Palaeontol. Afr., 15: 23—28.

Eriksson, K.A. and Truswell, J.F., 1974. Stratotypes from the Malmani Subgroup northwest of Johannesburg, South Africa. Trans. Geol. Soc. S. Afr., 77: 211—222.

Eriksson, K.A., McCarthy, T.S. and Truswell, J.F., 1975. Limestone formation and dolomitization in a Lower Proterozoic succession from South Africa. J. Sediment. Petrol., 45: 604—614.

Eriksson, K.A., Truswell, J. F. and Button, A., 1976. Palaeoenvironmental and geochemical models from an Early Proterozoic carbonate succession in South Africa. In: M.R. Walter (Editor), Stromatolites. Elsevier, Amsterdam, pp. 635—643.

Eugster, H.P. and I-Ming Chou, 1973. The depositional environments of Precambrian banded iron-formations. Econ. Geol., 68: 1144—1169.

Fockema, P.D., 1967. Crocidolite and Associated Rocks of the Kuruman Area in the Northern Cape. Ph.D. Thesis, Univ. Witwatersrand, Johannesburg, 189 pp. (unpubl.).

Francke, M., 1897. Rapport over de Malmanie-goudvelden. Jaarrapport van den Staatsmyningenieur over het jaar 1897, Aanhangsel B, Government Printer, Pretoria.

Frankel, J.J., 1940. Notes on some of the minerals in the Black Reef Series. Trans. Geol. Soc. S. Afr., 43: 1—8.

Frankel, J.J., 1958. Manganese ores from the Kuruman district, Cape Province, South Africa. Econ Geol., 53: 577—597.

Genis, J.H., 1961. The Genesis of the Blue Amphibole Asbestos of the Union of South Africa. Ph.D. Thesis, Univ. Cape Town, 146 pp. (unpubl.).

Genis, J.H., 1964. The formation of crocidolite asbestos. In: S.H. Haughton (Editor), The Geology of some Ore Deposits of Southern Africa, Vol. II. Geol. Soc. S. Afr., Johannesburg, pp. 571—578.

Gibson, W., 1892. The geology of the gold-bearing and associated rocks of the southern Transvaal. J. Geol. Soc. London, 58: 404—435.

Grobbelaar, W.S. and Beukes, N.J., in press. The Bishop and Glosam Manganese Mines and the Beeshoek Iron Ore Mine of the Postmasburg area. New Ore Deposits Volume, Geol. Soc. S. Afr., Johannesburg.

Grobler, N.J. and Botha, B.J.V., 1976. Pillow-lavas and hyaloclastite in the Ongeluk Andesite Formation in a road-cutting west of Griquatown, South Africa. Trans. Geol. Soc. S. Afr., 79: 53—57.

Hall, A.L., 1905. The geology of Pretoria and Neighbourhood: An explanation of the geological map of the environs of Pretoria. Mem. Geol. Surv. Transvaal, 1, 55 pp.

Hall, A.L., 1908a. On contact metamorphism in the Pretoria Series of the Lydenburg and Zoutpansberg districts. Trans. Geol. Soc. S. Afr., 11: 1—24.

Hall, A.L., 1908b. Note on contemporaneous igneous rocks in the Pretoria Series of the Lydenburg and Rustenburg districts. Trans. Geol. Soc. S. Afr., 11: 47—54.

Hall, A.L., 1909. On contact metamorphism in the Western Transvaal, Trans. Geol. Soc. S. Afr., 12: 119—138.

Hall, A.L. 1910. The geology of the Pilgrims Rest Gold Mining District. Mem. Geol. Surv. Transvaal, 5, 158 pp.

Hall, A.L., 1911. The geology of Sekukuniland. An explanation of sheet 8 (Sekukuniland). Geol. Surv. S. Afr., 42 pp.

Hall, A.L., 1912. The geology of the Murchison Range and district. Mem. Geol. Surv. S. Afr., 6, 186 pp.

Hall, A.L., 1913. The geology of the country south-west of Lydenburg. An explanation of Sheet 11 (Lydenburg). Geol. Surv. S. Afr., 38 pp.

Hall, A.L., 1914. The geology of the Haenertsburg gold fields and surrounding country. An explanation of Sheet 13 (Olifants River). Geol. Surv. S. Afr., 62 pp.

Hall, A.L., 1918. The geology of the country round Belfast. An explanation of Sheet 16 (Belfast). Geol. Surv. S. Afr., 56 pp.

Hall, A.L., 1926. The manganese deposits near Postmasburg West of Kimberley. Trans. Geol. Soc. S. Afr., 29: 17—46.

Hall, A.L., 1930. Asbestos in the Union of South Africa. Geol. Surv. S. Afr., 5, 324 pp.

Hall, A.L. and Humphrey, W.A., 1910. The geology of the country round Zeerust and Mafeking including the Malmani Goldfields and the lead and zinc deposits south of Zeerust. An explanation of Sheet 5 (Zeerust) and Sheet 6 (Mafeking). Geol. Surv. Transvaal, 47 pp.

Hamilton, Jo., 1977. Sr isotope and trace element studies of the Great Dyke and Bushveld mafic phase, and their relation to early Proterozoic magma genesis in southern Africa. J. Petrol., 18: 24—52.

Hammerbeck, E.C.I., 1970. On the genesis of lead-zinc and fluorspar deposits in the southwestern Marico district. Transvaal. Ann. Geol. Surv. S. Afr., 8: 103—110.

Hanekom, H.J., 1966. The Crocidolite Deposits of the Northern Cape Province. D.Sc. Thesis, Univ. Pretoria, South Africa, 213 pp. (unpubl.).

Harding, R.R., Crockett, R.N. and Snelling, N.J., 1974. The Gaborone Granite, Kanye Volcanics and Ventersdorp Plantation Porphyry, Botswana: geochronology and review. Inst. Geol. Sci., Rep. No. 74/5, 25 pp.

Harger, S.H., 1934. An early geological map by Carl Mauch. Trans. Geol. Soc. S. Afr., 37: 1—4.

Harington, J.S., 1962. Occurrence of oils containing 3 : 4-benzpyrene and related substances in asbestos. Nature, 193: 43—45.

Harington, J.S. and Cilliers, J.J. Le R., 1963. A possible origin of the primitive oils and amino acids isolated from amphibole asbestos and banded ironstone. Geochim. Cosmochim. Acta, 27: 411—418.

Hatch, F.H., 1904. The geology of the Marico District, Transvaal. Trans. Geol. Soc. S. Afr., 7: 1—6.

Hatch, F.H. and Corstorphine, G.W., 1905. The Geology of South Africa. MacMillan, London, 389 pp.

Hattingh, E., Snyman, C.P. and Förtsch, E.B., 1979. Enkele aspekte van die kalksilikaatgesteente noordoos van Pretoria. Trans. Geol. Soc. S. Afr., 82: 231—242.

Haughton, S.H., 1969. Geological History of Southern Africa. Geol. Soc. S. Afr., Johannesburg, 535 pp.

Hiemstra, S.A. and Van Biljon, W.J., 1959. The geology of the Upper Magaliesberg Stage and the Lower Bushveld Complex in the vicinity of Steelpoort. Trans. Geol. Soc. S. Afr., 62: 239—255.

Holmes, G.G., 1904a. The geology of a part of Bechuanaland west of Vryburg. Trans. Geol. Soc. S. Afr., 7: 130—132.

Holmes, G.G., 1904b. The Pretoria Series in the Marico District. Trans. Geol. Soc. S. Afr., 7: 167—174.

Holmes, G.G., 1905. The geology of a part of the Rustenburg District. Trans. Geol. Soc. S. Afr., 8: 1—6.

Horwood, C.B., 1910. Notes and analyses of typical Transvaal rocks. Trans. Geol. Soc. S. Afr., 13: 29—53.

Humphrey, W.A., 1908. The geology of Rustenburg and the surrounding district. An explanation of Sheet 4 (Rustenburg). Geol. Surv. Transvaal, 22 pp.

Humphrey, W.A. and Kynaston, H., 1911. The geology of the country lying northwards from Zeerust. An Explanation of Sheet 9 (Marico). Geol. Surv. S. Afr., 30 pp.

Humphrey, W.A. and Kynaston, H., 1914. The geology of the Pilandsberg and the surrounding country. An explanation of Sheet 12 (Pilandsberg). Geol. Surv. S. Afr., 32 pp.

Irwin, M.L., 1965. General theory of epeiric clear water sedimentation. Bull. Am. Assoc. Pet. Geol., 49: 445—459.

Jansen, H., 1975. Precambrian basins on the Transvaal craton and their sedimentological and structural features. Trans. Geol. Soc. S. Afr., 78: 25—34.

Jenkyns, H.C., 1974. Origin of red nodular limestones (Ammonitico Rosso, Knollenkalke) in the Mediterranean Jurassic: a diagenetic model. In: K.J. Hsü en H.C. Jenkyns (Editors), Pelagic Sediments on Land and under the Sea. Spec. Publ. Int. Assoc. Sedimentologists, 1, pp. 249—272.

Jennings, M., in press. The Middelplaats manganese ore deposit, northern Cape Province. New Ore Deposits Volume, Geol. Soc. S. Afr., Johannesburg.

Kent, L.E., Russell, H.D. and Van Rooyen, D.P., 1943. Fluorspar in the Union of South Africa and South West Africa. Bull. Geol. Surv. S. Afr., 14, 69 pp.

Key, R.M., 1976. Radiometric age determinations on rocks from Botswana up to the end of 1975. Bull. Geol. Surv. Botswana. 6., 16 pp.

Kirkman, H.L., 1930. Some notes on crocidolite and amosite occurences in the Union. Trans. Geol. Soc. S. Afr., 33: 13—18.

Klemm, D.D., 1979. A biogenetic model of the formation of the banded iron formation in the Transvaal Supergroup, South Africa. Miner. Deposita, 14: 381—385.

Klopp, A.A.C., 1978. The Metamorphosed Sediments of the Pretoria Group and the Associated Rocks Northwest of Zeerust, Western Transvaal. M.Sc. Thesis, Univ. Pretoria, South Africa, 87 pp. (unpubl.).

Krauskopf, K.B., 1957. Separation of manganese from iron in sedimentary processes. Geochim. Cosmochim Acta, 12: 61—84.

Krumbein, W.C. and Garrels, R.M., 1952. Origin and classification of chemical sediments in terms of pH and oxidation-reduction potentials. J. Geol., 60: 1—33.

Kupferberger, W., 1927. The fluorspar, lead and zinc deposits of the Western Transvaal. Trans. Geol. Soc. S. Afr., 30: 5—56.

Kupferburger, W., Boardman, L.G. and Bosch, L.G., 1956. New considerations concerning the manganese ore deposits in the Postmasburg and Kuruman areas, northern Cape Province, Union of South Africa. Papers Int. Geol. Congr. Mexico, Vol. II, Africa, pp. 73—87.

Kynaston, H., 1907. The geology of the country surrounding Pretoria. An explanation of Sheet no. 1. Geol. Surv., Transvaal, 38 pp.

Kynaston, H., 1929. The geology of the country surrounding Pretoria. An explanation of Sheet no. 1 (new series) revised by L.J. Krige and B.V. Lombaard. Geol. Surv. S. Afr., 48 pp.

Kynaston, H. and Humphrey, W.A., 1920. The geology of the northern portions of the districts of Marico and Rustenburg. An explanation of Sheet no. 14 (Witfontein) and Sheet no. 15 (Crocodile Pools). Geol. Surv. S. Afr., 38 pp.

Kynaston, H. and Mellor, E.T., 1912. The geology of the country round Warmbaths and Nylstroom including the Rooiberg tinfields. An explanation of Sheet no. 10 (Nylstroom). Geol. Surv. S. Afr., 50 pp.

Kynaston, H., Mellor, E.T. and Hall, A.L., 1911. The geology of the country round Potgietersrust. An explanation of Sheet 7 (Potgietersrust). Geol. Surv. S. Afr., 64 pp.

La Berge, G.L., 1966. Pyroclastic rocks in South African iron-formations. Econ. Geol., 61: 572—581.

Lichtenstein, H., 1812. Reisen im Südlichen Afrika in den Jahren 1803, 1804, 1805 und 1806. 2 volumes. Salfeld, Germany.

Litherland, M. and Malan, S.P., 1973. Manganiferous stromatolites from the Precambrian of Botswana. J. Geol. Soc. London, 129: 543—544.

Lombaard, B.V., 1931. The geology of the north-eastern Pretoria district and adjoining country. An explanation of Sheet no. 18 (Moos River). Geol. Surv. S. Afr., 48 pp.

MacGregor, I.M., Truswell, J.F. and Eriksson, K.A., 1974. Filamentous algae from 2300 m.y. old Transvaal dolomite. Nature, 247: 538—540.

Malherbe, S.J., 1970. Flat-pebble conglomerates in the Dolomite Series in the northern Cape Province. Ann. Geol. Surv. S. Afr., 8: 89—94.

Martini, J.E.J., 1976. The fluorite deposits in the Dolomite Series of the Marico district, Transvaal, South Africa., Econ. Geol., 71: 625—635.

Martini, J.E.J., 1979. A copper-bearing bed in the Pretoria Group in the northern Transvaal. In: A.M. Anderson and W.J. Van Biljon (Editors), Some Sedimentary Basins and Associated Ore Deposits of South Africa. Geol. Soc. S. Afr., Spec. Publ., 6: 65—72.

Mauch, C., 1870. Karl Mauch's Reisen im Inneren von Süd Afrika. Peterman's Geographische Mitteilungen, 16: 1—8, 92—103, 139—142.

McLaren, C.H., 1974. Die Geologie van 'n Gebied Suid van Plooysburg, Noord-Kaapland. M.Sc. Thesis, Univ. Orange Free State, Bloemfontein, 257 pp. (unpubl.).

Mellor, E.T., 1921. The geology of the country surrounding Johannesburg. An explanation of Sheet 12 (Johannesburg). Geol. Surv. S. Afr., 45 pp.

Minnitt, R.C.A., Button, A. and Kable, E.J.D., 1973. The gold content of Pre-Malmani argillaceous sediments in the Transvaal Supergroup, northeastern Transvaal. Inf. Circ., Econ. Geol. Res. Unit., Univ. Witwatersrand, Johannesburg, 82, 11 pp.

Moffat, R., 1858. Report of a survey of a portion of the Orange River eastward of Little Namaqualand. Parliamentary Report G1 — 1858, Cape Town.

Molengraaff, G.A.F., 1898, Report of the state geologist of the South African Republic for the year 1897. Trans. Geol. Soc. S. Afr., 4: 119—147.

Molengraaf, G.A.F., 1904. Geology of the Transvaal. Constable, Edinburgh, 90 pp.

Nel, T., 1929. The geology of the Postmasburg manganese deposits and surrounding country. An explanation of the geological map. Geol. Surv. S. Afr., Spec. Publ., 7, 109 pp.

Nel, L.T., 1935. The geology of the Klerksdorp-Ventersdorp area. An explanation of the geological map. Geol. Surv. S. Afr., 168 pp.

Nel, L.T. and Jansen, H., 1957. The geology of the country around Vereeniging. An explanation of Sheet 62 (Vereeniging). Geol. Surv. S. Afr., 90 pp.

Nel, L.T., Frommurze, H.F., Willemse, J. and Haughton, S.H., 1935. The geology of Ventersdorp and adjoining country. An explanation of Sheet 53 (Ventersdorp). Geol. Surv. S. Afr., 102 pp.

Nel, L.T., Truter, F.C. and Willemse, J., 1939. The geology of the country around Potchefstroom and Klerksdorp. An explanation of Sheet no. 61 (Potchefstroom) Geol. Surv. S. Afr., 156 pp.

Ortlepp, R.J., 1964. Nsutite (battery grade manganese dioxide) from the Western Transvaal. Trans. Geol. Soc. S. Afr., 67: 149—160.

Papenfus, J.A., 1964. The Black Reef Series within the Witwatersrand basin with special reference to its occurrence at Govenment Gold Mining Areas. In: S.H. Haughton (Editor), The Geology of some Ore Deposits of Southern Africa, Vol. II. Geol. Soc. S. Afr., Johannesburg, pp. 191—218.

Peacock, M.A., 1928. The nature and origin of the amphibole asbestos of South Africa. Am. Mineral., 13: 241—285.

Penning, W.H., 1885. A sketch of the gold-fields of Lydenburg and De Kaap in the Transvaal. Q. J. Geol. Soc. London, 41: 569—590.

Penning, W.H., 1891. A contribution to the geology of the southern Transvaal. Q. J. Geol. Soc. London, 47: 451—461.

Perry, E.C., Tan, F.C. and Morey, G.B., 1973. Geology and stable isotope geochemistry of the Biwabik Iron Formation, northern Minnesota. Econ. Geol., 68: 1110—1125.

Potgieter, C. de W., 1973. Die Geologie van 'n Gebied Suid van Douglas, Noord Kaapland. M.Sc. Thesis, Univ. Orange Free State, Bloemfontein (unpubl.).

Potgieter, G.J.A. and Visser, J.N.J., 1978. Character of the basal beds of the Griqualand West sequence, west of Prieska. Trans. Geol. Soc. S. Afr., 81: 369—371.

Reineck, H.E. and Singh, I.B., 1972. Genesis of laminated sand and graded rhythmites in storm-sand layers of shelf mud. Sedimentology, 18: 123—128.

Reinecke, L. and McClure, L., 1934. Variations in the quality of amosite asbestos at Penge, Transvaal. Trans. Geol. Soc. S. Afr., 36: 29—39.

Reinecke, L. and Stein, W.G.A., 1929. Ore bodies of the Pilgrims Rest goldfield. Trans. Geol. Soc. S. Afr., 32: 63—86.

Rhodes, R.C., 1971. Palaeocurrents in the Pretoria Group north of Marble Hall, Transvaal. Ann. Geol. Surv. S. Afr., 9: 119.

Rogers, A.W., 1906a. Geological survey of parts of Hay and Prieska, with some notes on Herbert and Barkly West. Annu. Rep. 1905, Geol. Comn. Cape of Good Hope, pp. 141—204.

Rogers, A.W., 1960b. The Campbell Rand and Griquatown Series in Hay. Trans. Geol. Soc. S. Afr., 9: 1—9.
Rogers, A.W., 1906c. The glacial beds in the Griquatown Series. Rep. S. Afr. Assoc. Adv. Science, pp. 261—265.
Rogers, A.W., 1907. Geological survey of parts of Bechuanaland and Griqualand West. Ann. Rep. 1906, Geol. Comn. Cape of Good Hope, pp. 7—86.
Rogers, A.W., 1908. Geological survey of parts of Vryburg, Kuruman, Hay and Gordonia. Annu. Rep. 1907, Geol. Comn. Cape of Good Hope, pp. 9—122.
Rogers, A.W., 1925. The geological structure of the Union. An explanation of the geological map of the Union of South Africa on the scale of one to a million. Geol. Surv. S. Afr., Spec. Publ., 34 pp.
Rogers, A.W., 1937. The pioneers in South African geology and their work. Trans. Geol. Soc. S. Afr., Annexure to Vol. 39, 130 pp.
Rogers, A.W. and Du Toit, A.L., 1909a. Report on the geology of parts of Prieska, Hay, Britstown, Carnavon and Victoria West. Annu. Rep. 1908, Geol. Comn. Cape of Good Hope, pp. 9—110.
Rogers, A.W. and Du Toit, A.L., 1909b. An Introduction to the Geology of Cape Colony. Longmans Green and Co., London, 491 pp.
Roy, S., 1981. Manganese Deposits. Academic Press, London, 458 pp.
Rust, I.C., 1961. Note on turbidite structures in the Pretoria Series. Trans. Geol. Soc. S. Afr., 64: 99.
S.A.C.S., 1980. Stratigraphy of South Africa. Part 1 (Compiler L.E. Kent). Lithostratigraphy of the Republic of South Africa, South West Africa/Namibia, and the Republics of Bophuthatswana, Transkei and Venda. Handb. Geol. Surv. S. Afr., 8, 690 pp.
Schweigart, H., 1965. Genesis of the iron ores of the Pretoria Series, South Africa. Econ. Geol., 60: 269—298.
Schwellnus, J.S.J., Engelbrecht, L.N.J., Coertze, F.J., Russel, H.D., Malherbe, S.J., Van Rooyen, D.P. and Cooke, R., 1962. The geology of the Olifants River Area. Transvaal. An explanation of Sheets 2429B (Chuniespoort) and 2430A (Wolkberg). Geol. Surv. S. Afr., 87 pp.
Shatskiy, N.S., 1964. On manganiferous formations and metallogeny of manganese. I. Volcanogenic-sedimentary manganiferous formations. Int. Geol. Rev., 6: 1030—1056.
Siever, R., 1971. Low temperature geochemistry of silicon. In: K.H. Wedepohl (Editor), Handbook of Geochemistry, V. II-7, section 14. Springer-Verlag, Heidelberg.
Smit, C.A., 1973. Die Geologie van 'n Gebied rondom Koopmansfontein, Noord-Kaapland, met Spesiale Verwysing na die Serieë Swartrif en Dolomiet. M.Sc. Thesis, Univ. Orange Free State, Bloemfontein (unpubl.).
Smith, J.D. and Hopkins, T.S., 1972. Sediment transport on the continental shelf off Washington and Oregon in light of recent current measurements. In: D.J.P. Swift, D.B. Duane and O.H. Pilkey (Editors), Shelf Sediment Transport: Process and Pattern. Dowden, Hutchinson and Ross, Stroudsburg, Penn., pp. 143—180.
Söhnge, P.G., 1977. Timing aspects of the manganese deposits of the northern Cape Province (South Africa). In: D.D. Klemm and H.J. Schneider (Editors), Time- and Strata-bound Ore Deposits. Springer-Verlag, Heidelberg, pp. 115—122.
Stanton, R.L., 1972. Ore Petrology. McGrawHill, New York, N.Y., 713 pp.
Stauffer, P.H., 1967. Grainflow deposits and their implications, Santa Ynez Mountains, California. J. Sediment. Petrol., 37: 487—508.
Stear, W.M., 1977. The stratigraphy and sedimentation of the Pretoria Group at Rooiberg, Transvaal. Trans. Geol. Soc. S. Afr., 80: 53—65.
Stow, G.W., 1874. Geological notes upon Griqualand West. Q. J. Geol. Soc. London, 30: 581—680.
Strauss, C.A., 1964. The iron ore deposits in the Sishen area, Cape Province. In: S.H.

Haughton (Editor), The Geology of some Ore Deposits in Southern Africa, Vol. II. Geol. Soc. S. Afr., Johannesburg, pp. 393—403.

Stromeyer, F. von, and Hausmann, J.FL., 1831. Krokydolith vom Kap. Gött. Gel. Anz., 3: 1593—1594.

Swiegers, J.U., 1938. The Black Reef Series in the Klerksdorp and Randfontein areas. Trans. Geol. Soc. S. Afr., 41: 177—191.

Swiegers, J.U., 1939. Gold, carbon, pyrite and other sulphides in the Black Reef. Trans. Geol. Soc. S. Afr., 42: 35—45.

Swiegers, J.U., 1948. The gold deposits of the Pilgrims Rest gold mining district, Transvaal. Trans. Geol. Soc. S. Afr., 51: 81—132.

Taljaardt, J.J., 1979. Major manganese ore fields, Republic of South Africa. Rep., SAMANCOR, Johannesburg, 7 pp. (unpubl.).

Tankard, A.J., Jackson, M.P.A., Eriksson, K.A., Hobday, D.K., Hunter, D.R. and Minter, W.E.L., 1982. Crustal Evolution of Southern Africa. Springer-Verlag, New York, N.Y., 523 pp.

Thord-Gray, I., 1905. Notes on the geology of the Lydenburg goldfields. Trans. Geol. Soc. S. Afr., 8: 66—81.

Toens, P.D., 1966. Precambrian dolomite and limestone of the northern Cape Province. Mem. Geol. Surv. S. Afr., 57, 109 pp.

Trendall, A.F. and Blockley, J.G., 1970. The iron-formations of the Hamersley Group, Western Australia, with special reference to the associated crocidolite. Bull. West. Aust., Geol. Surv., 119, 353 pp.

Truswell, J.F., 1967. A critical review of stratigraphic terminology as applied in South Africa. Trans. Geol. Soc. S. Afr., 70: 81—116.

Truswell, J.F., 1970. An introduction to the Historical Geology of South Africa. Purnell, Cape Town, 167 pp.

Truswell, J.F., 1977. The Geological Evolution of South Africa. Purnell, Johannesburg, 218 pp.

Truswell, J.F. and Eriksson, K.A., 1972. The morphology of stromatolites from the Malmani Dolomite north-west of Johannesburg, South Africa. Trans. Geol. Soc. S. Afr., 75: 99—110.

Truswell, J.F. and Eriksson, K.A., 1973. Stromatolitic associations and their palaeo-environmental significance: A re-appraisal of a Lower Proterozoic locality from the northern Cape Province, South Africa. Sediment Geol., 10: 1—23.

Truswell, J.F. and Eriksson, K.A., 1975. A palaeoenvironmental interpretation of the Early Proterozoic Malmani Dolomite from Zwartkops, South Africa. Precambrian Res., 2: 277—303.

Truter, F.C., 1949. A review of volcanism in the geological history of South Africa. Trans. Geol. Soc. S. Afr., 52: 29—89.

Truter, F.C., Wasserstein, B., Botha, P.R., Visser, D.J.L., Boardman, L.G. and Paver, G.L., 1938. The geology and mineral deposits of the Oliphants Hoek area, Cape Province. An explanation of Sheet no. 173, Oliphants Hoek. Geol. Surv. S. Afr., 144 pp.

Tyler, N., 1978. A Stratigraphic Analysis of the Pre-Chuniespoort Group Strata around the Makoppa Dome, West-central Transvaal. M.Sc. Thesis, Univ. Witwatersrand, Johannesburg, 206 pp. (unpubl.).

Tyler, N., 1979a. Stratigraphy, geochemistry and correlation of the Ventersdorp Supergroup in the Derdepoort area, west-central Transvaal. Trans. Geol. Soc. S. Afr., 82: 133—147.

Tyler, N., 1979b. The stratigraphy of the Early-Proterozoic Buffalo Springs Group in the Thabazimbi area, west-central Transvaal. Trans. Geol. Soc. S. Afr., 82: 215—226.

Vajner, V., 1974a. The tectonic development of the Namaqua mobile belt and its foreland in parts of the Northern Cape. Bull. Precambrian Res. Unit, Univ. Cape Town, 14, 201 pp.

Vajner, V., 1974b. The Doringberg fault and its relation to the post-Waterberg deforma-Reneke). Geol. Surv. S. Afr., 76 pp.
Van Biljon, S. 1936. Limestone in the upper part of the Pretoria Series. Trans. Geol. Soc. S. Afr., 39: 45—76.
Van Biljon, W.J., 1964. The chrysotile deposits of the Eastern Transvaal and Swaziland. In: S.H. Haughton (Editor), The Geology of Some Ore Deposits of Southern Africa, Vol. II. Geol. Soc. S. Afr., Johannesburg, pp. 625—669.
Van der Merwe, J., 1973. Die Geologie van 'n gebied om Campbell, Noord-Kaapland, met Spesiale Verwysing na die Serieë Swartrif en Dolomiet. M.Sc. Thesis, Univ. Orange Free State, Bloemfontein (unpubl.).
Van Eeden, O.R., De Wet, N.P. and Strauss, C.A., 1963. The geology of the area around Schweizer-Reneke. An explanation of Sheet 2724B (Pudimoe) and 2725A (Schweizer-Reneke). Geol. Surv. S. Afr., 76 pp.
Van Niekerk, C.B., 1968. The Suitability of Extrusive Rocks for U-Pb Radiometric Dating. Ph.D. Thesis, Univ. Cape Town, South Africa (unpubl.).
Van Niekerk, C.B. and Burger, A.J., 1964. The age of the Ventersdorp System, Ann. Geol. Surv. S. Afr., 3: 75—86.
Van Niekerk, C.B. and Burger, A.J., 1978. A new age for the Ventersdorp acidic lavals. Trans. Geol. Soc. S. Afr., 81: 155—163.
Van Rooyen, D.P., 1947. Sekere pre-Transvaal rotse noordoos van Potgietersrus. Trans. Geol. Soc. S. Afr., 50: 63—70.
Van Wyk, J.P., 1980. Die Geologie van die Gebied Rooinekke-Matsap-Wolhaarkop in Noord-Kaapland met Spesiale Verwysing na die Koegas Subgroep, Transvaal-Supergroep. M.Sc. Thesis, Rand Afrikaans University, Johannesburg, 159 pp. (unpubl.).
Venter, F.A., 1934. The geology of the country between Springs and Bethal. An explanation of Sheet no. 51 (Bethal). Geol. Surv. S. Afr., 87 pp.
Vermaas, F.H.S., 1952. The amphibole asbestos of South Africa. Trans. Geol. Soc. S. Afr., 55: 199—229.
Verwoerd, W.J., 1964. Die geologiese struktuur van die Krokodilrivierfragment. Trans. Geol. Soc. S. Afr., 66: 49—66.
Visser, D.J.L., 1944. Stratigraphic features and tectonics of portions of Bechuanaland and Griqualand West. Trans. Geol. Soc. S. Afr., 47: 197—254.
Visser, D.J.L., 1954. Deposits of manganese ore on Rooinekke and neighbouring farms, District Hay. Trans. Geol. Soc. S. Afr., 57: 61—75.
Visser, D.J.L. (Compiler), 1958. The geology and mineral deposits of the Griquatown area, Cape Province. An explanation of Sheet 175 (Griquatown). Geol. Surv. S. Afr., 72 pp.
Visser, H.N. and Verwoerd, W.J. (Compilers), 1960. The geology of the country north of Nelspruit. An explanation of Sheet 22 (Nelspruit). Geol. Surv. S. Afr., 128 pp.
Visser, H.N., Spies, J.J., Fourie, G.P., Viljoen, J.J., Söhnge, A.P.G. and Venter, F.A., 1961. Die geologie van die gebied tussen Middelburg en Cullinan, Transvaal. Toeligting van Blad 2528D (Bronkhorstspruit) en Blad 2529C (Witbank). Geol. Surv. S. Afr., 73 pp.
Visser, J.N.J., 1969. 'n Sedimentologiese Studie van die Serie Pretoria in Transvaal. D.Sc. Thesis, Univ. Orange Free State, Bloemfontein, 263 pp. (unpubl.).
Visser, J.N.J., 1971. The deposition of the Griquatown glacial member in the Transvaal Supergroup. Trans. Geol. Soc. S. Afr., 74: 187—199.
Visser, J.N.J., 1972. The Timeball Hill Formation at Pretoria — A prograding shoreline deposit. Ann. Geol. Surv. S. Afr., 9: 115—118.
Visser, J.N.J. and Grobler, N.J., 1972. The transition beds at the base of the Dolomite Series in the northern Cape Province. Trans. Geol. Soc. S. Afr., 75: 265—274.
Von Backström, J.W. (Compiler), 1960. Die geologie van Rustenburg en die omliggende gebied. Toeligting van Blad 4 (Rustenburg). Geol. Surv. S. Afr., 93 pp.
Von Backström, J.W., 1963. Septarian concretions in the Lower Griquatown Stage of the

Pretoria Series near Kuruman, Cape Province. Ann. Geol. Surv. S. Afr., 2: 78—88.
Von Dessauer, A., 1909. Some shales in, and observations on the dolomites of Pilgrims Rest. Trans. Geol. Soc. S. Afr., 12: 78—81.
Wagner, P.A., 1917. Asbestos. S. Afr. J. Indust., 1: 251—270.
Wagner, P.A., 1920. Note on the nature and origin of the Crocodile River iron deposits. Trans. Geol. Soc. S. Afr., 23: 118—129.
Wagner, P.A., 1921. Report on the Crocodile River iron deposits. Mem. Geol. Surv. S. Afr., 17, 65 pp.
Wagner, P.A., 1927. The geology of the north-eastern part of the Springbok Flats and surrounding country. An explanation of Sheet 17 (Springbok Flats). Geol. Surv. S. Afr., 104 pp.
Wagner, P.A., 1928. The iron deposits of the Union of South Africa. Mem. Geol. Surv. S. Afr., 26, 268 pp.
Welch, R.H., 1969. The Nature and Origin of the Banded Ironstone and Some Crocidolite Deposits of the Pretoria Series in the Kuruman District, Northern Cape Province. Unpubl. M.Sc. Thesis, Rand Afrikaans University, Johannesburg, 179 pp.
Wessels, J.T., 1967. Teorie bewys daar is meer erts op Sishen. Yskornuus, Desember, pp. 2—7.
Willemse, J. and Bensch, J.J., 1964. Inclusions of original carbonate rocks in gabbro and norite of the eastern part of the Bushveld Complex. Trans. Geol. Soc. S. Afr., 67: 1—87.
Wilson, J.L., 1975. Carbonate Facies in Geologic History. Springer-Verlag, Heidelberg, 471 pp.
Wilson-Moore, C., 1896. Some observations on the geology of the Sabie Valley. Trans. Geol. Soc. S. Afr., 2: 131—141.
Wybergh, W.J., 1920. The limestone resources of the Union, Vol. II. Mem. Geol. Surv. S. Afr., 11, 149 pp.
Wybergh, W.J., 1925. The economic geology of Sabie and Pilgrims Rest. Mem. Geol. Surv. S. Afr., 23, 124 pp.
Wybergh, W.J. and Du Toit, A.L., 1918. The limestone resources of the Union, Vol. I. Mem. Geol. Surv. S. Afr., 11, 122 pp.
Young, R.B., 1906. The calcareous rocks of Griqualand West. Trans. Geol. Soc. S. Afr., 9: 57—66.
Young, R.B., 1932. The occurrence of stromatolitic or algal limestones in the Campbell Rand Series, Griqualand West. Trans. Geol. Soc. S. Afr., 35: 29—36.
Young, R.B., 1933. Conditions of deposition of the Dolomite Series. Trans. Geol. Soc. S. Afr., 36: 121—135.
Young, R.B., 1934a. A comparison of certain stromatolitic rocks in the Dolomite Series of South Africa with modern algal sediments in the Bahamas. Trans. Geol. Soc. S. Afr., 37: 153—162.
Young, R.B., 1934b. Alterations effected by solutions in the limestones of the Dolomite Series. Trans. Geol. Soc. S. Afr., 37: 163—169.
Young, R.B., 1940a. Further notes on algal structures in the Dolomite Series. Trans. Geol. Soc. S. Afr., 43: 17—22.
Young, R.B., 1940b. Note on an unusual type of concretionary structure in limestones of the Dolomite Series. Trans. Geol. Soc. S. Afr., 43: 23—25.
Young, R.B., 1943. The domical-columnar structure and other minor deformations in the Dolomite Series. Trans. Geol. Soc. S. Afr., 46: 91—104.
Young, R.B., 1945. Nodular bodies in the Dolomite Series. Trans. Geol. Soc. S. Afr., 48: 43—47.
Young, R.B. and Mendelsohn, E., 1948. Domed algal growths in the Dolomite Series of South Africa, with associated fossil remains. Trans. Geol. Soc. S. Afr., 51: 53—62.
Zietsman, A.L., 1964. The Geology of the Sabie-Pilgrims Rest Goldfield. Unpubl. M.Sc. Thesis, Univ. Orange Free State, Bloemfontein, 84 pp.

Chapter 5

THE KRIVOY ROG BASIN

YA. N. BELEVTSEV, R. YA. BELEVTSEV and R.I. SIROSHTAN

INTRODUCTION

The sediments which were laid down in the Krivoy Rog basin now form a band of ferruginous rocks 2—7 km wide, which extends 85 km north—south along the Ingulets, Saksagan and Zheltaya rivers, in the central part of the Ukrainian crystalline shield. The area now has low relief, and the crystalline rocks are exposed as cliffs along river valleys and deep ravines incised into the superficial cover of Phanerozoic limestones, sandstones, and loams.

The Krivoy Rog iron ores were discovered late in the 18th century (1781) when V. Zuev described "iron slate" on the Ingulets river banks. Commercial mining of iron ore in the Krivoy Rog basin started in 1875. At present in the basin there are twelve mines producing high-grade ore and five which produce concentrates from low-grade iron-formation. Their combined annual output is about 130 million tons of ore and concentrate. About 2 billion tonnes of iron ore have been mined in the Krivoy Rog basin since mining began. The reserves of high-grade ores have been estimated down to a depth of 1500 m and those of low-grade magnetite ores to a depth of 500 m.

HISTORY OF GEOLOGICAL RESEARCH

S.O. Kontkevich (1880) first described the stratigraphy of the basin. He suggested the three-unit stratigraphic sequence of the Krivoy Rog series which has been used since, with some variations. Later, P.P. Pyatnitsky (1898) studied the petrography of the Krivoy Rog rocks and gave a two-unit stratigraphic sequence of the Krivoy Rog series.

A Geological Committee was formed in 1896 to study the basin. At first the research was headed by A.S. Mikhalsky, and then, after 1904, by A.V. Faas. A.S. Mikhalsky (1908) concluded from his analysis of the geological data that the structure of the basin was a complex open syncline, or synclinorium, rather than a deep trough. Further research revealed in the Krivoy Rog area a synclinorium with its central trough several kilometres deep.

From 1916 onwards I.I. Tanatar performed extensive research on the Krivoy Rog rocks. He developed (Tanatar, 1916, 1923, 1939) the concept of a magmatic origin of the ores and iron-formations of the basin. In 1932 there appeared a voluminous work (Svitalsky et al., 1932) which gave the most complete, to that date, description of the geological structure, petrography and genesis of the Krivoy Rog rocks. The authors divided the Krivoy Rog series into a three-unit sequence. Accepting only one ferruginous jaspilite bed, they explained multiple outcrops of these rocks by isoclinal folding with fractures and multiple thrusts. In 1933—1938 L.I. Martynenko and Yu. G. Gershoig suggested a new stratigraphic sequence which was called double-bedded, according to which two ferruginous and two slate horizons were distinguished.

From 1939 till 1958 a group of geologists from Krivoy Rog headed by Ya. N. Belevtsev compiled detailed geological maps of the Krivoy Rog basin and proposed a new stratigraphic scheme for the Krivoy Rog series, in which seven slate and seven ferruginous horizons were distinguished in the middle ("iron ore") suite. This stratigraphic sequence is now employed in all the mines.

In 1946 a paper by P.M. Kanibolotsky was published dealing with the petrography of the Krivoy Rog basin. This author was the first to predict the possible role of metamorphism in the formation of the high-grade iron ores. The structure of the ore fields and tectonic pattern of the basin were studied for many years. From 1938 till 1979 a large group of mine geologists headed by N.P. Semenenko, Ya. N. Belevtsev and G.V. Tokhtuev performed structural mapping of numerous mining horizons down to a depth of 900 m. Their work made it possible to study in detail the folding, fracturing, and cleavage of rocks and ores. During 1954—1962 geologists of the Ukrainian SSR Academy of Sciences, Dniepropetrovsk Institute of Mines and Krivoy Rog Geological Survey under the guidance of Ya. N. Belevtsev completed work which made stratigraphy, tectonics, and genesis of the Krivoy Rog basin ores more precise. The exploration results are presented in three monographs (Belevtsev, 1957, 1959, 1962). They deal with the detailed geological structure, ore deposits and genesis of iron ores. The basic ideas developed from work performed have been applied in the mining industry and in geological surveys.

The bibliography on the Krivoy Rog basin geology exceeds 1000 titles.

GEOLOGICAL STRUCTURE OF THE BASIN

General

The Ukrainian crystalline shield lies in the central part of the Ukrainian Republic. It extends from east to west for 1000 km, its width varying from 150 km in the east to 350 km in the west. The shield is composed of Archaean and Proterozoic complexes overlain by the Cenozoic sedimentary cover.

The Archaean rocks, with isotopic ages ranging from 3.5×10^9 to 2.8×10^9 years are represented by various granitoids, migmatites and relics of metabasites, ultrabasites, and to a lesser extent of supracrustal rocks. The Archaean rocks were formed at the pre-geosynclinal stage of geological history. Basic and ultrabasic volcanism prevailed, with only restricted sedimentation. Primary sedimentary-volcanogenic formations were intensively granitized and transformed into granitoids and metamorphic rocks.

The Proterozoic rocks, between 2.7×10^9 and 1.8×10^9 years in age, are represented by metamorphic rocks, including gneiss, crystalline schist, metasandstone, metaconglomerate, marble and ferruginous-siliceous rock with varying degrees of granitization. The lithology and facies of these rocks suggest a primary geosynclinal origin.

The structure of the shield is very complicated and results from several episodes of folding and fracturing. The Archaean folds are mainly aligned north-westwards, while the Proterozoic ones run approximately north—south. The whole shield is divisible into fault-bounded blocks which have undergone substantial relative vertical displacement.

Ferruginous-siliceous rocks in the Ukrainian shield form north—south bands or zones, often discontinuous. These appear in Fig. 5-1 and are named as follows: Odessko—Belotserkov, Krivoy Rog—Kremenchug, Pridnieper, Belozero—Orekhov, and Priazov. They comprise three types of iron-formation: sedimentary, of the Superior type, common in Krivoy Rog—Kremenchug and Belozero—Orekhov bands; sedimentary-volcanogenic, occurring in the Odessko—Belotserkov band, in the Priazov, and the Pridnieper areas; and volcanogenic, common in the Pridnieper area.

Stratigraphy

The Krivoy Rog basin forms the southern part of the Krivoy Rog—Kremenchug trough, which is marginal to a large Proterozoic geosyncline composed of rocks of the Krivoy Rog series; these lie with angular unconformity on the eroded surface of the Archaean rocks of the Pridnieper block (see Fig. 5-1).

Two Precambrian complexes, of Archaean and Lower Proterozoic age, define the structure of the basin. The Archaean complex is composed of plagioclase granite and migmatite (Saksagan) with numerous relics of metabasites and ultrabasites and rare gneisses. The rocks of this complex form the structural basement of the Krivoy Rog basin, and belong to the older Archaean Pridnieper block. The Proterozoic complex consists of the Krivoy Rog series rocks, represented by conglomerate, metasandstone, various slates and shales, iron-formation and jaspilite. Below the basal unconformity of the Krivoy Rog series the Archaean granites and migmatities show paleo-weathering, and were also later deformed during the total folding of the Krivoy Rog basin.

The Krivoy Rog series (K) includes all the known ferruginous rocks of the Krivoy Rog basin and is divided, from the lowest upwards, into the following five suites: New Krivoy Rog (K_0), or "ancient Krivoy Rog"; Skelevat (K_1), or lower suite; Saksagan (K_2), or middle suite; Gdantsev (K_3^1), or "aboveore" suite; Gleevat (K_3^2), or upper suite.

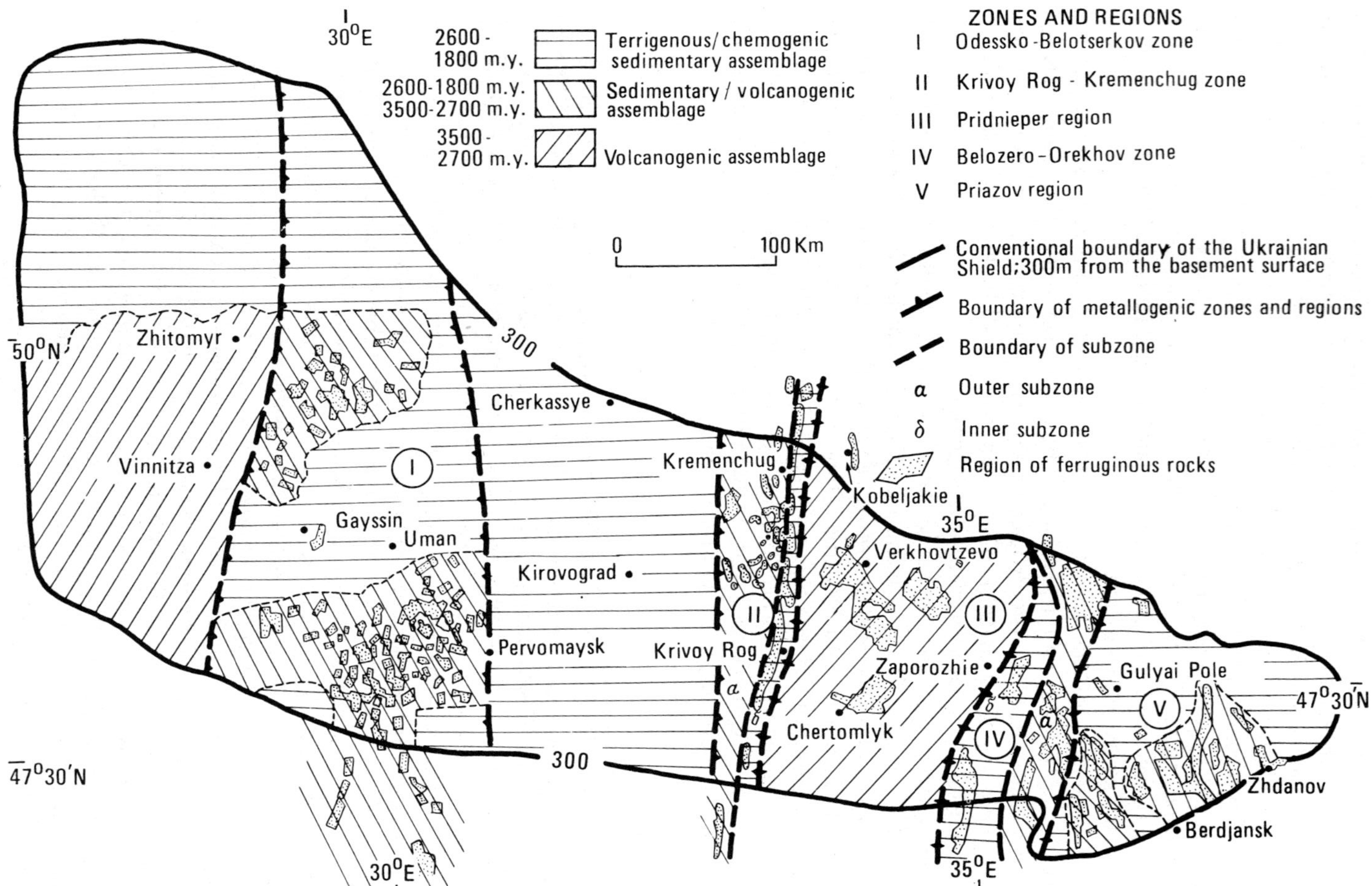

Fig. 5-1. Distribution of metallogenic zones, regions, and localities with ferruginous rocks in the Ukrainian Shield.

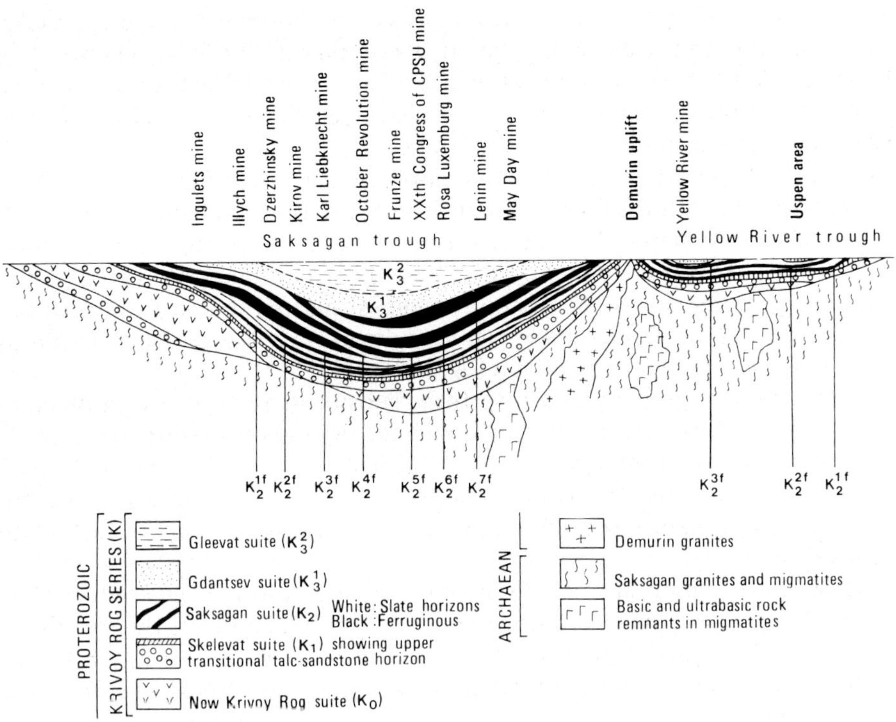

Fig. 5-2. Distribution of suites and beds of the Krivoy Rog series (longitudinal projection).

The New Krivoy Rog suite (K_0) immediately overlies the basement of the basin. Amphibolite is the most abundant rock in the sequence; amphibole and quartz-biotite schists as well as metasandstone are of subordinate significance. White quartzite and metaconglomerate locally occur in the lower part. Amphibolites with an amygdaloidal texture are quite common. The thickness of the suite varies from 80 to 2000 m. Amphibolites of K_0 lie almost everywhere at the base of the Krivoy Rog series. They form an outer frame of the Krivoy Rog structures at their boundary with the Archaean plagioclase granite, and represent the metamorphosed basic lavas which unconformably rest on the old Saksagan plagioclase granites and migmatites.

The Skelevat (lower) suite (K_1) unconformably overlies an eroded surface of amphibolite and plagiogranite. It is divided, from below upward, into three horizons: the arkose-quartzite horizon, consisting of interbedded metasandstones, which locally grade into metaconglomerates, with quartz-sericite schists and quartzites; the phyllitic horizon, composed of quartz-mica phyllite-like schists often enriched with black graphite; and the carbonate-talc-sandy horizon represented by talc, carbonate-chlorite-talc schists, sometimes with interbeds of talcose meta-sandstones and conglomerates. This horizon is transitional from the Skelevat suite, which consists predominantly of clastic rocks, to the overlying Saksagan suite, which is composed chiefly of chemogenic rocks. The talc-bearing rocks are shown by their chemical composition to be derived from effusive products of ultrabasic rocks. The total thickness of the Skelevat suite ranges between 50 and 300 m.

The Saksagan (middle) suite (K_2) is composed of ferruginous-siliceous rocks and represents an alternation of thick iron-formation horizons (beds) with horizons of various slates. The number and thickness of ferruginous and slate horizons vary along the strike and dip. The most complete sequence of the suite, corresponding to the structurally deepest part of the basin, consists of seven ferruginous horizons and seven slate horizons. The greatest thickness of the middle suite reaches 1400 m (Fig. 5-2).

The Saksagan suite is divided into three subsuites: the lower iron ore subsuite comprising the two lowest slate and ferruginous horizons, the middle slate subsuite represented by the third and fourth slate horizon and the third ferruginous horizon, and the upper iron ore subsuite consisting of the fourth, fifth, sixth and seventh ferruginous and the fifth, sixth and seventh slate horizons (Table 5-I).

Beds of chert, which are usually more abundant towards the contact with the ferruginous horizons, are common in slate horizons of all the subsuites. An internal lithological regularity is characteristic of most ferruginous horizons. It consists of a greater abundance of carbonate and silicate quartzites in the upper and lower parts of each horizon, with magnetite and essentially hematite iron-formation confined to the central part. This authigenic-mineralogical regularity is believed to reflect a primary sedimentary regularity in the depositional environment.

The Gdantsev (aboveore) suite (K_3^1) rests unconformably on the Saksagan suite. At the base there occur sedimentary breccias of iron-formation suggesting a significant break after the Saksagan suite sedimentation. Two horizons may be distinguished, based on lithology: the lower (K_3^{1-1}) horizon composed of metasandstones, chlorite and chloritoid schists, conglomerate and chlorite-magnetite ores; and the upper (K_3^{1-2}) horizon consisting of chlorite slate, quartz-mica and graphite schists. The total thickness of the Gdantsev suite reaches 850 m.

The uppermost Gleevat (upper) suite (K_3^2) is found in the axial part of the Krivoy Rog synclinorium. Three horizons may be distinguished, their total thickness being 3500 m. At the base there occur metaconglomerates, metasandstones and quartz-biotite schists of the first horizon. The second horizon is composed of dolomites and quartz-graphite schists with subordinate beds of metasandstones. The third horizon is represented by sandstones with thin beds of quartz-biotite schists. The Krivoy Rog series is intruded by dolerite and granite, neither of which is shown on Fig. 5-2.

Dolerite dykes, trending approximately east—west, are common and in some places are 30 to 35 m thick. The dykes cut across all the rocks and ores

TABLE 5-I

Stratigraphy of the Krivoy Rog series

	Suite	Sub-suite	Horizon - bed name	Symbol	Lithology
			Microcline granites Dolerite dykes		
Krivoy Rog series	Gleevat (upper) K_3^2		Quartz-feldspathic schist and sandstone	K_3^{2-3}	Quartz-biotite schist and metasandstone
			Carbonate rocks	K_3^{2-2}	Dolomite and dolomitized limestone
			Quartz-graphite schist	K_3^{2-1}	Quartz-graphite schist, quartz-biotite-chlorite schist
					Metaconglomerate and metasandstone
	Gdantsev (above-ore) K_3^1		Sandy shale	K_3^{1-2}	Quartz-mica, quartz-chlorite, graphite schist
			Metasandstone, quartzite and conglomerate	K_3^{1-1}	Metasandstone, sandy shale, conglomerate quartzite. Magnetite-chlorite ores
			Unconformity		
	Saksagan (middle) K_2	Upper iron ore	Seventh ferruginous	K_2^{7f}	Amphibole-magnetite, chlorite-carbonate-amphibole-magnetite iron-formation and chlorite-biotite schist
			Seventh slate	K_2^{7s}	Amphibole-chlorite-biotite schist
			Sixth ferruginous	K_2^{6f}	Magnetite and martite red-white-banded jaspilite, chlorite-magnetite iron-formation
			Sixth slate	K_2^{6s}	Amphibole-chlorite schist and chert with characteristic scattered magnetite crystals
			Fifth ferruginous	K_2^{5f}	Magnetite (martite), hematite red-blue-banded jaspilite
			Fifth slate	K_2^{5s}	Chlorite-sericite-quartz schist
			Fourth ferruginous	K_2^{4f}	Chlorite-amphibole-magnetite, carbonate-magnetite iron-formation
		Middle slate	Fourth slate	K_2^{4s}	Graphite-sericite, graphite-biotite, sericite-magnetite iron-formation
			Third ferruginous	K_2^{3f}	Chlorite-magnetite iron-formation, chlorite-quartz schist
			Third slate	K_2^{3s}	Quartz-graphite (carbonaceous) shale, slate, sericite-biotite schist, chert
		Lower iron ore	Second ferruginous	K_2^{2f}	Magnetite-martite jaspilite, hematite-magnetite-chlorite iron-formation
			Second slate	K_2^{2s}	Biotite-chlorite schist and carbonate-bearing chert
			First ferruginous	K_2^{1f}	Magnetite-martite jaspilite, chlorite-amphibole schist
			First slate	K_2^{1f}	Quartz-sericite and chlorite-carbonate schist
			Unconformity		
	Skelevat (lower) K_1		Talc-carbonate sandy	K_{1-2}	Chlorite-talc schist, carbonate-talc, serpentine-talc rock, sandstone, conglomerate.
			Phyllitic	K_1^2	Sericite, muscovite, mica-staurolite phyllite
			Arkose-quartzite	K_1	Quartz arkose metasandstone, conglomerate and quartzite
			Unconformity		
	New Krivoy Rog (ancient) K_0			K_0	Biotite, hornblende, hornblende-epidote amphibolite, biotite schist, metasandstone, quartzite
			Great unconformity		
			Archaean metabasite and meta-ultrabasite, plagioclase granite and migmatite (Saksagan)		

of the Krivoy Rog series, without significant displacement of the adjacent rocks.

Granites cutting across the Krivoy Rog series or in contact with them are found in several places in the northern region of the basin, where the grade of metamorphism resulted in amphibolite and granulite facies. Pink or red microcline granites crossing old plagioclase migmatites and granites are common. A body of microcline granite, several kilometres long and 0.5 to 0.7 km thick was traced in the Gdantsev suite north of the May Day mine. There is no significant alteration of the Krivoy Rog series rocks along the contact of these granites, although they are believed to be of later age.

Tectonic framework

Deformation of the whole complex of the Krivoy Rog metamorphic series resulted in the formation of a complicated north—south folded zone or synclinorium which rests on the Archaean basement.

The Krivoy Rog synclinorium is much affected by major faults, block uplifts, and subsidences. As a result, in many places it is represented only by second-order folds which reflect the general synclinorium pattern. The Krivoy Rog synclinorium is most completely preserved in the central part of the Krivoy Rog basin in the area of the city of Krivoy Rog. Here it is expressed as a series of large folds complicated by a higher-order folding and numerous fractures (Fig. 5-3).

The Main (Krivoy Rog) syncline lies in the central part of the synclinorium. In the east it abuts on to the Saksagan anticline, and in the west on to the Tarapako—Likhmanov anticline. The Main syncline is an open-type fold plunging northwards, in which the limbs lie at an angle of between 50° and 80° at the keel. It is a complex structure, and consists of the East Ingulets and West Ingulets synclines, separated by the Ingulets anticline. These folds, in their turn, are complicated by higher-order folds. The Main syncline is closed and exposed to the south of the city of Krivoy Rog. Northwards it plunges beneath a thick cover of the Gdantsev and Gleevat suites (Fig. 5-3).

All the folded structures of the Krivoy Rog synclinorium plunge northwards at 18—20°. From an analysis of sedimentary thicknesses it is possible to establish a transverse uplift of the synclinorium near the October Revolution and Frunze mines (Fig. 5-9). Farther north the northerly plunge continues and the basin is deepest (6—7 km) in the area of the Lenin mine. Closure of the synclinorium takes place in the Annovsky district. The absence of western limbs in many synclinal structures of the Krivoy Rog synclinorium (Saksagan, Likhmanov, Annovsky and Zheltaya Reka) justifies another interpretation of the general Krivoy Rog structure: it may be regarded as a large flexure which developed in the area of the city of Krivoy Rog and was overfolded eastwards, with a general western dip. It should be mentioned that similar one-limbed structures resembling flexures occur in the Belozero and Kremenchug regions

pp. 221—222

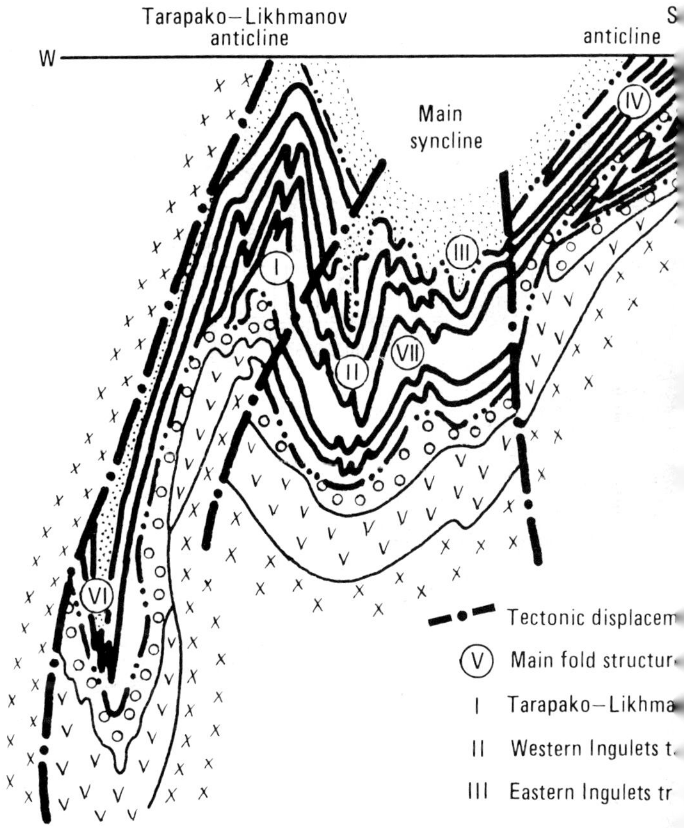

Fig. 5-4. Structural cross-section of the Krivoy Rog basin.

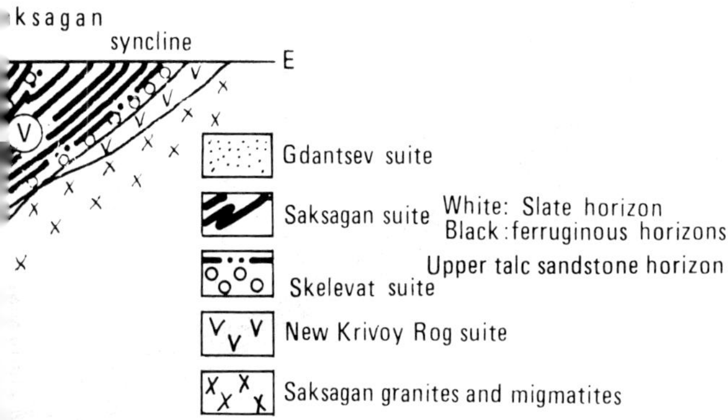

ent IV Saksagan anticline

s V Saksagan syncline

nov anticline VI Likhmanov syncline

ough VII Sovet anticline

ugh

of the Ukraine, in the Starooskol area of the Kursk magnetic anomaly, in the Mesabi range and in other iron-ore provinces of the world.

The eastern limb of the Main syncline is composed of the Saksagan anticline and Saksagan syncline, which together form the structure of the Saksagan ore area. Both folds occur in north-northeasterly trending structures. They are overfolded east-southeastwards, as a result of which the axial surfaces of these folds dip west-northwestwards at angles ranging from 35 to 80°. The common limb of these structures is almost completely destroyed by the longitudinal and conformably dipping Saksagan thrust. Thus the folds are one-limbed: the Saksagan syncline retained its eastern limb, while the Saksagan anticline preserved its western limb. The Saksagan syncline plunges northwards at angles of 12—22°. This results in exposure of its core to the south of the Dzerzhinsky mine. In the Kirov mine it is at a depth of 1000 to 1300 m and in the Lenin mine at a depth of 5 to 6 km. The Saksagan anticline extends from the southern boundary of the city of Krivoy Rog to the Frunze mine, where it is overlain by the Gdantsev suite. Within the general monoclinal trend of the Saksagan syncline limb, gentle transverse bends or flexures have formed through later transverse folding. The iron ore deposits of the Saksagan region (Fig. 5-4) are confined to these bends.

The western limb of the Krivoy Rog synclinorium consists of two large folds: the Tarapako—Likhmanov anticline and the Likhmanov syncline.

The Tarapako—Likhmanov anticline is an open fold with a moderate curve and steeply dipping limbs. As distinct from other folds of the Krivoy Rog basin, it is an upright fold with a vertical axial surface and limbs extending westwards and eastwards. The limbs of the anticline dip at angles mainly from 45 to 70°, sometimes becoming vertical. The crest of the fold plunges at 10 to 15° in a northwesterly direction. Higher-order folding is common within the Tarapako—Likhmanov anticline. Longitudinal thrusts and transverse faults and minor displacements are also abundant. The Western Annovsky band of ferruginous rocks is an extension of the western limb of the synclinorium to the north.

The Likhmanov syncline is a compressed fold with an undeveloped western side. Higher-order folds are abundant within it. The core of the fold consists of Gdantsev suite rocks, the crest plunges northwards at an angle ranging from 12 to 35°.

In the northern part of the Saksagan area, at the May Day mine, the outcrop of the Krivoy Rog rocks turns northwesterly. North of this mine the Saksagan outcrop area of the rocks passes into the Eastern Annovsky band of ferruginous quartzites, which corresponds to the eastern side of the synclinorium. The narrow (about 1—2 km), complicated, Zheltaya Reka syncline borders the northern margin of the basin (Fig. 5-5).

Minor displacements along fracture lines are abundant in the region and superimpose a pattern of small blocks or scales on the main structure pattern. The basin structure was most affected by: (a) longitudinal subconcordant

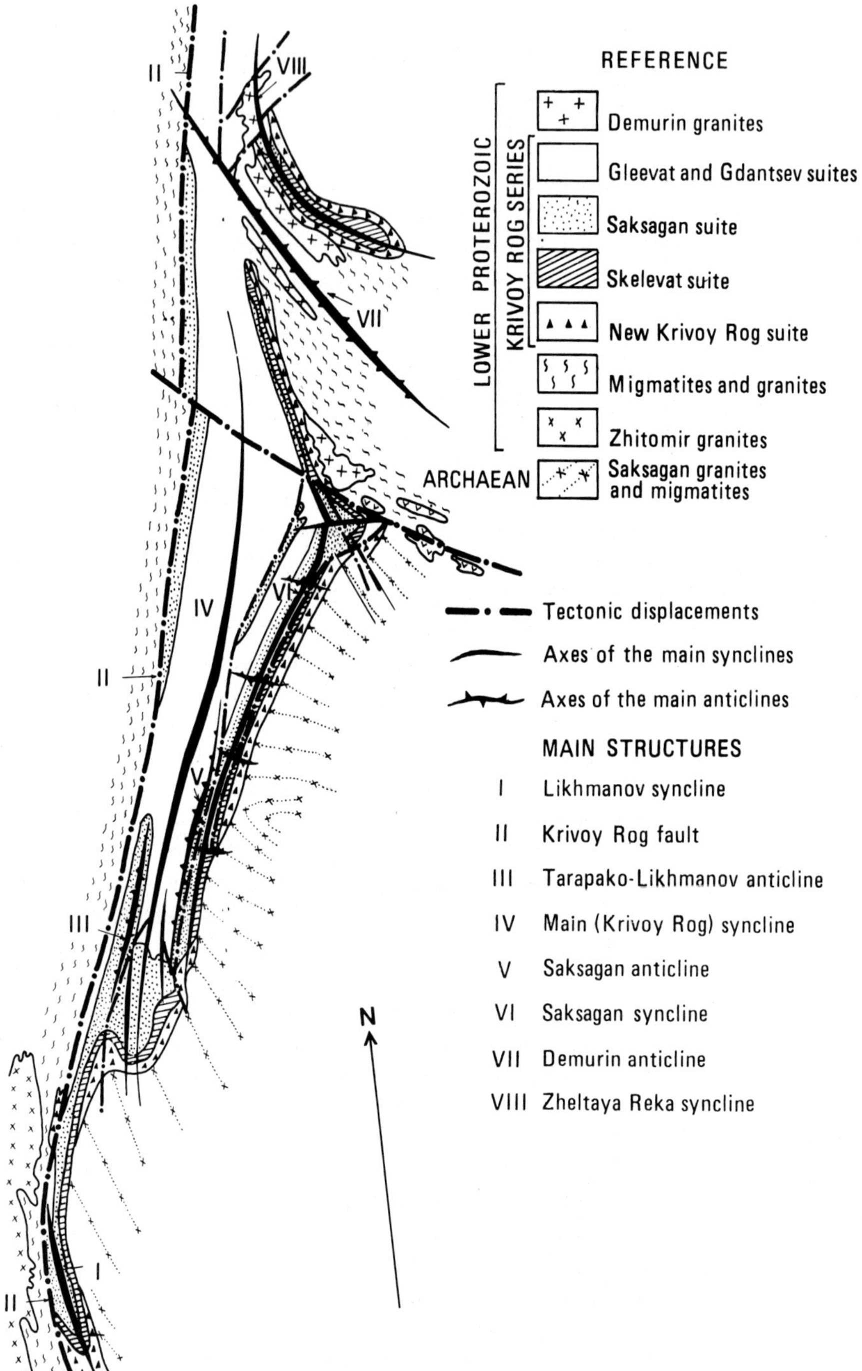

Fig. 5-5. Structural map of the Krivoy Rog basin.

faults and thrusts; (b) diagonal displacements of the thrust and overfault types; (c) transverse fault-type dislocations marked by insignificant rock displacement.

Among the first group of displacements the Krivoy Rog (Western) fault is the best studied. It is a part of the deep Krivoy Rog—Kremenchug fault along which the granitoids of the Ingulets area are thrust over the rocks of the Krivoy Rog basin (Fig. 5-5). The second group of displacements include the Tarapakov, Saksagan, Diagonal, Skelevat and other faults. The third group of dislocations are the most numerous in the basin, but they are small.

The large folded structures defining the main tectonic setting are associated with smaller folds or plications of various orders, down to a very small scale. The pattern of minor folding, despite its apparent discordance, is similar to that of the large folds, which suggests their common origin. Two types of small folds and plications may be distinguished: (a) approximately north—south compressed or closed folds with a tight elongate core, which resemble isoclinal folds (Fig. 5-6). This type of fold is common in the finely banded rocks, such as the iron-formations and jaspilites; (b) transverse open folds which form transverse bands and flexures of the beds or small folds or plications within individual beds (Fig. 5-7). The limbs of these folds have a moderate dip, the cores are open, and they are usually asymmetric. Such folds are most frequent in ore bands. For this reason they are known among Krivoy Rog geologists as ore folds or open folds. Intense cleavage of one or several of the beds involved is common, and leads to an increase in permeability of the rocks; as a result there often occur cleavage cracks filled with ore.

Fig. 5-6. Tight north-trending folds in iron-formation. Dark = iron oxides; light = silica.

Fig. 5-7. Transverse open folding in iron-formation. Dark = iron oxides; light = silica.

THE IRON-FORMATIONS

Mineral composition and texture

The iron formation includes two large groups of rocks: (a) iron-formation proper; and (b) metapelite.

Iron-formation

This category includes all banded quartz-silicate, quartz-silicate-ferruginous and quartz-ferruginous rocks consisting of variously interlayered combinations of quartz, silicate and iron oxide minerals. The quartz (chert) beds have a cryptocrystalline structure. The iron oxide layers are composed of magnetite, martite, hematite, goethite and dispersed hematite and the silicate bands consist of chlorite, sericite, amphibole, biotite and quartz. The iron oxide interbeds have a growth-oriented, network growth-oriented and non-oriented structure. The schist interbeds have a lepidoblastic and nematoblastic structure.

All these rocks have a banded texture. The character of the banding makes it possible to divide them into three groups: fine-banded (bands 1—3 mm wide), medium-banded (3—10 mm) and coarse-banded (more than 10 mm).

These rocks are also divisible into three groups on the basis of composition, silicate iron-formation, oxide-silicate iron-formation, and oxide iron-formation and jaspilite (Table 5-II).

Silicate iron-formation consists of chert (quartz) beds, sometimes with a small amount of iron carbonate, alternating with beds of iron silicate and in places with aluminosilicate and carbonate. The total thickness of chert interbeds usually amounts to between 50 and 80% of the total thickness of the rocks. Rock with less than 30% of chert is classed as schist. Silicate iron-formation containing more than 15% of iron oxide interbeds is classified as oxide-silicate iron-formation.

In the oxidized zone iron carbonate is replaced by dispersed hematite and goethite; quartz and silicates (chlorite-biotite) are decomposed, resulting in the formation of goethite and kaolinite.

Oxide-silicate iron-formation consists of quartz, iron and magnesium silicates, rarely of aluminosilicate, and iron oxide minerals, either granular or dispersed magnetite, hematite, and goethite. Three types of band are common: chert, or quartz bands; iron oxide, consisting variously of magnetite, hematite or martite; and silicate bands made up mainly of chlorite, amphibole or dispersed hematite.

When oxidized, the quartz-siderite interbeds are partially or completely substituted by dispersed hematite and become dark red to black. Dispersed hematite and goethite substitute for silicate bands. The rock is called martite-dispersed-hematite iron-formation.

Oxide iron-formation and jaspilite consist of quartz and the iron oxide minerals: magnetite, hematite and martite. Three types of bands are distin-

TABLE 5-II

Classification of iron-formation

Silicate iron-formation	Oxide-silicate iron-formation	Oxide iron-formation and jaspilite
1. Sericite	1. Chlorite-magnetite	*Oxide iron-formation*
2. Sericitic-chlorite	2. Martite-dispersed-hematite	1. Magnetite or martite-magnetite
3. Chlorite	3. Amphibole-magnetite	2. Martite-dispersed-hematite
4. Dispersed-hematite	(a) cummingtonite-magnetite	*Jaspilite*
5. Goethite-dispersed-hematite with martite crystals	(b) actinolite-tremolite-magnetite	1. Martite
6. Amphibole-chlorite	(c) cummingtonite-magnetite riebeckitized	2. Magnetite
7. Chlorite-biotite	(d) cummingtonite-magnetite chloritized	3. Martite-hematite
	4. Chlorite-siderite-magnetite	
	5. Sericite-chlorite-martite	
	6. Aegirine-hematite	

guished: iron oxide-free or chert (quartz) bands; iron oxide bands composed of iron oxides and an insignificant amount of quartz; and mixed bands composed of iron oxide minerals and quartz.

Jaspilite is the name used in the Krivoy Rog basin for a fine-banded rock consisting of iron oxide minerals and quartz, which form iron oxide, oxide-free, and mixed bands. Jaspilite is characterised by: (a) complete or almost complete absence of silicate minerals (chlorite, amphibole, biotite, etc); (b) fine-banded texture, with the bands not more than 1—3 mm thick; (c) total thickness of oxide and mixed bands exceeding that of quartz bands.

Oxide iron-formation is a quartz-iron oxide banded rock containing not more than 5—10% of silicates. Mixed bands are relatively rare, and do not significantly affect the composition of the rock. Coarse banding, and the prevalence of the oxide-free bands over the iron oxide bands, are the main distinctive features of oxide iron-formation as compared with jaspilite.

At their upper and lower boundaries jaspilites are bounded by oxide iron-formation, which lie in turn against ferruginous oxide-silicate iron-formation, and then slate. Thus, oxide iron-formation is often a transitional rock from jaspilite to oxide-silicate iron-formation.

Metapelite

Slates of the Krivoy Rog series middle suite are variable in both composition and appearance. Slates are commonly completely recrystallized rocks derived from both clastic and chemogenic quartz-clay and siliceous-ferruginous-silicate sediments. However, in places they contain unchanged primary sedimentary material.

Chemical composition

Iron-formation

Besides these major elements — iron and silica — the Krivoy Rog ferruginous rocks contain various minor elements. The minor elements in ferruginous rocks and ores are commonly sulphur, aluminium, calcium, potassium, sodium, phosphorus, carbon, and magnesium, each rarely exceeding one tenth of a per cent. Some of them (phosphorus, sulphur, calcium) may have significance in stratigraphic correlation.

Metapelite

Slates of the Saksagan suite are divided on the basis of chemical composition into the following three isochemical types which reflect the composition of the primary constituents: (1) aluminosilicate; (2) ferrosilicate; and (3) magnesiosilicate (Table 5—III). The degree of rock metamorphism determines the fabric and mineral composition in each group. The chemical composition of these rocks are included in Table 5-IV.

The most important chemical components of the rocks are as follows: SiO_2, ranging from 37.5% in jaspilites to 60% in metapelite; Al_2O_3, from absent in

TABLE 5-III

Classification of pelitic rocks

By chemical composition	By degree of metamorphism			
	Microschists	Slates	Phyllitic schist	Crystalline schist
Alumosilicate schist	Clay	Sericite schist	Muscovite-sericite schist	Mica schist
Ferruginous-silicate schist	"Paint-rock"	Sericite-chlorite Chlorite Chlorite-biotite Sericite-biotite Chlorite-carbonate Magnetite-mica-chlorite	Biotite Chlorite-biotite Chlorite Chlorite-amphibole (garnet and kyanite) Mica-amphibole Amphibole-magnetite and hematite Garnet-sericite-chlorite	Chlorite-chloritoid Staurolite-mica Garnet-chlorite Biotite-amphibole (sometimes with garnet) Garnet-amphibole Amphibole (cummingtonite- and riebeckite-cummingtonite) Biotite-plagioclase gneiss
Magnesium-silicate schist	Talc china clay	Talc with magnetite and carbonate crystals Talc-carbonate Talc-chlorite	Talc-carbonate Talc-chlorite Talc-actinolite Talc-serpentine	Actinolite with a low content of talc Actinolite-serpentine

Quartz is not mentioned in the names of pelitic rocks as it is always present.

TABLE 5-IV

Chemical composition of metapelitic rocks and iron-formations

Oxides	1	2	3	4	5	6	7	8
SiO_2	56.95	59.86	54.29	51.14	48.73	57.20	46.74	37.55
Al_2O_3	14.73	12.35	5.24	5.10	7.72	0.36	—	0.53
Fe_2O_3	5.89	9.55	16.01	15.30	7.51	40.75	51.12	57.18
FeO	5.92	7.53	17.59	17.79	27.84	0.74	0.98	1.48
TiO_2	0.68	0.34	0.02	0.03	0.04	Trace	0.07	Trace
MnO	0.03	0.04	0.06	0.04	0.07	0.03	0.02	0.03
CaO	0.02	Trace	1.54	0.90	0.69	0.40	0.30	0.60
MgO	6.30	2.63	2.27	1.40	3.51	0.18	0.36	0.06
P_2O_5	0.15	0.14	0.08	0.13	0.12	0.15	0.10	0.17
SO_3	0.19	0.20	0.61	0.42	0.30	0.08	0.08	0.21
Na_2O+K_2O	3.72	2.77	0.81	0.43	0.16	—	—	—
H_2O at 105°	0.23	0.25	0.63	0.83	1.96	0.15	0.09	0.12
Ignition Loss	5.26	4.49	1.50	6.66	0.81	0.49	0.01	0.24
Total	100.25	100.15	100.02	100.17	99.45	100.53	99.87	99.17

1. Quartz-chlorite-sericitic schist, K_2^{1s}, the Saksagan river outcrop.
2. Chlorite-quartz-biotite schist, K_2^{2s}, the Saksagan river outcrop near the Artem Mine.
3. Iron-formation representing low-grade ore, K_2^{5s}, the Artem mine.
4. Chlorite-magnetite iron-formation, K_2^{4f}, the October Revolution mine.
5. Martite-chlorite iron-formation, K_2^{4f}, the Dzerzhinsky mine.
6. Hematite-magnetite iron-formation, K_2^{6f}, the Artem mine.
7. Martite jaspilite, K_2^{5f}, the Dzerzhinsky mine.
8. Martite jaspilite, K_2^{5f}, Gleevat ravine.

jaspilites to 15% in metapelites, Fe_2O_3 + FeO, from 59% in jaspilites to 10% in metapelites; CaO, from 0.06% in jaspilite to 6.0% in metapelites; Na_2O + K_2O, from absent in jaspilites and oxide iron-formation to 3.7% in metapelites.

Stable isotope data

Data on sulphur, oxygen and carbon isotopes in the ferruginous rocks of the Krivoy Rog basin have been studied by Tugarinov and Grinenko (1965), Chukhrov et al. (1968), Chukhrov et al. (1969), Belevtsev et al. (1969), Belevtsev and Koptyukh (1974), Lugovaya (1976) and Belevtsev et al. (1978, etc).

From numerous analyses a wide variation of oxygen isotope composition has been established for magnetite, hematite and martite sampled in different types of iron ores, ferruginous rocks, slates and zones of metamorphism. The lowest $\delta^{18}O$ values are observed in martites of martite ores (from −6 to 0‰) and of martite iron-formation (from −2 to +6‰). Values of $\delta^{18}O$ in magnetite from magnetite ores fall within the range from 0 to +25‰; the highest figures (from +18 to +25‰) are more pronounced in magnetites from metasomatic magnetite ores. The oxygen isotope ratio in magnetite from unoxidized oxide iron-formation and jaspilite varies from +2.4 to +18‰. This ratio in the magnetite from chlorite slates of the Saksagan suite varies from +1 to +20‰. The oxygen-18 isotope content in iron oxides increases from martite ores and martite-hematite iron-formation to magnetite ores, magnetite iron-formation and chlorite slates.

The oxygen-18 content in magnetite increases with the grade of metamorphism, namely, from greenschist to amphibolite and granulite facies. Magnetite of the greenschist facies has $\delta^{18}O$ mean about 2.5‰. In rocks of the amphibolite facies this ratio rises to 6—8‰ and in rocks of the granulite facies it grows up to 12—18‰.

An analysis of the data makes it possible to establish various $\delta^{18}O$ values for magnetite and hematite from iron ores of different genetic types. The difference between $\delta^{18}O$ values of magnetite ore and the enclosing ferruginous rocks is small for residual metamorphic ores formed at the progressive stage of the greenschist facies of metamorphism when silica was removed and rocks compressed. Magnetite from metamorphic magnetite ores formed under iron-magnesium metasomatism during the amphibolite and granulite stage of metamorphism shows a varying content of heavy oxygen, smaller or greater than that in enclosing rocks ($\delta^{18}O$ ranging from 8—10 to 21‰). Oxidized ores are characterized by extremely low values of $\delta^{18}O$ (from —6 to 0‰).

The variations of $\delta^{18}O$ in the rocks and ores of the Krivoy Rog basin are governed by the primary content of sediments, facies of metamorphism, and superimposed processes of metasomatism and oxidation.

The $^{34}S/^{32}S$ ratio has been studied in pyrite, which is widely distributed in all ores and rocks of the Krivoy Rog basin. Based on the geological setting and genetic type two morphological varieties of pyrite are distinguished: (1) banded pyrite, in which either massive or disseminated pyrite forms either continuous or discontinuous thin layers in the rock parallel to the bedding; and (2) veins of various composition containing pyrite or composed purely of pyrite crossing the stratification (Fig. 5-8).

Data on isotopic sulphur ratios in pyrite are set out in Table 5-V. It is possible to draw some conclusions from these results.

The isotopic content of sulphur in banded pyrite clearly reflects the conditions of formation of rocks from different suites of the Krivoy Rog series and is consistent in each suite in all the regions of the basin. The lowest values of $^{34}S/^{32}S$ in this variety of pyrite occur in the Saksagan suite ($\overline{x}\delta^{34}S$ = —6.1 to —1.9‰); they are slightly higher in the Skelevat suite immediately below ($\overline{x}\delta^{34}S$ = —0.8 to +0.5‰). The Gdantsev and Gleevat suites are characterized by the highest $^{34}S/^{32}S$ ratio ($\overline{x}\delta^{34}S$ = +9.2 to +23.7‰). Values of $\delta^{34}S$ are extremely high in pyrites from carbonaceous and carbonate metamorphic rocks (to +35.7‰).

Fig. 5-8. Bedded and cross-cutting pyrite in iron-formation. White = pyrite; black and grey = iron oxide and oxide-free bands in iron-formation.

The isotopic composition of sulphur in veins is similar to that of banded pyrite of the same rock or slightly lower. A considerable difference between the sulphur isotopic composition of the pyrites from metasomatic amphibole-magnetite ores ($\bar{x}\,\delta^{34}S = -10.2‰$) and martite (oxidized) residual-metamorphic ores ($\bar{x}\,\delta^{34}S = +14.6‰$) reflects different geological environments of their formation.

The $^{13}C/^{12}C$ ratio has been studied for all rocks of the Krivoy Rog series (Table 5-VI). The wide range of $\delta^{13}C$ (from -36.3 to $+6.3‰$) is best explained as the result of various geological environments of the rock formation and the presence of carbon of two types (organic and carbonate). The $\delta^{13}C$ values for most rocks fall into the field of biogenic carbon ($\bar{x}\delta^{13}C = -10.7$ to $-31.9‰$); magnetite ores of the Gdantsev suite and dolomitized limestones and carbonate rocks of the Gleevat suite ($\bar{x}\delta^{13}C = +2.0‰$), the carbon of which is very close to those of recent marine sediments are the exceptions.

TABLE 5-V

Isotopic data of sulphide sulphur in rocks of the Krivoy Rog series*

Suite	Enclosing rocks and forms of sulphide (in brackets)	$\delta^{34}S$, ‰		
		Min.	Max.	Mean
Gleevat (K_3^2)	Biotite-chlorite-graphite slates (banded)	+11.7	+23.9	+20.4
	Carbonate rocks (banded)	+17.2	+35.7	+23.7
	Biotite-graphite-quartz slates (vein)	−9.5	+5.8	−0.4
Gdantsev (K_3^1)	Quartz-biotite metasandstones (banded)	+7.3	+12.6	+9.2
	Breccia of iron-formation (banded)	−5.5	+30.5	+11.6
Saksagan (K_2)	Silicate-carbonate-magnetite, magnetite and hematite iron-formation and jaspilite (banded)	−10.7	−3.7	−6.1
	Magnetite-hematite iron-formation (vein)	−12.2	+1.0	−6.5
	Quartz-biotite-sericite-chlorite slates (banded)	−3.1	0	−1.9
	Amphibole-magnetite ores (vein)	−14.0	−3.9	−10.2
	Martite ores (banded)	+7.2	+31.4	+14.6
Skelevat (K_1)	Metaconglomerates (banded)	−1.8	+2.9	+0.5
	Metasandstones (banded)	−2.8	+2.1	−0.3
	Phyllite slates (banded)	−5.7	+4.9	−0.8
New Krivoy Rog (K_0)	Schistose amphibolite (impregnation)	—	—	+1.0

* According to F.I. Zhukov's sulphur isotope data.

Sedimentological synopsis

The following are the main points that need to be made, in a review paper of this length, concerning the sedimentation of the Krivoy Rog iron-formations:

(1) The iron-formation comprises rhythmic alternations of beds and horizons of ferruginous rocks (jaspilite and oxide iron-formation with pelitic (slate, schist) beds now composed of biotite, chlorite, quartz, amphibole and other metamorphic minerals. It lies on a clastic suite including conglomerate and is overlain by a thick clastic schist-carbonate suite.

(2) Each couplet of ferruginous and slate horizons represents a sedimentational microcycle, marked by the following rock sequence: aluminosilicate slate—iron-silicate slate—oxide iron-formation or jaspilite—iron-silicate slate and aluminosilicate slate. The middle part of each slate horizon consists of

TABLE 5-VI

Isotopic composition of carbon from the Krivoy Rog rocks and ores*

Suite	Rock	$\delta^{13}C$,‰ (total)		
		Min.	Max.	Mean
Skelevat	Phyllite schist	—	—	23.5
Saksagan	Quartz-sericite schist	−16.3	−27.8	−22.0
	Quartz-chlorite-biotite schist	−11.3	−17.4	−14.4
	Silicate-carbonate-magnetite iron-formation	−13.0	−13.4	−13.2
	Magnetite iron-formation	−9.2	−11.7	−10.7
Gdantsev	Magnetite ore	+1.2	+0.9	+1.1
	Sedimentary breccia	−19.5	−20.1	−19.8
Gleevat	Quartz-graphite-sericite-biotite schist ($\delta^{13}C$ organic)	−30.0	−36.3	−31.9
	Dolomitized limestone and carbonate rock	+6.3	−3.4	+2.0

* According to F.I. Zhukov's data.

aluminosilicate slate, and the middle part of each ferruginous horizon includes oxide iron-formation or jaspilite composed of magnetite, hematite and quartz. When a cycle of sedimentation is well developed, a gradual transition from a slate horizon to a ferruginous one is clear. Belevtsev (1947) has shown that this transition corresponds to one from clastic sedimentation, through mixed clastic and chemical sedimentation, to chemical sedimentation (oxide iron-formation and jaspilite).

(3) A well established zonation of authigenic minerals in the iron-formation suggests a consistent intergradation of various facies types of ferruginous rocks (Belevtsev and Skuridin, 1957; Plaksenko, 1969). In accordance with this zonation the primary ferruginous facies in the normal facies profile of the formation appear in the following order from the shore line into the deeper part of the basin: slate (clay and mud) facies; ferrous facies (low-grade ore iron-silicate metapelites with magnetite and siderite); ferric oxide—ferrous oxide facies (sideroplesite-magnetite and sideroplesite-chlorite-magnetite hornfels); ferrous oxide—ferric oxide facies (magnetite iron-formation); oxide facies (magnetite-hematite and hematite jaspilite and iron-formation). The change of ferruginous facies along the profile from the shore line into the deeper part of the basin is associated with a decreasing content of organic matter in sediments. N.M. Strakhov (1947) considers that such authigenic mineral zonation of quartz-ferruginous rocks is formed when iron is introduced into the basin as hydroxide colloids. It corresponds to that observed in recent oceans.

(4) A consistency in the association of iron and silica, in the ratio of one to the other, and in minor elements in rock-forming minerals in the facies profile, which is in good agreement with an ideal sedimentary profile, bears witness to the leading role of solutions in migration of primary components.

(5) The small-scale stratigraphic continuity of the ferruginous rocks is often broken, and there appear a variety of lenticular and cross-bedded structures, and some local brecciation. Some bands in the iron-formation have been removed, and the banding is discontinuous. All these features are common in other sedimentary rocks.

(6) Free carbon persists in the ferruginous rocks, in schist particularly. An average content of C_{free} in the Krivoy Rog iron-ore suite is 0.29%, and that of CO_2 is of about 4.10%, being in some horizons 0.52 and 10.2%, respectively. A clear correlation is established between the C_{free} content and the amount of terrigenous material in sediments. Data from X-ray diffraction, electron microscopy and infra-red spectroscopy make it possible to show the dependence of the relict graphite state of aggregation and structural ordering on the grade of regional metamorphism. This suggests that amorphous carbonaceous matter was buried syngenetically with the rocks. Remnants of blue-green algae and coralloidal invertebrates have also been detected in the Krivoy Rog carbonate rocks (Kalyaev and Snezhko, 1973).

The primary sediments have been variously affected by metamorphic processes. Petrographic data show complete recrystallization of sediments in various metamorphic and ultrametamorphic facies. Recrystallization transformed primary sediments into crystalline schists, gneisses, and migmatites; and ferruginous rocks into jaspilites and iron-formation. Metamorphic changes, however, did not much influence the primary stratigraphic relations of the rocks, which are typical of those resulting from various known sedimentation processes.

METAMORPHISM

General description

Rocks of the Krivoy Rog synclinorium are characterized by Early Proterozoic sedimentary-volcanogenic formations which are underlain by Archaean granites forming a structural basement. Both high-grade and low-grade iron ores are related to metamorphism of iron formation within the Early Proterozoic rocks.

Studies of geological and metallogenic relations in the Krivoy Rog basin lead to the conclusion that the character and extent of iron ore mineralization are governed by physical and chemical conditions of metamorphism. This account deals with the conditions of metamorphism and spatial arrangement of metamorphic facies in the basin. For this purpose, mineral parageneses of aluminous (metapelite) rocks, which are the most responsive to changes in metamorphic conditions, have been mainly studied. Data from deep drilling in the Krivoy Rog basin (to 2500 m) make it possible to evaluate both horizontal and vertical gradients of metamorphism.

The Krivoy Rog series was long considered to represent an evenly and slightly metamorphosed sequence, with the Archaean rocks underlying the synclinorium having been metamorphosed to a higher grade. But the studies of P.M. Kanibolotsky (1946) and then N.P. Semenenko, A.P. Nikolsky, V.N. Kobzar, and R.Ya. Belevtsev proved a distinct metamorphic zonation up to, but bare reaching, the amphibolite facies.

The best studied rock assemblages of the central (Saksagan) region include mainly chlorite, almandine-chlorite, biotite-sericite, and chloritoid phyllite and schist and chlorite-epidote-actinolite amphibolite which points to a greenschist facies of regional metamorphism. Higher-temperature muscovite-andalusite-staurolite, cummingtonite-biotite-garnet schist, plagioclase microgneiss and amphibolite with bluish-green hornblende and oligoclase-andesine occurring commonly in the southern and northern regions are associated with the muscovite-almandine-andalusite-staurolite subfacies, which is related to the epidote-amphibolite facies by Sobolev (1970). The highest temperature muscovite-microcline-sillimanite gneiss of the sillimanite-muscovite subfacies also related to the epidote-amphibolite facies, occurs in the central Annovsky band in the northern part of the basin. Migmatized metapelite gneiss with rare concordant leucocratic granite or pegmatite veins has been found in this band. It should be noted that migmatization is of extremely rare occurrence in the rocks of the Krivoy Rog series, which is taken as evidence for the absence of the amphibolite facies of metamorphism. Association of muscovite with quartz is common over the whole of the Krivoy Rog basin.

In the study of metamorphic facies and subfacies of the rocks in the basin, transitions were found along the strike. The Saksagan region is characterised by rocks of the greenschist facies. Northwards, metapelite schists become coarse-grained, and the features of primary sedimentary (clastic grains and thin-layered textures) and volcanic (porphyritic, ophitic structures, and amygdaloidal textures) rocks are less pronounced.

Rocks of each metamorphic facies or subfacies crop out over a substantial area, while the boundaries between them are rather sharp, as is typical of the zonation resulting from progressive regional metamorphism. Each subfacies, separated by isograds, corresponds to a metamorphic zone. Three metamorphic zones can be distinguished: almandine, staurolite, and sillimanite-muscovite (Fig. 5-3).

The almandine zone occupies most of the Saksagan region, except the Dalny-Zapadny bands and the area to the northwest of them, where metamorphism reached the staurolite zone. This latter zone is also general in the Likhmanov syncline and over the greater part of the Annovsky band (southern and eastern). The sillimanite-muscovite zone is present only in the northwestern part of the Annovsky band. Metamorphism is less developed in the southern part of the Saksagan band (from the Frunze mine to the Dzerzhinsky mine), where almandine garnet is somewhat rarer than in the other parts of the almandine zone. This area may be referred to the almandine-biotite subzone.

In exploratory drillholes of the central region to depths of 2500—2800 m, slates with chlorite and chloritoid have been found, suggesting a lack of distinct vertical zoning.

Although the accuracy of location of the isograds is not everywhere uniform, and is usually about ± 1—2 km, they quite clearly cross the Krivoy Rog synclinorium and pass into the granitoids of the basement. Analysis of the pattern of metamorphic zonation shows that the granitoids of the basement have been influenced by metamorphism. In the almandine zone the granitoids consist predominantly of foliated plagioclase granites and plagioclase-rich migmatites derived from metabasites, with epidote, greenish biotite, actinolitic hornblende, oligoclase-albite, chlorite and sericite. In the staurolite and sillimanite-muscovite zones the plagioclase granites are usually remigmatized and converted into polymigmatites with veins and patches of pink leucocratic microcline-plagioclase granites, pegmatites, or metasomatic microcline. Epidote is rare, chlorite is absent, amphibole is represented by bluish-green hornblende and plagioclase by oligoclase-andesine.

Rocks of the almandine zone of the Krivoy Rog basin contain the following mineral* associations:

$Gar_{94.6} + Ch_{69.6} + Cum_{73} + Mt$;
$Chd_{84.2} + Ch + Mt + Qz$;
$Chd + Bi_{60} + Mu + Qz + Gph$;
$Chd + Ch_{75} + Bi_{75} + Mu + Qz$;
$Gar_{96.0} + Bi_{74.2} + Ch_{72.0} + Mu + Qz$;
$Gar_{95.3} + Cum_{77.2} + Bi + Ch + Qz$;
$Gar_{91.2} + Bi_{61.4} + Qz \pm H$;
$Gar_{95.6} + Bi_{74.2} + Qz$;
$Gar_{91.5} + Bi_{64.3} + Qz + Gph \pm Mu \pm Ch$.

The contents of the spessartite and grossular components in almandine garnet are 3.7 ± 3.1 and 9.6 ± 2.6 mol per cent, respectively. The temperature

* Symbols of minerals: And-andalusite, Ant-anthophyllite, Bi-biotite, Ch-chlorite, Chd-chloritoid, Cpx-clinopyroxene, Cor-cordierite, Cum-cummingtonite, Fa-fayalite, Gph-graphite, Gar-garnet, Hb-hornblende, Hyp-hypersthene, Mt-magnetite, Mu-muscovite, Or-orthoclase, Pl-plagioclase, Qz-quartz, Sil-sillimanite, St-staurolite. Subscripts show, for each mineral, its total Fe content as a percentage of Fe + Mg; when these are derived from chemical analyses they are shown to the first decimal place, and when based on refractive indices they are shown as whole numbers. Subscripts for plagioclase represent anorthite content.

of metamorphism in this zone according to the garnet-biotite, geothermometer data of Perchuk (1970) and Thompson (1976) ranges from 430 to 550°C, with values in the range 450—520°C most common.

The following mineral associations are representative of the staurolite zone:

$Gar_{88.6-93.3} + Bi_{57.2-63.7} + St_{82.6-83.8}$ + And + Mu + Qz + Gph;
Qz + $Pl_{40} + Bi_{46.5} + Gar_{85.5}$;
Qz + Mu + Gph + $St_{74.3} + Bi_{41.5}$ + And;
$Gar_{89} + Bi_{49} + Cum_{63}$ + Qz;
Cor + Ant + Qz + Bi.

The average contents of the spessartite and grossular components in the garnets are 3.25 ± 2.3 and 7.5 ± 2.0 mol per cent, respectively. The temperatures of garnet-biotite equilibrium in the staurolite zone on the basis of geothermometry data are 510—600°C, the range of 530—590°C being most common (Perchuk, 1970; Thompson, 1976).

In the sillimanite-muscovite zone the association of garnet and microcline appears for the first time. Typical associations are as follows:

$Gar_{83.5} + Bi_{57.8}$ + Sil + Mu + Or + Pl_{17} + Qz + Gph;
$Gar_{90} + Bi_{70} + Cum_{87}$ + Qz ± Mt;
Hb + Cpx + Pl + Bi + Qz;
$Gar_{87.1} + Bi_{54.5} + Hb_{60.6} + Pl_{50}$ + Qz;
$Cum_{85.2} + Hyp_{87.1} + Fa_{97.0}$ + Mu + Qz;
$Gar_{75.4} + Cpx_{38.0} + Pl_{45}$ + Or + Gph + $Bi_{46.9}$ + Qz;
$Gar_{91.1} + Bi_{70.4} + Cum_{73}$ + Qz.

Garnets of this zone are commonly enriched in manganese to a maximum of 20 per cent of the spessartite molecule. The temperature of garnet-biotite equilibrium in the sillimanite-muscovite zone ranges from 580°C to 630°C.

The total metamorphic pressure of the Krivoy Rog basin as evaluated from Perchuk's (1970) experimental equilibria on the staurolite-garnet and clinopyroxene-garnet geobarometers, ranges from 3 to 6 kbar, with most determinations within the 4—5 kbar range. The metamorphic zonation superimposed on the folded structure of the Krivoy Rog synclinorium, and the discordance of the isograds with tectonic boundaries make it possible to evaluate the isobaric metamorphism in the Krivoy Rog basin. It appears that the temperature of metamorphism bears no direct relationship to the lithostatic load caused by the weight of rocks studied.

Based on the foregoing facts, several conclusions can be drawn:

(1) The sedimentary-volcanogenic rocks of the Krivoy Rog series, which include the iron-formation units, were regionally and progressively metamorphosed in the Early Proterozoic orogeny ($2000 \pm 300 \times 10^6$ years). The Archaean granitoids of the basement were also metamorphosed during this stage.

(2) The Early Proterozoic metamorphism in the Krivoy Rog basin is reflected in a regional metamorphic zonation which transects the Krivoy Rog

Series at an acute angle, and extends into granitoids of the basement. Almandine, staurolite, and sillimanite-muscovite metamorphic zones can be distinguished. No distinct pattern of vertical zonation related to depth has been observed, to a depth of 2500 m.

(3) The lateral metamorphic zonation in the Krivoy Rog basin is temperature-dependent, but isobaric. The temperature of metamorphism, as in the whole of the central part of the Ukrainian shield, increases from south to north and from east to west (from eugeosyncline to miogeosyncline) and at the same time also rises in the narrower parts of the Krivoy Rog synclinorium.

Thermobarometric data

A large number of gas—liquid inclusions in the ferruginous rocks of the Krivoy Rog basin and other regions of the Ukrainian Shield have been studied recently (Belevtsev and Tereshchenko, 1979). Quartz from bands in iron-formation and jaspilite from iron ore deposits with various degrees of metamorphism has attracted most attention. Quartz from cross-cutting veins was investigated as well. In order to gain some further insight into the origin of the rocks fluid inclusion studies were carried out on garnet, calcite, pyroxene, and amphibole of oxide iron-formation, slates, and other rocks.

The highest temperatures of homogenization in the Krivoy Rog basin (Table 5-VII) are characteristic of the rock inclusions in the northern (408—468°C) and southern (452—469°C) ore regions; the lowest temperatures correspond to inclusions in the rocks of the central (Saksagan) ore region (346—400°C). Homogenization in all primary inclusions of the Krivoy Rog rock results in a liquid phase. The shape of the inclusions is isometric, regular and bipyramidal. The inclusions are usually single, while groups of 2—3 inclusions

TABLE 5-VII

Homogenization temperature of gas-liquid inclusions in minerals of ferruginous rocks of the Krivoy Rog basin*

Region of the basin	Homogenization temperature of inclusions (°C)		Homogenization temperature of carbon dioxide (CO_2) primary inclusions (°C)	
	Quartz from bands	Quartz from veins	Banded quartz	Quartz from veins
Central	346—400	325—357	27—30	26—29
Southern	452—469	336—402	26—28	23—31
Northern	408—468	324—391	28—29	28—30

* Analytical data obtained by S.I. Tereshchenko.

located chiefly in the centres of quartz grains are less abundant. The size of the inclusions varies from fractions of a micron to 5 microns.

Thermobarometric data for the Krivoy Rog rocks indicate greenschist facies for the central region, and amphibolite and partly granulite facies for the northern and southern regions.

The mode of occurrence, state of aggregation, type of homogenization and setting of the primary inclusions in banded quartz are similar to those of inclusions in quartz from veins. However, the homogenization temperature of primary inclusions in vein quartz is lower than that of quartz of bands of the enclosing rocks (see Table 5-VII). Here again, homogenization produces a liquid phase. The quartz veins common in quartz-rich rocks were formed from residual metamorphic solutions.

In addition to primary gas-liquid inclusions, band and vein quartz both contain a large number of secondary and late-secondary inclusions which upon homogenization pass into a liquid phase over a wide range of temperatures (from 120 to 400°C) and pressures. The secondary inclusions occur along a rather complicated system of healed fissures and show great diversity.

Carbon dioxide is the main component of the metamorphic solutions. Thermobarometric data show that the water/carbon-dioxide ratio varies over a wide range in the ore regions of the Krivoy Rog basin and the role of carbon dioxide is not the same for rocks of different metamorphic grades. The greatest quantity of carbon dioxide in inclusions is observed in the rocks of the southern and northern regions. Carbon dioxide and water/carbon dioxide inclusions make up about 90% of all inclusions and they occur in practically all minerals studied. Carbon dioxide inclusions comprise only about 10% of the inclusions in rocks of the central region.

The inclusions in garnets from the rocks of the granulite facies of metamorphism are composed of pure carbon dioxide. Primary carbon dioxide inclusions are characterized by an ideal shape of the negative host-crystal and homogenization temperatures ranging from 10 to 20°C, which corresponds to a 0.856—0.776 g/cm^3 density of carbon dioxide.

The pyroxene of the rocks of the granulite facies contains inclusions of brine-melt consisting mainly of a solid phase and solutions of high concentration. The solid phase dissolves at a temperature of 680—720°C. In addition, pyroxene has primary inclusions of pure carbon dioxide with homogenization temperatures between 12° and 17°C and a density of 0.851—0.761 g/cm^3.

The results of the studies of inclusions in minerals of the ferruginous rocks metamorphosed to the granulite stage (quartz, garnet, pyroxene) suggest that carbon dioxide was of primary importance in comparison with rocks of the amphibolite and greenschist facies. Calculated from density data, the carbon dioxide pressure in the primary inclusions of minerals in the granulite facies ranges from 4000 to 5000 bar, and in rocks of the greenschist and amphibolite facies from 1500 to 3000 bar.

Based on geological evidence, and studies of the fluid inclusions in minerals, thermobarometric conditions for the formation of the Krivoy Rog ferruginous rocks are as follows:

(1) Metamorphism in the greenschist facies progressed at a temperature of 320—400°C and a pressure of 1500—2500 bar, and the phase in the fluid inclusions is characterized by a negligible content of carbon dioxide.

(2) Metamorphism in the amphibolite facies occurred at a temperature of 470—340°C and a pressure of 200—2500 bar, and carbon dioxide appears as a component of about 90% of the fluid inclusions.

(3) Metamorphism of the granulitic facies is visualized at temperatures from 460 to 730°C and pressures from 4200 to 4400 bar. In carbon dioxide solutions the density of carbon dioxide amounts to 0.796—0.851 g/m^3.

The obtained data on temperature and especially on pressure are somewhat lower than those for corresponding metamorphic facies accepted in petrology (Sobolev, 1970).

GENETIC MODEL FOR PRECAMBRIAN BANDED IRON-FORMATIONS

Physicochemical investigations show that conditions of simultaneous precipitation of iron and silica are extremely restricted. Only a favourable combination of many factors in specific places of the Earth's surface and during certain periods of geological history may therefore be expected to lead to chemical precipitation of ferruginous-siliceous sediments uncontaminated by terrigenous and volcanogenic material. A genetic model for the chemical deposition of iron formation is suggested on the basis of new experimental and estimated thermodynamic data (Belevtsev and Mel'nik, 1976).

Environmental conditions of iron migration and precipitation

Fe^{2+} and Fe^{3+} ions are the prevailing ionic forms for iron migration, and Fe^{3+} migrates in colloidal form as well. An analysis of geochemical, experimental and estimated data (Mel'nik, 1973) shows that migration of iron oxide in colloidal or ionic form is closely constrained. A considerable amount of the Fe^{3+} ion may exist only in very acid solutions (pH = 0—2), and an increase in pH to 2—4 causes hydrolysis and precipitation of the insoluble $Fe(OH)_3$ hydroxide. The existence of strong acid solutions in the Precambrian crust of weathering cannot be accounted for by physicochemical processes even if there was a very high content of carbon dioxide in the atmosphere. Acid thermal waters of the type which commonly occur in regions of present volcanic activity were probably no less abundant in the Precambrian. These waters are commonly rich in ferrous, rather than ferric, iron. When volcanic waters reach the surface, the Fe^{2+} ion is oxidized to its trivalent state. The migration of iron is sharply limited, and iron precipitates in the form of hydroxide due to its reaction with rocks, dilution by meteoric waters and the buffering effect of carbonate and silicate systems in the marine waters.

Experimental data show that colloidal ferric iron is unstable in the presence of electrolytes, especially the SO_4^{2-} ion. The migration of iron in colloidal form was, probably, less widespread in Precambrian than in Phanerozoic time, as there was no organic matter in the weathered land surface or its associated water (there was no terrestrial vegetation during the Precambrian).

Therefore, migration of iron in the presence of free oxygen is not likely to have taken place. Still less probable is the incursion of considerable amounts of Fe^{3+} iron into the basin of chemogenic accumulation beyond the areas of terrigenous sedimentation. The pH factor and the gradients of electrolyte concentration (Mg^{2+}, Na^{2+}, Fe^{2+}) acted as geochemical barriers and restricted iron precipitation in the form of hydroxide to the near-shore areas.

When free oxygen is absent, the ability of iron to migrate increases sharply; this is shown in thermodynamic diagrams by an expansion of the stability field of the Fe^{2+} ion. In slightly acid and even neutral solutions retention of appreciable amounts of ferrous iron in solution is quite possible. From such solutions iron precipitated mainly as carbonates or oxides, depending on the changing CO_2 pressure and pH, and the increase in Eh associated with the evolution of the atmosphere, hydrosphere and biosphere. The interdependent factors pH and P_{CO_2} acted as geochemical barriers for the precipitation of iron in the form of carbonate. The chemogenic formation of siderite could only take place in oxygen-free evolutionary stages, but periodic fluctuations in P_{CO_2} permitted precipitation. A fundamental reason for the massive precipitation of iron in the form of oxides and hydroxides was a sharp increase in redox potential.

Migration and precipitation of silica

The solubility of silica in the form of the ionic monomer $Si(OH)_4^0$, within pH values ranging from 2 to 10, is 80—100 mg/l, and does not depend on the acidity of the solution. Silica derived from the weathered crust and volcanic sources accumulated in Precambrian basins, which were practically free of organic matter. It is known that silica can be removed from saturated or undersatured solutions by biological activity.

In the colloidal form, the migration of SiO_2 and its accumulation in chemogenic sediments are greatly increased. A volcanogenic source of colloidal silica is the most probable. Thermal waters of present-day volcanic regions are acid, and contain about 200—300 mg/l SiO_2, rarely 900 mg/l (Zelenov, 1972). These figures exceed many times the "equilibrium" concentration required for the precipitation of amorphous silica. On the basis of Mel'nik's experimental data (Mel'nik et al., 1973) such ion-colloidal solutions are quite stable in acid and slightly acid environments (pH $<$ 4—5), with mean SiO_2 contents of about 100—1000 mg/l. Such solutions could readily transport silica over great distances. Geochemical barriers which caused the precipitation of colloidal silica are suggested by these studies. They are: (a) pH gradients, orginating inevitably from mixing acid volcanic and slightly acid or neutral fluvial or marine waters and (b) gradients of electrolyte concentrations, mainly Mg^{2+}, Na^{+}, and possibly Fe^{2+}.

Formation of banded ferruginous-siliceous sediments

It is possible now to summarize a wide range of geological and experimental data and to present a generalized physicochemical model of the formation of banded ferruginous-siliceous sediments.

A biochemical variant of the sedimentary-volcanogenic hypothesis is preferred. It is assumed that during the oxygen-free stage of hydrospheric and atmospheric history all iron as Fe^{2+} ion produced from weathering of the crust and volcanic sources was accumulated in the slightly acid environment of the ancient basins. Simultaneously there occurred an accumulation of ionic silica up to "equilibrium" concentrations.

When iron and silica concentrations reached high values, the inflow of oversaturated volcanic solutions caused the precipitation of iron and silica. The oldest (Archaean) iron formations closely associated with volcanics ("Algoma type") precipitated from oversaturated solutions formed by mixing acid thermal waters with oceanic waters saturated with carbon dioxide but devoid of free oxygen. Temperature, pH gradients, concentration of carbon dioxide and electrolytes all acted as geochemical barriers.

The redox conditions of Proterozoic time were greatly influenced by biogenic factors. Eh increased and Fe^{2+} was oxidized to Fe^{3+} with consequent precipitation of iron oxides and hydroxides. A peak of iron-formation deposition occurred during the Proterozoic. Any connection of the Krivoy Rog Superior type iron-formations with volcanism is quite remote. A volcanogenic source for silica and for part of the iron, and a normal process of accumulation of ferruginous-siliceous sediments in the marine, comparatively shallow, basin are suggested.

Periodic sharp changes of the redox conditions (Eh gradients) in the zone of photosynthesis were the main cause of iron precipitation. It is suggested that significant phytoplankton colonies, whose relicts are preserved in the ferruginous rocks, grew locally in the ancient basins, at specific water depths (5—10m) sufficient to protect the living organism from ultraviolet radiation, and at an optimum distance from the shoreline.

Periodic bursts of intense phytoplankton "blooming" in the Precambrian led not to an increase of oxygen in the atmosphere, but to the oxidation of Fe^{2+} iron to Fe^{3+} in the water, causing precipitation of iron in the form of insoluble hydroxide. A considerable release of oxygen under photosynthesis was accompanied by an intense increase of biogenic mass and CO_2 absorption, which in its turn, led to a local decrease in P_{CO_2} and a rise of pH. The resulting $Fe(OH)_3$ precipitation was accompanied by the accumulation of carbonate. A reducing zone was present in deep parts of the basin, where iron precipitates were dissolved. This fact accounts for the absence of great accumulations of ferruginous rocks in regions of deep water (Belevtsev and Mel'nik, 1976).

Coagulation of colloidal silica began in near-shore zones, where clots of amorphous SiO_2 precipitated with terrigenous material or were carried away by currents. The zone of maximum chemogenic precipitation of SiO_2 was at some distance from the shore and did not coincide spatially with the zone of maximum accumulation of argillaceous sediments which were later transformed into slates. This displacement of zones is explained by a spatial separation of the main geochemical barriers (pH and electrolyte concentrations) which caused the precipitation of colloidal silica (Belevtsev and Mel'nik, 1976). With increasing distance from the shoreline discrete cherty layers appear in terrigenous sediments of clayey composition; their number and thickness increase towards the zones of sedimentation of purely chemogenic rocks. The banding of cherty pelitic sediments (cherty slates and iron-poor cherts) is explained by the periodicity of sedimentation of the terrigenous components. After passing the near-shore geochemical barriers the solutions contain almost no colloidal silica, and in the deep zones of the basins the intensity of chemogenic accumulations of silica sharply decreased.

Cyclic (seasonal) precipitation of iron compounds in relatively shallow basins, accompanied by constant sedimentation of amorphous silica, accounts for the character of banding and variety of textures, structures and mineral associations of iron formations. Rapid sedimentation of dense particles of iron hydroxide, and slow accumulation of amorphous masses of silica represent the kinetic factors that governed the distinct separations of interbeds. Iron carbonates did not precipitate as fast as iron hydroxides and this led to the formation of peculiar siderite-cherty interlayers in the succession. It should be noted that biological and chemical precipitation of iron occurred not only in relatively deep zones but also in near-shore areas. But in this case terrigenous sedimentation dominated and neither the iron components nor the amorphous silica formed separate layers. A considerable amount of iron in almost all slates related to iron-formations is taken as indirect evidence for this.

When the main episode of iron-formation deposition was completed, the chemogenic sedimentation of carbonate rocks, composed of dolomite and calcite, was possible. Free oxygen appeared in equilibrium with the atmosphere after the complete oxidation of Fe^{2+}, and the consequent precipitation of all iron accumulated in the hydrosphere during the early stages of the Earth's evolution.

Stabilization of sediments and formation of authigenic iron minerals (siderite, hematite, magnetite, silicate and sulphide) came to an end during the diagenesis stage. During regional dynamothermal metamorphism of the ferruginous-siliceous rocks many structural and textural features, mineral compositions, mineral associations and separate minerals were preserved, and the chemical composition of the sediments changed little. Later silica was removed and iron redistributed during circulation of metamorphic solutions and this led to the formation of the high-grade iron ores.

IRON ORE DEPOSITS

Two types of iron ore are distinguished in the Krivoy Rog basin: high-grade ores, with 46—70% Fe content, utilized in metallurgical processes without preliminary treatment, and low-grade ores, or oxide iron-formation and jaspilite, with Fe contents ranging from 15 to 35%, requiring preliminary upgrading.

The high-grade deposits are mostly concentrated within jaspilite and oxide iron-formation of the middle suite. They are composed of ore beds, thick hinge deposits, shoots and pockets. Ore bodies are confined to folded and combined folded and faulted structures, where they form groups or clusters making up major deposits.

The deposits of the basin lie in three ore fields — southern, central (Saksagan), and northern — which are characterized by various geological and structural conditions, mineral composition and genetic characteristics of the ores.

The southern ore field is situated in the southern part of the basin and extends from the Ingulets mine to the town of Krivoy Rog. It is characterized by bedded and lenticular deposits of iron-mica-magnetite and chlorite-magnetite ore confined to the upper iron ore subsuite of the Krivoy Rog series (Table 5-I).

The Saksagan ore field is in the central part of the basin, extending from the Dzerzhinsky mine to the Lenin mine. Massive and porous martite and loose goethite-hematite-martite and goethite-hematite ores are common and form rather complex deposits among which there often occur ore columns and stocklike and bedded ore deposits morphologically related to the thick keel of the Saksagan geosyncline (Fig. 5-9).

The northern ore field, in the northern part of the basin, contains massive amphibole-magnetite and hematite-magnetite ores, confined to complex folded and fractured block structures.

The Saksagan ore field, producing about 85% of the Krivoy Rog basin output, has the greatest commercial importance.

Four types of ore occur in the basin:

(1) Massive magnetite and siliceous-magnetite ores, confined to combined fold-fault structures in rocks substantially affected by Mg + Fe and Fe metasomatism.

(2) Martite and martite-hematite ores, forming shoots and sheets. They occur in complexly folded areas of ferruginous rocks. Magnetite ores grade

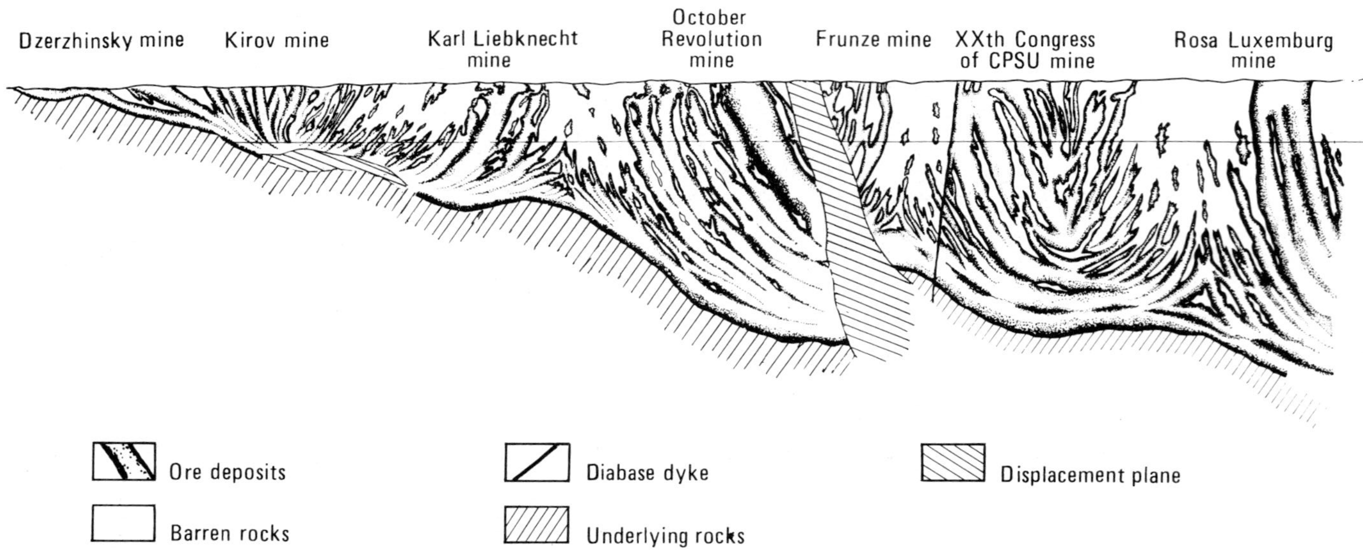

Fig. 5-9. Projection of high-grade iron ore deposits of the Saksagan syncline on an approximately vertical and north—south plane, showing their distribution in relation to structure.

into martitic ores at either shallow or appreciable depths within one vein or deposit.

(3) Soft, hydrated ores represented by goethite-hematite-martite, goethite-hematite varieties. They are common within the weathering zone, or in narrow deep oxidation zones.

(4) Low-grade ores, composed of (a) magnetite jaspilite and iron-formation and (b) oxidized martite and goethite-hematite iron-formation and slate.

Ores of the second type (martite and martite-hematite) constitute the bulk of all the Precambrian iron ore deposits. They are common in the Saksagan ore field. Ores of the first and third types are of restricted occurrence and of local significance. The basin is rich in low-grades ores, from which so far only magnetite has been concentrated.

Magnetite and silicate-magnetite ores of the first type are typical of the deposits of the northern ore field and also occur in the southern ore field. They occur among amphibole-magnetite iron-formation and slate of the Krivoy Rog middle (Saksagan) suite. These deposits are confined to zones of metasomatism where rocks have been converted to magnetite-amphibole compositions.

The metasomatic character of ore formation governed the ore body morphology and was controlled by two factors: (a) the degree of tectonic processing of the rock (folding and faulting) which provided passages for the ore-forming fluids; and (b) the widespread occurrence of amphibole, magnetite-amphibole slate, and jaspilite, which are lithologically favourable for metasomatism. Ore bodies occur in structural concordance with the enclosing metamorphic rocks and are confined to zones of metasomatism.

Martite and martite-hematite ores of the second type are common. They make up numerous deposits of the Saksagan ore field; each deposit is made up of several shoots either on the limb of a syncline or in its hinge (see Fig. 5-9). Martite ore deposits are found in oxide iron-formation and jaspilite of horizons K_2^{4Fe}, K_2^{5Fe} and K_2^{6Fe} of the upper iron ore subsuite of the Saksagan suite. Horizon K_2^{5Fe}, composed of jaspilite, embraces the largest number of deposits. Its iron content varies between 42 and 77%. The total area of ore deposit area of this horizon is more than 70% of the total of the deposits of the whole ore field of the Saksagan region.

The development of ore deposits in various horizons is highly variable. The average mineralization coefficient for all ferruginous rocks of the basin is equal to 0.04, but varies from 0 to 0.9 for different horizons. Thickening and thinning can be observed along the strike in all, or almost all, horizons simultaneously. As a result of this, transverse mineralization bands were developed, and may be traced in two, three, or even four, neighbouring ferruginous horizons. Such transverse bands of ore are confined to displacement in the parallel bedding of one or several iron-formation and jaspilite bands forming gentle flexures. The flexures are arranged in moderately dipping transverse folds 100—1000 m wide and 100—150 m high.

Ore deposits in the Saksagan region are closely associated in space and shape with transverse zones of deformation, and do not occur in unfolded rock. Ore deposits are not found in localities devoid of folded rock between mines. In these places magnetite bearing rocks or lean ore beds are abundant.

Loose or soft hydrated ores of the third type are characteristic of the central (Saksagan) and in part of the southern regions. They form separate (goethite-hematite) or complex (goethite-hematite-martite) deposits. Goethite, dispersed hematite, and martite are the basic ore minerals. These ores are confined to oxidation zones in silicate-oxide iron-formation and ferruginous slates. Hydrated ores occur in deep oxidation zones and in the old zone of weathering.

Ore deposits have been studied in mines 800—1000 m deep and surveyed in drillholes with a depth range of 1500—3000 m. This has permitted for the first time an estimate of the depth and character of the oxidation zones. Linear zones extend commonly to a depth of 1300—1400 m, and in places to 2900 m. The lower boundary of oxidation is not determined. Deep oxidation in the Saksagan region coincides with ore belts along transverse zones of folding.

Comparison of the compositions of massive and porous ore makes it evident that porous ores were formed by almost complete solution and removal of quartz. The silica was not replaced by newly formed minerals such as chlorite, carbonate, and goethite. This resulted in a porosity change from 4—5% for massive ores to 25—30% for porous ores. The quartz content, in the process of formation of porous ores, decreased from 15—25% for compact ores to 0.5—8% for porous ores. Magnetite is almost completely oxidised to martite, consistent with a reduction of ferrous iron to 0.6—0.7%.

The main characteristics of the Krivoy Rog ores are: their mineral compositions are analogous to those of the enclosing rocks; the ores contain the same set of chemical elements as the enclosing rocks; ore deposits are confined to folded and faulted structures; metasomatic processes greatly influenced formation of ores; there is no alteration adjacent to ore; there is a close relation in space and time between the formation of ore and folding; there is no zonality of mineral associations within the ore; ore deposits show no spatial or time relation to intrusive rocks.

Low-grade ore occurs throughout the basin and is represented by magnetite jaspilites and iron-formation as well as by martite and goethite-hematite schists. Magnetite ores mined in the basin are located between high-grade ore deposits or either southwards or northwards from them. These low-grade ore bodies vary in thickness and length. Their grade depends on the primary iron content and the magnetite grain size, which is in itself dependent on conditions of metamorphism. The deposits are usually large, with reserves of many hundreds of millions to several billion tons.

The genesis of iron ores of the Krivoy Rog basin is considered as a natural historical process of iron accumulation consisting of successively developing sedimentation, metamorphic and supergene processes. The sedimentation and diagenesis of ferruginous and siliceous material which formed the basis of all the ferruginous rocks represent the earliest stage of the iron accumulation process. The Archaean crystalline rocks — metabasites, ultrabasites, gneiss, migmatites and granites — were the source of the initial material of the iron oxide suite. The sedimentation occurred in the synclinal environment of the Krivoy Rog—Kremenchug subsyncline.

The second phase of iron concentration in the rocks is related to the dynamothermal metamorphism which resulted from the formation of the fold structures in the Krivoy Rog basin. This phase is associated with the formation of the bulk of high-grade and low-grade ores of the Krivoy Rog basin.

The folding, flow, and inter-layer movement of ferruginous-siliceous sediments caused heating and circulation of metamorphic solutions, which in turn were responsible for the migration of iron, silica, manganese, sodium, calcium and aluminium, and recrystallization of the rocks, with the appearance of new mineral associations.

At the same time ferruginous-siliceous sediments were converted to iron-formation (jaspilites) and slates. In places where folds (mainly transverse) and fissure zones developed, and where metamorphic solutions circulated intensely, transport of rock components occurred, in the first place iron and silica. Under tectonic compression, in certain parts of ferruginous rocks, quartz became unstable, was dissolved and removed from compression zones. As a final result, these sites favoured formation of residual high-grade metamorphic ores which are common in greenschist facies regions in the basin.

A second stage is related to ore-forming Mg and Fe metasomatism characteristic of the amphibolite and granulite facies. This stage is associated with the formation of metasomatic hematite-magnetite ores.

The third stage of ore formation and alteration is related to supergene alteration of deep zones of ferruginous rock and formation of high-grade iron ore by oxidation. Supergene processes caused considerable removal of silica, and compact magnetite was converted into soft ore. These ores show a chemical relationship to their host rocks, so that martite ores were formed in jaspilites, goethite-hematite-martite ores in silicate-ferruginous hornfels, goethite-hematite ores in ferruginous-silicate schists, and so on.

REFERENCES

Belevtsev, R. Ya., 1970. Metamorphic zonation of the Krivoy Rog basin. Geol. Zhurn., 30 (4): 25—38 (in Russian).

Belevtsev, Ya. N., 1947. Deposition of the rocks of the Krivoy Rog suite. Sov. Geologiya (Soviet Geology), 23: 44—53 (in Russian).

Belevtsev, Ya. N., (Editor-in-Chief), 1957. Geological Development and Iron Ores of the Krivoy Rog Basin. Gosgeoltekhizdat, Moscow, 280 pp. (in Russian).

Belevtsev, Ya. N., (Editor-in-Chief), 1959. Genesis of Iron Ores of the Krivoy Rog Basin. Izd. Akad. Nauk Ukrain. S.S.R., Kiev, 308 pp. (in Russian).

Belevtsev, Ya. N., (Editor-in-Chief), 1962. Geology of Krivoy Rog Iron Ore Deposits (2 volumes). Izd. Akad. Nauk Ukrain. S.S.R., Kiev, 448 pp. (in Russian).

Belevtsev, Ya. N. and Koptyukh, Yu. M., 1974. Characteristics of the formation of Precambrian iron-formations from the evidence of sulphur isotopic composition of sulphides. Geol. Zhurn., 34 (3): 41—48 (in Russian).

Belevtsev, Ya. N., Lugovaya, I.P. and Mel'nik, Yu. P., 1969. Isotopic composition of oxygen of ore minerals of ferruginous rocks of Krivoy Rog. In: Problemy obrazovaniya zhelezistykh porod dokembriya (Problems of the Formation of the Precambrian Iron Formations). Izd. Naukova Dumka, Kiev, pp. 271—279 (in Russian).

Belevtsev, Ya. N. and Mel'nik, Yu. P., 1976. Biogeochemical-accumulation model for the formation of Precambrian iron ore formations. MGK (Int. Geol. Cong.), XXV sessiya, Dokl. Sov. Geol., Nauka, Moscow, pp. 67—78 (in Russian).

Belevtsev, Ya. N. and Skuridin, S.A., 1957. History of formation of the rocks of the Krivoy Rog series. In: Geologicheskoe stroenie u zheleznye rudy Krivorozhskovo basseyna (Geological development and iron ores of the Krivoy Rog basin). Gosgeoltkhizdat, Moscow, pp. 88—103 (in Russian).

Belevtsev, Ya. N. and Tereshchenko, S.I., 1979. Thermobarometric conditions of formation of rocks of iron ore formations of the Ukrainian shield. In: Osnovye parametry prirodnykh protsessov endogennogo rudoobrazovaniya. T.I. Fizikokhimicheskaya evoluytsiya rudnoobrazuyushchikh sistem. Medno-nikelevye, zeheleznorudnye, molibdenovye mestorozhdeniya. (Basic parameters of natural endogenic processes of ore formation. Vol. I. Physico-chemical evolution of ore-forming systems. Copper-nickel, iron ore, molybdenum deposits). Nauka, Sibirskoe otdelenie, Novosibirsk, pp. 166—171 (in Russian).

Belevtsev, Ya. N., Zhukov, F.I., Skobelev, V.M. and others, 1978. Characteristics of the formation of the Precambrian rocks of the Krivoy Rog iron ore basin from the evidence of sulphur isotopic composition of sulphides. Geol. Zhurn., 38 (1): 1—19 (in Russian).

Chukhrov, F.V., Vinogradov, V.I. and Yermilova, L.P., 1968. On the question of sulphur isotope fractionation in the Proterozoic. Izv. Akad. Nauk S.S.S.R. Ser. Geol. (Proc. Acad. Sci. U.S.S.R., Geol. Ser.). 11: 3—11 (in Russian).

Chukhrov, F.V., Yermilova, L.P. and Vinogradov, V.I., 1969. On the isotopic composition of sulphur as an indicator of the possibility of some geochemical processes in the older Precambrian. Izv. Akad. Nauk S.S.S.R. Ser. Geol. (Proc. Acad. Sci. U.S.S.R., Geol. Ser.). 9: 50—60 (in Russian).

Cloud, P.E. and Licari, G.R., 1968. Microbiotas of the banded iron formations. Proc. Nat. Acad. Sci. U.S.A., 61 (3): 779—786.

Kalyaev, G.I. and Snezhko, A.M., 1973. New data on the stratigraphic position of the Krivoy Rog Series. Geol. Zhurn., 33 (6): 16—28 (in Russian).

Kanibolotskiy, P.M., 1946. Petrogenesis of the Rocks and Ores of the Krivoy Rog Iron Ore Basin. Izd. Akad. Nauk Ukrain. S.S.R., Chernovtsy, 312 pp. (in Russian).

Kobzar', V.N., 1963. On the stratigraphic and structural position of the metamorphic rocks of the Western Annovskiy belt of northern Krivoy Rog. Geol. Zhurn., 23 (1): 65—73 (in Russian).

Kontkevich, S.O., 1880. Geological description of the environs of Krivoy Rog. Gornyy Zhurn., 1 (in Russian).

La Berge, G.L., 1967. Microfossils and Precambrian iron formations. Geol. Soc. Am. Bull., 78 (3): 331—342.

Lugovaya, I.P., 1976. Characteristics of the isotopic composition of oxygen of some genetic types of iron ores of the Precambrian of the Ukraine. Geol. Rudn. Mestorozhd., 6: 59—67 (in Russian).

Mel'nik, Yu. P., 1973. Physico-chemical Conditions of Deposition of Precambrian Iron-formations. Nauk. Dumka, Kiev, 287 pp. (in Russian).

Mel'nik, Yu. P., Drozdovskaya, A.A. and Vorobyeva, K.A., 1973. New experimental and calculated data on the conditions of deposition of Precambrian ferruginous-siliceous sediments. Geol. Zhurn., 33 (2): 12—23 (in Russia).

Mikhalsky, A.S., 1908. On some basic quastions of Krivoy Rog geology. In: Sbornik neizdannykh trudov A.S. Mikhal'skogo. Trudy Geologicheskogo Komiteta. Novaya ceriya (Collection of the unpublished works of A.S. Mikhalsky. Works of the geological committee. New Series), 32: 3—60 (in Russian).

Nikol'skiy, A.P., 1960. Geological-metallogenic Sketch of the Eastern Part of the Ukraine Shield. VSEGEI, Novaya seriya, 162 pp. (in Russian).

Perchuk, L.L., 1970. Equilibrium of Rock-forming Minerals. Nauka, Moscow, 391 pp. (in Russian).

Plaksenko, N.A., 1969. Particulars of the palaeogeographic setting of the formation of the ferruginous-siliceous sediments of the Kursk series and questions on the theory of Precambrian iron-ore deposition. In: Problemy obrazovaniya zhelezistykh porod dokembriya (Problems of formation of Precambrian iron-formations). Izd. Naukova Dumka, Kiev, pp. 11—27 (in Russian).

Pyatnitskiy, P.P., 1898. Studies of the crystalline schists of the steppe belt of southern Russia. Trudy Obshchestva ispyto prirody pri Khar'kovskom un-te (Works of the Society for the Investigation of Nature at Kharkov University), 32 pp. (in Russian).

Pyatnitskiy, P.P., 1925. Genetic relationships of the Krivoy Rog ore deposits. Vol. 1. Iron-formations and jaspilites. Trudy Inst. Priklad. Mineralogii i Petrografii (Trans. Inst. Econ. Mineral Petrography), Kharkov, 17: 42 pp. (in Russian).

Semenenko, N.P., 1966. Metamorphism of Mobile Zones. Izd. Naukova Dumka, Kiev, 298 pp. (in Russian).

Sobolev, V.S. (Editor-in-Chief), 1970. Facies of Metamorphism. Izd. Nedra, Moscow, 432 pp. (in Russian).

Strakhov, N.M., 1947. Iron ore facies and their analogs in the history of the Earth. Izd. Akad. Nauk S.S.S.R., Moscow 276 pp. (in Russian).

Svital'skiy, N.I., Polovinkina, Yu. G., Dubyaga, Yu. G., Lisovskiy, A.L., Muzylev, S.V., Dubrova, B.S. and Rabinovich, F.K., 1932. The Krivoy Rog Iron Ore Deposits. Trudy. Vses. Geol.-Razv. Ob'edinen: NKTP (Trans. All-Union Geol.-Prosp. Soc.), 153: 283 pp. (in Russian).

Tanatar, I., 1916. Some considerations on the Krivoy Rog ores and the quartzites enclosing them. Yuzhniy Inzhener (Southern Engineer), pp. 7—8 (in Russian).

Tanatar, I.I., 1923. The genesis of the Krivoy Rog iron ores and the quartzites containing them. Gornyy Zhurn., 7 (in Russian).

Tanatar, I.I., 1939. The geochemical characteristics of Bolshoy Krivoy Rog in connection with the question of genesis of its ores. Trudy XVII MGK (Int. Geol. Congr.), 1.

Thompson, A.B., 1976. Mineral reactions in pelitic rocks, Parts I and II. Am. J. Sci., 4: 401—454.

Tugarinov, A.I. and Grinenko, V.A., 1965. Conditions of deposition of Lower Proterozoic formations according to data on variations in sulphur isotopic compositions in sulphides. In: Problemy geokhimii (Problems of Geochemistry). Izd. Nauka, Moscow, pp. 193—203 (in Russian).

Zelenov, K.K., 1972. Volcanoes as Sources of Ore-forming Components of Sedimentary Layers. Nauka, Moscow, 214 pp. (in Russian).

Chapter 6

IRON-FORMATION IN FOLD BELTS MARGINAL TO THE UNGAVA CRATON

G.A. GROSS and I.S. ZAJAC

INTRODUCTION

Lake Superior type iron-formation is distributed with folded Proterozoic sedimentary and volcanic rocks in a sequence of basins which surround the Ungava craton in the eastern part of the Superior Province of the Canadian Shield. The Labrador—Quebec geosyncline is the largest of these fold belts and extends along the entire eastern margin of the Ungava craton for more than 1200 km. The Cape Smith fold belt continues along the northern margin of the craton, the Belcher and Nastapoka Islands fold belt along the western side, the Sutton Lake homocline appears to be a western continuation of the Belcher fold belt, and the Albanel—Temiscamie homocline occurs on the southeast side of the craton. The Gunflint, Mesabi, Cuyuna, Gogebic, Marquette, Crystal Falls—Iron River and Menominee iron ranges of the Penokean fold belt in Ontario, Minnesota, Wisconsin, and Michigan may be remnants of basins that were located on the southeastern edge of the same craton before it was disrupted by later tectonic events that formed the Lake Superior basin (Fig. 6-1).

The fold belts marginal to the Ungava craton consist of thick sequences of shale, dolomite, chert breccia, quartzite, iron-formation and black shale that were deposited on the older crystalline rocks in basins and embayments on the continental shelf. The shelf sediments interfinger in their offshore extensions with greywacke, turbidites and intermediate to basic and ultramafic rocks that form prominent volcanic belts along the margins of the craton.

The iron-formation, because of its distinctive lithology, continuity over great distances and consistent stratigraphic position in the major basins has been used as a horizon marker for regional correlation of stratigraphy in single sedimentary basins and between fold belts on the craton margins. Isotopic dates indicate that the iron-formation in the various basins was deposited between 2400 and 1800 m.y. ago and may have formed a continuous stratigraphic unit on the coastal shelves around the edge of the craton with thicker sections marking deeper depositional basins and embayments. It is considered unlikely that deposition of the iron and silica beds was contemporaneous in all of the wide range of basin environments and more than one horizon of

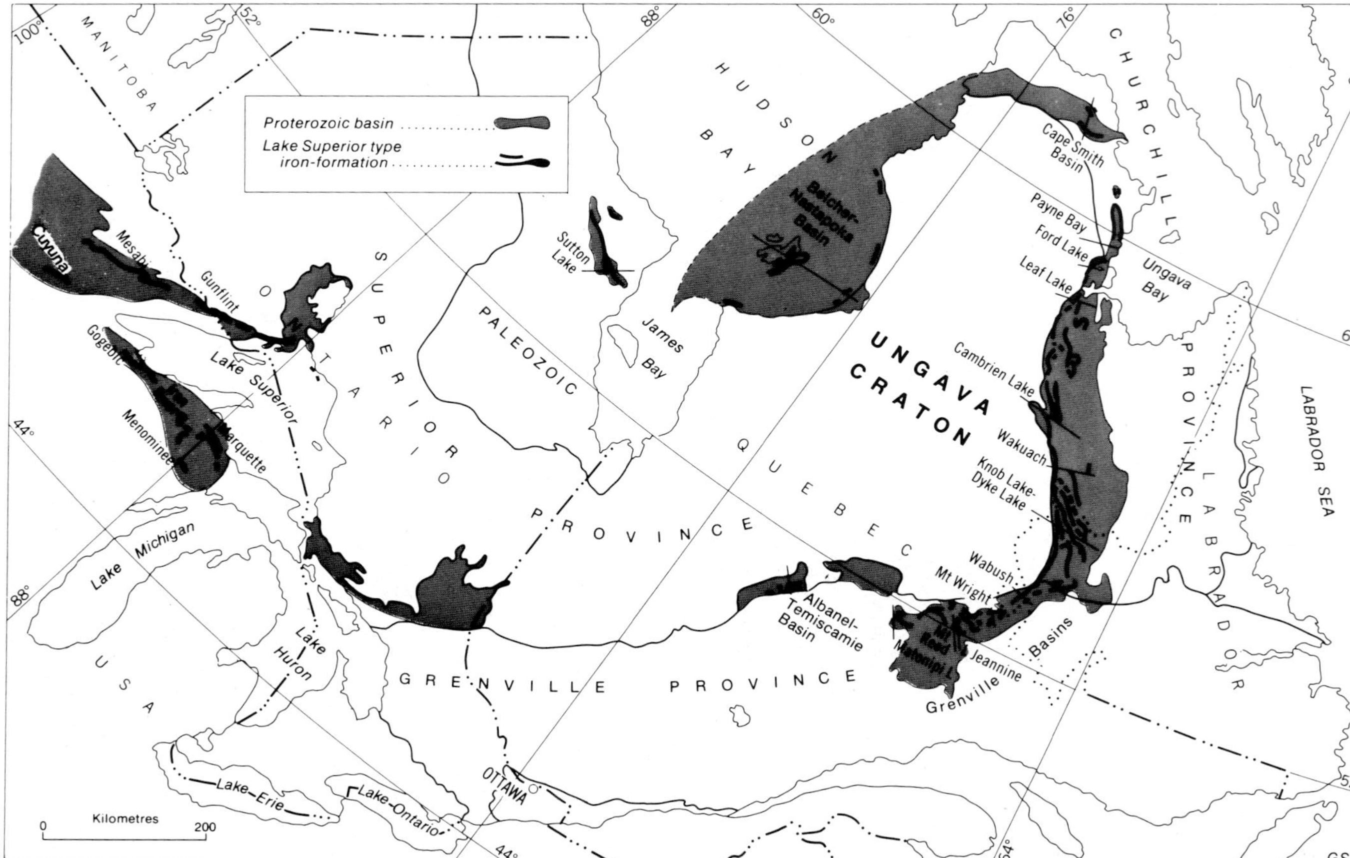

Fig. 6-1. General distribution of iron-formation in fold belts marginal to the Ungava craton.

iron-formation has been identified in the Belcher basin and in parts of the Labrador—Quebec geosyncline. Deposition of iron-formation took place in local basins on the craton shelves as the tectonic-volcanic arc systems developed along the craton margins. Extensive effusive and fumarolic activity associated with these systems are believed to have been major hydrothermal sources of iron and silica.

The iron-formation merges from basin to basin in the Labrador—Quebec fold belt to form a continuous stratigraphic unit more than 1200 km long and it is probably the most continuous single iron-formation in the world. It was deposited along the margins of the craton in separate local basins with distinctive sequences of quartzite, dolomite, and shale near basin shorelines and with other clastic facies, greywacke, turbidites, tuff and volcanic rocks futher offshore on the craton shelves. The thickest sections in the Labrador—Quebec and Belcher—Nastapoka belts occur adjacent to major accumulations of volcanic rock and ultramafic intrusions in the offshore areas. The Wabush basin to the south (Fig. 6-1) appears to be an exception but the corresponding volcanic belt in this area may have been uplifted and eroded during the later Grenville orogeny.

Characteristic structural patterns of the fold belts include low-dipping homoclines of quartzite and iron-formation that lie unconformably on the Archean gneisses, granulites and granitoid rocks of the craton margins and the exposed unconformable contacts along the western margin of the Labrador—Quebec fold belt may trend parallel to the original shorelines of the depositional basins. In some places the younger sedimentary rocks have been thrust over the basement on low-angle thrust planes with the development of imbricate or occasionally nappe structures. These marginal structures pass outward from the craton to broad open folds that are deformed in their crestal parts by complex isoclinal folds and faults developed by thrusting and tectonic transport directed toward the craton. Thick accumulations of mafic, ultramafic and intermediate rocks of volcanic and intrusive origin appear to be related to major structural breaks in the volcanic belt and most are distributed offshore beyond the miogeosyncline part of the craton shelf. Outer boundaries of the Labrador—Quebec geosyncline offshore from the craton are marked by structural breaks at the edge of the fold belt and by highly metamorphosed basin rocks that appear as remnants amid the outlying gneisses and schists.

The southeast border of the Ungava craton is truncated by the Grenville orogenic belt which transects the southwestern extension of the Labrador—Quebec geosyncline (Fig. 6-2). The iron-formation and associated shelf sediments continue southwest into the Grenville terrain for more than 100 km where they are highly metamorphosed, complexly folded and form isolated structural segments. Metamorphism of the iron-formations around the craton varies from subgreenschist to greenschist facies except where lower to upper amphibolite facies are found within the Grenville Province; in part of the Albanel—Temiscamie basin adjacent to the Grenville front; in a part of the

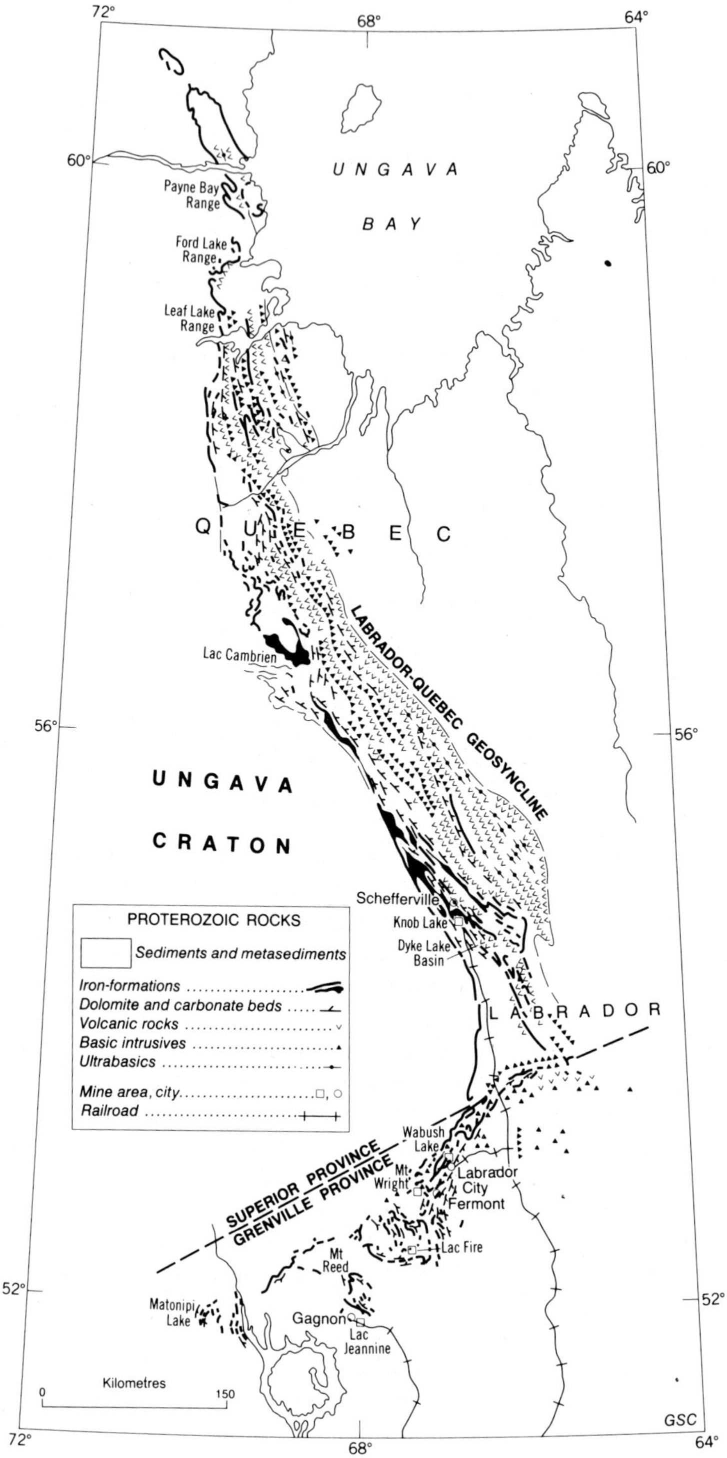

Fig. 6-2. Distribution of iron-formation in the Labrador—Quebec geosyncline.

Labrador—Quebec belt that lies along the west side of Ungava Bay; and in the Cape Smith belt.

Information on the interrelationship of the different basins marginal to the Ungava craton remains fragmented although understanding of individual basins has advanced considerably in the past two decades. Cross-sections through prominent marginal basins in Fig. 6-3 illustrate some of their principal geological features.

The Labrador—Quebec fold belt has received the most attention and has been studied more extensively than other marginal belts. The variation and diversity in detailed geology from basin to basin throughout this belt requires considerably more detailed work and documentation on a uniform scale before some questions regarding the genesis of the iron-formation can be answered and satisfactory generalizations made about its depositional environment. The relative position of major lithological groups of rocks in the Labrador—Quebec fold belt is shown in Table 6-I and a selection of sectional diagrams across this belt illustrate prominent stratigraphic and structural features (Fig. 6-4).

The Knob Lake basin including the Schefferville mine area in its central part is described in some detail in this paper as most of the specialized research has been carried out in this basin area. It exhibits remarkable variation and diversity in the development of primary sedimentary facies and depositional features in the iron-formation which reflect important changes in environmental conditions during the chemical precipitation of the siliceous iron bearing sediments.

The iron-formations in the Circum-Ungava fold belts provide very large iron ore resources of three main genetic types. The first type, located in the Knob Lake basin, consists of earthy hematite-goethite ore derived by secondary enrichment processes from the various lithological facies of iron-formation protore. Oxidation of the iron and leaching of the siliceous minerals took place under the action of deeply circulating groundwater that left large residual masses of iron-oxide minerals in the folded iron-formation. The second genetic type of iron ore consists of highly metamorphosed oxide facies of iron-formation located in the Grenville Province and west of Ungava Bay. Textural changes in the iron-formation involving recrystallization and enlargement of the mineral grains and segregation of the iron and silica constituents in discrete particles have improved the quality and amenability of this iron-formation as a source of high-quality iron ore concentrate in the Grenville Province. Iron-formations that are not highly metamorphosed but amenable to processing and concentration of the iron minerals constitute the third major type of iron ore resource. These fine-grained cherty iron-formations, comparable to the taconite ores of the Lake Superior region, are widely distributed in the Circum-Ungava belt but iron ore has not been produced from them to date.

Very little was known about the geology of the Ungava region prior to this century and only limited reconnaissance was carried out before systematic exploration and mapping programs were initiated by mining companies and the Geological Survey of Canada between 1946 and 1949. Incidental geographical information was acquired by the fur traders, and the iron occurrences were first mentioned between 1866 and 1870 by Louis Babel, a missionary. A.P. Low of the Geological Survey of Canada recognized the Labrador—Quebec fold belt as a major geological feature and anticipated the economic significance of the iron-formations during his exploration of the Ungava region between 1893 and 1895. Because of the remoteness of the region and difficult travel conditions little further geological information was gained before mineral exploration parties visited the Belcher Islands between 1914 and 1918 and reported on their geology (Flaherty, 1918; Moore, 1918; Young, 1922). W.F. James and J.E. Gill discovered iron deposits with material of ore quality near Knob Lake in the Central Labrador—Quebec belt in 1929 and visited the Wabush Lake area further south in 1933. Large concessions of land in the central and southern part of the belt covering the Knob Lake basin were granted to mineral exploration companies in 1933 in Labrador and in 1941 in Quebec.

The Geological Survey of Canada initiated systematic mapping on a scale of 1 : 250,000 around Knob Lake in 1949 with the study of a cross section of the Knob Lake basin (Harrison, 1952; Harrison et al., 1972); and eventually this work was extended to the southwest to cover part of the iron-formation in the Grenville province. More detailed mapping, 1 : 50,000 scale, was carried out by the Quebec Department of Natural Resources in selected areas in the northern, central and southern Grenville parts of the belt. The mining companies were in the vanguard of systematic geological study and investigation during the development of the region with their detailed mapping, and mineral evaluation studies. Their encouragement and sponsorship of research related to the development of the mines, carried out in cooperation with government agencies and universities, has brought a wealth of data on the depositional environment and genesis of the ore deposits.

The first map, scale 1 : 1,000,000, showing the distribution of the iron-formations and related rocks throughout the Labrador—Quebec fold belt and its continuation in the Grenville Province was based on government maps and detailed data from the mining companies and prepared in the Geological Survey's project on the geology of iron deposits in Canada (Gross, 1961b). This was a preliminary step in the preparation of an Economic Geology Series report (Gross, 1968), in which the distribution and detailed stratigraphy of the iron-formation and the various types of iron deposits throughout the geosynclinal belt were described in some detail.

Geological investigations were continued throughout the 1960's resulting in a number of important papers dealing with the mineralogy, metamorphism and sedimentary environment of the iron-formation. The work of Zajac (1974), on the stratigraphy and mineralogy of the Sokoman Formation documented the depositional environment for various facies of iron-formation in the Knob Lake basin and established typical stratigraphic sections for reference in the study of other shelf basins around the Ungava craton.

The geology of the Circum-Ungava geosyncline was reviewed at some length by Baragar, Bergeron, Dimroth and Jackson in a symposium on basins and geosynclines of the Canadian Shield (Dimroth et al., 1970), and the general conditions were considered under which the iron-formation and the Kaniapiskau Supergroup rocks were deposited. There is extensive literature on the Labrador-Quebec geosyncline that includes maps, memoirs, bulletins and papers from the federal and provincial government programs, detailed information in scientific and mining journals and in unpublished university theses. The bibliography included here is but a selection from the extensive literature available. Correlation studies have not made the fullest use of the wealth of detailed data available and there are conspicuous gaps in the systematic documentation of the mineralogy, petrology and sedimentary features of the iron-formation in individual basins. In view of this situation many of the interpretations concerning the environment for deposition of the iron-formations published previously are considered to be tentative.

DESCRIPTION OF BASINS

Stratigraphic sections for major basins marginal to the Ungava craton illustrate some of the characteristic features as well as differences in local sedimentary conditions (Figs. 6-3 and 6-4).

The Belcher—Nastapoka basin

The stratigraphy of the Belcher—Nastapoka basin has been correlated by Jackson (1960) and his diagrammatic reconstruction of the central part (Dimroth et al., 1970) is modified in Fig. 6-3. Arkose and sandstone overlying Archean basement rocks in the near shore areas give way westward and offshore to a thick sequence of stromatolitic dolomite that has pink to red argillite interbedded in its basal parts and tuff and red shale near the top. The lower quartzite and dolomite unit is overlain by a volcanic unit 900 m thick comprising mostly aphanitic amygdaloidal basaltic lava, feldspar porphyry and associated gabbro that extends over most of the basin. A lean iron-formation member, up to 165 m thick, consisting of ferruginous jasper interbedded with argillite, greywacke and sandstone overlies the lower volcanic rocks.

The lower iron-formation members are succeeded by the Nastapoka Group composed in the east near the old shoreline of quartzite and varicoloured ar-

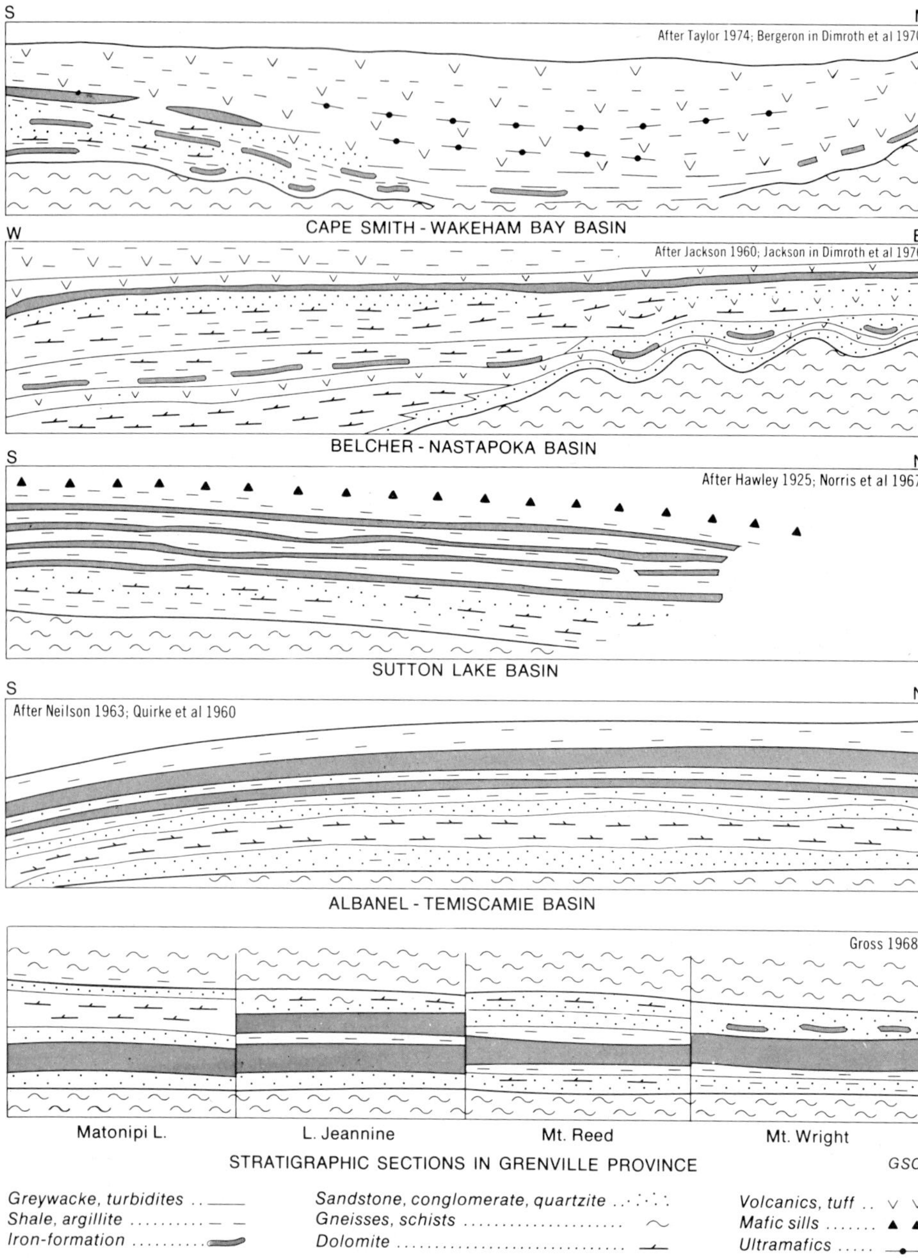

Fig. 6-3. Diagrammatic sections showing the relative position of major lithological units in basins marginal to the Ungava craton.

gillite which give way westward to a thick complex succession of interbedded dolomite, varicoloured shale and tuff, and to local volcanic members in the eastern and upper parts of the basin.

Stratigraphic correlation in the Belcher—Nastapoka basin presents difficult problems because of the insular distribution of exposed stratigraphic sections and possible thrust faults. An alternative hypothesis has been presented recently by Chandler and Schwarz (1980) for correlating the stratigraphy in the eastern part of the basin. Their proposed revisions show a different correlation of key volcanic members and imply that the lower iron-formation member which is probably much thinner than previously reported may occur in the Nastapoka Group rather than in the underlying Richmond Gulf Group that may be significantly older.

The overlying Belcher Group includes sandstone and shale in the east and a continuous dolomite member in the west that are overlain by sandstone and conglomerate at the base of the Kipalu iron-formation. This iron-formation is a continuous unit, 60—120 m thick, highly varied in composition and facies development throughout the basin. It is made up of cherty oolitic hematite and magnetite, jasper, carbonate, iron-silicate and ferruginous shale facies interlayered with varicoloured argillites, tuff and greywacke. Carbonate facies are most prominent in the near shore eastern parts of the basin. Sections of cherty carbonate iron-formation that are relatively free of clastic sediment are usually a few metres thick and reach a maximum of about 60 m.

The iron-formation is overlain by a thick sequence of massive pillowed basalt succeeded by interlayered greywacke, sandstone and argillite. The shelf rocks are intruded by gabbro-diabase sills and dykes. Most of the sedimentary and volcanic shelf rocks were deposited in a miogeosyncline environment with intermittent exposure at the surface giving rise to red beds in many parts of the sedimentary sequence. A more typical assemblage of eugeosynclinal rocks is believed to lie further westward under Hudson Bay. Unstable tectonic conditions over a broad shelf area have given rise to many local depressions in the basin floor where sedimentary facies and stratigraphy, especially in the iron-formation, vary considerably from basin to basin.

The Cape Smith—Wakeham Bay basin

The Cape Smith—Wakeham Bay fold belt extends westward across the northern tip of the Ungava craton and contains two distinctive groups of rocks. The lower Povungnituk Group resting unconformably on the Archean crystalline gneisses contains a thin sequence of iron-formation and black shale associated with quartzite, dolomite, calcareous arkose, mica and chlorite-amphibole schists with interlayered grey-green pillowed basalt intruded by thick gabbro sills. The Chukotat Group unconformable above the Povungnituk Group is composed mainly of pillowed tholeiitic and komatiitic basalts intruded by gabbro and ultramafic sills and contains graphitic slate, tuff, quartzite, conglomerate and black chert in its lower part. The rocks of this belt, especially the upper group, appear to have been deposited in a distal shelf or eugeosyncline environment but are believed to correlate with small remnant basins of iron-formation in the northern part of the Labrador—Quebec belt that were deposited in shallow water and closer to shore.

The Albanel Lake—Temiscamie River basin

The Temiscamie iron-formation is part of the Mistassini Group of rocks located immediately north of the Grenville Province boundary in central Quebec. The group includes a thick succession of dolomitic limestone, conglomerate, quartzite, iron-formation, slate, argillite, greywacke and tuffaceous shale and forms a basin structure about 160 km long and 40 km wide. Basin rocks dip gently southeast and are truncated in the southeast by a fault zone along the edge of the Grenville Province where they are folded and deformed and the rank of metamorphism increases from greenschist facies common throughout the basin to amphibolite facies near the fault zone. Isotopic dates on the iron-formation of 1.3 b.y. (Quirke et al., 1960) and 1.78 b.y. (Fryer, 1972) are thought to be affected by metamorphism that preceded the Grenville orogeny. The Mistassini Group was probably deposited contemporaneously with rocks of the main Labrador—Quebec basin.

The older metasedimentary, metavolcanic, gneissic and granitic rocks are overlain unconformably by conglomerate, arkose, sandstone and greywacke of the Papaskwasati Group. It is conformable with the overlying dolomite limestone of the Albanel Formation in the lower part of the Mistassini Group which is 2000—2500 m thick and characterized by fossil cryptozoan or algal-like structures, anthraxolite and interbedded sandy or argillaceous dolomite. It is overlain by a conglomerate, sandstone and quartzite sequence that is 6—15 m thick and grades upward into iron-formation.

The Temiscamie iron-formation forms a monoclinal structure that dips gently to the southeast and extends southwest for 56 km between Lake Albanel and a fault zone along the Temiscamie River. It is up to 215 m thick and consists of silicate-carbonate facies in the lower part, oxide facies in the middle and carbonate-silicate facies in the upper parts. It has been divided into six lithological units (Quirke et al., 1960; Neilson, 1963) which in ascending order include the:

- "lower argillaceous iron-silicate member", 3—12 m thick;
- "lower sideritic chert and iron-carbonate member", 6—30 m thick, containing stilpnomelane and minnesotaite;
- "magnetite chert member", 20—60 m thick, average thickness 45 m, with prominent oolitic texture and containing variable amounts of magnetite, hematite and siderite, of principal economic interest and similar to taconite ore of the Lake Superior region;
- "upper argillite member", 1.5—14 m thick, that is dark-green to brown

or black, consisting largely of ankerite, siderite, and stilpnomelane;
- "magnetite-minnesotaite-carbonate member", about 30 m thick in the eastern part of the basin; and
- "upper sideritic chert member" about 90 m thick, with prominent oolitic and granular texture that is comparable to the lower siderite chert member but contains less minnesotaite.

The average thickness of the Temiscamie Formation is 137 m; however, the various members are not clearly defined because of interbedding and transitions from one facies to the next and the distribution of lithological facies within the basin is not well known.

The iron-formation is overlain conformably by the Kallio Formation composed of argillite, black slate, and greywacke with abundant carbon and pyrite present in the finer clastic material.

Basins in the Grenville Province

Segments of a highly metamorphosed and complexly folded iron-formation within the Grenville Province extend northeastward from Albanel Lake to Wabush Lake for a distance of 480 km (Fig. 6-1). These structural segments appear to mark a number of small local depositional basins for iron-formation along the southeast shoreline of the Ungava craton. Primary features in the iron-formation and associated metasediments have been largely destroyed by deformation during at least two periods of orogeny; however, the associated quartzite, metadolomite, graphitic schists, and gneisses indicate shelf type sediments that were probably similar to those in the larger basins in the main belt to the north. The stratigraphic sequence and lithology of the metasediments associated with the iron-formation differ considerably from segment to segment indicating a marked variation in depositional conditions. The iron-formation at Mount Wright, Lac Jeannine, and Fire Lake is principally a quartz-hematite facies associated with quartzite. Iron-formation overlying quartzite in the Wabush Lake area consists of a Lower silicate-carbonate unit, a Middle oxide facies and an Upper silicate-carbonate unit.

In most other areas hematite, magnetite and silicate facies are associated with quartzite, dolomite and other metasedimentary rocks. The stratigraphic succession is not consistent and the position of the iron-formation varies in relation to metadolomite and quartzite members. The distribution of different facies of iron-formation and associated metasediments was shown by Gross (1968), with preliminary work on the reconstruction of depositional basins.

Basins in the Labrador—Quebec geosyncline

The fold belt along the eastern margin of the Ungava craton is made up of two separate assemblages of rocks; those on the western side of the fold belt

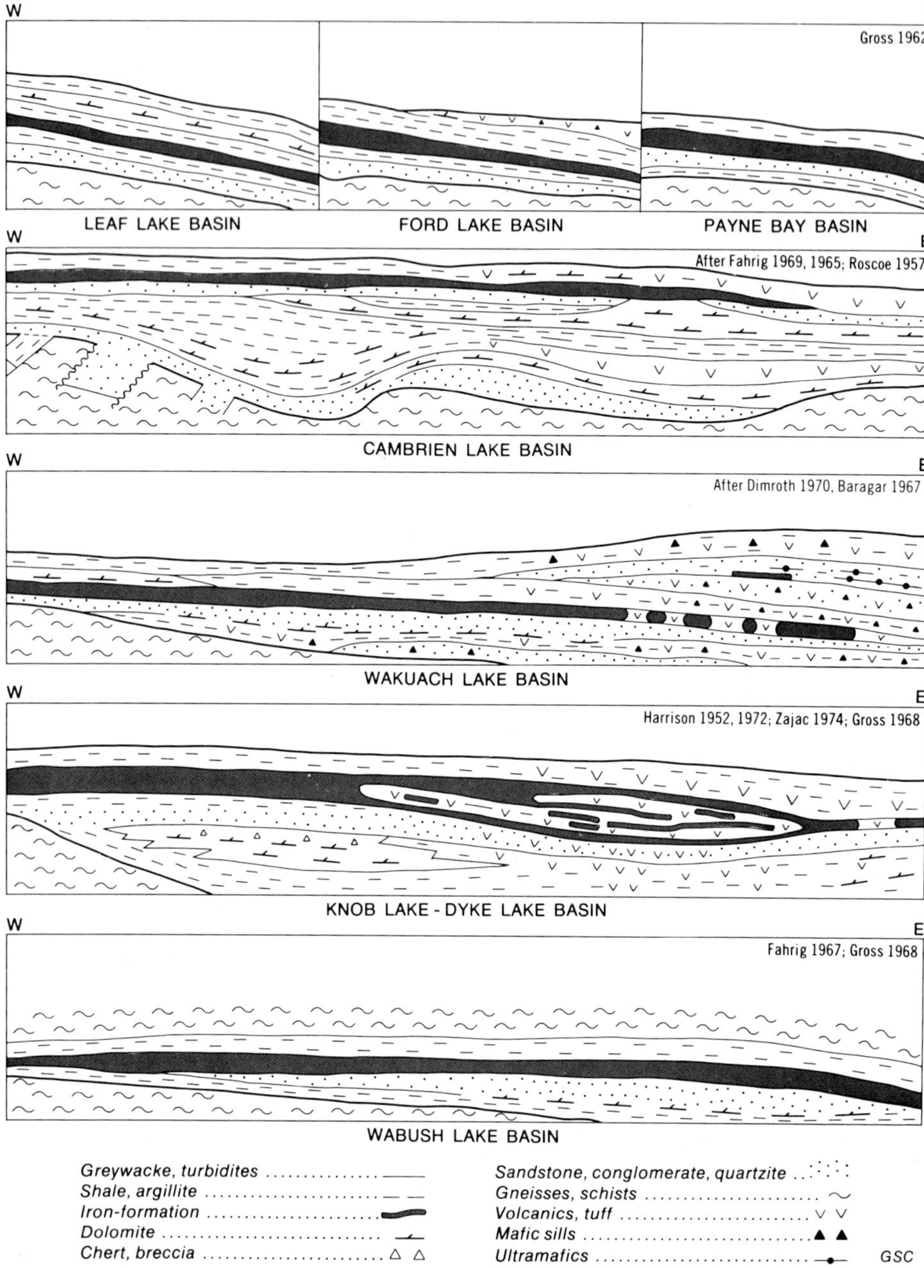

Fig. 6-4. Diagrammatic sections showing the relative position of major lithological units in basins of the Labrador—Quebec geosyncline.

are a typical continental shelf group deposited in a miogeosynclinal environment and those on the eastern side are a eugeosynclinal group composed of greywacke, argillite, turbidites, basic volcanic rocks, and ultrabasic intrusive assemblages (Fig. 6-2). The iron-formation was deposited mainly with the shelf sediments in a number of interconnected basins. The stratigraphic successions in some of the larger depositional basins on the craton shelf are shown in Fig. 6-4 where the iron-formation and associated quartzite and dolomite are thicker and have local distinctive characteristics. Major basins identified along the belt from south to north are located around Wabush Lake, Dyke Lake, Knob Lake, Wakuach Lake, Lac Cambrien, Leaf Lake, Ford Lake, and Payne bay (Fig. 6-2). Rocks of the Kaniapiskau Supergroup are unconformable above older Precambrian granite, granodiorite, and gneisses along the west side of the belt and on the east side geosynclinal rocks are in fault contact with granite, gneisses, hypersthene granites and amphibolites and are in part derived from geosyncline rocks. Table 6-I prepared by Frarey and Duffell (1964), gives a generalized stratigraphic succession for the central part of the fold belt.

The sedimentary and volcanic rocks in this belt were derived from two major source areas. The quartzite, dolomite, and arkose deposited in the west came from the adjacent craton area and form a typical miogeosynclinal, shallow-water succession of continental shelf sediments. Similar rocks in the lower part of the succession southeast of the main group of volcanic rocks in the southern part of the belt were probably derived from a source area lying to the east. The other major source area, a volcanic belt that extended along the eastern part of the geosyncline contributed considerable tuff and clastic material in the argillites and greywackes, as well as extrusive and intrusive rocks. Much of the silica and iron deposited in the cherty iron-formations was probably derived from this volcanic belt and deep-seated fissure systems along its western margin. Because of the different sources of sediment there is a marked change in the rock successions from west to east in all parts of the geosyncline with interfingering of the two groups in its central and eastern parts.

The *Wishart quartzite*, *Sokoman iron-formation*, and *Menihek slate* occur in ascending order throughout the western part of the belt. Dolomite and chert breccia members are present below the quartzite around Knob Lake and in some places in the central and southern part of the belt. Volcanic rocks are interbedded with the iron-formation in the Dyke Lake area. Further north in the Wakuach Lake area abundant greywacke, pyroclastic and volcanic material is interbedded in the succession and obscures the relationship of quartzite and dolomite formations with the shelf rocks in the west (Baragar, 1967; Dimroth 1978; Dressler, 1979).

Sections around Lac Cambrien differ from those in the southern region mainly by the presence of a major group of argillites, quartzites and conglomerates below the iron-formation and dolomite above it. East of Lac Cam-

TABLE 6-I

Table of formations, Central Labrador — Quebec Geosyncline; after Frarey and Duffell (1964)

Era	Supergroup	Group	Formation	Lithology and remarks
			Shabogamo Gabbro	Diabasic olivine gabbro, coarse-grained norite, anorthositic gabbro, hypersthene-augite-plagioclase gneiss
			Sims	Quartzite, grit, conglomerate (flat lying)
			Unconformity	
		Montagnais	Retty Peridotite	Serpentinized peridotite; pyroxenite sills may be older than Wakuach Gabbro
			Wakuach Gabbro	Gabbro, metagabbro, glomeroporphyritic gabbro ("leopard rock"), diorite
			Intrusive Contact	
		Doublet	Willbob	Basalt, metabasalt, flow breccia, minor sediments
			Thompson Lake	Quartzite, greywacke, shale, argillite, conglomerate, intercalated basalt
Proterozoic			Murdoch	Agglomerate, breccia, tuff, basalt, minor sediments
		Knob Lake	Menihek	Carbonaceous slate and shale, quartzite, greywacke; basic volcanic rocks; minor dolomite and chert
			Purdy	Dolomite, minor argillaceous beds
	Kaniapiskau		Sokoman	Iron-formation; intercalated basic volcanic rocks; ferruginous slate, slaty iron-formation, black and brown slate, carbonaceous shale
			Wishart	Feldspathic quartzite, arkose, minor chert, greywacke and slate, intercalated basic volcanic rocks
			Fleming	Chert breccia, minor lenses of shale and slate
			Denault	Dolomite, limestone and cherty facies, fragmental dolomite
			Attikamagen	Green, red, grey and black shales, slate, graphitic slates, phyllites and argillites, intercalated basic flows
			Seward	Grit, arkose, conglomerate, white or pink quartzite, greywacke, acidic flows

brien the statigraphic succession up to the top of the iron-formation is similar to that in the Knob Lake area and the main facies members in the iron-formation are present in the same order of succession in both areas. In the Leaf Lake basin thin quartzite and slate members are present below the iron-formation, a distinctive dolomite member lies above it, and lavas are present within it near the western margin of the belt.

Wishart quartzite with lesser amounts of arkose and conglomerate overlies Precambrian gneisses and granite with angular unconformity along the western edge of the belt except where the iron-formation has been thrust over the gneisses west of Wakuach Lake. West of Lac Cambrien the main beds of quartzite and iron-formation are underlain by thick beds of argillite, quartzite and dolomite, which overlie 1500 m of feldspathic quartzite, arkose and conglomerate that filled deep westerly trending troughs in the basement. The Wishart quartzite below the iron-formation in the Knob Lake area is less than 30 m thick near the western margin of the belt but increases to twice this thickness in the centre of the basin.

A thick succession of slate and greywacke-argillite known as the *Attikamagen Formation* in the Knob Lake region, occurs below the iron-formation and the Wishart quartzite throughout the geosyncline. This basal unit is very much like the upper Menihek slate in composition but carbon-rich beds are not as abundant and much of it is light greenish grey with occasional maroon, yellow or brown beds. In the Knob Lake area the unit thickens from 30 m near the western margin of the basin to more than 365 m in the central part. To the east these lower slate beds are interlayered with volcanic rocks and basic sills and cannot be distinguished from the folded and faulted argillites and slates of the Menihek Formation. Thin beds of dolomite, quartzite, conglomerate, chert, and tuff occur in the Attikamagen slate in the southern part of the area associated with fine-grained clastic material of volcanic origin.

A number of dolomite members are found below and above the quartzite iron-formation units. The *Denault dolomite* of the Knob Lake area and the dolomite member in the Wabush Lake area lie above the lower slate-argillite formation (Attikamagen). Three dolomite members may be present in the Wakuach Lake area: one below the lower slate, one below the main quartzite unit, and one above the iron-formation. A thick dolomite formation (the Abner) occurs in the succession of argillite rocks above the iron-formation in the Koksoak River area in the north. Dolomite beds of uncertain stratigraphic position are present in a few places along the eastern margin of the belt and some occur below the iron-formation in the central-eastern area.

Chert breccia of the *Fleming Formation* lies above the Denault dolomite and below the Wishart Quartzite in the western part of the Knob Lake basin. The lower part, directly overlying dolomite, is composed of angular laminated grey chert fragments and brecciated dolomite embedded in a carbonate matrix that grades upward into a colloform dense chert matrix, and the main upper part consists of chert fragments embedded in quartzite. This lenticular unit is

TABLE 6-II

Description of iron-formation facies, French Mine (after Gross, 1968), for chemical analyses see Table 6-III.

	B523 Silicate-Carbonate	B524 Lower Red Cherty	B525 Pink Cherty
Location	1.6 km northwest of French Mine on northwest slope of Pete Signal Hill	76 m north of ore loading station north side of French Mine	107 m northeast of loading station north of French Mine
Thickness sampled	15.2 m	12.2 m	7.6 m
Megascopic description	Mainly thin bedded (1.9—5.1 cm) with beds composed of laminae 0.6 cm or less thick. Dull olive-green to grey with khaki or brown cast. Deep orange-red to orange-brown on weathered surface. Very fine grained, some beds strongly magnetic. Very rich in iron-silicate minerals. Few cherty beds. Uniform section	Lensy banded red jasper and grey-blue hematite chert, stubby lenses, laminae and nodules 0.6—2.5 cm thick, give rock a thin-bedded appearance, fractures and breaks in slabs 10.1—15.2 cm thick, medium-sized chert granules in finer-grained matrix, some in grey to brownish red blotches and patches. Hematite is dense blue-black in iron-rich beds, and where disseminated in chert bands gives a pink to brown colour. Some fine-grained specular hematite	Fairly uniform thin-banded pinkish chert with disseminated blue hematite interbanded with blue-grey hematite-rich bands. Beds and slabby fragments 1.9—5.1 cm thick composed of wavy laminae, laminae less distinct than in lower red facies, some beds of brownish chert. Differs mainly from lower red in colour. Some 0.6 cm thick laminae composed of coarse granules
Microscopic description	A thin-banded, dense, felty mass of minnesotaite, a few secondary veins of minnesotaite. Granular texture preserved in some bands, granules sheared and distorted in others; considerable brown stain. Very little free quartz or carbonate in sections examined	Composed almost exclusively of hematite and chert. Granular to oolitic texture. Jasper is fine-grained chert with disseminated dusty red hematite. Chert granules rimmed by coarser-grained hematite that is recrystallized. Some patches of coarser-grained quartz in matrix to granules and centres of many granules are selectively recrystallized to coarser quartz	Coarse granular or oolitic to nodular texture. Small oolites not abundant. Cherty texture over large areas interrupted by patches of coarse-grained quartz and crystalline hematite, minor brown iron oxide

up to 90 m thick northwest of Knob Lake, and thin beds of chert breccia are found with dolomite and iron-formation in other parts of the belt.

The maroon to grey-black slate beds above the Wishart quartzite at the base of the iron-formation vary in thickness from less than 3 mm to more than 3 cm. Thin ferruginous chert layers interbedded with fine clastic carbon-bearing layers in the upper part of the unit mark the transition from slaty clastic beds to cherty silicate-carbonate iron-formation. The ferruginous beds in the *Ruth slate member* mark the beginning of abundant iron deposition in the region and this member is now included in the Sokoman iron-formation.

The Sokoman Formation, composed of a variety of complex lithological facies of iron-formation, underlies the greater part of the geosyncline. The thickness and order in which the silicate, carbonate and oxide facies occur may vary from basin to basin but iron-silicate and iron-carbonate members are present in the lower part of the iron-formation throughout the belt. The

B526 Grey Cherty	B527 Brown Cherty	B528 Upper Red Cherty	B529 Grey Upper Cherty
122 m northeast of loading station, French Mine	152 m northeast of loading station, French Mine	168 m northeast of loading station, French Mine	152 m east of loading station, French Mine
15.2 m	6.1 m	18.3 m	20.4 m
Bands and zones vary in colour from pinkish grey to grey to brown. Predominantly thin banded (0.64 cm), but much is crudely laminated to lensy, or forms thicker (5.1—15.3 cm) massive beds. Some wavy to lenticular banded iron-rich and leaner cherty beds are fairly well differentiated. Weakly magnetic in places	Lenticular thin banding 7.6—10.2 cm of brownish grey to pinkish jasper iron-formation. Considerable variation in high ferruginous bands from blue to brown laminae and lenses. Coarse granular texture in most beds with nodes and stubby lenses (1.3 cm thick) of pink and brown jasper	Thick massive beds prevalent up to 30.5 cm thick with gradational patches of blue to grey-pink iron-rich beds interspersed with banded, lenticular and nodular jasper, 1.3—2.5 cm thick. Magnetite- and hematite-rich lenses abundant in jasper. Some coarse granular to nodular material. Generally pinkish blue to dark grey with abundant red jasper	Grey-green magnetite carbonate chert with blue to brown hematite-goethite-rich beds. Spotty distribution of carbonate in grey chert and iron-oxide beds. Blue hematite beds with metallic lustre and granular textured jasper beds dispersed in the lower part of the member. Considerable blue and brown iron oxide uniformly disseminated in grey-green chert. Magnetite-rich beds in places
Much medium-grained quartz with numerous fine-grained cherty granules. Granular to oolitic texture present in some beds. Considerable hematite crystallized in fine discrete grains, much goethite present in grains, secondary bands or stringers, and as brown stains	Fairly pure oxide facies of hematite, goethite and chert. Jasper nodules consist of fine chert with minor hematite dust. Ferruginous beds have coarser recrystallized hematite and quartz and secondary goethite. Similar to grey cherty facies	Granular textured chert hematite and magnetite, with some goethite, minor minnesotaite. About half of silica recrystallized to coarse chert, remainder is fine-grained chert. Much of coarser chert in centres of granules. Minnesotaite in cherty patches. Some hematite altered to goethite	Predominantly hematite and goethite in coarse chert. Fine chert in centres of granules set in matrix of coarse chert. Grains of hematite border granules with some disseminated in the matrix. Brown iron oxide replaces some hematite grains and much is derived from siderite. Minor iron-silicate content distributed in grey-green chert

oxide facies comprising the main central part of the formation are succeeded by various silicate, carbonate or lean chert facies in its upper part. The Sokoman iron-formation is more than 170 m thick in the Knob Lake area and it is rarely less than 30 m thick along the western margin of the belt. It thickens in the central part of the area and apparently pinches out to the east where it is structurally deformed and its distribution cannot be defined by mapping.

The formation is composed mainly of thin-banded ferruginous chert layers with oolitic or granular texture, and the metamorphic equivalent of such material. Oxide facies composed of hematite, magnetite and chert are the most abundant but iron-silicate facies composed of minnesotaite, stilpnomelane, and occasionally greenalite in chert and siderite occur persistently in its basal parts. Iron-formation at French Mine is described in Table 6-II and chemical analyses of surface samples of the lithological units are given in Table 6-III. Separate from the main iron-formation unit are other thin bands in ferruginous slate in the Lac Cambrien area and a thin member in the volcanic rocks east of Murdock Lake.

Very fine-grained, thinly laminated black slate, containing considerable carbon and pyrite, known as the *Menihek Formation* in the Knob Lake area, overlies the iron-formation conformably. In places cherty iron-formation beds are interbanded with the black slate and form a transitional zone between the two stratigraphic units. Dolomite lenses occur in the slate in some areas. The black slate marks the beginning of a thick succession of slate, argillite, and greywacke that increases in thickness to the east, where it is interbedded with lava flows and gabbro sills. Lithologically similar rocks are found above the iron-formation in nearly all parts of the geosyncline.

Volcanic rocks in the south-central part of the belt are derived from effusive centres in the Dyke Lake area. Basic lavas lie between the quartzite and the

TABLE 6-III

Chemical analyses (%) of chip samples of iron-formation facies, French Mine (for descriptions see Table 6-II).

	Sample No.						
	B523	B524	B525	B526	B527	B528	B529
SiO_2	49.41	41.42	48.16	51.24	43.77	49.01	56.49
Al_2O_3	0.68	0.79	0.53	0.42	0.42	0.37	0.37
Fe_2O_3	16.34	54.49	46.96	41.97	49.85	44.50	38.10
FeO	24.19	1.35	1.50	3.25	2.27	3.65	1.99
CaO	0.02	0.00	0.01	0.00	0.00	0.00	0.00
MgO	2.95	0.37	0.31	0.62	0.37	0.19	0.00
Na_2O	0.03	0.08	0.03	0.02	0.02	0.03	0.02
K_2O	0.07	0.01	0.01	0.01	0.01	0.01	0.01
H_2O+	5.20	0.98	2.04	2.10	2.54	1.94	2.42
H_2O-	0.38	0.06	0.04	0.05	0.05	0.02	0.03
TiO_2	0.00	0.00	0.00	0.00	0.00	0.00	0.00
P_2O_5	0.08	0.04	0.04	0.03	0.04	0.05	0.04
MnO	0.65	0.02	0.02	0.02	0.03	0.03	0.03
CO_2	0.22	0.02	0.02	0.06	0.02	0.04	0.04
S	0.05	0.05	0.03	0.00	0.00	0.00	0.00
C	0.15	0.12	0.10	0.08	0.13	0.04	0.03
Total	100.42	99.80	99.80	99.87	99.52	99.88	99.57

Analyst G.A.Bender, Geological Survey of Canada.
Spectrographic Analyses. Composition of all samples within these ranges: Co 0.1—1.0%; Ni $<$ 0.01%; Ti $<$ 0.01%; Cr 0.01—0.1%; Cu 0.01%; Ba $<$ 0.01%; B not detected.
Analyst W.F. White, GSC; B523 - Silicate-carbonate facies; B524 - Lower red cherty facies; B525 - Pink cherty facies; B526 - Grey cherty facies; B527 - Brown cherty facies; B 528 - Upper red cherty facies; B529 - Grey upper cherty facies (from Gross, 1968).

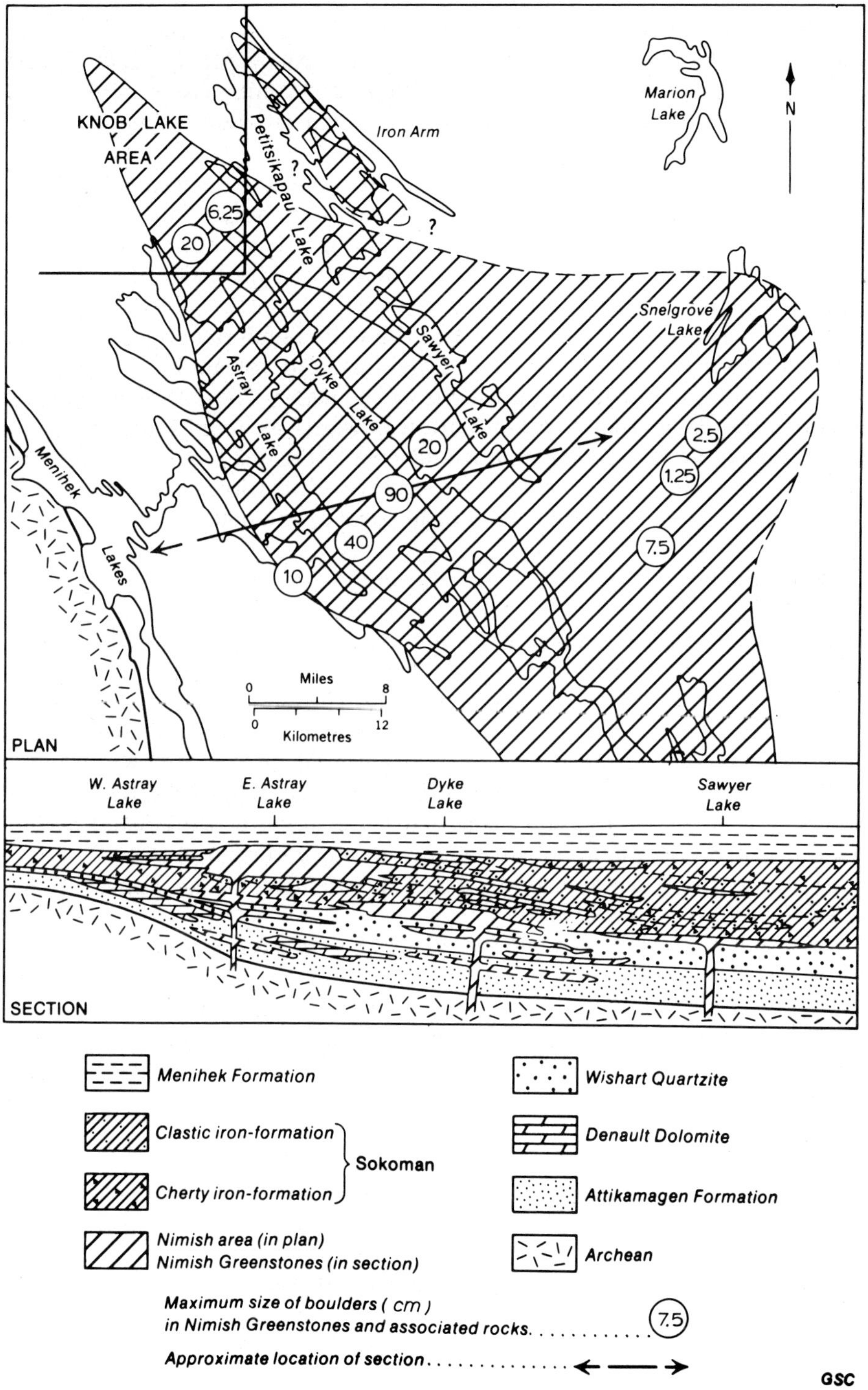

Fig. 6-5. Distribution of Nimish greenstones, from Zajac (1974), adapted from Sauve (1953).

iron-formation a few miles southeast of Knob Lake, within the iron-formation around Dyke Lake, and in its upper parts east of Astray Lake. Basalt flows, pillow lavas, pyroclastic rocks, and flows containing jasper pebbles are interbedded with other sedimentary rocks throughout the southeastern part of the area.

A group of volcanic and sedimentary rocks along the eastern side of the belt consists of basic pyroclastics interlayered with basalt flows and greywacke, quartzite, slate and conglomerate. The group is about 600 m thick east of Knob Lake but thickens to the north around Murdock Lake where it includes a thin band of iron-formation.

Much of the central and eastern part of the belt is underlain by thin gabbro sills that are conformable with the sedimentary and volcanic rocks and are considered to be the intrusive equivalents of the latter. These tholeiitic intrusions show trends in their differentiation towards extreme enrichment in iron with negligible alkali enrichment and characteristically a low potassium and strontium content. Serpentinized peridotite intrusions occur with the volcanic rocks in the east and north-central parts of the belt.

The Knob Lake basin

This basin area in the central western part of the geosyncline belt extends from 54°30′ to about 55°00 N latitute. The iron-formation continues north through the adjacent Wakuach lake area from 55°00′ to 55°30′N latitude. The Dyke Lake basin overlaps the southeastern part of the Knob Lake basin and the general relationship of the Nimish volcanic rocks to the Knob Lake sedimentary group is illustrated in Fig. 6-5. A cross section of the Knob Lake basin (Fig. 6-4), extending southeast from the mine area at Schefferville shows the general stratigraphic succession before deformation. Most of the detailed studies have been carried out in the vicinity of the mines in the Schefferville and Knob Lake area.

The Sokoman Formation includes all of the iron-formation and related stratigraphic units between the Wishart quartzite and the Menihek Formation. It was divided into ten members by Zajac (1974) that are commonly grouped as Lower, Middle, and Upper iron-formation. The general distribution of different lithological facies of iron-formation in these members is illustrated in Fig. 6-6.

Lower iron-formation

Members I, II and III — mainly black shale and slate, and silicate-carbonate facies.

Member I, the *Ruth slate*, is 25 to 45 m thick and grades upward from black slate into carbonate, silicate, oxide, or mixed facies of iron-formation. The relative position and distribution of facies units in this member are shown in

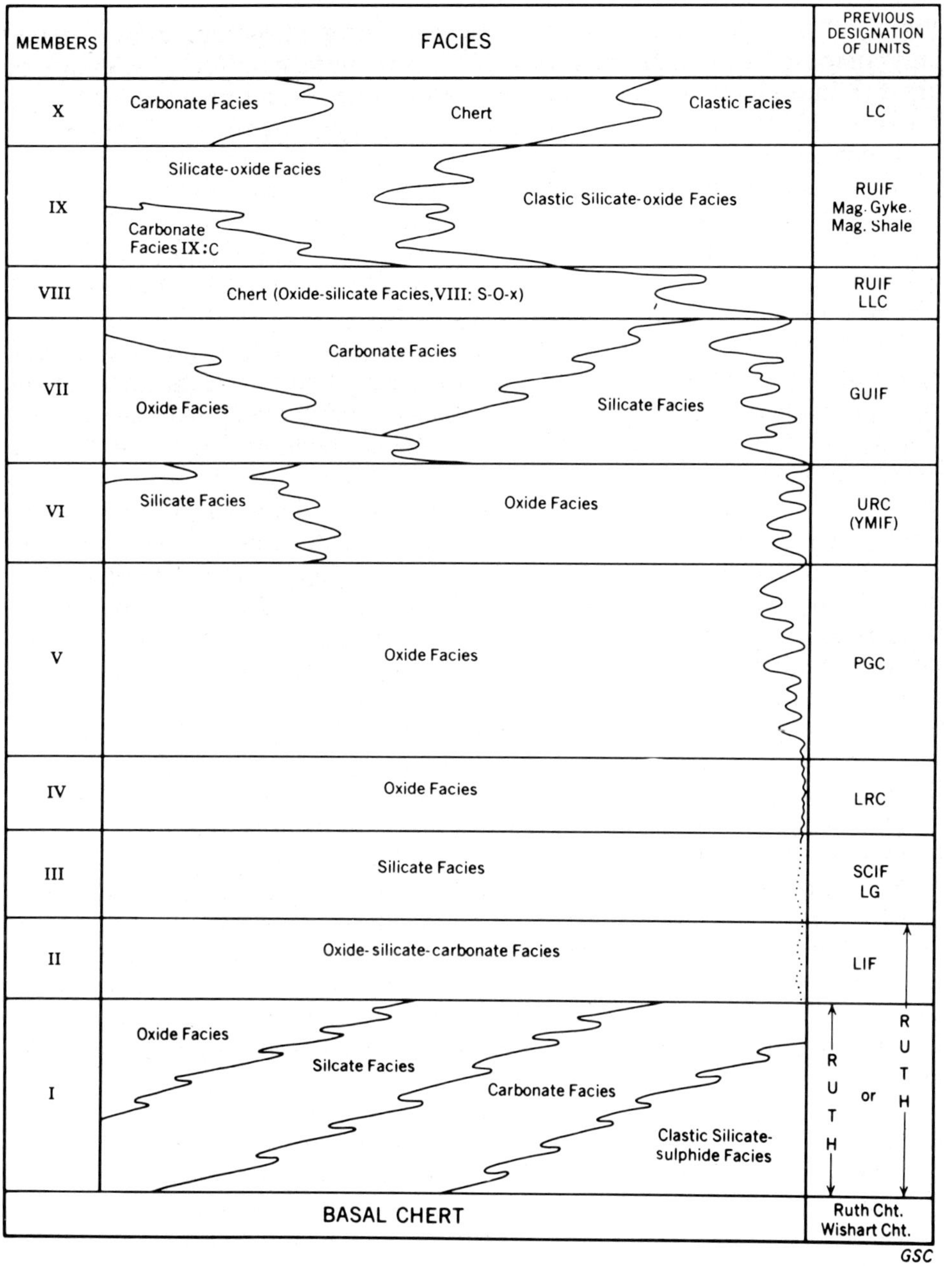

Fig. 6-6. Subdivisions of the Sokoman Formation, from Zajac (1974).

Fig. 6-7. The clastic silicate-sulphide facies is fine grained, dark-grey to black, and composed of opaque mixtures of chlorite, potassium feldspar, quartz or chert, minute feathery flakes of iron-silicate minerals and fine-grained pyrite. Laminated silty clastic layers show graded bedding and are interlayered with dark banded chert. Light-greenish to dark-grey massive beds of chloritic tuff, usually less than 5 cm thick, are dispersed through the lower 12 m of the member and are similar in texture and composition to Nimish volcanic rocks to the southeast. The carbonate facies, composed of thin alternating layers of laminated chert and siderite with shaly parting parallel to the bedding, weathers to various shades of yellow and brown.

The silicate facies is normally thin-bedded green to grey-green and composed of varying amounts of chert, minnesotaite, siderite and magnetite. Beds rich in minnesotaite weather to a distinctive bright yellow-orange colour. The silicate facies has thicker and more irregular bedding than the carbonate facies and consists of minnesotaite granules, 0.3—2.0 mm in size, cemented by chert of variable grain size. The oxide facies in the northwest part of the basin contains distinctive bright-red jaspilite and pisolitic pebble conglomerate.

Prominent features of the jaspilite are alternating layers, 5—25 mm thick, of grey iron oxide and red chert in which wavy laminae give the appearance of cross-bedding. Iron oxide-rich bands 1 cm or less thick, composed of magnetite and minor amounts of hematite, pinch and swell and enclose lenticular beds and pods of jasper. Twenty to seventy per cent of the jasper beds are composed of spherites 30—45 μm in diameter that consist of spherical patches of clear microcrystalline chert outlined by dusty rims of hematite and inclusions of magnetite. Grey cherty beds in the jaspilites are coarse grained, thicker bedded, contain cherty iron-oxide granules and siderite but lack the thin laminae and spherites commonly found in other jaspilites.

A distinctive conglomerate restricted to the upper part of the oxide facies consists of tabular, well-rounded, and irregular to subangular jaspilite fragments, 1—6 cm in size, enclosed in a pink to grey chert matrix. Many fragments have outer concentric rims or shells composed of chert and iron oxide. The monolithic composition and pisolitic textures are typical of the conglomerate and small scale crossbedding and poorly developed graded bedding are present locally.

Member II, a thin unit in the basin, is typically laminated, evenly bedded, nongranular, chert-carbonate facies containing small quantities of minnesotaite and silicates. A granular silicate-oxide facies is common in the western parts of the basin and a well-banded carbonate-silicate facies is typical in its deeper eastern parts.

Member III, silicate-carbonate iron-formation or SCIF, is 9—27 m thick, and the distribution of four prominent subfacies is shown in Fig. 6-8. The stilpnomelane-siderite facies in the east-central part of the basin is thin banded,

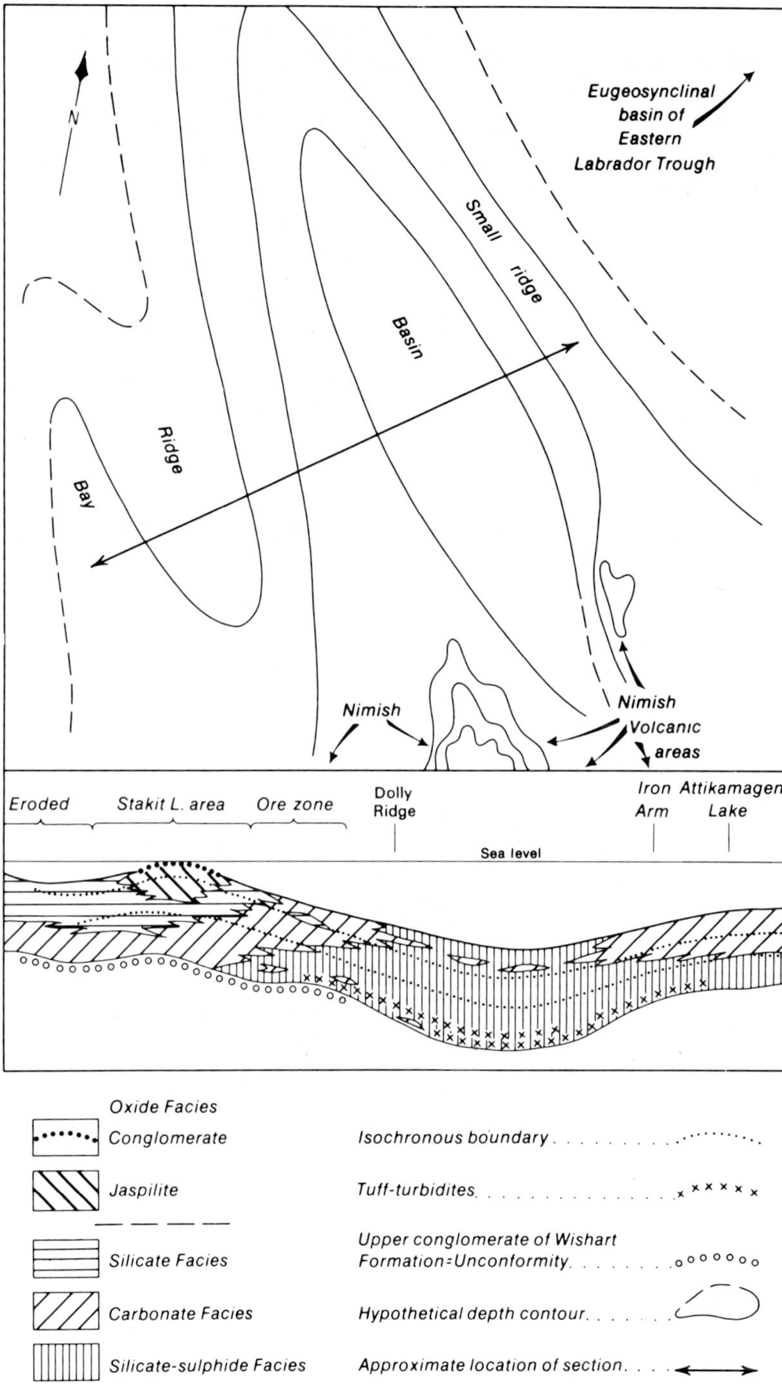

Fig. 6-7. Interpretation of the depositional environment of Member I, from Zajac (1974).

light to dark green, fine grained and composed of felt-like intergrowths of stilpnomelane, chert, siderite, and minnesotaite, with minor magnetite and ilmenite in fine elongate particles. The siderite-minnesotaite subfacies consists of nongranular, thin, uniform alternating layers of chert siderite and minnesotaite with magnetite in its upper parts. The minnesotaite subfacies, typical of the SCIF member, has irregular lenticular beds that are thicker in the western part of the basin and show evidence of reworking. Minnesotaite is interlaminated with chert and occurs in felt-like masses, radiating intergrowths and granules. Thin layers of magnetite forming 10—30% of the subfacies are most common in its upper parts. The minnesotaite-magnetite facies in the western part of the basin is composed of thin irregular wavy beds and microcrystalline greenalite is dispersed in granules and elongate particles in the

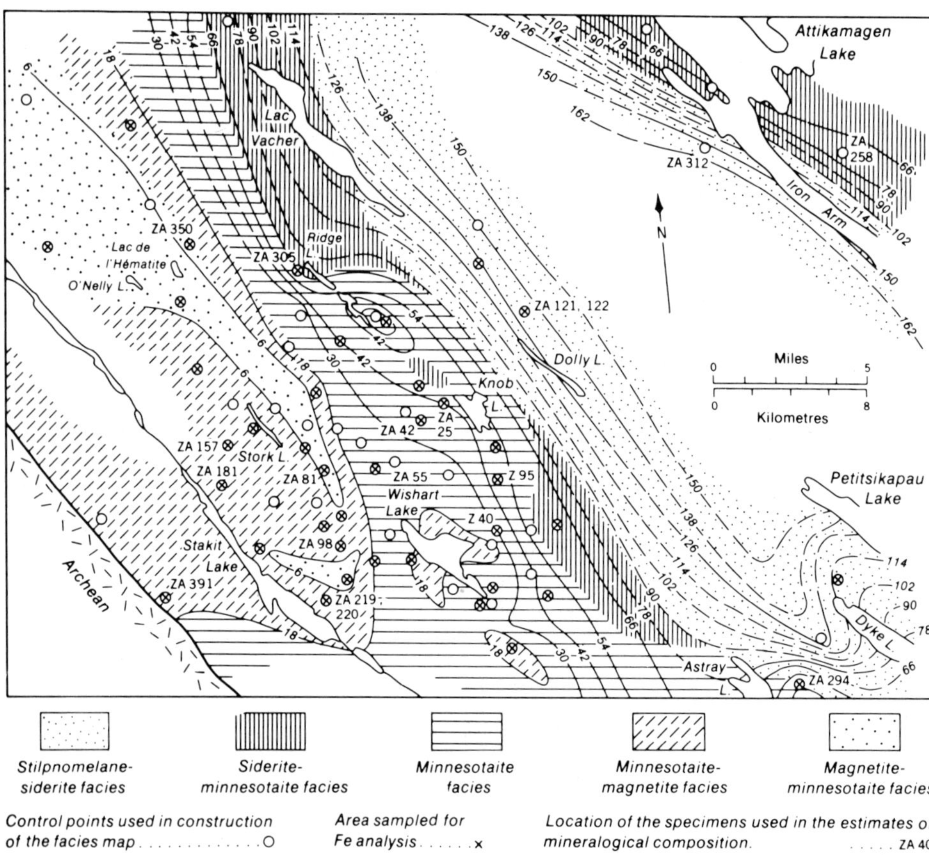

Fig. 6-8. Facies map of Member III. The contours indicate the average number of beds per foot of the iron-formation, from Zajac (1974).

chert. Granules are abundant and crossbedding is found in beds 5—10 cm thick.

Middle iron-formation

Members IV, V, VI — mainly oxide facies. Member IV is made up of the Lower Red Cherty (LRC) unit, Member V of Pink and Grey Cherty (PGC) and Brown Chert (BC), and Member VI of the Upper Red Cherty (URC) and Yellow Middle Iron-Formation (YMIF). The Lower and Upper Red Cherty units are distinctive stratigraphic markers in the central part of the basin; however, the oxide facies varies locally in colour, thickness and texture of individual beds, in the nature of granules and oolites, the development of

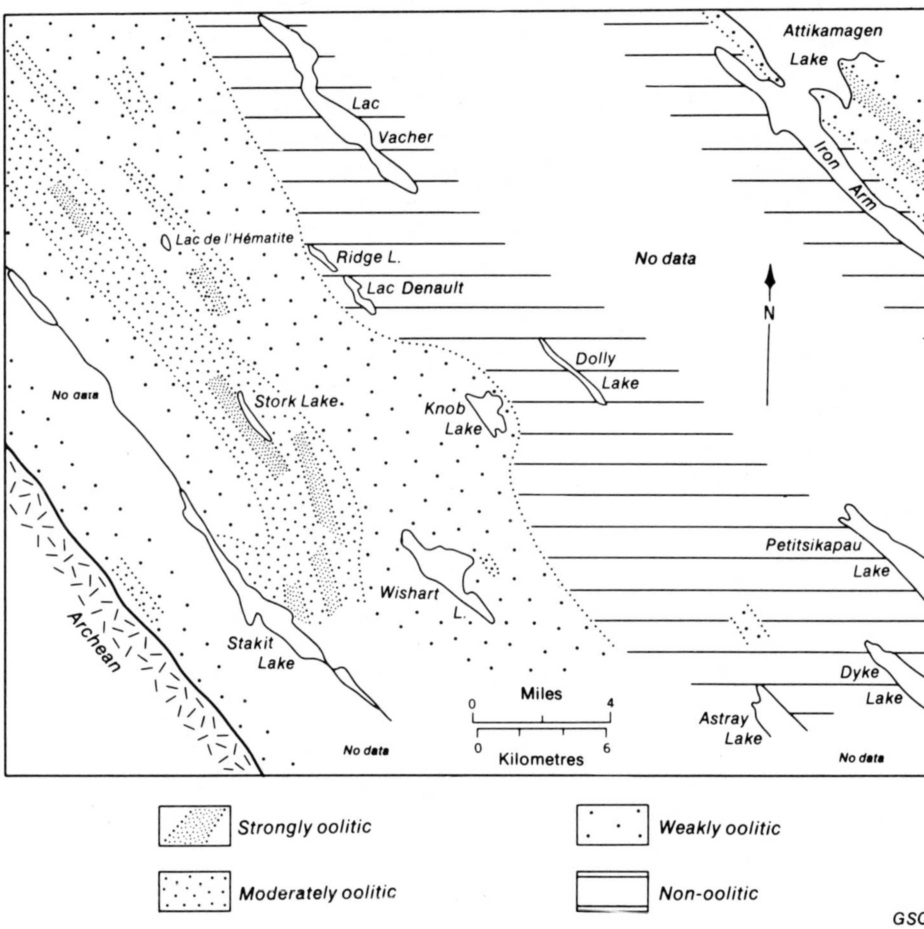

Fig. 6-9. Distribution of oolitic rocks in the upper part of Member IV, from Zajac (1974).

soft sediment deformation, and in the character of the intraformational conglomerates and breccia.

Member IV, the Lower Red Cherty unit, thickens from 7 m in the west to 15 m or more in the central part of the basin. It consists mainly of thin irregular jasper layers and massive metallic-looking iron-oxide beds and lenses that are abundant in its upper parts. Granules and oolites which vary greatly in composition and physical character are a prominent textural feature in the western part of the LRC unit and their distribution is shown in Fig. 6-9.

Member V, essentially an oxide facies with prominent granular to oolitic textures, is composed of thin to thick massive beds of jaspilite, pink to grey chert, hematite and magnetite with minor disseminated ankerite or minnesotaite. Conglomerate beds consisting of cherty fragments that were broken and deformed when still in a semi-plastic state are common. Massive pink and red jasper beds containing clastic material occur in the southeast; thicker intraformational conglomeratic facies are prominent throughout the western part; and clastics derived from the Dyke Lake volcanic area are distributed throughout its central part (Fig. 6-10). Granules of chert which vary in texture, size and composition are abundant in the conglomerate beds, oolites and oolite fragments are found in most beds, and cross bedding is developed locally.

Member VI, is composed of the Upper Red Cherty (URC) oxide facies, and the Yellow Middle Iron-Formation (YMIF) silicate facies. It ranges in thickness from 3 m along the eastern and western margins of the basin to over 15 m in the central part. The URC facies forms an important marker unit consisting of interbedded granular grey to pink chert and metallic grey iron-oxide layers in beds 15—45 cm thick. Conglomerate beds made up of lenses and angular or well-rounded fragments of red jasper less than 5 cm in size are interspersed throughout the member and give it distinctive colour and character. Lenses of minnesotaite and patches of ankeritic carbonate are common accessory constituents.

The YMIF is developed in the Wishart Lake area where it forms a thin-bedded, chert-minnesotaite-magnetite unit. Some of the thicker irregular beds have granular textures, platy fragments, lenses of broken beds and small-scale crossbedding.

Upper iron-formation

Members VII, VIII, IX, X — a mixed group of shale, clastic, lean chert, oxide, silicate and carbonate facies.

Member VII, mainly Grey Upper iron-formation, is 9 to 27 cm thick and varies greatly in composition (Fig. 6-11). It is irregularly bedded, grey to pink, and composed of chert, magnetite, silicates and hematite in the eastern and central part of the basin; to the southeast it is thin to medium bedded, grey to purplish brown and argillaceous, composed of chert, iron oxides, iron silicates (stilpnomelane) and scattered grains of feldspar and detrital quartz; beds of carbonate, silicate and oxide facies are interfingered in the west and a silicate facies is dominant in the southwest.

The carbonate facies composed of medium to coarse-grained granular siderite forms thick irregular beds with minor crossbedding. The silicate facies is mostly thick-bedded and composed of granules of fibrous or felty minneso-

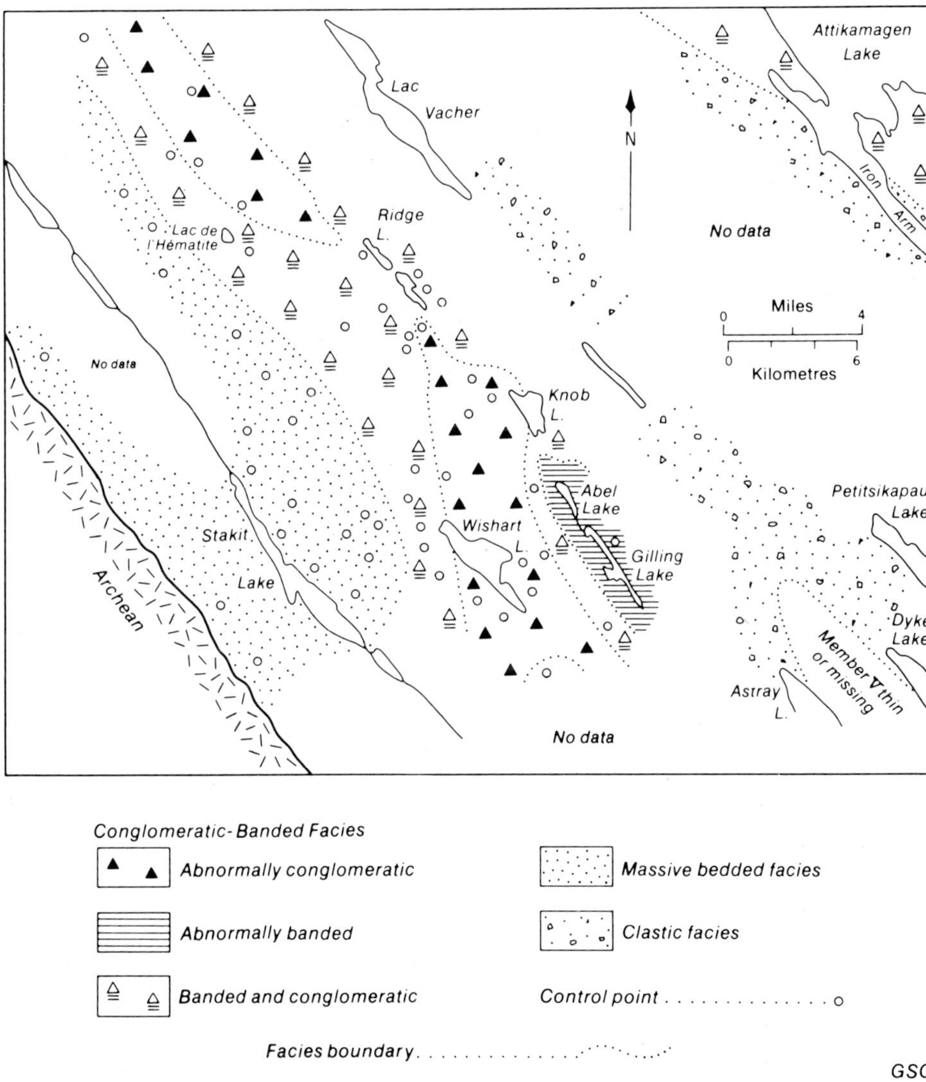

Fig. 6-10. Facies map of Member V, from Zajac (1974).

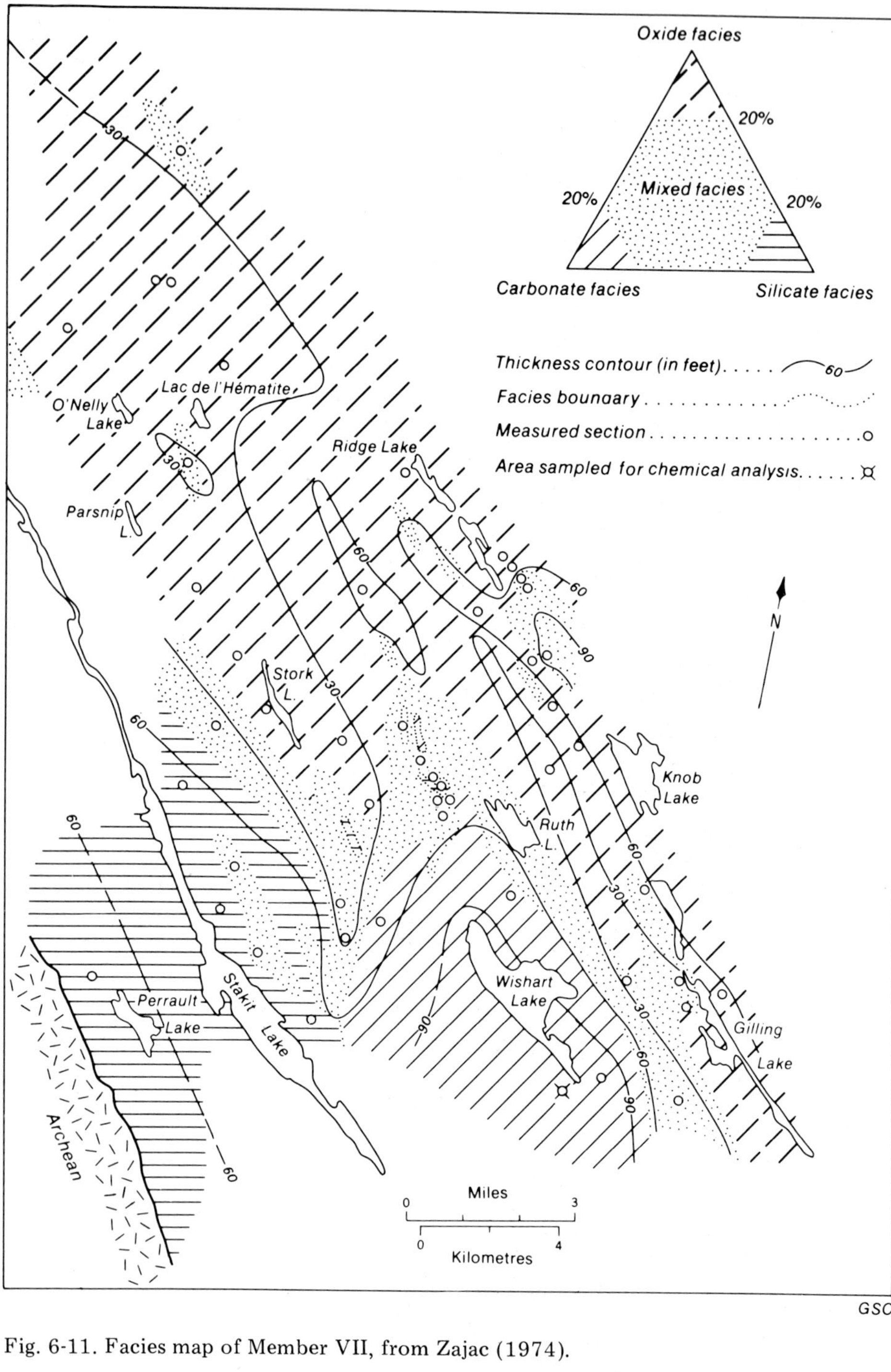

Fig. 6-11. Facies map of Member VII, from Zajac (1974).

taite, siderite and chert and euhedral crystals of magnetite. The oxide facies varies from grey to pink; massive beds range in thickness from 0.5 to 1.5 m and have pitted surfaces where carbonate has weathered; jasper fragments are conspicuous in the lower parts, and small-scale crossbedding is outlined by concentrations of iron-oxide minerals. Pebbly beds and lenses of chert rimmed by iron oxides are prominent in the more siliceous parts of the member. Thin beds and lenses of chert-siderite, chert-minnesotaite, or chert-iron oxide are interlayered in the lower part of the member and a variety of granules, pebbles, fragments and lenses are intermixed in single beds.

Member VIII is 1.5 to 2.4 m thick in the western part of the basin and up to 7.5 m thick in the central area. It varies in colour from blotchy green to purplish red, has a vitreous waxy appearance, and is composed of granular chert with disseminated minnesotaite, stilpnomelane, ankerite, siderite, and magnetite grains, dusty hematite and spherites of greenalite. Oolitic and conglomeratic cherts are distributed mainly in the purple red hematitic jaspers. In the southeast it consists of massive- to thin-bedded hematitic jasper interbedded with greywacke and shale. Algal structures are well developed in the upper parts of the member especially in the southwest part of the basin.

Member IX, 7.5 m thick in the northwest to 60 m in the southeast around Dyke Lake, is composed of intermixed clastic and nonclastic beds, mainly greywacke, shales, chert, and silicate-oxide and carbonate facies of iron-formation. Carbonate facies in the southwest part of the basin has alternating beds of greenish grey chert, chert-siderite and siderite with fine- to coarse-grained spheroidal to lobate granules of chert and siderite in a matrix of chert with irregular patches of stilpnomelane and siderite.

Silicate-oxide facies vary throughout the basin from a nonclastic subfacies prevalent in the northwest to a clastic subfacies in the southeast composed of cherty silicate-oxide beds, detrital feldspar and quartz, fragments of volcanic rock, and clasts derived from a variety of rocks. The greywacke and shales of the clastic silicate-oxide facies are greenish brown to purplish or grey, thin bedded and have variable amounts of stilpnomelane, minnesotaite, and chlorite associated with magnetite, chert, or feldspar; they are crossbedded in places and contain thin conglomerate beds with jasper pebbles. Twenty percent of the facies consists of clastic grains of titaniferous magnetite with overgrowths of titanium-free magnetite that were most likely derived from Nimish volcanic rocks, and titanium-free magnetite grains derived from cherty iron-formation. The nonclastic silicate-oxide facies is interbedded with thinly banded shale, greywacke and greenish grey chert. Cherty beds, composed of dense granules of magnetite and iron-silicate are interlayered with shaly fissile beds of stilpnomelane, magnetite and chert, that are coarse grained, sandy or conglomeratic and often crossbedded.

Member X, at the top of the Sokoman is 6 to 15 cm thick, and consists of bedded grey to black chert with subordinate shale and siderite-rich beds. Microcrystalline chert, and granules are outlined by fine-grained greenalite, minnesotaite and dusty amorphous carbon. Thin beds of siderite are prominent in the western part of the basin and dark clastic carbon bearing shales are typical in its eastern parts.

The Menihek Formation, overlying the Sokoman iron-formation in the Knob Lake basin consists of grey to black shale, argillite, and microgreywacke with pyritic dolomite and cherty horizons developed locally. The Sokoman and Menihek rocks are generally conformable with a disconformity of bedding noted in only a few places.

DEPOSITIONAL ENVIRONMENTS

The iron-formations around the Ungava craton display similar lithological and depositional features from basin to basin suggesting a common genetic model for their development. Their deposition on the shelves of the Ungava craton probably coincided with offshore tectonic and volcanic activity that was related to deep-seated fault and rift systems that developed between 2400 and 1900 m.y. ago during the separation of the craton from other plates or continental land masses. Thick sequences of iron-formation were deposited in the troughs, shallow lagoonal basins and tidal flats located between the shores of the craton and the surrounding offshore volcanic belts and tectonic ridges marginal to its shelves. The deposition and accumulation of these distinctive siliceous iron-rich chemical sediments were dependent on a large number of interrelated tectonic, physical, chemical and probably organic factors that controlled the depositional environment.

Except for the clastic and some mixed facies, the iron-formation is believed to have originated as chemical and colloidal precipitates of silica and iron oxide, carbonate or silicate that were deposited as amorphous gelatinous ooze and muds, in beds, lenses, thin laminae and delicate microbands and by accretion of spheroidal granules, pisolites, oolites and nodules. The primary chemical sediment was altered in texture and form by diagenetic processes, crystallized as chert and iron minerals and deformed and recrystallized by later metamorphic processes. Evidence for changes in the bulk chemical composition of individual beds or layers or their component granules, oolites, microbands or matrix has not been demonstrated in any of the facies; however, mineralogical and textural changes involving recrystallization during diagenesis and subsequent metamorphism were extensive in some of the carbonate and silicate rich facies.

The shallow basins, tidal flats, lagoons and coastal embayments in which the iron-formation formed were located over relatively stable tectonic areas of the shelf where epeirogenic movement and adjustments controlled basin depth and configuration. The stable craton apparently was reduced to low topographical relief and was the source of fine-grained clastic sediment, carbonate and silica that formed dolomite reefs and chert beds that are interspersed with the shales, greywacke and sandstones. The orthoquartzite, arkose and conglomerate were derived through sorting and extensive reworking of sandstone and clastic sediments in the nearshore areas.

The tectonic ridges and volcanic belts on the outer margins of the craton slopes probably formed offshore barriers that separated the coastal basins from the deeper ocean and at times restricted the circulation of currents and the intermixing of sediment carried by coastal and deep-ocean currents. During deposition of the iron-formation chemical precipitation prevailed with very little clastic sediment being introduced from the craton or intrabasinal offshore ridges, except for the occasional influx of tuff and pyroclastic material from contemporaneous vulcanism.

High-energy sedimentary environments were characteristic of the coastal basins throughout their depositional history. Circulation within the basins was generally open and unrestricted in the shallow surface waters and more limited or restricted in local deeper depressions on basin floors. Tidal, wave or current action are conspicuously recorded in the iron-formation by channeling, scour and fill structures, crossbedding, brecciation, and deformation of bedding and layering in the partially consolidated siliceous ooze and sediment. The depth of water fluctuated but was generally shallow with intermittent exposure of the soft sediment. Extensive reworking of thin beds and the unevenness and irregularity of bedding and laminations indicates an undulating sediment surface of low relief that was subjected at times to intensive turbulence and wave action.

The nature and origin of the oolites has been discussed by Zajac (1974) and Gross (1968) and typical oolitic textures in oxide facies from the northwest part of the basin are shown in Fig. 6-12. Granules and oolites have developed through processes of chemical precipitation and accretion with possibly some biogenic influence that is not clearly demonstrated or understood. The energy level in the depositional basin was a critical factor in determining whether microbands or oolites formed. Where water was agitated and turbulent in high-energy parts of the basin, precipitation and accretion of iron and silica took place around nucleii and formed oolites whereas thin laminae and microbands formed in quiet undisturbed water. Concepts proposed for a replacement origin of the iron-formation emphasized by Dimroth and Chauvel (1973) are not considered by the writers to be applicable for any of the facies studied in detail in the Knob Lake basin. The textural evidence cited by them for replacement processes are interpreted by the writers as crystallization of components within laminae, oolites and granules during diagenesis and metamorphism of the beds.

Different physical and chemical environments that existed in the basins around the Ungava craton during deposition of the iron-formation are clearly reflected in well defined lithological facies. Conditions favorable for development of the oxide, carbonate, silicate and sulphide facies of iron-formation as defined by Krumbein and Garrels (1952) and James (1954) and adopted by Goodwin (1956), Gross (1968) and Zajac (1974) are directly applicable in the interpretation of facies environments around the Ungava craton. However, some inferences as to depth of water and proximity to shorelines are less certain and need to be evaluated in local situations. Many interrelated factors determine the kind and location of facies development in a sedimentary basin. Probably the most significant are size and shape of the basin, pattern and degree of circulation, influx and preservation of organic material, influence of organisms, input and chemistry of land drainage or hydrothermal solutions related to fumarolic and volcanic activity, composition of the atmosphere, depth of water and proximity to shore. The complex interrelationships of these factors control the locations within a basin where different kinds of facies may develop.

Conditions in the Knob Lake basin varied from euxinic to highly oxidizing as shown by the thick accumulation of carbon-bearing black shales, red shales, and siderite, silicate and oxide facies of iron-formation. The carbon in the shales and iron-formation was derived from organic sources and the fine-

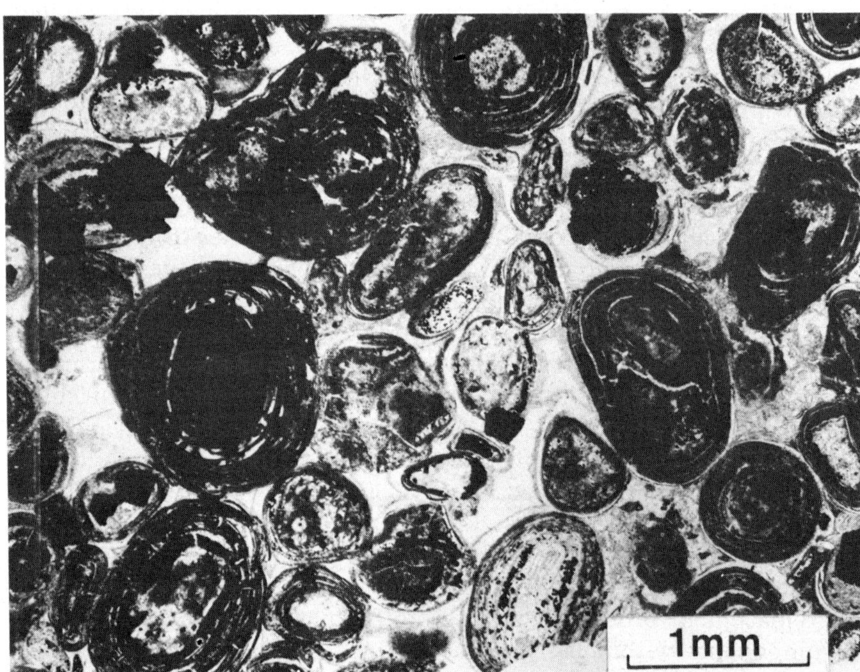

Fig. 6-12. Typical oolitic textures in oxide facies of iron-formation from the northwestern part of the Knob Lake basin.

grained clastic material came from adjacent land areas and from volcanic sources offshore. If the iron and silica were derived from the land mass a gradual build-up and increase of these constituents in the offshore sediments would be expected as the land was reduced to low relief. Deposition of silica and iron coincided with volcanic and tectonic activity along the offshore ridges and these constituents are believed to be derived from hydrothermal springs and from effusive and fumarolic sources related to vulcanism in these ridges as proposed by Gross (1965, 1968). The iron-formations have lower alumina and titanium content than would be expected if they were derived from lateritic regoliths but have minor element contents that are comparable with those found in material from hydrothermal sources. If crustal temperatures in tectonic belts were higher in Proterozoic time as generally believed, the solution, transportation and dispersion of silica in the basins would have been facilitated greatly and may account for the more extensive development of iron-formation during this period.

Deposition of iron-formation in the Knob Lake basin

The Attikamagen slate, Denault dolomite, Fleming chert breccia and Wishart quartzite below the Sokoman iron-formation represent normal shelf sediments with no distinctive characteristics that might herald the major period when chemical sedimentation exceeded clastic deposition. After deposition of the Wishart quartzite low-energy euxinic conditions were established over the basin area and laminated carbon-rich muds of the Ruth slate or Member I of the iron-formation began to accumulate in its deeper parts (Fig. 6-8), along with the introduction of tuff and some turbidites near the volcanic centres to the east and southeast. The first indication of iron-formation deposition is seen in the pyrite-bearing shales in the lower part of Member I designated as clastic silicate-sulphide facies. Reducing euxinic conditions prevailed as the transition from clastic to predominantly chemical sedimentation took place with the development of carbonate facies in both the western and eastern parts of the basin. Silicate facies were deposited over cherty carbonate beds and a red jaspilite oxide facies formed over a narrow offshore ridge in the western part of the basin. The distribution of granules, intraformational breccias and conglomerates reflects an increase in energy levels to the west with considerable reworking of the jaspilite and cherts in shallow water near shore and over ridges.

Lower-energy conditions were more uniform over the basin during the deposition of the laminated evenly bedded chert carbonate facies of Member II. Granular silicate-oxide facies in the west suggests shallow agitated water conditions while banded carbonate-silicate facies developed in deeper parts of the basin in the east.

Moderate to more strongly reducing conditions throughout the basin are recorded in Member III with distribution of magnetite-bearing facies in the west, and minnesotaite-rich silicate facies and siderite facies in its central parts. Shallower water, moderately reducing and higher-energy environments in the western coastal area are recorded in thicker beds by the extensive development of granules, uneven to wavy laminations and banding, and reworking of beds (Fig. 6-8). Stilpnomelane-siderite facies are extensively developed in the eastern deeper parts of the basin where tuff and fine-grained volcanoclastic material are abundant.

A fairly abrupt but transitional change took place from the reducing environment prevalent during deposition of the Lower iron-formation to shallow water, strongly oxidizing and higher-energy conditions that characterized the deposition of red pink, grey, and brown cherty oxide beds of the Middle iron-formation.

Agitated to turbulent shallow-water conditions were common over the basin as vast quantities of iron and silica were deposited in the form of granules, oolites, gelatinous muds and ooze in massive or thinly laminated beds on uneven undulating surfaces. Conditions varied greatly and deposition of the iron and silica was intermittent with occasional rhythmic precipitation in thinly laminated beds and in the delicate concentric shells of oolites. The distribution of strongly oolitic to non-oolitic jasper facies of Member IV shown in Fig. 6-9 suggests a transition from high-energy conditions in the western coastal areas of the basin to less disturbed and probably deeper water in its central part.

The depth of water appears to have fluctuated considerably and circulation or current patterns altered periodically. High-energy conditions in the basin are recorded in the extensive reworking and sorting of granular and oolitic material, in development of scour and fill structures, crossbedding, and wavy discontinuous layers. More violent turbulence and intensive wave action periodically broke up partly consolidated beds producing fragmented layers, intraformational breccias and conglomerates that were recemented in a matrix of granules, oolites and siliceous muds.

During the deposition of Member V banded conglomerate beds developed over a high along the central western part of the basin floor (Fig. 6-10), and more clastic material was introduced in the southeast and central parts of the basin from the Nimish volcanic area. Well-oxygenated shallow-water conditions prevailed over much of the basin during deposition of fragmental beds of oxide facies iron-formation in Member VI, except in a depression or subsidiary basin in the Wishart Lake area in the southwest where thin-bedded minnesotaite facies were developed under moderately reducing low-energy conditions.

Facies distribution of Member VII shown in Fig. 6-11, suggests deepening of the water in the south and southwest parts of the basin where carbonate facies were deposited in the Wishart Lake area and silicate facies along the western margin of the basin. A relatively thin oxide member is distributed over the northern basin with many lenses of mixed silicate, carbonate and oxide facies that formed in local depressions on the basin floor. A considerable amount of clastic sediment was introduced to form the argillaceous mixed facies in the eastern part of the basin and carbonate and silicate facies developed in the southwest under reducing conditions. High-energy levels, persisted throughout the basin during the deposition of all facies. Thin layers of different composition interfinger, and granules, pebbles and fragments of contrasting mineral composition are mixed in single beds indicating erratic disturbances and changes in facies development.

A shallow, oxidizing to mildly reducing environment prevailed during the widespread deposition of oolitic jasper beds and pebble conglomerates of Member VIII. Deeper-water reducing conditions continued to mark the Wishart subsidiary basin where oolites, conglomerate beds and hematite are absent. Algal structures along the western margin of the basin indicate clean shallow sunlit water while other parts of the basin were somewhat deeper and subject to the occasional influx of fine clastic muds, especially toward the end of the period.

There was a general deepening of the basin during the deposition of Member IX with reducing conditions permitting further accumulation of carbonate facies in the subsidiary Wishart basin and deposition of mixed silicate-oxide facies throughout the remainder of the basin. A marked increase in the influx of coarser clastic material from the Nimish volcanic area is recorded in widely dispersed shale and greywacke beds that interfinger with silicate-carbonate iron-formation in the deeper eastern part of the basin.

The last stages of deposition of the Sokoman iron-formation took place in deeper-water and reducing environments with carbonaceous black to grey chert being characteristic of Member X. Siderite is more widely distributed in the southwest where magnetite is scarce and hematite and jasper are absent. Clastic sediment in thin shales and greywacke was contributed from islands and shallow parts of the basin near volcanic centres of the Dyke Lake and Astray Lake areas. As the sea gradually engulfed the islands the influx of clastic material ceased and typical thin-banded, lean, black, non-oolitic cherts of the upper Sokoman represent the final stages in the great episode of iron and silica deposition during which iron-formation up to 170 m thick was accumulated. The increasing depth of the transgressing sea during Menihek time enabled several hundred feet of black carbonaceous shale to accumulate with pyritic zones distributed in its lower members.

Iron-formation deposition around the Ungava craton

Iron-formation was deposited in chains of predominantly shallow interconnected silled basins along the coast of the Ungava craton where circulation was restricted by tectonic and volcanic ridges on their outer margins. The basins developed on tectonically metastable shelf areas between the stable craton and mobile eugeosynclinal belts that surrounded the craton. Active vulcanicsm contemporaneous with the deposition of the iron-formation was distal to the shallow-water and coastal areas where the thickest sections of iron-formation with the highest iron content were deposited. Fumaroles and hydrothermal springs situated along deep fractures and fault systems in the shelf and platform areas that were closely related to active volcanic centres are considered to be the principal sources of iron and silica for the iron-formation.

The order in which different facies of iron-formation occur varies in detail from basin to basin and within individual basins but a predominance of silicate-carbonate facies in the Lower iron-formation, oxide facies in the Middle, and silicate-carbonate and lean chert facies in the Upper iron-formation are typical in most parts of the Circum-Ungava belt. The mineralogy and texture of the iron-formation are highly variable with oxide, silicate, and carbonate facies being most abundant. Sulphide facies are represented exclusively, it seems, by carbonaceous and pyritic shales. The oxide facies are noted for the abundance of granules, irregular bedding, oolites, intraformational conglomerates and breccias and soft sediment deformation features.

The coincidence of numerous favorable factors that controlled the deposition and accumulation of the iron-formations appears to have been repeated occasionally during the earth's history but never under identical circumstances. Iron-formation deposition marginal to the Ungava craton was distinctive in having many common features and continuity in the chain of basins that extended along the craton shelf for more than 3000 km.

REFERENCES AND SELECTED BIBLIOGRAPHY

Baragar, W.R.A., 1960. Petrology of basaltic rocks in part of the Labrador Trough; Geol. Soc. Am. Bull., 71: 1589—1644.

Baragar, W.R.A., 1967. Wakuach Lake Map-Area, Quebec—Labrador (230) Geol. Surv. Can., Mem. 344.

Bayley, R.W., and James, H.L., 1973. Precambrian iron-formations of the Unites States. Econ. Geol., 68: 934—959.

Béland, R. and Auger, P.E., 1958. Structural features of the northern part of the Labrador Trough. Trans. R. Soc. Can., Sec. 4, 52: 5.

Bell, R., 1878. Nastapoka Island, Hudson Bay Mining Division, Northwest Territories. Geol. Surv. Can., Rep. Progress, 1877—1878: 15c—18c.

Bérard, J., 1957. Bones Lake area, New Quebec. Que., Dep. Mines, Prelim. Rep. 342.

Bérard, J., 1958. Finger Lake area, New Quebec. Que., Dep. Mines, Prelim. Rep. 360.

Bérard, J., 1959. Leaf Lake area, New Quebec. Que., Dep. Mines, Prelim. Rep. 384.

Bergeron, R., 1954. Gérido Lake area, New Quebec. Que., Dep. Mines, Prelim. Rep. 291.

Bergeron, R., 1955. Thévenet Lake area (west part), New Quebec. Que., Dep. Mines, Prelim. Rep. 311.

Bergeron, R., 1956. Harveng Lake area (west half), New Quebec. Que., Dep. Mines, Prelim. Rep. 320.

Bergeron, R., 1957a. Brochant—De Bonnard area, New Quebec. Que., Dep. Mines, Prelim. Rep. 348.

Bergeron, R., 1957b. Preliminary Report on Cape Smith—Wakeham Bay Belt, New Quebec. Geol. Surv. Branch. Dep. Mines, Que., Can., Prelim. Rep. 355.

Bergeron, R., 1959. Preliminary Report on Povungnituk Range area. New Quebec. Geol. Surv. Branch, Dep. Mines, Que., Can.; Prelim. Rep. 392.

Blais, R.A., 1959. L'Origine des minerals crétacés du gisement de fer de Redmond, Labrador. Nat. Can., 86: 265—200.

Blais, R.A. and Stubbins, J.B., 1962. The role of mine geology in the exploitation of the iron deposits of the Knob Lake range, Canada Soc. Min. Eng., March: 15—23.

Chakraborty, K.L., 1963. Relationship of anthophyllite, cummingtonite and mangano-cummingtonite in the metamorphosed Wabush iron-formation, Labrador. Can. Mineral., 7 (5): 738—750.

Chakraborty, K.L., 1966. Ferromagnesian silicate minerals in the metamorphosed iron-formation of Wabush Lake and adjacent areas, Newfoundland and Quebec. Geol. Surv. Can. Bull. 143.

Chandler, F.W. and Schwarz, E.J., 1980. Tectonics of the Richmond Gulf area, Northern Quebec — a hypothesis. In: Current Research, Part C. Geol. Surv. Can., Pap., 80-1C: 59—68.

Chauvel, J.J. and Dimroth, E., 1974. Facies types and depositional environment of the Sokoman iron formation, Central Labrador Trough, Quebec, Canada. J. Sediment. Petrol., 44 (2): 299—327.

Choubersky, A., 1957. The operations of the Iron Ore Company of Canada. Trans. Bull. Can. Inst. Min. Metall., 67 (2).

Clarke, P.J., 1960. Normanville area. Que., Dep. Mines, Prelim. Rep. 413.

Clarke, P.J., 1967. Gras Lake—Felix Lake Area, Saguenay County. Que., Dep. Na. Resour., Geol. Rep. 129.

Dimroth, E., 1965. Geology of Otelnuk Lake area, New Quebec Territory. Que., Dep. Na. Resour., Prelim. Rep. 532.

Dimroth, E., 1968. Sedimentary textures, diagenesis, and sedimentary environment of certain Precambrian ironstones. N. Jahrb. Geol. Palaontol., Abh. 130: 247—274.

Dimroth, E., 1970. Evolution of the Labrador Geosycline. Geol. Soc. Am. Bull., 81: 2717—2742.

Dimroth, E., 1977. Facies models-5. Models of physical sedimentation of iron formations. Geosci. Can., 4 (1) 23—30.

Dimroth, E., 1978. Labrador Trough area between latitudes 58°30′ and 56°30′. Dir. Gén. Rech. Géol. Minér., Minist. Richesses Naturelles, Que., Can. Rap. Geol. 193.

Dimroth, E. and Chauvel, J.J., 1973. Petrography of the Sokoman Iron Formation in part of the central Labrador Trough. Geol. Soc. Am. Bull., 84: 111—134.

Dimroth, E. and Dressler, B., 1978. Metamorphism in the Labrador Trough. In: J.A. Fraser and W.W. Heywood (Editors), Metamorphism in the Canadian Shield. Geol. Surv. Can., Pap., 78—10: 215—236.

Dimroth, E., Baragar, W.R.A., Bergeron, R. and Jackson, G.D., 1970. The filling of the Circum-Ungava geosyncline. In: A.J. Baer (Editor), Symposium on Basins and Geosynclines of the Canadian Shield. Geol. Surv. Can., Pap., 70—40: 45—157.

Donaldson, J.A., 1959. Marion Lake, Quebec—Newfoundland. Geol. Surv. Can., Map 17—1959.

Donaldson, J.A., 1960. Geology of the Marion Lake Area, Quebec—Labrador. Ph.D. Thesis, The Johns Hopkins University, Baltimore, Md., (unpubl.).

Donaldson, J.A., 1966. Marion Lake map-area, Quebec—Newfoundland (23 I/13). Geol. Surv. Can., Mem. 338.

Donaldson, J.A., 1963. Stromatolites in the Denault Formation Marion Lake, Coast of Labrador, Newfoundland. Geol. Surv. Can., Bull. 102.

Dorf, E., 1959. Cretaceous flora from beds associated with rubble iron-ore deposits in the Labrador Trough. Geol. Soc. Am. Bull., 70:1591.

Douglas, R.J.W. (Editor), 1970. Geology and Economic Minerals of Canada. Geol. Surv. Can., Econ. Geol., Rep. No. 1, 5th ed., 838 pp.
Dressler, B., 1979. Région de la Fosse du Labrador, (56°30′—57°15′). Service de L'Exploration Géologique, Ministère des Richesses Naturelles, Rap. Géol., 195.
Duffell, S. and Roach, R.A., 1959. Mount Wright, Quebec—Newfoundland. Geol. Surv. Can., Map 6-1959.
Dugas, J., 1970. Metallic mineralization in part of the Labrador Trough. Que., Dep. Nat. Resour., Mines Branch, Spec. Pap. 5.
Eade, K.E., 1952. Unknown River (Ossokmanuan Lake, east half), Labrador, Newfoundland. Geol. Surv. Can., Pap. 52—9.
Eade, K.E., 1966. Fort George River and Kaniapiskau River (West Half) map-areas, New Quebec. Geol. Surv. Can., Mem. 339.
Eade, K.E., Stevenson, I.M., Kranck, S.H. and Hughes, O.L., 1959. Nichicun-Kaniapiskau, New Quebec. Geol. Surv. Can., Map 56-1959.
Eugster, H.P. and I-Ming Chou, 1973. The depositional environments of Precambrian banded iron-formations. Econ. Geol., 68: 1144—1168.
Evans, J.L., 1978. The Geology and Geochemistry of the Dyke Lake area (parts of 23J/8, 9), Labrador. Miner. Dev. Div., Dep. Mines Energy, Gov. Newfoundland and Labrador, Rep. 78-4.
Fahrig, W.F., 1951. Griffis Lake (west half), Quebec. Geol. Surv. Can., Pap. 51-23.
Fahrig, W.F., 1955. Lac Herodier, New Quebec. Geol. Surv. Can., Pap. 55-1.
Fahrig, W.F., 1956a. Lac Herodier (east half), New Quebec. Geol. Surv. Can., Pap. 55-37.
Fahrig, W.F., 1956b. Cambrian Lake (west half), New Quebec. Geol. Surv. Can., Pap. 55-42.
Fahrig, W.F., 1957. Geology of certain Proterozoic rocks in Quebec and Labrador. R. Soc. Can., Spec. Publ. 2.
Fahrig, W.F. 1965. Lac Herodier, Quebec. Geol. Surv. Can., Map 1146A.
Fahrig, W.F., 1967. Shabogamo Lake map-area, Newfoundland-Labrador and Quebec, 23G, E 1/2. Geol. Surv. Can., Mem. 354.
Fahrig, W.F., 1969. Lac Cambrien (west half), Quebec. Geol. Surv. Can., Map 1223A.
Flaherty, R.J., 1918. The Belcher Islands of Hudson Bay; Their Discovery and exploration. Geograph. Rev., v. 5 (6): 433—458.
Frarey, M.J., 1952. Willbob Lake, Quebec and Newfoundland. Geol. Surv. Can., Pap. 52-16.
Frarey, M.J., 1961. Menihek Lakes, Newfoundland and Quebec. Geol. Surv. Can., Map 1087A.
Frarey, M.J., 1967. Willbob Lake and Thompson Lake map-areas, Quebec and Newfoundland (230/1 and 23 0/8). Geol. Surv. Can., Mem. 348.
Frarey, M.J. and Duffell, S., 1964. Revised stratigraphic nomenclature for the central part of the Labrador Trough. Geol. Surv. Can., Pap. 64-25.
French, B.M., 1973. Mineral assemblages in diagenetic and low-grade metamorphic iron-formation. Econ. Geol., 68: 1063—1074.
Fryer, B.J., 1972. Age determinations in the Circum-Ungava geosyncline and the evolution of Precambrian banded iron-formations. Can. J. Earth Sci., 9 (6): 652—663.
Fryer, B.J., 1977a. Trace element geochemistry of the Sokoman iron formation. Can. J. Earth Sci., 14: 1598—1610.
Fryer, B.J., 1977b. Rare earth evidence in iron-formations for changing Precambrian oxidation states. Geochim. Cosmochim. Acta, 41: 361—367.
Garrels, R.M., 1960. Mineral Equilibria at Low Temperature and Pressure. Harper and Brothers, New York, N.Y., 254 pp.
Garrels, R.M., Perry, E.A. Jr., and Mackenzie, F.T., 1973. Genesis of Precambrian iron-formations and the development of atmospheric oxygen. Econ. Geol., 68: 1193—1179.
Gastil, G.R. and Knowles, D.M., 1960. Geology of the Wabush Lake area, Southwestern Labrador and Eastern Quebec, Canada. Geol. Soc. Am. Bull., 71: 1243—1254.

Gastil, G.R. et al., 1960. The Labrador geosyncline. Int. Geol. Congr., XXI session, Norden, pt. 9.
Gélinas, L., 1958a. Gabriel Lake area (west half), New Quebec. Que., Dep. Mines., Prelim. Rep. 373.
Gélinas, L., 1958b. Thévenet Lake area (east half), New Quebec. Que., Dep. Mines, Prelim. Rep. 363.
Gélinas, L., 1960. Gabriel Lake area (east half) and the Fort Chimo area (west part), New Quebec. Que., Dep. Mines, Prelim. Rep. 407.
Gélinas, L. and Bergeron, R., 1962. Geology of northern Ungava. Precambrian, 35 (3): 20—25.
Gill, J.E., Bannerman, H.M. and Tolman, C., 1937. Wapussakatoo Mountains, Labrador. Bull., Geol. Soc. Am. v. 48: 567—585.
Goldich, S.S., 1973. Ages of Precambrian banded iron-formation. Econ. Geol., 68: 1125—1134.
Goodwin, A.M., 1956. Facies relations in the Gunflint iron-formation. Econ. Geol., 51 (6): 565—595.
Green, B.A., 1974. An outline of the geology of Labrador. Miner. Dev. Div., Dep. Mines and Energy, Prov. Newfoundland, Inf. Circ. 15.
Gross, G.A., 1951. A Comparative Study of three Slate Formations in the Ferriman Series in the Labrador Trough. M.A. Thesis, Queen's Univ., Kingston, (unpubl.).
Gross, G.A., 1955. The Metamorphic Rocks of the Mount Wright and Matonipi Lake Areas of Quebec. Ph.D. Thesis, Univ. Wisconsin, (unpubl.).
Gross, G.A., 1959. Metallogenic map, Iron in Canada. Geol. Surv. Can., Map 1045A-M4.
Gross, G.A., 1960. The iron ranges and current developments in New Quebec and Labrador, Canada. 21st Annu. Mining Symposium, Univ. Minnesota.
Gross, G.A., 1961a. Metamorphism of iron-formations and its bearing on their beneficiation. Trans. Can. Inst. Min. Metall., 64: 24—31.
Gross, G.A., 1961b. Iron-formations and the Labrador geosyncline. Geol. Surv. Can., Pap. 60-30.
Gross, G.A. 1962. Iron deposits near Ungava Bay, Quebec. Geol. Surv. Can., Bull. 82.
Gross, G.A. 1965. Geology of Iron Deposits in Canada; Volume 1, General Geology and Evaluation of Iron Deposits. Geol. Surv. Can., Econ. Geol. Rep. No. 22.
Gross, G.A., 1967. Geology of Iron Deposits in Canada; Volume II, Iron Deposits in the Appalachian and Grenville Regions of Canada. Geol. Surv. Can., Econ. Geol. Rep. No. 22.
Gross, G.A., 1968. Geology of Iron Deposits in Canada; Volume III, Iron Ranges of the Labrador Geosyncline. Geol. Surv. Can., Econ. Geol. Rep. No. 22.
Gross, G.A., 1972. Primary features in cherty iron-formation. Sediment. Petrol., 7: 241—261.
Gross, G.A., 1973. The depositional environment of principal types of Precambrian iron-formation. In: Genesis of Precambrian Iron and Manganese Deposits. Proc. Kiev Symposium, 1970, Unesco Earth Sciences, 9.
Gross, G.A., 1980. A classification of iron-formation based on depositional environments. Can. Mineral., 18 (2): 215—222.
Gross, G.A. and McLeod, C.R., 1980. A preliminary assessment of the chemical composition of iron-formations in Canada. Can. Mineral., 18 (2): 223—229.
Gross, G.A., Glazier, W., Kruechl, G., Nichols, L. and O'Leary, J., 1972. Iron ranges of Labrador and northern Quebec 24th Int. Geol. Congr., Guideb. Field Excursion A55.
Gustafson, J.K. and Moss, A.E., 1953. The role of geologists in the development of the Labrador Quebec iron ore districts. Paper presented at Am. Inst. Min. Met., Los Angeles, Calif., U.S.A.
Harrison, J.M., 1952. The Quebec-Labrador iron belt, Quebec and Newfoundland. Geol. Surv. Can., Pap. 52-20.

Harrison, J.M., Howell, J.E. and Fahrig, W.F., 1972. A geological cross-section of the Labrador miogeosyncline near Schefferville, Quebec, Geol. Surv. Can., Pap. 70-37.

Hawley, J.E., 1925. Sutton Lake area, Patricia District, Ontario. Ont., Dep. Mines, Rep., 34 (7).

Henderson, E.P., 1959. A glacial study of Central Quebec-Labrador, Geol. Surv. Can., Bull. 50.

Hofmann, H.J. and Jackson, G.D., 1969. Precambrian (Aphebian) microfossils from Belcher Islands, Hudson Bay. Can. J. Earth Sci., 6 (5): 1137—1144.

Jackson, G.D., 1960. Belcher Islands, Northwest Territories. Geol. Surv. Can., Pap. 60-20.

James, H.L., 1954. Sedimentary facies of iron-formation. Econ. Geol., 49: 235—293.

Kearey, P. and Halliday, D.W., 1976. The gravity field of the Central Labrador Trough, Northern Quebec, with map No. 162 — Lac Nachicapau — Central Labrador Trough. Earth Physics Branch, Energy, Mines and Resources, Canada.

Klein, C., Jr., 1966. Mineralogy and petrology of the metamorphosed Wabush iron-formation, southwestern Labrador. J. Petrol., 7 (2): 246—305.

Klein, C., Jr., 1973. Changes in mineral assemblages with metamorphism of some banded Precambrian iron-formations. Econ. Geol., 68: 1075—1088.

Klein, C., Jr., 1978. Regional metamorphism of Proterozoic iron-formation, Labrador Trough, Canada. Am. Mineral., 63: 898—912.

Klein, C., Jr. and Fink, R. P., 1976. Petrology of the Sokoman Iron Formation in the Howells River area, at the western edge of the Labrador Trough. Econ. Geol., 71: 453—487.

Knowles, D.M. and Gastil, G.R., 1959. Metamorphosed iron-formation in southwestern Labrador. Trans., Can. Inst. Min. Metall., 62: 265—272.

Kranck, S.H., 1959. Chemical Petrology of Metamorphic Iron-formations and Associated Rocks in the Mount Reed Area in Northern Quebec. Ph.D. Thesis, Mass. Inst. Technol., Cambridge, Mass.

Kranck, S.H., 1961. A study of phase equilibrium in a metamorphic iron-formation. J. Petrol., 2 (2): 137—184.

Krishnan, T.K. and Oertel, G., 1980. Aspects of strain history in folded sediments from the Schefferville mining district, Labrador Trough, Canada. Tectonophysics, 64: 33—46.

Krumbein, W.C. and Garrels, R.M., 1952. Origin and classification of chemical sediments in terms of pH and oxidation-reduction potentials. J. Geol., 60: 1—33.

Lesher, C.M., 1978. Mineralogy and petrology of the Sokoman Iron Formation near Ardua Lake, Quebec. Can. J. Earth Sci., 15 (4): 480—500.

Low, A.P., 1896. Report on exploration in the Labrador Peninsula along Eastmain, Koksoak, Hamilton, Manikuagan and portions of other rivers. Geol. Surv. Can. Annu. Rep. 1895, Rep. L.

Low, A.P., 1902. Report on an exploration of the East Coast of Hudson Bay. Geol. Surv. of Can., Publ. No. 778.

Low, A.P., 1903. Report on the geology and physical character of the Nastapoka Islands, Hudson Bay. Geol. Surv. Can., Publ. No. 819.

Lowdon, J.A., 1960. Age determinations by the Geological Survey of Canada, Report 1, isotopic ages. Geol. Surv. Can., Pap. 60-17.

Lowdon, J.A. (Compiler), 1961. Age determinations by the Geological Survey of Canada, Report 2, isotopic ages. Geol. Surv. Can., Pap. 61-17.

Macdonald, R.D., 1960. Iron deposits of Wabush Lake, Labrador. Min. Eng., Oct. 1960.

Markun, C.D. and Randazzo, A.F., 1980. Sedimentary structures in the Gunflint Iron Formation, Schreiber Beach, Ontario. Precambrian Res., 12: 287—310.

Moore, E.S., 1918. The iron-formation on Belcher Islands, Hudson Bay, with special reference to its origin and its associated algal limestone. J. Geol., 26: 412—438.

Mueller, R.F., 1960. Compositional characteristics and equilibrium relations in mineral assemblages of a metamorphosed iron-formation. Am. J. Sci., 258: 449—497.

Murphy, D.L., 1959. Mount Wright area. Saguenay Electoral District. Que., Dep. Mines, Prelim. Rep. 380.

Neilson, J.M., 1953. Albanel Area, Mistassini Territory. Geol. Surv. Branch, Dep. Mines, Que., Can., Geol. Rep. 53.

Neilson, J.M., 1963. Lake Albanel Iron Range, Northern Quebec. Trans. Can. Inst. Min. Metall., 66: 21—27.

Norris, A.W., Sanford, B.V. and Bostock, H.H., (Compilers), 1967. Geology, Hudson Bay Lowlands, Manitoba, Ontario, Quebec and District of Keewatin. Geol. Surv. Can., Map 17-1967, Pap. 67—60.

Phillips, L.S., 1958. Tuttle Lake area, Saguenay Electoral District. Que., Dep. Mines, Prelim. Rep. 377.

Phillips, L.S. 1959. Peppler Lake area (east half), Saguenay Electoral District. Que., Dep. Mines, Prelim. Rep. 401.

Quirke, T.T., Goldich, S.S., and Krueger, H.W., 1960. Composition and age of the Temiscamie iron-formation, Mistassini Territory, Quebec, Canada. Econ. Geol., 55: 311—326.

Roach, R.A. and Duffell, S., 1968. The pyroxene granulites of the Mount Wright map-area, Quebec-Newfoundland. Geol. Surv. Can., Bull. 162.

Roach, R.A. and Duffell, S., 1974. Structural analysis of the Mount Wright map-area, southernmost Labrador Trough, Quebec, Canada. Geol. Soc. Am. Bull., 85: 947—962.

Roscoe, S.M., 1957. Cambrian Lake (east half), New Quebec. Geol. Surv. Can., Pap. 57-6.

Rose, E.R., 1955. Manicouagan Lake-Mushalagan Lake area, Quebec. Geol. Surv. Can., Pap. 55-2.

Sauvé, P., 1953. Clastic Sedimentation during a Period of Volcanic Acitivity, Astray Lake, Labrador. M.Sc. Thesis, Queen's University, Kingston, Ont., Can., (unpubl.).

Sauvé, P., 1955. Gérido Lake area (east half), New Quebec. Que. Dep. Mines, Prelim. Rep. 309.

Sauvé, P., 1956a. Leopard Lake area (east half), New Quebec. Quebec Dep. Mines, Prelim. Rep. 325.

Sauvé, P., 1956b. De Freneuse Lake area (west half), New Quebec. Quebec Dep. Mines, Prelim. Rep. 332.

Sauvé, P., 1957. De Freneuse Lake area (east half), New Quebec. Quebec Dep. Mines, Prelim. Rep. 358.

Sauvé, P., 1959. Leaf Bay area, New Quebec. Que., Dep. Mines. Prelim. Rep. 399.

Schwellnus, J.E.G., 1957. Ore Controls in Deposits of the Knob Lake Area, Labrador Trough. Ph.D. Thesis Queen's Univ., Kingston, Ont. (unpubl.).

Sérvice des Gîtes menéraux, 1971. Iron in Quebec. Gouvernement du Quebec, Ministère des Richesses Naturelles, Direction Générale des Mines. Spec. Pap. 12.

Sinclair, A.J., 1960. Georget Lake area (east half). Que. Dep. Mines, Prelim. Rep. 414.

Stevenson, I.M., 1968. A geological reconnaissance of Leaf River map-area, New Quebec and Northwest Territories. Geol. Surv. Can., Mem. 356.

Stubbins, J.B., Blais, R.A. and Zajac, I.S., 1961. Origin of the soft iron ores of the Knob Lake Range. Trans. Can. Inst. Min. Metall., 64: 37—52.

Taylor, F.C., 1974. Reconnaissance geology of a part of the Precambrian Shield, northern Quebec and northwest Territories. Geol. Surv. Can., Pap. 74-21.

Waddington, G.W., 1960. Iron ore deposits of the Province of Quebec. Que., Dep. Mines, Prelim. Rep. 409.

Wahl, W.G., 1953. Temiscamie River Area, Mistassini Territory; Geological Surveys Branch, Department of Mines, Quebec, Canada, Geological Report 54.

Westra, L., 1978. Metamorphism in the Cape Smith-Wakeham Bay area north of 61°N, New Quebec. In: J.A. Fraser and W.W. Heywood (Editors), Metamorphism in the Canadian Shield. Geol. Surv. Can., Pap., 78—10: 237—244.

Wynne-Edwards, H.R., 1960. Michikamau Lake (west half), Quebec-Newfoundland. Geol. Surv. Can., Map 2-1960.

Young, G.A., 1922. Iron-bearing rocks of Belcher Islands, Hudson Bay. Geol. Surv. Can., Summary Rep., 1921, Part E.

Zajac, I.S., 1974. The stratigraphy and mineralogy of the Sokoman Formation in the Knob Lake area, Quebec and Newfoundland. Geol. Surv. Can., Bull. 220.

Chapter 7

THE NABBERU BASIN OF WESTERN AUSTRALIA

A.D.T. GOODE, W.D.M. HALL and J.A. BUNTING

INTRODUCTION

The Early Proterozoic Nabberu Basin lies along the northern margin of the Archaean Yilgarn craton in central Western Australia. Extensive granular iron-formations of the Superior type are developed in the basin, in distinct contrast to the laminated (Algoman-style) iron-formations within the Early Proterozoic Hamersley Basin to the north. Both Early Proterozoic basins were developed on the edge of the stable granite-greenstone basements of the Yilgarn and Pilbara cratons and on the inferred projection of the complex "mobile" Western Gneiss Belt which would appear to separate the cratons (Goode, 1981; Fig. 7-1).

Although rocks within the Nabberu Basin had been described previously, it was not until 1973 that the full extent and significance of the basin was recognized following publication of the Bureau of Mineral Resources regional aeromagnetic maps of Nabberu and Stanley (see Hall and Goode, 1978).

A considerable amount of work has followed since 1973, and mapping of the 1:250,000 sheet areas covering the basin has recently been completed by the Geological Survey of Western Australia. The basin has been divided into an eastern Earaheedy Sub-basin, a central Glengarry Sub-basin and a western Padbury Sub-basin on tectonic and stratigraphic grounds (Fig. 7-2). Partly because of deep weathering and lack of outcrop in critical areas and the general lack of detailed work, there are various interpretations on a number of points, particularly the correlation between the sub-basins (especially the Earaheedy and Glengarry sub-basins), the regional relationship of sequences developed in the Padbury sub-basin and the broader relationships with the Hamersley Basin (see discussion by Goode, 1981).

Much of the description that follows is confined to the Earaheedy Sub-basin, as it is here that the iron-formations are most extensively developed and have been better studied. The iron-formations in the Padbury Sub-basin (Macleod, 1970; Barnett, 1975; Hall and Goode, 1978; Gee, 1979a) have not received the same detailed attention as those in the Earaheedy Sub-basin.

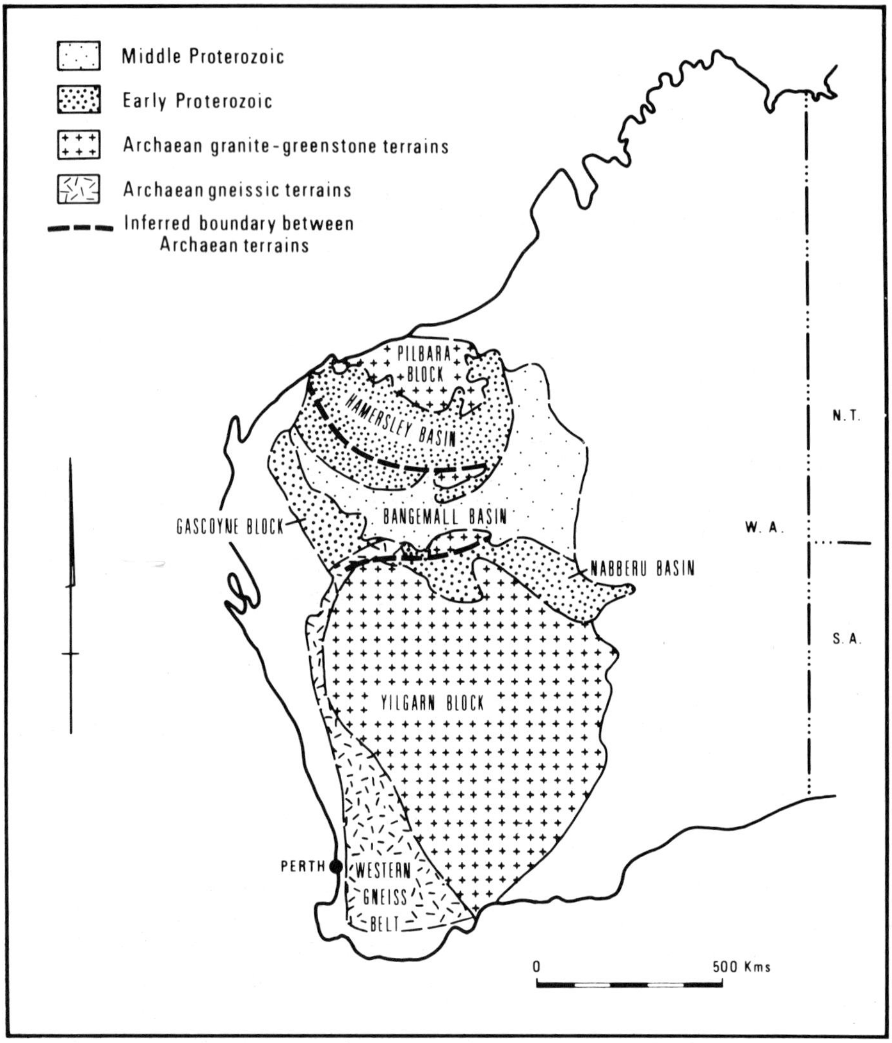

Fig. 7-1. Location of the Nabberu Basin in relation to other major tectonic subdivisions of the Western Australian Shield.

DOCUMENTATION OF THE BASIN

The first geological investigation of the area was undertaken by Talbot (1910, 1920) who recognized the unconformities between the Middle Proterozoic Bangemall ("Nullaginian" of Talbot) and Earaheedy Groups, and between the Early Proterozoic rocks and the Archaean rocks of the Yilgarn Block. However, he mistakenly identified another unconformity, correspond-

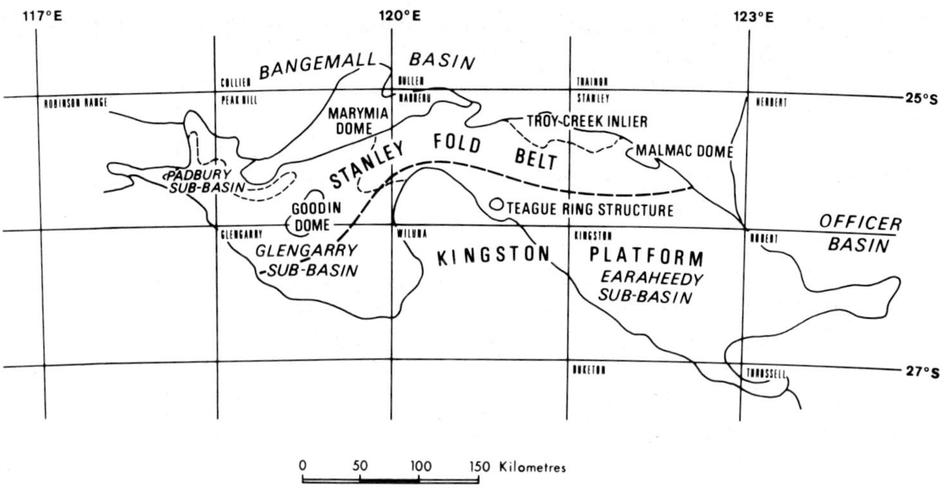

Fig. 7-2. Generalized map of the Nabberu Basin (also showing the location of individual 1 : 250,000 map sheets).

ing to the change from folded to unfolded rocks at the southern edge of the Stanley Fold Belt. This led to the belief that the southern undeformed rocks south of this line were equivalent to the "Nullagine" rocks.

Talbot's interpretation remained on State Geological maps until the 1966 edition, in which all the Proterozoic rocks in the eastern part of the area were included in the Middle Proterozoic Bangemall Group, while those in the west were included with the Archaean basement. This interpretation was in vogue (Sofoulis and Mabbutt, 1963; Daniels and Horwitz, 1969; MacLeod, 1970; Sanders and Harley, 1971) until the major uncomformity between the Bangemall Basin and Nabberu Basin was rediscovered by Hall and Goode (1975) and Horwitz (1975a). These two studies recognized the significance of the granular iron-formations and compared them with the Superior-type iron-formations of North America. This was further emphasized by Walter et al. (1976) who described microfossils from the Frere Formation that were comparable with those of the Gunflint Iron Formation, and by Goode and Hall (1976) who presented evidence for a shallow-water origin for the iron-formations.

The original stratigraphic subdivisions of Hall and Goode (1975) were modified slightly and defined by Hall et al. (1977). A full description of the basin was presented by Hall and Goode (1978).

Mapping at 1:250,000 scale by the Geological Survey of Western Australia (GSWA) and Bureau of Mineral Resources (BMR) initially skirted around the basin. In the west Barnett (1975) defined the Padbury Group by modifying the subdivisions of Macleod (1970). Parts of the eastern margin of the basin were mapped in 1972—1973 by Jackson (1978), Bunting et al. (1978), and Bunting and Chin (1975). Following Jackson's work, glauconite from the basal Yelma Formation was dated at about 1700 m.y. (Preiss et al., 1975), and Preiss (1976) described stromatolites from the Yelma and Windidda Formations.

Systematic mapping of the basin by the GSWA was carried out between 1975 and 1977, and the results published as a series of explanatory notes (Bunting, 1977; Commander et al., 1979; Bunting et al., 1979; Elias and Bunting, 1979; and Elias et al., 1979). Bunting et al. (1977) presented a preliminary synthesis and summary of the work (a more comprehensive bulletin is in preparation) while Gee (1979a) described new formations in the Glengarry Group and redefined the Padbury Group.

DESCRIPTION OF THE BASIN

General information

The Nabberu Supergroup rests unconformably on Archaean rocks along the northern margin of the Yilgarn craton, and now covers an area of approximately 60,000 km^2. A maximum exposed sedimentary thickness of about 6000 m occurs in the Earaheedy Sub-basin, about 7000 m in the Glengarry Sub-basin and about 5000 m in the Padbury Sub-basin.

The present boundaries do not represent the original depositional extent of the basin and a few outliers of Proterozoic sedimentary rocks to the south were almost certainly once part of it. On its western and northwestern sides the Nabberu Basin becomes increasingly affected by tectonism in which the Archaean basement is involved. In the north, sediments of the Nabberu Basin are unconformably covered by or faulted against Middle Proterozoic rocks. Their northern extent beneath these younger rocks is not known. In the east the basin is covered by sediments of the Officer Basin (principally Early Permian sediments). It is assumed that the Nabberu Basin continues for a considerable distance under the Officer Basin.

Stratigraphy

Sediments of the Nabberu Basin form the Nabberu Supergroup, which is subdivided into the Padbury, Glengarry and Earaheedy Groups, corresponding to the three major sub-basins which comprise the basin.

The Glengarry Group will not be described here in detail; its constituent formations have been described by Bunting et al. (1977), Elias et al. (1979), and Gee (1979a). The group consists of two facies — a thick (7000 m) trough facies in the north containing greywacke, shale, arkose, quartz arenite, and mafic volcanic rocks, and a thinner shelf facies in the south consisting of a basal quartz arenite overlain by shale, marl, and thin carbonate beds.

The redefined Padbury Group unconformably overlies the Glengarry Group according to Gee (1979a) although previously the lower Padbury Group and the Glengarry Group were regarded as equivalent. Lithological similarity of the iron-formations and stratigraphic compatability suggest that the Padbury Group may be correlated with part of the Tooloo Subgroup of the Earaheedy Group, although the similarity may only reflect similar depositional conditions.

Most of the iron-formations in the Padbury Group are contained in the *Robinson Range Formation*, which is at least 3000 m thick. This consists of two major units of iron-formation separated by magnetite-bearing hematite-sericite shale. The lower unit is a laminated (Algoman or Hamersley style) banded iron-formation, but the upper unit is a granular iron-formation with discontinuous bedding on outcrop scale. The granular iron-formation is characterised by lenses of granular and oolitic chert 10—20 mm thick, more continuous bands of the same thickness of red jasper, and beds of clastic ironstone up to 1 m thick. Clasts in the ironstone include spherical peloids of chert 0.5 mm across, hematitic shale, green chloritic shale, chert and specular hematite up to 10 mm across, and larger fragments of jasper. All clasts were derived from the immediate sedimentary environment.

Laminated iron-formations up to 250 m thick and 50 km in lateral extent also occur in the *Horseshoe Formation*, part of the Padbury Group of Barnett (1975) but regarded as part of the (older?) Glengarry Group by Gee (1979a). Iron-formation units up to 40 m thick are intercalated with quartz-chlorite-magnetite shale. The iron-formations thin laterally to less than 1 m thick.

The Earaheedy Group (Fig. 7-3), which is the main topic of the following discussion, is divided into an older Tooloo Sub-group and a younger Miningarra Sub-group (Table 7-I) on the basis of a disconformity in the eastern part of the sub-basin. Each sub-group represents a cycle comprising an initial transgressive phase and a terminal regressive phase. The disconformity, which is marked by a period of emergence from the dominantly marine sequence, does not occur in the northern and northwestern parts of the sub-basin, where, in probably deeper-water sediments, there is a continuous sedimentary sequence dominated by shales.

The *Frere Formation*, which contains the bulk of the iron-formations in the Earaheedy Sub-basin, forms a series of low hills and ranges. It conformably overlies the clastics and dolomites of the Yelma Formation and is in turn conformably overlain by the limestones and shales of the Windidda Formation in the southeast, and the fine clastics of the Wandiwarra Formation in the north and west (Fig. 7-4).

Shallow dips in the south of the sub-basin and strong folding in the north and the west, together with generally poor exposure, make it difficult to calculate the thickness. However, detailed work in the western Frere Range suggests that the formation is approximately 1300 m thick in that area.

The Frere Formation consists typically of interbedded iron-formation

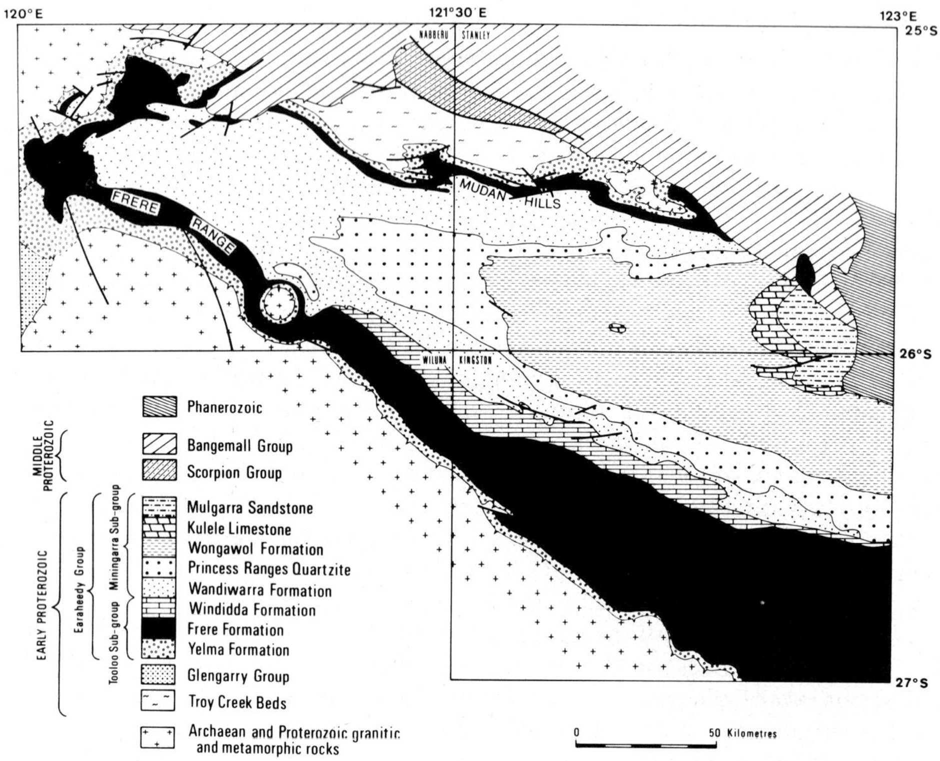

Fig. 7-3. Geological map of the Earaheedy Sub-basin.

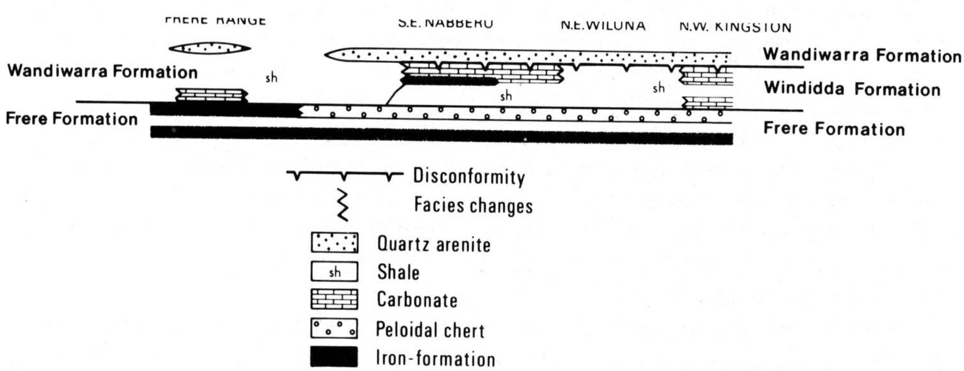

Fig. 7-4. Schematic stratigraphic relationships between the Frere, Windidda and Wandiwarra Formations.

and shale, the thickness of individual iron-formation units varying from a few centimetres to 180 m (Fig. 7-5). The ratio of iron-formation to shale increases northwest along the southern margin of the sub-basin from 1 : 50 in the southeast Kingston sheet to 1 : 10 in the northeast Wiluna sheet to 2 : 1

TABLE 7-I

Stratigraphy of the Earaheedy Group (after Hall and Goode, 1978)

Formation	Maximum thickness (m)	Lithologies	Comments
Miningarra Sub-Group:			
Mulgarra Sandstone	100+	quartz sandstone, minor shale, limestone	cross-bedding, slumps, mud pellets, minor load casts, glauconite
Kulele Creek Limestone	300	limestone, minor sandstone, shale	stromatolites, oolites, intraformational conglomerates
Wongawol Formation	1500	fine micaceous sandstone, minor siltstone, limestone	festoon cross-bedding, ripple marks, slump rolls, scour channels, mud pellets, rare mud cracks
Princess Ranges Quartzite	200—600	quartz sandstone, siltstone, minor grit	cross-bedding, glauconite, mud pellets
Wandiwarra Formation	350—(1500)	fine micaceous sandstone, shale	cross-bedding, minor glauconite, ripple marks, mud pellets
Tooloo Sub-Group:			
Windidda Formation	800	limestone (ankeritic, chamositic at base), shale, minor sandstone, chert	stromatolites, intraformational conglomerates
Frere Formation	1300	granular iron-formation, chert, siltstone, minor carbonates, laminated iron-formation. Dolomite at base	local stromatolites, oolites, cross-bedding
Yelma Formation	0—1000+	sandstone, siltstone, minor conglomerate, iron-formation, carbonate	cross-bedding, ripple marks, stromatolites; glauconite; thickens and becomes more clayey to north and west; gypsum pseudomorphs in dolomite

in the Frere Range at the western end of the sub-basin. Along the northern edge of the sub-basin the ratio is more consistent at about 1 : 1, decreasing only slightly to the east.

Superimposed on the increasing ratio of iron-formation to shale is an increase in the total iron content of both iron-formation and shale. In the eastern part of the Kingston sheet most iron-formation members are represented by ferruginous chert containing between 1 and 10% iron. In the northeast of the Wiluna sheet it is estimated that most iron-formations contain between 10 and 20% iron, while in the western and northern Nabberu sheet they contain over 20% iron. Secondary enrichment in many cases has increased the iron content to over 50%, and values of up to 65% have been recorded.

Rocks of the Frere Formation comprise mainly granular iron-formation and shale with minor laminated and shaly iron-formation, chert and carbonate. The shale units contain various fine-grained, terrigenous rocks including siltstone, sandy siltstone, fissile shale and massive mudstone. In the northern deformed part of the sub-basin these rocks have a well-developed cleavage and are slates and phyllites. In the southeast, siltstone units are the dominant lithology and are typically well-laminated, cream, buff, red, brown or purple rocks with common small-scale cross lamination.

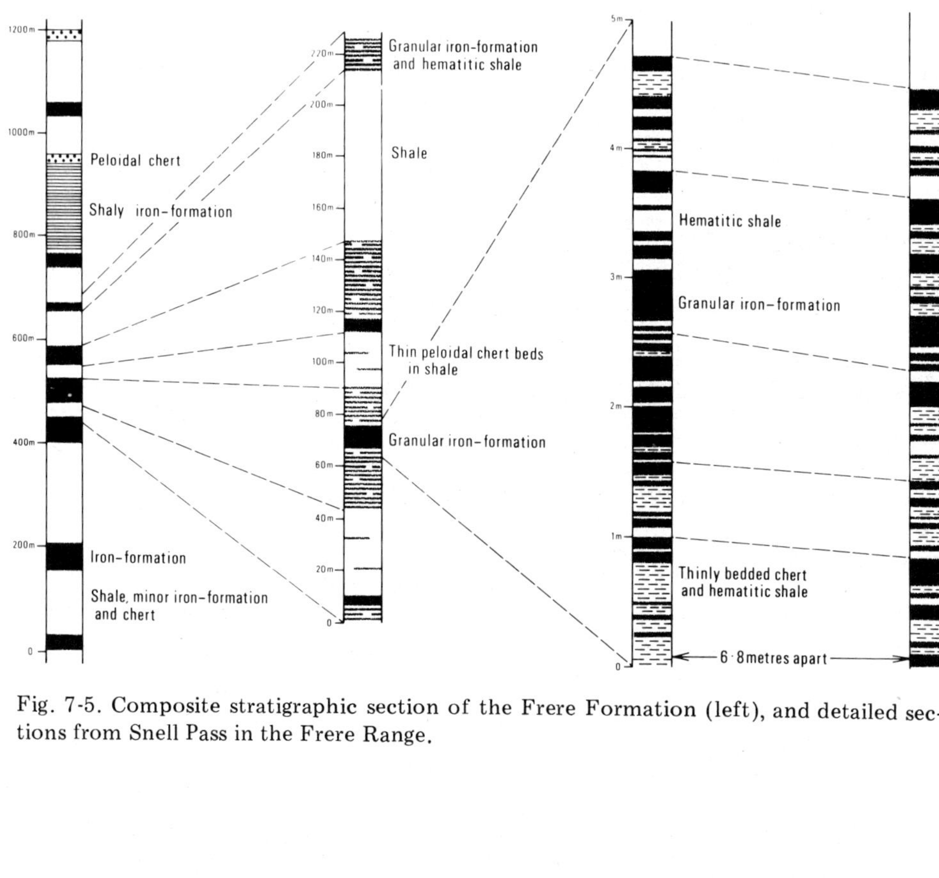

Fig. 7-5. Composite stratigraphic section of the Frere Formation (left), and detailed sections from Snell Pass in the Frere Range.

On a regional scale, iron-formation of the Nabberu Basin is concentrated into distinct units which may be informally termed members. These contain a significant amount of shale, much of it hematitic. At the scale of a measured section the iron-formation members are seen to contain variable amounts of shale, and thus thinner units of predominantly granular iron-formation can be distinguished. Some parts of a member may contain only 10% granular iron-formation, while others contain up to 95%, the remainder being hematitic shale or shaly iron-formation.

Granular iron-formation (GIF, formerly also called pelletal or intraclastic iron-formation) is the most characteristic rock type of the Frere Formation. Individual GIF beds attain a thickness of 1 m but are generally between 5 and 30 cm thick. Beds are typically wavy, lenticular, internally unbanded and seldom persist laterally for more than a few tens of metres (Fig. 7-6a). Local small-scale scouring is observed. GIF beds are separated by hematitic shale which varies from a few millimetres to several metres thick.

The granular texture comprises clasts set in a cement or matrix of chert or chalcedony. The clastic fraction is usually peloidal but may be oolitic, oncolitic or an intraclastic breccia. The peloids are rounded, subspherical grains of either jasper (ferruginous chert), platy hematite and chert, or, in the more recrystallized areas, granular quartz and martite. The ooliths contain concentric layers of jasper and chert, in part altering to platy hematite. Both peloids and ooliths range from 1 to 2 mm in diameter. Most peloidal iron-formation beds contain some breccia fragments, which can be iron-rich or cherty. These are usually angular and form tabular fragments several centimetres long, although smaller ones may be rounded and grade into peloids. An unusual bed of oolitic iron-formation, near Camel Well in the northeast Wiluna sheet, contains stromatolites, oncolites and intraclastic breccia derived from them. The oncolites contain microfossils comparable to forms in the Gunflint Iron Formation in North America (Walter et al., 1976).

Laminated iron-formation (LIF) is virtually restricted to the northern side of the sub-basin, although some thin, microbanded jasperoidal chert beds occur in the hematitic shale portions of the GIF on the southern side. The most conspicuous LIF occurs at the top of the Frere Formation in the north. It is about 50 metres thick and contains alternating chert and iron oxide mesobands between 2 cm and 20 cm thick (Fig 7-6b).

Shaly iron-formation along the southern side of the sub-basin occurs as thin beds interlayered with GIF, and as a thick member without GIF beds near the top of the formation in the west (Illagie Iron-Formation Member). The principal rock type is a dark grey, or purple, shaly iron-formation with bedding on a scale of 1 m or less. Bedding is defined by variations in iron and clay content. A fine lamination or microbanding is present and is laterally continuous over several metres (Fig. 7-7a). It is not clear whether the shaly iron-formation pinches out eastwards or

Fig. 7-6. a. Interbedded lenticular granular iron-formation and shales of the Frere Formation, Frere Range. b. Laminated (banded) iron-formation from the top of the Frere Formation, Mudan Hills.

passes by a facies change into the more typical GIF-shale association.

Non-ferruginous chert may be massive or peloidal. Massive chert is usually grey-green, cryptocrystalline and forms rare beds and lenses up to 10 cm thick within non-ferruginous siltstones in the southeast of the sub-basin. Also in the southeast, non-ferruginous peloidal chert beds 1—2 m thick are separated by great thicknesses of sandstone and shale. Textures are similar to those in the granular iron-formation. A distinctive, green peloidal chert forms an important marker unit at the top of the Frere Formation in the Kingston sheet and extends for a distance of 75 km. It is about 50 m thick and displays an irregular, thick bedding which in places is lensoid. Texturally the chert is very similar to the peloidal chert previously described.

The Frere Formation contains carbonate lenses in at least four localities. The largest of these is near Simpson Well in the western Nabberu sheet where a lens 35 m thick occurs about 500 m above the base of the formation. The carbonate is pale grey to white and recrystallized to a sparry calcite. It contains stromatolites similar to those in the upper part of the Yelma Formation near Sweetwater's Well (Hall and Goode, 1978; Grey, 1979).

The basal few metres of the overlying Windidda Formation, transitional to the Frere Formation, consists of brown ankeritic carbonates with interbedded, pelletal ferruginous cherts, pelletal chlorites, black sulphidic cherts and stromatolitic horizons. Sedimentary chlorite with a chamositic-thuringitic composition occurs as very fine-grained material on bedding planes and algal laminations, and as reworked pellets within carbonates and cherts. Rare detrital chlorite is also present, and some laminated chamosite has been found finely intergrown with chert. Apatite, ankerite, chalcopyrite and sphalerite also occur in some of the thin, dark chert bands. The alternation of chert and dolomite in this basal unit results in an irregular, wavy layering. There is some evidence in the field that dolomite has been replaced by silica to form the chert, giving a form of podding which has been subsequently modified by differential compaction.

The iron-formations

The iron formations in the Nabberu Basin are significant in that they represent the first area of extensive, granular (Superior-type) iron-formations to be recognized outside North America. They contrast strikingly with the laminated iron-formations of both the Hamersley Basin and the Archaean cratons of Western Australia.

Texture

The terminology used in describing the textural elements of the granular iron-formation is based on that of Dimroth and Chauvel (1973) who made the analogy between iron-formation and limestone textures, and adapted the limestone terminology of Folk (1959) for use with iron-formation. A fundamental subdivision can be made into allochems and orthochems.

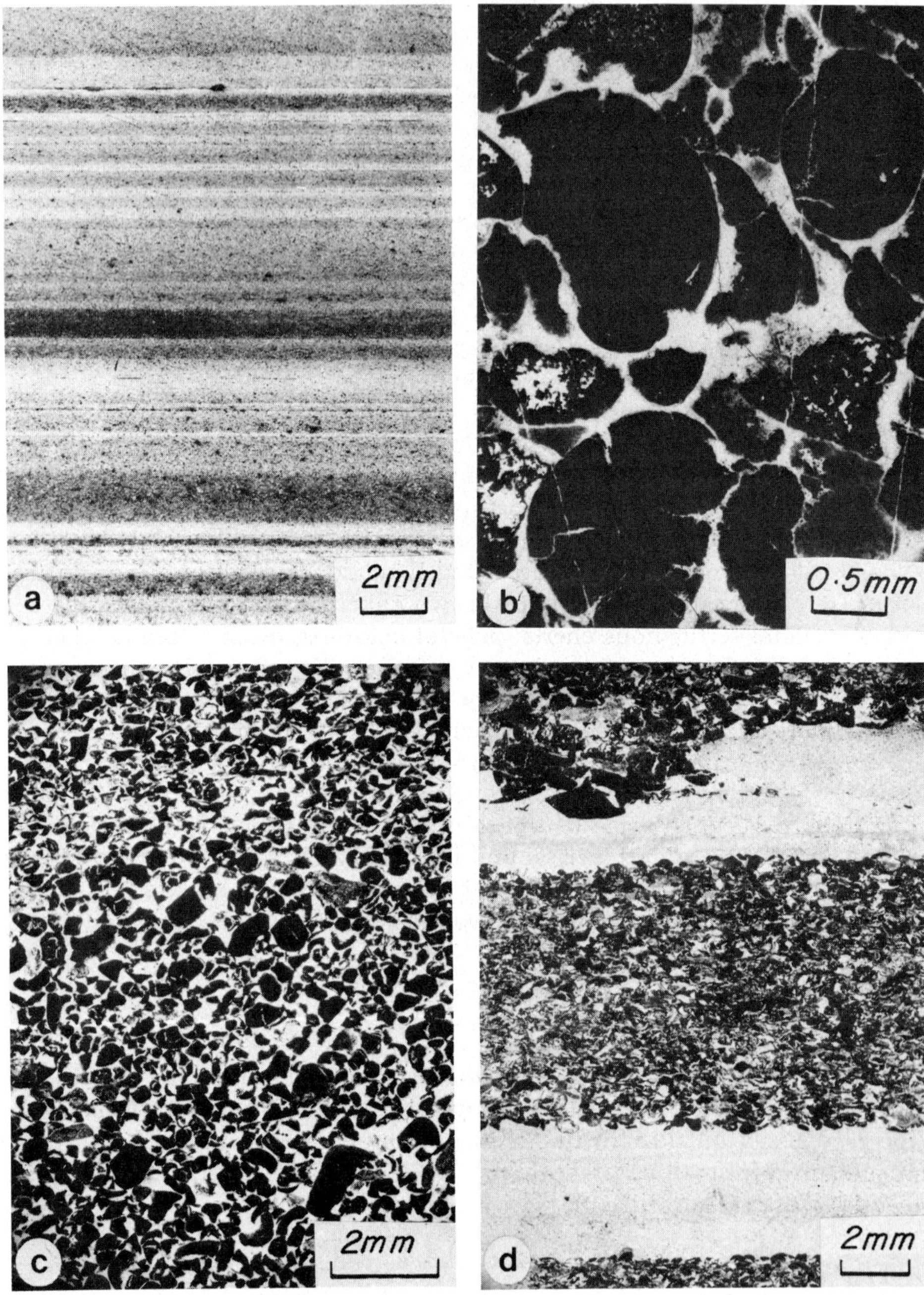

Fig. 7-7. a. Laminated (shaly) iron-formation, Illagie Iron-Formation Member, Frere Range. b. Rounded to sub-angular peloids of ferruginous chert in cherty matrix. c. Angular peloids of ferruginous chert in cherty matrix. d. Alternating mesobands of peloidal ferruginous chert/chert (with rare large chert intraclast) and microbanded chert.

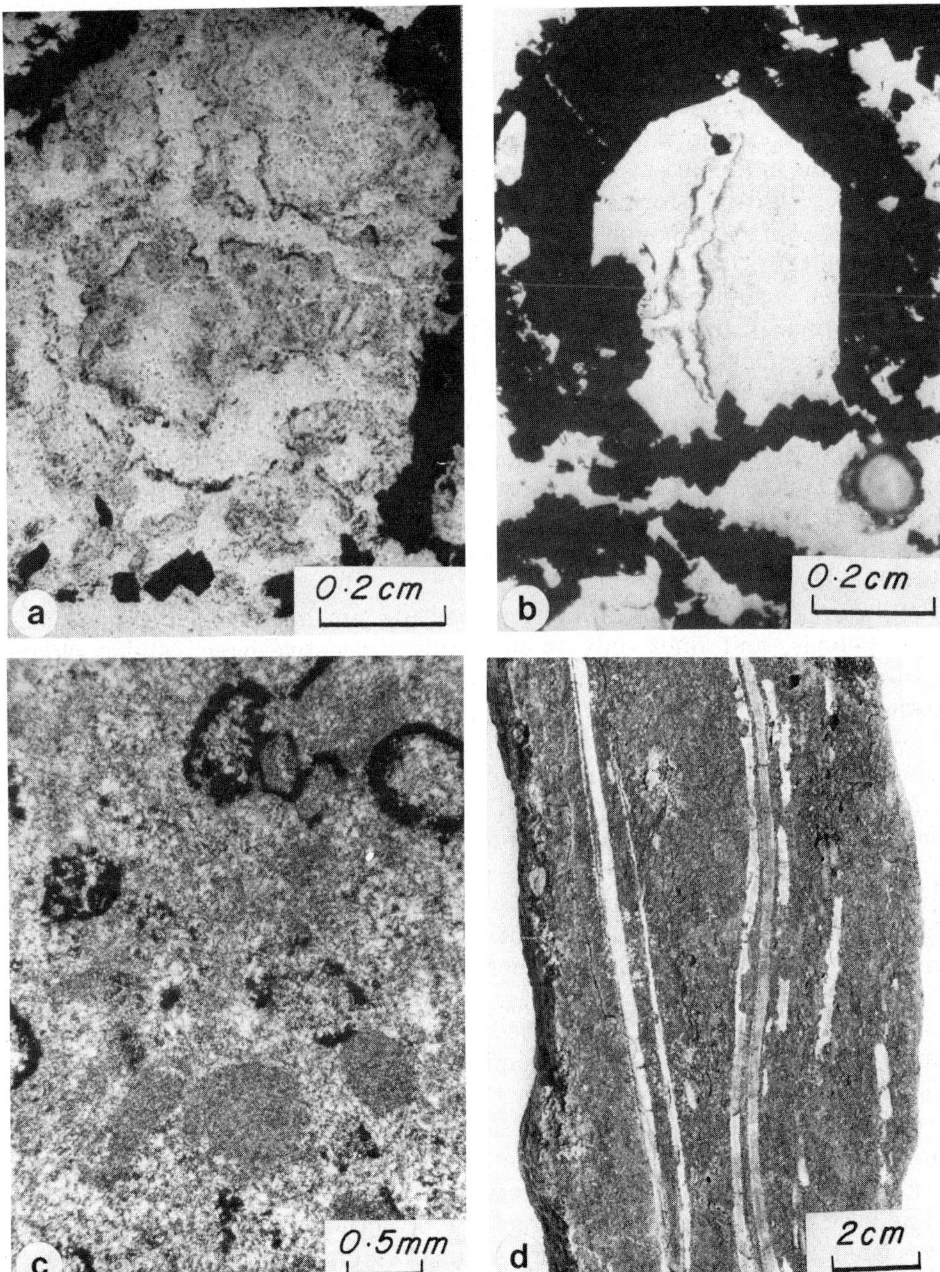

Fig. 7-8. a. Shrinkage (syneresis) cracks filled by clear, non-dusty quartz in rounded chert intraclast. Note euhedral magnetite on peloid rim. b. Textural relic of shrinkage crack in single authigenic quartz crystal within magnetite-rimmed peloid. c. Rounded peloids of chert, and chert rimmed by magnetite, in cherty matrix. Crossed polars. d. Thin chert bands alternating with granular iron-formation containing large, tabular chert intraclasts (from top of the Frere Formation, Frere Range).

Allochems represent the particulate (clastic) fraction of iron-formations, and consist of that chemically derived material formed within the basin of deposition but redeposited elsewhere in the basin.

(1) *Peloids:* rounded to rarely angular chert or ferruginous chert grains up to 2 mm in diameter (Fig. 7-7b, c). They are by far the most abundant allochem in the iron-formation. The peloids are generally sub-spherical in shape (Fig. 7-7b), although in some units they may be irregularly tabular and curved (Fig. 7-7d) as a result of soft-sediment compaction.

In some instances peloids are inhomogeneous, generally comprising iron-rich rims (usually magnetite) and chert cores (Fig. 7-8a—c). Shrinkage or syneresis cracks filled with clear, non-dusty quartz are common in individual peloids (Fig. 7-8a); authigenic quartz crystals up to 0.5 mm in size also rarely contain texturally preserved shrinkage cracks (Fig. 7-8b).

The mineralogy of peloids in unmetamorphosed rocks is largely a simple mixture of chert and hematite, with gradations in proportion ranging from pure chert through jasperoidal cherts in which the peloids consist of hematite dust in chert to iron-rich rocks in which the peloids are entirely hematite (usually platy).

(2) *Intraclasts:* in this instance intraclasts are arbitrarily distinguished from peloids, and refer only to texturally distinctive large, tabular clasts up to 10 cm long and 4 cm thick (Fig. 7-8d). Intraclastic breccias form distinctive units, particularly in the upper part of the Frere Formation in the Frere Range. The breccias are characterised by a great diversity in clast lithologies representing all rock types within the Frere Formation. Predominant types are black chert, red jasperoidal chert and platy hematite. Many of the large intraclasts show internal microbanding and a few are compounds of earlier formed peloids.

(3) *Ooliths* and *pisoliths:* ooliths are particularly common in the ferruginous cherts along the southern margin of the basin. Where best developed they consist of multiple alternations of hematite and chert arranged in concentric laminae about a central nucleus (Fig. 7-9a). Pisoliths (diameter greater than 2 mm) are restricted to a single bed 0.2 to 0.5 m thick near Camel Well. Apart from their size and greater number of concentric laminae, the pisoliths are similar to the ooliths.

(4) *Oncolites* and *stromatolites:* these are restricted to the jasperoidal pisolite bed east of Camel Well. The oncolites form irregular bodies several centimetres across which display very fine concentric laminae (Fig. 7-9b). Some have cores of jasperoidal intraclastic breccia, and others are themselves brecciated. Stromatolitic beds a few centimetres thick are laterally continuous and display fine wavy lamination. Both oncolites and stromatolites are rich in algal microfossils (Walter et al., 1976) (Fig. 7-9c).

(5) *Terrigenous detritus:* detrital tourmaline, muscovite, chlorite and monocrystalline quartz are rarely observed.

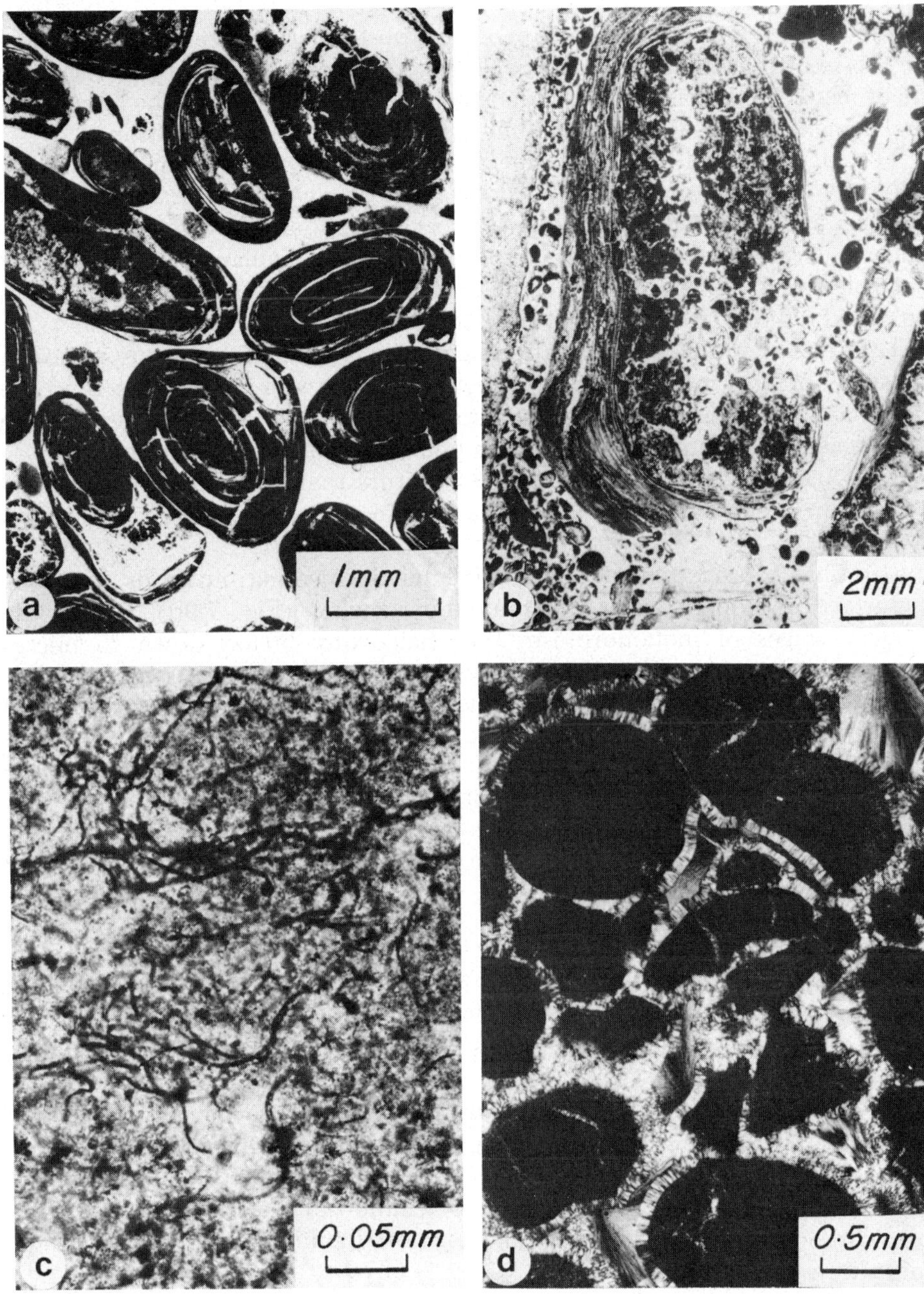

Fig. 7-9. a. Ferruginous chert ooliths, including rare multiple ooliths, in cherty matrix. b. Oncolite with algal layers coating intraclast core. Note large shrinkage crack in core filled with rounded peloids. c. Algal filaments (*Gunflintia minuta*) in ferruginous chert oncolite. d. Two stage chalcedonic cement between rounded ferruginous chert peloids; an inner columnar variety, and a younger coarse fan-aggregate variety. Crossed polars.

Orthochems represent that portion of the iron-formation directly precipitated at the site of formation of the rock. Where possible a distinction has been made between matrix (material deposited at the same time as the allochems) and cement (material precipitated in pore spaces during lithification).

(1) Matrix chert in the unmetamorphosed iron-formation and cherts typically forms a very fine-grained mosaic with an average grain size of between 1 and 5 microns (Fig. 7-8c). Matrix chert is commonly similar in appearance to chert within allochems, and probably had a similar origin as a gel-like precipitate or ooze. Much of the matrix chert in the iron-formation is coloured red due to the presence of hematite dust.

(2) Cement in the iron-formation and chert is generally siliceous, although in some of the more iron-rich varieties hematite or goethite is now the dominant cementing agent. Four principal types of siliceous cement are recognized, and commonly more than one is present in the same rock:

(a) Quartz with columnar (fibrous) texture that forms post-depositional rims on allochems (Fig. 7-9d). Some terminate in poorly developed crystal faces.

(b) Chalcedonic cement as radiating fans projecting outwards from the surface of allochems or columnar quartz rims (Fig. 7-9d). During the early stages of metamorphism the chalcedony breaks down to microcrystalline quartz and the boundaries of the fan become sutured.

(c) Quartz mosaic which represents the final in-filling of the pore space and occurs with both chalcedonic and columnar types. The quartz is usually equant and has an anhedral texture. Some allochems may be preferentially replaced by the same material (Fig. 7-10a, b).

(d) Microcrystalline quartz cement which is identical to matrix chert in texture, and can be distinguished only when it fills desiccation cracks in allochems.

Mineralogy

No diamond drill cores are available from the iron-formations, and all samples are from surface exposures. This limits the amount of mineralogical and chemical work that can be done because the extent of surface weathering and silicification cannot be accurately determined.

Oxide facies assemblages dominate the mineralogy, the most common being hematite-quartz and hematite-magnetite-quartz. Iron-rich carbonates, chamositic chlorite and sulphides locally form a distinctive assemblage in the transition from Frere Formation to Windidda Formation, but elsewhere iron carbonate, silicate and sulphide form only minor components of oxide-facies iron-formation.

Quartz: mostly present as a very fine-grained mosaic in matrix chert or cherty allochems. Grain size is usually less than 0.01 mm but increases with metamorphic grade. Silica cement has several forms, the most common being fibrous to columnar chalcedony.

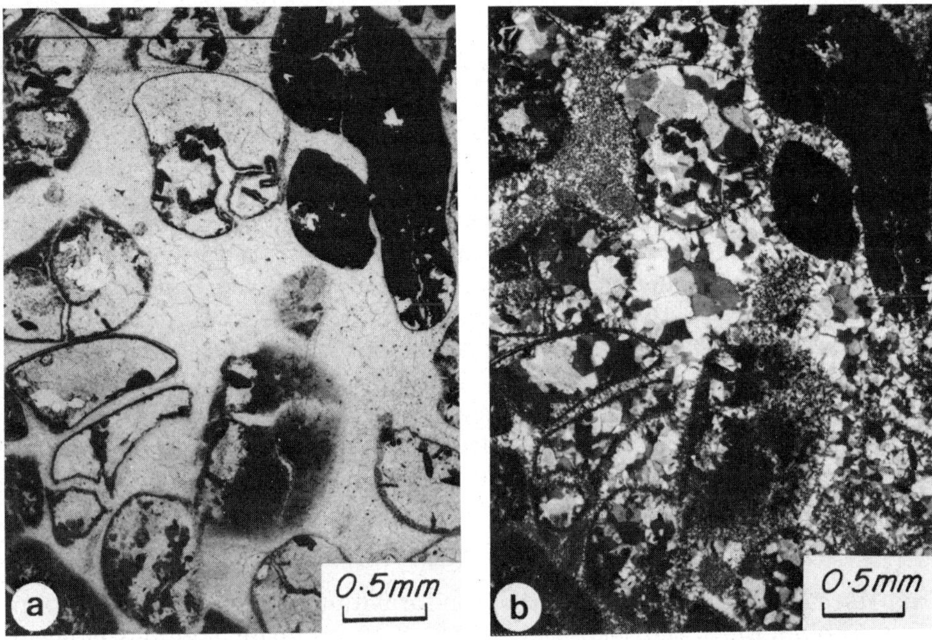

Fig. 7-10. Rounded chert and ferruginous chert peloids with shrinkage cracks. Note irregular presence of coarse-grained quartz in both interstices and allochems (b in crossed polars).

Hematite: in all the jasperoidal chert and iron-formation, hematite occurs as finely disseminated, sub-microscopic dust, giving the characteristic red colour. In peloidal rocks it forms dense aggregates of very small, almost opaque red platelets which are usually less than 30 microns across. These are concentrated within peloids, either filling them or forming irregular patches. In the centres of some peloids the hematite has recrystallised to a coarser type of specular (platy) hematite, which is visible in hand specimens as small metallic grey clusters. Hematite also occurs in the form of martite as a secondary replacement of coarse-grained, commonly octahedral magnetite.

Magnetite: in its unaltered form it seldom occurs in surface exposures of the Frere Formation. It is not present in the unmetamorphosed flat-lying iron formations in the southeast, and elsewhere it is largely replaced by hematite (the former presence of magnetite can be readily established by its ubiquitous octahedral habit). Individual crystals range from about 3 microns to 200 microns, but are mostly in the range 20—50 microns. The magnetite occurs as discrete octahedra or in irregular aggregates (Fig. 7-8b) which form patches within peloids and occasionally transgress boundaries.

Sulphide: traces of pyrite, chalcopyrite and sphalerite occur in some cherts associated with iron-formations, particularly in the transitional zone with the overlying Windidda Formation. Some limestone and dolomite units within the Frere Formation contain cubes up to 10 mm square of pyrite or limonitized pyrite.

Carbonate: iron-rich carbonate minerals are rare, and only at one locality is there a rock which approaches a carbonate-facies iron-formation. This is near Tooloo Bluff (central Kingston Sheet) where a thin bed of ankeritic carbonate occurs in the transition zone between the Frere and Windidda Formations. The carbonate minerals are a granular mixture of yellow-buff ferroan dolomite and brown ankerite. Varying proportions of brown ankerite define a crude 20—30 mm bedding, in a 0.5 m thick unit.

Rhombs of iron (?) carbonate occur rarely in GIF and more commonly in laminated and shaly iron-formations, mainly in the form of isolated crystals up to 100 microns across. Nowhere do they occur in more than accessory amounts.

Iron silicates: extremely rare in the iron-formations of the Frere Formation. Near Tooloo Bluff the transitional ankeritic carbonate contains elongate clots of a pale green chlorite which has properties consistent with its being thuringite. Elsewhere, phyllosilicate minerals occur only as very rare accessories, usually as small flakes only a few microns long. Most are colourless and too fine-grained for positive identification. Probable stilpnomelane has been identified at one locality. Greenalite has not been identified in any samples from the Frere Formation, although in some ferruginous chert the iron-oxide peloids have a dark-green tinge which may be finely disseminated, sub-microscopic greenalite. Small rosettes of possible minnesotaite occur in the oncolitic iron-formation. Nontronite has also been observed at one locality.

Structure

The Nabberu Supergroup and its Archaean basement were deformed into the Stanley Fold Belt during the late Early Proterozoic.The boundary between the fold belt and the essentially undeformed Kingston Platform (Fig. 7-2) is taken for convenience at the first appearance of slaty cleavage in pelitic rocks which broadly corresponds to the appearance of lower greenschist-facies assemblages.

The degree of deformation in the fold belt increases northwards and north-westwards. Folds are commonly asymmetric and overturned southwards, while cleavage and thrust faults are north dipping (see Hall and Goode, 1978, fig. 4). Both indicate that deformation was directed from north to south. Towards the west and north basement rocks are increasingly involved in the deformation. In the northwest large-scale fold interference patterns are dominant, and intrusive granites are present in the basement.

One of the most noticeable features of the Stanley Fold Belt is the abrupt change in trend in the western Earaheedy Sub-basin. This change in trend occurs immediately to the west of the aeromagnetic extension of a large north-northwest trending fault which displaces the Frere Formation in the eastern Frere Range (Fig. 7-3). In the south of the Earaheedy Sub-basin the Yelma Formation appears to thicken immediately to the west of this fault, suggesting that it was active during sedimentation. Another major fault with a similar trend lies about 30 km to the west, and is taken as the eastern boundary of the Glengarry Sub-basin. This fault clearly displaces the Frere Formation, but its effect on the thickness and facies within the Yelma Formation is unknown.

These faults are the extensions of major structures (e.g. Celia, Keith—Kilkenny Lineamants) in the Archaean of the Yilgarn Block. It is possible that they were also intermittently active over a considerable time span in the Proterozoic, controlling deposition within the Nabberu Basin and later influencing the deformation pattern in the zone between the Earaheedy and Glengarry Sub-basins.

Metamorphism

Regional changes in metamorphic grade accompany the variation in structural deformation of the Nabberu Supergroup and its Archaean basement (Hall and Goode, 1978, fig. 5). Mineral assemblages indicate a general increase in grade from essentially unmetamorphosed sediments on the Kingston Platform in the southeast to greenschist facies in the western part of the Nabberu Basin. Amphibolite and granulite facies assemblages in the western part of the Stanley Fold Belt are largely or totally within basement rocks, and although broadly consistent with a regional gradient associated with a major Early Proterozoic metamorphic episode (Hall and Goode, 1978; Muhling, 1980), they may have been inherited from an earlier Archaean event (Williams et al., 1978).

Six metamorphic zones, based mainly on the distribution of co-existing iron oxides, have been established in the oxide facies iron-formation in the Nabberu Basin (Fig. 7-11). The highest-grade zone is only found in Archaean iron-formations (Goode and Hall, in prep.) This zonation and the corresponding changes in quartz grain size reveal a subtle metamorphic variation in the low-grade and "unmetamorphosed" areas which is not otherwise apparent.

Quartz grain size

In the greenschist facies and higher-grade areas quartz grain sizes are generally uniform and consistent within any one specimen, although restraining and recrystallization associated with younger deformations or local structural features can give unusual variations in grain size. In the essentially unmetamorphosed areas grain sizes of quartz in the chert and iron-formations are much more irregular. Relatively coarse-grained quartz is developed in texturally or compositionally controlled areas in the granular iron-formations, and is assumed to correspond to similar features of diagenetic origin observed in carbonate rocks (c.f. Dimroth and Chauvel, 1973). However, very fine-grained areas of quartz are usually preserved in these rocks, and are interpreted as representing primary grain sizes as they form a continuous, gradational size range with the more metamorphosed cherts.

Quartz grain size consistently and regularly increases to the northwest across the Stanley Fold Belt, and is obviously related to metamorphic facies as delineated by other rock types. In the southeastern areas of the Nabberu Basin, furthest from the fold belt, quartz grain sizes are always less than 10 μm and generally less than 5 μm. Magnetite and platy hematite grain sizes vary in sympathy with quartz, but are less regular.

The increases in quartz grain size are considered to reflect the response of primary colloidal or early crystallized silica to increasing metamorphic grade (particularly due to temperature) following nucleation. Similar results in grain size variations have been reported from metamorphosed iron-

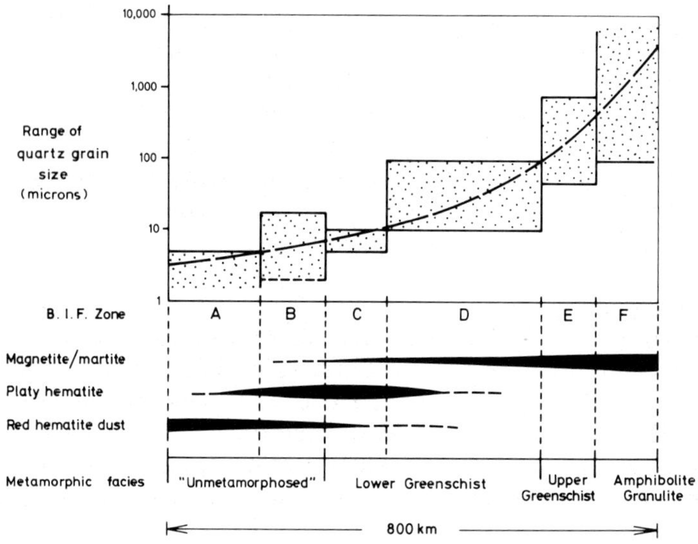

Fig. 7-11. Generalized variation in iron-oxide mineralogy and quartz grain size with metamorphic grade across the Stanley Fold Belt. The amphibolite-granulite assemblages are within probable Archaean iron-formations and may represent an older metamorphic event.

formations in Liberia (White, 1973), and from the Marquette Range in the Animikie Basin (James, 1955).

Iron-oxide assemblages

In the southeastern part of the Nabberu Basin very fine-grained, red hematite "dust" is the dominant to sole iron oxide present in the iron-formations. Towards the fold belt this dust is progressively replaced by fine-grained, grey "crystalline" hematite with a platy morphology. This platy hematite becomes coarser grained at higher grades.

Only rare and minor amounts of magnetite are present in the Kingston Platform, a feature supported by the regional aeromagnetic surveys which show a marked southeastwards weakening in the amplitude of the anomalies associated with the iron-formation. No magnetite has been found in the extreme southeast of the basin, illustrating that at extremely low metamorphic grades the iron-formations may be non-magnetic.

At progressively higher metamorphic grades, in the greenschist facies and above, red hematite dust and platy hematite are progressively replaced by magnetite. At upper greenschist facies and higher, magnetite or its oxidized equivalent martite is the only iron oxide observed.

Silicate mineralogy

Chamosite and possible minnesotaite have been recorded within the "unmetamorphosed", ultra low-grade zone, while metamorphic chlorite and biotite occur rarely in the upper greenschist facies iron-formations. Clinopyroxene, orthopyroxene and tremolite have been observed in Archaean iron-formations from the highest-grade areas. The pyroxenes contain exsolution lamellae in some instances, possibly similar to those in pyroxenes in high-grade iron-formations in the Lake Superior District, U.S.A. (Simmons et al., 1974; Bonnichsen, 1975).

Age

Sediments of the Nabberu Basin unconformably overlie Archaean rocks of the northern Yilgarn Craton. About 70 km south-southeast of Wiluna, ages of between 2634 ± 17 m.y. to 2481 ± 18 m.y. have been recorded from Archaean rocks by Roddick et al. (1976), while a further 70 km to the south, ages between 2718 ± 50 and 2474 ± 14 m.y. have been recorded by Cooper et al. (1978). Stuckless et al. (1981) record a Pb/Pb age of 2370 m.y. for alkali granite 130 km southeast of Wiluna. The youngest of these provides a probable maximum age for the inception of the Nabberu Basin.

To the north and east the Nabberu Basin is overlain by sediments of the Bangemall Basin in which deposition commenced about 1100 m.y. ago (Gee et al., 1976). Dolerite sills and plugs associated with the Bangemall

Group intrude Nabberu sediments in the east of the Earaheedy Sub-basin, and have been dated at 1050 m.y. by Preiss et al. (1975).

Samples of glauconite from basal Nabberu sediments in the extreme southwest of the basin give minimum K/Ar ages of around 1700 m.y., and minimum Rb/Sr ages of between 1590 and 1710 m.y. (Preiss et al., 1975). A comparable K/Ar age of 1685 m.y. was recorded by Horwitz (1975b) from glauconite at the base of the Wandiwarra Formation. Coarse-grained galena from a thin dolomite (upper Yelma Formation) has recently provided a Pb/Pb age of approximately 1700 m.y. Bunting et al. (1980) report a Rb/Sr age of about 1630 m.y. for an intrusive quartz syenite into the Teague Ring Structure. Intrusive granites in the Gascoyne Province in the north-western Stanley Fold Belt have been dated variably between 1700 and 1550 m.y. (Compston and Arriens, 1968; Williams et al., 1978).

DEPOSITIONAL ENVIRONMENT

Sedimentation in the Earaheedy Sub-basin began during a regional marine transgression over Archaean basement rocks from the north and west. The presence of glauconite in the basal clastics, the upward lithological variation in the Tooloo Sub-group from sand to silt to chemical sediments, and the occurrence of probable deeper water iron-formations at the top of the Frere Formation, indicate progressively deeper water conditions associated with trangression. The land surface on to which the Early Proterozoic sea transgressed was probably a peneplain (Allchurch and Bunting, 1976; Hall and Goode, 1978; Gee, 1979b), as indicated by the essentially flat nature of the unconformity surface combined with its largely linear outcrop trace over some 300 km, evidence of deep weathering in the source areas and the lack of thick, coarse detrital units (Hall and Goode, 1978).

The initial transgression deposited a thin cover of reworked strandline sand that now forms basal Yelma Formation, and blankets much of the sub-basin floor. As the transgression continued a mud-floored, shelf sea developed with local deposition of cryptalgal and stromatolitic carbonates, possibly in a partly evaporitic (littoral, sabkha?) environment.

The transgressive conditions continued during the deposition of the Frere Formation when chemical precipitates and fine-grained clastics were laid down on an extensive stable shelf (Fig. 7-12). The Frere Formation was deposited in a shallow sub-aqueous environment, probably marine, as indicated by ripple marks, scour structures and lenticular bedding. Clastic material is either absent (in the cherts and iron formations) or fine grained, indicating either remoteness from the source, a very low-lying, mature hinterland, a barred environment or arid conditions in the source areas. The cherts and iron-formations form a background of silica and iron precipitation punctuated by periods of fine-grained sediment influx.

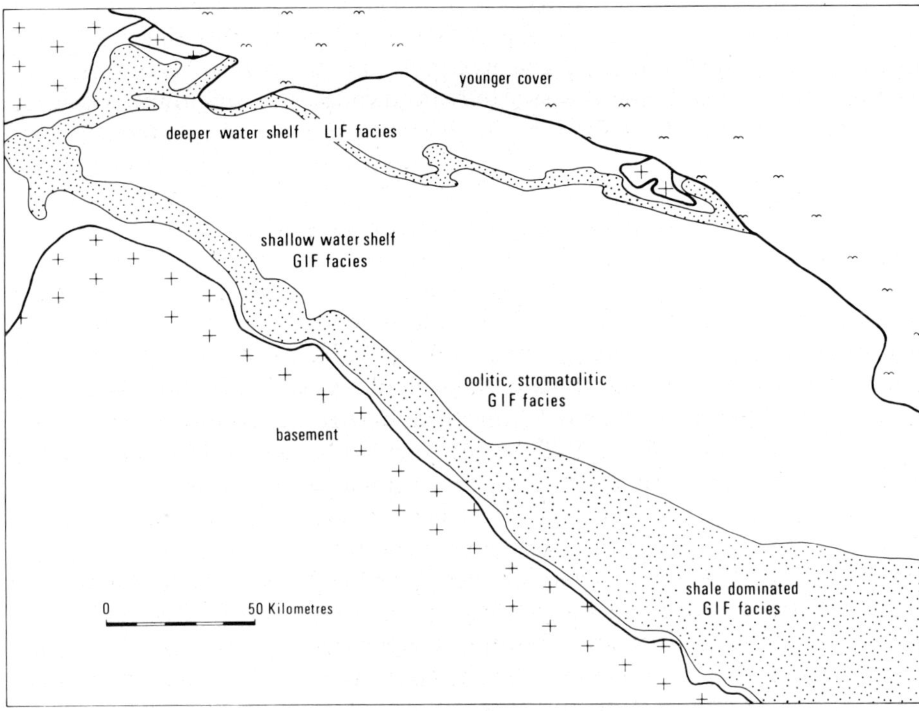

Fig. 7-12. Generalized distribution of iron-formation sedimentary facies within the Frere Formation, Earaheedy Sub-basin.

Regional lithological variation is a direct response to variation in the local environment of deposition. A higher clastic component (shale, siltstone, and fine-grained sandstone) in the southeast of the Earaheedy Sub-basin suggests proximity to the source area of the sediment. This association with a sediment-generating shore line is consistent with sedimentary relationships elsewhere in both the Tooloo and Miningarra Sub-groups.

The confined occurrence of laminated iron-formation to the western and northern parts of the sub-basin is another indication of the shallowing of the sub-basin to the southeast. Laminated iron-formation in the northern part of the sub-basin is also concentrated near the top of the formation, suggesting that water depth was increasing towards the end of iron-formation deposition. In these areas water depths were probably greater than 200 m (i.e. below wave base).

Facies changes within the iron-formation almost certainly reflect changes in water depth. While the laminated iron-formations formed below wave base, the granular iron-formations formed in shallower water on an extensive shelf, where wave action was constantly breaking up recently deposited iron-formation and chert. Brief periods of quiescence allowed thin, continuous chert bands to form (e.g. Fig. 7-7d), and it is notable that these lack iron, suggesting that much of the iron-rich allochems may have formed in deeper water before being transported shorewards into shallower water. As well as the southeast to northwest increase in the ratio of iron-formation to shale, there is an increase in the iron content of the iron-formation which further suggests that the source of the iron was to the northwest.

There is general agreement that granular iron-formations formed in shallow water, where storm action could disrupt the layers of silica and iron "hydroxide" gel precipitated during quieter periods. Storm activity rather than strong current activity (such as tidal currents) as the mechanism of disruption is based on the general lack of consistent current structures within the iron-formation. Small scale ripple cross-lamination is present in the interbedded shales, but the iron-formation contains only a few scours.

Storm waves also disrupted earlier formed shale and iron-formation and mixed these together to form the intraclastic breccias. Some of the finer intraclastic material was transported into shallower water and winnowed by normal wave activity, thus producing the better sorted peloidal iron-formation and chert. The peloids were deposited in low, shoal-like deposits which resulted in the lensoid and wavy bedding characteristic of these rocks. The finely laminated iron-formation requires deeper water (> 200 m, below wave base) in order to preserve the delicate, continuous microbanding from the effects of storms and bottom currents.

Oolitic and pisolitic iron-formations are concentrated in a specific zone (Fig. 7-12) and represent oolite shoals and banks with probable slight relief above the surrounding muds of the shelf. Biological activity may have helped to build up these shoals, particularly as coarser pisoliths are associated with oncolites, stromatolites and bacterial microfossils.

In the Nabberu Basin, at the present state of knowledge, there is little evidence as to the nature of the original precipitate apart from the fact that it was colloidal. Early diagenesis resulted in dewatering and lithification of the original precipitate by crystallization of quartz (as cryptocrystalline chert and interstitial cement) and iron oxides. That gel dewatering began early is indicated by the occurrence of small rounded peloids *within* larger syneresis cracks in oncolite cores (Fig. 7-9b). There is no evidence in the Frere Formation that the earliest formed iron oxide was other than hematite. It appears that iron silicates and carbonates were of only minor and very localized importance.

The regressive phase following the Frere Formation mainly affected the south of the sub-basin, where lagoonal or marsh conditions developed behind extensive carbonate banks. The lagoonal-algal flat facies deposited a thick sequence of muds and thin carbonates in quiet, shallow water subject to periodic storm activity. Seaward of the lagoonal environment, and off shore from the carbonate bank, the basin received a small but constant supply of fine-grained detritus. The culmination of the regressive phase of the Windidda Formation resulted in partial emergence of the carbonate platform accompanied by sub-aerial brecciation and vadose encrustation (Hall and Goode, 1978). The emergence seems to have occurred only in the southeastern part of the Earaheedy Sub-basin.

The Miningarra Sub-group was deposited during a second transgressive—regressive cycle in which shallow-marine conditions prevailed (see Hall and Goode, 1978).

DEPOSITIONAL FACIES MODEL

The transition from Ca-Mg carbonates to oxide facies iron-formations via a complex, usually thin, iron carbonate, silicate and rarely sulphide assemblage is becoming increasingly well known in classic Early Proterozoic sequences, e.g., Nabberu, Hamersley, Transvaal (see discussion by Goode, 1981). In some regions the iron carbonate—silicate zone is much thicker, e.g., Western Animikie Basin; Middleback Range, South Australia (Parker and Lemon, 1982). In the Animikie Basin and Labrador Trough the Ca-Mg carbonate zone is missing. The essential point of these observations is that iron-formations occupy a specific niche in the stratigraphic record, and are gradational with other sediments in the basin. The origin of iron-formations therefore does not appear to have a predominant extra-basinal influence.

An idealized sedimentary facies model based on the observed lateral and vertical facies relationships in the Tooloo sedimentary cycle is shown in Fig. 7-13. It is believed that this model has general application to other iron-formation basins. In reality, of course, deposition is much more complicated than this, being especially dependent on local environmental/palaeogeographic factors and on the interplay between clastic supply, volcanism and chemical precipitation.

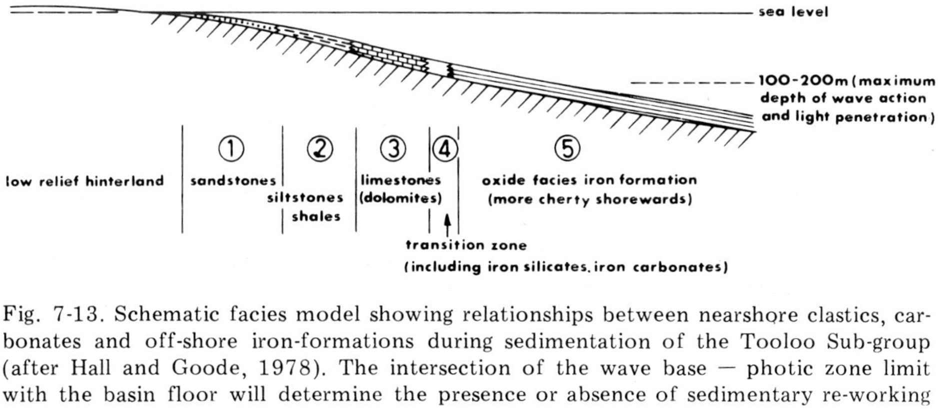

Fig. 7-13. Schematic facies model showing relationships between nearshore clastics, carbonates and off-shore iron-formations during sedimentation of the Tooloo Sub-group (after Hall and Goode, 1978). The intersection of the wave base — photic zone limit with the basin floor will determine the presence or absence of sedimentary re-working and photosynthetic benthonic organisms in the sediments.

This model proposes that the deposition of BIF occurs offshore from areas of carbonate deposition, and therefore BIF appears in any given stratigraphic column at a time of maximum transgression (note that in most Early Proterozoic sequences, BIF occurs near the top of any transgressive sequence prior to a subsequent erosional or regressive event). This relationship of course is difficult to reconcile with evaporitic theories of origin for BIF.

The close and transitional association with Ca-Mg carbonates suggests a common or related origin for these chemical sediments. Since it is generally agreed that Phanerozoic, and presumably Proterozoic, carbonates were precipitated as a result of organic activity, it would therefore seem reasonable to propose that iron-formations were also precipitated as a result of organic activity. Microfossils were certainly prolific at certain times in the Early Proterozoic (note Hamersley black shales with up to 17% free carbon). Dominantly benthonic? forms are known in several shallow-water iron-formations (Walter et al., 1976), but planktonic forms although suspected have yet to be confirmed.

The source of the iron is not definitely known. However, the increase in iron content towards deeper water in the Nabberu iron-formations suggests the source may have been the offshore ocean basins. If the early Archaean atmosphere had been reducing as has been suggested by many writers, over the span of Archaean time there could have been a huge build-up of soluble ferrous iron in the deep-ocean reservoirs from onshore weathering processes. The unique thickness and extent of broadly synchronous Early Proterozoic iron-formations may be due to the evolutionary advent of large quantities of algae and bacteria suitable for precipitating the iron-formations and carbonates, coupled with the development of large stable marine shelf environments adjacent to the deep-ocean (permanent?) reservoirs.

ACKNOWLEDGEMENTS

The authors are grateful to the staff of The Broken Hill Proprietary Co. Ltd. and the Geological Survey of Western Australia for their assistance and encouragement.

This paper is published with the permission of the General Superintendent Exploration of The Broken Hill Proprietary Co. Ltd. Previously unpublished Geological Survey data is published with the permission of the Director, Geological Survey of Western Australia.

REFERENCES

Allchurch, P.D. and Bunting, J.A., 1976. The Kaluweerie Conglomerate: a Proterozoic fluviatile sediment from the Yilgarn Block. West. Aust., Geol. Surv., Annu. Rep., 1975: 83—87.

Barnett, J.C., 1975. Some probable Lower Proterozoic sediments in the Mount Padbury area. West. Aust., Geol. Surv., Annu. Rep., 1974: 52—54.

Bonnichsen, B., 1975. Geology of the Biwabik Iron Formation, Dunka River area, Minnesota. Econ. Geol., 70: 319—340.

Bunting, J.A., 1977. Explanatory notes on the Kingston 1 : 250,000 Geological Sheet, Western Australia: West. Aust., Geol. Surv., Rec., 1977/5.

Bunting, J.A. and Chin, R.J., 1975. Explanatory notes on the Duketon 1 : 250,000 Geological Sheet. West. Aust., Geol. Surv., Rec., 1975/7.

Bunting, J.A., Commander, D.P. and Gee, R.D., 1977. Preliminary synthesis of Lower Proterozoic stratigraphy and structure adjacent to the northern margin of the Yilgarn Block. West. Aust., Geol. Surv., Annu. Rep., 1976: 43—48.

Bunting, J.A., Jackson, M.J. and Chin, R.J., 1978. Explanatory notes on the Throssell 1 : 250,000 Geological Sheet. Aust. Bur. Miner. Resour., Canberra.

Bunting, J.A., Brakel, A.T. and Commander, D.P., 1979. Explanatory notes on the Nabberu 1 : 250,000 Geological Sheet, Western Australia. West. Aust., Geol. Surv., Rec., 1978/12.

Bunting, J.A., de Laeter, J.R. and Libby, W.G., 1980. Evidence for the age and cryptoexplosive origin of the Teague Ring Structure, Western Australia. West. Aust., Geol. Surv., Annu. Rep., 1979: 81—85.

Commander, D.P., Muhling, P.C. and Bunting, J.A., 1979. Explanatory notes on the Stanley 1:250,000 Geological Sheet, Western Australia. West. Aust., Geol. Surv., Rec., 1978/15.

Compston, W. and Arriens, P.A., 1968. The Precambrian geochronology of Australia. Can. J. Earth Sci., 5: 561—583.

Cooper, J.A., Nesbitt, R.W., Platt, J.P. and Mortimer, G.E., 1978. Crustal development in the Agnew region, Western Australia, as shown by Rb-Sr isotopic and geochemical studies. Precambrian Res., 7: 31—59.

Daniels, J.L. and Horwitz, R.C., 1969. Precambrian tectonic units of Western Australia. West. Aust., Geol. Surv., Annu. Rep., 1968: 37—38.

Dimroth, E. and Chauvel, J.J., 1973. Petrography of the Sokoman Iron Formation in part of the central Labrador Trough, Quebec, Canada. Geol. Soc. Am. Bull., 84: 111—134.

Elias, M. and Bunting, J.A., 1979. Explanatory notes on the Wiluna 1 : 250,000 Geological Sheet, Western Australia. West. Aust., Geol. Surv., Rec., 1978/10.

Elias, M., Bunting, J.A. and Wharton, P.H., 1979. Explanatory notes on the Glengarry 1 : 250,000 Geological Sheet, Western Australia. West. Aust., Geol. Surv., Rec., 1979/3.

Folk, R.L., 1959. Practical petrographic classification of limestones. Bull. Am. Assoc. Pet. Geol., 43: 1—38.

Gee, R.D., 1979a. The geology of the Peak Hill area. West. Aust., Geol. Surv., Annu. Rep., 1978: 55—62.

Gee, R.D., 1979b. Structure and tectonic style of the Western Australian Shield. Tectonophysics, 58: 327—369.

Gee, R.D., de Laeter, J.R. and Drake, J.R., 1976. Geology and geochronology of altered rhyolite from the lower part of the Bangemall Group near Tangadee, Western Australia. West. Aust., Geol. Surv., Annu. Rep., 1975: 112—117.

Goode, A.D.T., 1981. Proterozoic geology of Western Australia. In: D.R. Hunter (Editor), Precambrian of the Southern Hemisphere. Elsevier, Amsterdam, pp. 105—203.

Goode, A.D.T., and Hall, W.D.M., 1976. Shallow water banded iron-formations from the Nabberu Basin, Western Australia. 25th Int. Geol. Congr., Sydney, (Abstr.), p. 162.

Goode, A.D.T., and Hall, W.D.M., in prep. Metamorphism of banded iron-formations from the Nabberu Basin, Western Australia.

Grey, K., 1979. Preliminary result of biostratigraphic studies of Proterozoic stromatolites in Western Australia. West. Aust., Geol. Surv., Rec., 1979/22.

Hall, W.D.M. and Goode, A.D.T., 1975. The Nabberu Basin: a newly discovered Lower Proterozoic basin in Western Australia. Geol. Soc. Aust., 1st Aust. Geol. Convention, "Proterozoic Geology" (Abstr.), pp. 88—89.

Hall, W.D.M. and Goode, A.D.T., 1978. The Early Proterozoic Nabberu Basin and associated iron-formations of Western Australia. Precambrian Res., 7: 129—184.

Hall, W.D.M., Goode, A.D.T., Bunting, J.A. and Commander, D.P., 1977. Stratigraphic terminology of the Earaheedy Group, Nabberu Basin. West. Aust. Geol. Surv. Annu. Rept., 1976: 40—43.

Horwitz, R.C., 1975a. The southern boundaries of the Hamersley and Bangemall Basins of sedimentation. Geol. Soc. Aust., 1st Aust. Geol. Convention, "Proterozoic Geology" (Abstr.), p. 91.

Horwitz, R.C., 1975b. Provisional geologic map at 1 : 2,500,000 of the northeast margin of the Yilgarn Block, Western Australia. CSIRO Min. Res. Lab. Rep., FP-10.

Jackson, M.J., 1978. Explanatory notes on the Robert 1 : 250,000 Geological Sheet. Aust. Bur. Miner. Resour.

James, H.L., 1955. Zones of regional metamorphism in the Precambrian of Northern Michigan. Bull. Geol. Soc. Am., 66: 1455—1488.

Macleod, W.N., 1970. Explanatory notes on the Peak Hill 1 : 250,000 Geological Sheet. West. Aust., Geol. Surv.

Muhling, J.R., 1980. Evolution of the high-grade gneiss complex at Errabiddy, North West Yilgarn Block, Western Australia. 2nd Int. Archaean Symposium, Perth, (Abstr.), pp. 34—35.

Parker, A.J. and Lemon, N.M., 1982. Reconstruction of the Early Proterozoic stratigraphy of the Gawler Craton, South Australia. J. Geol. Soc. Aust., 29: 221—238.

Preiss, W.V., 1976. Proterozoic stromatolites from the Nabberu and Officer Basins, Western Australia, and their biostratigraphic significance. South Aust., Geol. Surv., Rep. Invest. 47.

Preiss, M.V., Jackson, M.K., Page, R.W. and Compston, W., 1975. Regional geology, stromatolite biostratigraphy and isotopic data bearing on the age of a Precambrian sequence near Lake Carnegie, Western Australia. Geol. Soc. Aust., 1st Aust. Geol. Convention, "Proterozoic Geology", (Abstr.), pp. 92—93.

Roddick, J.C., Compston, W. and Durney, D.W., 1976. The radiometric age of the Mount Keith Granodiorite, a maximum age estimate for an Archaean greenstone sequence in the Yilgarn Block, Western Australia. Precambrian. Res., 3: 55—78.

Sanders, C.C. and Harley, A.S., 1971. Hydrogeological reconnaissance of part of the Nabberu and East Murchison Mining Areas. West. Aust., Geol. Surv., Annu. Rep., 1970: 23—27.

Simmons, E.C., Lindsley, D.H. and Papike, J.J., 1974. Phase relations and crystallization sequence in a contact-metamorphosed rock from the Gunflint Iron Formation, Minnesota. J. Petrol., 15: 539—565.

Sofoulis, J., and Mabbutt, J.A., 1963. Geology of the Wiluna-Meekatharra area. In: Lands of the Wiluna-Meekatharra Area, Western Australia. CSIRO Land Res. Series, 7(4): 93—106.

Stuckless, J.S., Bunting, J.A. and Nkomo, I.T., 1981. U-Th-Pb systematics of some granitoids from the northeastern Yilgarn Block, Western Australia and implications for uranium source potential. J. Geol. Soc. Aust., 28: 365—375.

Talbot, H.W.B., 1910. Geological observations in the country between Wiluna, Halls Creek and Tanami. West. Aust., Geol. Surv., Bull., 39.

Talbot, H.W.B., 1920. The geology and mineral resources of the Northwest, Central and Eastern Divisions, between longitude 119° and 122°E and latitude 22° and 28°S: West. Aust., Geol. Surv., Bull., 83.

Walter, M.R., Goode, A.D.T. and Hall, W.D.M., 1976. Microfossils from a newly discovered Precambrian stromatolitic iron formation in Western Australia. Nature, 261: 221—223.

White, R.W., 1973. Progressive metamorphism of iron formation and associated rocks in the Wologizi Range, Liberia. U.S. Geol. Surv. Bull., 1302.

Williams, S.J., Elias, M., and de Laeter, J.R., 1978. Geochronology and evolution of the eastern Gascoyne Province and adjacent Yilgarn Block. West. Aust., Geol. Surv., Annu. Rep., 1977: 50—56.

Chapter 8

PART A. A CONTRIBUTION ON THE CHEMICAL COMPOSITION OF PRECAMBRIAN IRON-FORMATIONS

R. DAVY

INTRODUCTION

In 1954 James published his classic paper on the facies of iron-formations, and followed this, in 1966, with his major compilation "The chemistry of iron-rich rocks". In this later paper he discussed the iron minerals present in iron-rich sedimentary rocks, the composition of the different facies of iron-formation, the distribution of deposits in time and the chemistry of iron in natural waters. He included data on the composition of Precambrian iron-formations.

Since 1966 there have been extensive investigations of iron-formations, their distribution, stratigraphy and origin. A great number of papers contain geochemical analyses, but many factors detract from the value of these analyses as contributions to a comprehensive understanding of the composition of these rocks, and so the chemical composition of iron-formations remains inadequately known.

Many authors have drawn attention to compositional similarities between iron-formations from different regions, but have not sufficiently appreciated the difficulties involved in sampling these extremely variable rocks. The dramatic changes in composition that characterize all scales of banding in iron-formations vitiate attempts to assess the overall composition of whole formations from analyses of hand specimens of short lengths of core. Thin, individual bands may conform to "end-member" compositions of James' facies whereas adjacent bands conform to other "end-member" compositions or to mixed facies. In addition, many reported analyses have been of rocks that have been altered by oxidation, leaching or metamorphism, or by a combination of these processes. Apparent "facies" changes can result from such alteration.

The main objective in attempting to obtain a comprehensive picture of the composition of iron-formations and of the range of variation in composition of their component parts is to be able to use this information in understanding their genesis and how they have been altered by subsequent processes, including such economically important processes as the formation of iron ores. It is necessary, therefore, to recognize what is original, to concentrate on rocks that are fresh, and to exclude rocks that contain such evidence of oxidation as martite and goethtite. No matter how fresh an iron-formation may appear to be, it will always have been subjected to diagenetic changes during consolidation, and it may be difficult to distinguish these changes from the subsequent effects of low-grade metamorphism. More extreme metamorphism may lead to chemical redistributions that progressively obscure the nature of the original rock. A quantitative assessment of these changes can only be made by comparing analyses of rocks from identical stratigraphic horizons.

This paper is an attempt to bring together data that are acceptable as contributions to the comprehensive knowledge of the composition of Precambrian iron-formations. Much more information is needed, particularly analyses of adequate samples of whole formations from major deposits around the world, and also detailed analyses of the composition of individual bands at all scales of banding.

Isotope and rare-earth-element data are discussed elsewhere in this volume. General studies of the composition of minerals within iron-formations are excluded.

SYSTEMATIC STUDIES

Analyses of thin, single bands or layers

There are few analyses of very thin (submillimetric) layers in iron-formations.

Limited numbers of analyses are available of mesobands from the Puolanka area of Finland (Laajoki and Saikkonen, 1977) and from the Dales Gorge Member of the Brockman Iron Formation in the Hamersley Group of Western Australia (Trendall and Blockley, 1970). Representative analyses are given in Table 8A-I.

Trendall and Blockley illustrated variations in mesoband composition of essentially unmetamorphosed iron-formation in three ways:

(a) Variations within mesobands of one facies

The data of Table 8A-I columns 1 to 4 are all of oxide-facies mesobands. However, the composition of each of these facies subtypes is very different and reflects the differing proportions of individual minerals. Chemically there are major differences in total Fe, SiO_2, CaO, CO_2, P_2O_5, and Fe^{III}/Fe^{II}. The content of SiO_2 is generally antipathetic to that of CO_2, but the relationship is by no means consistent. Although no analyses are given, Trendall and Blockley mention mesobands which contain much more iron than, for example, the chert matrix sub-type of column 4. These authors refer to magnetite mesobands and indicate a continuous gradation between these and "chert matrix".

TABLE 8A-I

Selected mesoband analyses. Data for columns 1—10 from Trendall and Blockley, 1970, table 12, pp. 139—142. Data for columns 11—12 from Laajoki and Saikkonen, 1977, table 7, p. 85.

	1	2	3	4	5	6	7	8	9	10	11	12
SiO_2	74.71	48.41	66.08	35.74	30.03	45.26	24.76	32.49	92.98	80.69	91.61	35.14
Al_2O_3	0.19	0.16	0.19	0.81	2.87	1.69	1.88	2.42	0.12	0.08	0.09	0.62
Fe_2O_3	0.46	14.86	1.60	35.43	4.86	8.24	5.69	4.92	0.20	0.63	1.61	24.66
FeO	8.99	19.00	8.68	19.71	30.01	21.60	32.93	29.44	2.12	6.37	3.64	26.94
MgO	2.08	3.44	3.28	2.41	8.79	7.55	8.55	8.70	0.45	0.99	0.45	3.69
CaO	3.39	0.55	6.45	1.31	0.17	0.79	0.04	0.05	1.29	3.17	0.05	3.44
Na_2O	0.03	0.07	0.09	0.19	0.54	0.55	0.40	0.52	0.04	0.04	0.00	0.07
K_2O	0.08	0.17	n.d.	0.12	1.11	1.17	0.86	1.08	0.08	0.18	0.02	0.04
H_2O^+	0.57	0.81	0.21	0.58	3.69	5.52	3.42	3.86	0.23	0.76	0.40	1.20
H_2O^-	n.d.	0.06	0.12	0.12	1.03	0.72	0.81	1.02	0.13	0.18	0.00	0.04
CO_2	9.22	12.57	12.90	3.68	16.34	6.73	19.44	15.35	2.27	5.92	1.59	0.66
TiO_2	0.01	0.06	0.09	0.05	0.35	0.13	0.10	0.09	0.03	0.02	0.00	0.05
P_2O_5	0.01	0.05	0.45	0.09	0.08	0.10	0.28	0.31	0.12	0.46	0.04	2.16
MnO	0.04	0.06	0.04	N.D.	N.D.	0.07	0.15	0.14	0.05	0.16	0.01	0.09

Columns 1 — 4. Different mesoband types from the same facies: 1 — Coarsely microbanded chert; 2 — Finely microbanded chert: 3 — Podded chert; 4 — Chert matrix. (B.I.F. 2. macroband. Dales Gorge Member, Wittenoom from Trendall and Blockley, 1970, table 12, cols. 1—4, p. 139.)

Columns 5—8. Single mesoband analyzed in 4 different holes in Wittenoom area, B.I.F. 2 macroband. Dales Gorge Member mineralogy: stilpnomelane-siderite: 5 — DDH 28 Mesoband "about 10 mm thick"; 6 — DDH 33; 7 — DDH 40; 8 — DDH 46. 6—8 thickness not stated. Trendall and Blockley, 1970, table 12, cols. 8—12, p. 140.

Columns 9—10: Different mesobands of same lithological type. Chert mesoband in chert-siderite. S13 macroband 37 cm apart DDH 51 Wittenoom: 9 — Depth 98.45 m (319 ft 8.5 in); 10 — Depth 98.09 m (318 ft 6 in). Trendall and Blockley, 1970, table 12, cols. 17 and 18, p. 141.

Columns 11—12. Mesoband analyses, Puolanka area, Finland: 11 — Chert mesoband; 12 — Magnetite-amphibole-chert mesoband. Laajoki and Saikkonen, 1977, table 7, cols. 1 and 2, p. 85.

n.d. = not detected.

N.D. = not determined.

(b) Variations within the same mesobands

The composition of any given mesoband may vary from place to place. The data for Table 8A-I columns 5 to 8 reflect these variations in a specific stilpnomelane-siderite mesoband, sampled from 4 drill holes no more than 10 km apart. A feature of these data is the relative consistency of MgO when compared with the widely fluctuating (inverse) proportions of SiO_2 and CO_2.

(c) Variations within mesobands of similar lithological subtypes

Table 8A-I, columns 9 and 10 represent mesobands of the same subtype, some 37 cm apart in the same core and show significant differences in the content of most components.

The amount of variation shown by these analyses of thin bands show how unwise it is to extrapolate from hand specimen analyses to the bulk composition of a significant thickness of iron-formation.

Laajoki and Saikkonen (1977) provide "typical" compositions of a few mesoband types from amphibolite-facies iron-formation from the Puolanka area, Finland (Table 8A-I, columns 11 and 12).

No systematic study of the internal chemical features within individual layers or of lateral and vertical variations between individual bands has yet been undertaken.

Analyses of thick, compound bands or layers (macrobands)

The Dales Gorge Member of the Brockman Iron Formation (Hamersley Group), Western Australia has been divided by Trendall and Blockley (1970) into 17 macrobands of cherty banded iron-formation (BIF) separated by 16 thinner S macrobands. The BIF macrobands are composed of cherty banded iron-formation, with only minor silicates and carbonates: the S macrobands are of phyllosilicate and interbanded carbonate and chert. Each macroband is composed of many individual layers (mesobands and submesobands).

Analyses of macrobands BIF 12-16, S6 (part) and S12-16 in the Wittenoom area have been made by Trendall and Pepper (1977), and BIF 0-16 and S1-16 in the Paraburdoo area have been analysed by Ewers and Morris (1981). Paraburdoo and Wittenoom are approximately 130 km apart. Some comparisons are given in Table 8A-II. Comparisons of analyses of handspecimen-sized samples with complete sections through macrobands (Table 8A-II) illustrate the difficulties of collecting small samples which are representative of the maroband as a whole. However, there are overall similarities between the macrobands at the two localities.

In the Dales Gorge Member successive BIF macrobands have similar compositions (with no wild fluctuations) and there are no clear-cut temporal trends (Ewers and Morris, 1981; Table 8A-II). However, there are slight differences in composition (spatially), with the BIF macrobands at Wittenoom containing slightly higher total Fe, Na_2O and K_2O than those at Paraburdoo.

That there are more compositional differences between successive S macrobands is due partly to a wider variation in the proportion of carbonate minerals. However, the changes are not systematic. Some S macrobands are markedly unusual. For example, at Paraburdoo, macroband S15 has more than twice the CaO content (9.04%) of any other S band, and S9 contains almost

TABLE 8A-II

Comparison of analyses of portions of macrobands with the whole macroband and a comparison of the same macrobands at two localities for the Dales Gorge Member (Values in percentage)

	BIF 12 Macroband			S13 Macroband			
	1	2	3	4	5	6	7
Thickness (m)	0.04	8.73	8.5	0.075	<0.02	2.04	1.38
SiO_2		45.33	43.66	29.60	80.69	43.63	44.67
Al_2O_3		0.12	0.09	0.13	0.08	2.69	1.66
Fe_2O_3	42.42	30.13	44.51(a)	2.71	0.63	4.93	31.39(a)
FeO	22.37	15.10		36.58	6.37	22.15	
MgO	4.50	2.37	3.54	3.51	0.99	4.34	6.59
CaO	2.06	1.56	1.86	0.89	3.17	2.66	1.79
Na_2O		0.03	0.01	0.08	0.04	0.07	0.02
K_2O		0.13	0.01	0.23	0.18	2.01	0.09
H_2O+	0.77	0.69		1.02	0.76	2.90	
H_2O-	0.12	0.07		0.05	0.18	0.68	
CO_2	2.18	4.27	5.11	24.75	5.92	12.48	13.47
TiO_2		0.02	0.00	0.01	0.02	0.12	0.07
P_2O_5	0.86	0.20	0.18	0.36	0.46	0.17	0.19
MnO		0.04	0.04	0.27	0.16	0.26	0.31
C						0.02	
S			0.05				0.15
FeS_2	n.d.	0.06		n.d.		0.93	
Fe as Fe_2O_3	67.03	46.74	44.51	42.95	7.64	29.30	31.29

BIF 12 macroband:

1 — DDH 51 Wittenoom, partial analysis of mixed group of Calamina cyclothem. Trendall and Blockley, 1970, table 11, col. 7, p. 134.

2 — Wittenoom, complete macroband. Trendall and Pepper, 1977, pp. 9—10.

3 — Paraburdoo, complete macroband. Ewers and Morris, 1981, p. 1934.

S 13 macroband:

4 — DDH 51 Wittenoom, chert siderite. Trendall and Blockley, 1970, table 11, col. 8, p. 135.

5 — DDH 51 Wittenoom, chert siderite, single mesoband. Trendall and Blockley, 1970, table 12, col. 18, p. 141.

6 — Wittenoom, complete macroband, Trendall and Pepper, 1977, pp. 9—10.

7 — Paraburdoo, complete macroband, Ewers and Morris, 1981, p. 1934.

n.d. = not detected.

(a) = Fe as Fe_2O_3.

ten times more K_2O (1.76%) than the next highest band (0.21%, S16) (Ewers and Morris, 1981).

Differences between the chemical composition of the Dales Gorge Member at Wittenoom and at Paraburdoo are more pronounced in the S macrobands. Al_2O_3, CaO, Na_2O and K_2O are higher overall at Wittenoom in bands S12—16 than in the comparable bands at Paraburdoo, with K_2O being consistently one order of magnitude higher. Conversely, the S macrobands at Paraburdoo are relatively richer in SiO_2, total Fe and MgO (Ewers and Morris, 1981).

Profiles constructed by Ewers and Morris (1981, fig. 5) show that although the lithological boundaries between the S and BIF macrobands are apparently sharp, the BIF zones immediately above and below the S macrobands show a gradational change in certain components. They suggest that iron and silica precipitation continued during S macroband deposition, but that depositional conditions were changed by addition of pyroclastic material (see also Ewers, this volume). Certainly some S macrobands contain volcanic shards (La Berge, 1966). However, very little other clastic material has been recognized.

At Mount Tom Price, 60 km from Paraburdoo, both BIF and S macrobands are markedly different chemically and mineralogically from equivalent bands at Paraburdoo. At Mount Tom Price both CO_2 and Mn are more abundant in BIF than in S macrobands, the reverse of the situation at Wittenoom and Paraburdoo. Total Fe (as Fe_2O_3) and MgO are higher, and SiO_2 is lower at Mount Tom Price than at Paraburdoo. These differences are ascribed by Ewers and Morris (1981) to metasomatic changes and a higher metamorphic grade in the rocks at Mount Tom Price relative to those at Paraburdoo. These authors consider that the primary deposits, particularly those of the BIF macrobands, were probably chemically similar at all three localities. The metamorphic grade increases from Paraburdoo to Wittenoom to Tom Price, but nowhere have the rocks been metamorphosed beyond the lower greenschist facies.

Lateral variations in iron-formation

Apart from the studies of Gruner (1946, pp. 54—66) in the Biwabik Iron Formation of the Mesabi Range in Minnesota, there has been little systematic consideration of lateral variations in iron-formations. Of the more recent papers, only Ewers and Morris (1981) have been able to compare the composition of stratigraphically equivalent sequences of iron-formation. This work was discussed in the previous section. The significance of that work is limited since comparisons have only been possible between 2 localities for any given band. Thus macrobands S6 (in part), S12—16 and BIF 12—16 have been studied at Wittenoom and Paraburdoo, and macrobands S4 (in part), S5 and BIF 4—5 at Paraburdoo and Mount Tom Price.

Other authors have made even more restricted comparisons: data have been obtained from the same iron-formation but no attempt has been made to obtain stratigraphically equivalent samples. For example, as part of a more general paper, Gole and Klein (1981) include mean values and the ranges of composition for each major component of samples from iron-formations of the Labrador Trough. These samples were taken from both the unmetamorphosed iron-formations north of the Grenville Front and the metamorphosed (amphibolite-grade) iron-formations south of the Grenville Front. Assuming that the metamorphic processes were essentially isochemical except for loss of H_2O and CO_2, they recalculated analyses on a volatile-free basis in an attempt to take into account the effects of metamorphism. Their data were gained by averaging results from small samples and the comparison merely emphasizes the wide range in composition of the analysed samples.

Separation of chemical and clastic source material

One vital aspect of the study of iron-formations is the identification of components which are immobile or which maintain constant proportions during all post-depositional processes. Such components may reflect "primary" source material.

Ewers and Morris (1981) identify 2 groups of components in the S macrobands of the Dales Gorge Member of the Brockman Iron Formation: those characteristic of chemical sediments (CaO, MgO, MnO and CO_2) and those characteristic of "shales" (Al_2O_3, TiO_2, K_2O). Although it has been suggested that the last group were chemically precipitated (Trendall and Blockley, 1970), the presence of shard bands (La Berge, 1966) and clastic material in at least one S macroband at Paraburdoo suggests a composite origin. Analysis of 57 samples from the S macrobands, each representing at least 250 mm of continuous core, showed that Al_2O_3 and TiO_2 varied in a linear manner with a correlation coefficient of +0.96. Data for Al_2O_3 and TiO_2 for the S macrobands at Wittenoom (Trendall and Pepper, 1977) also plot close to the regression line established by Ewers and Morris. In contrast, a plot of Trendall and Blockley's data for individual mesobands (Trendall and Blockley, 1970, pp. 139—141) from Wittenoom reveals no particular relationship between Al_2O_3 and TiO_2 and indicates either that a different factor is controlling the distribution of these oxides at this scale, or that there may have been later, short-range redistribution of these oxides. A plot of Al_2O_3 and TiO_2 for both mesoband and "hand-specimen" data for the Puolanka rocks of Laajoki and Saikkonen (1977) also suggests a linear relationship but with different ratios. These data are expressed in Fig. 8A-1.

Ewers and Morris (1981) also observed a linear relationship between TiO_2 and Zr in the same set of 57 samples from Paraburdoo, and speculated on the cause. They concluded that the source of the material could be pyroclastics from intermediate igneous rocks (cf. Pearce and Cann, 1971), though they also drew attention to similar data from sedimentary clays or slates (e.g., Rosler and Lange, 1972), both of which have similar TiO_2/Zr ratios.

If the sources of the TiO_2, Zr, and Al_2O_3 at Paraburdoo were intermediate igneous rocks, the higher TiO_2:Al_2O_3 ratio in Finland might suggest a more mafic igneous source. Unfortunately there is not enough minor or trace-element data available to pursue this line of thought.

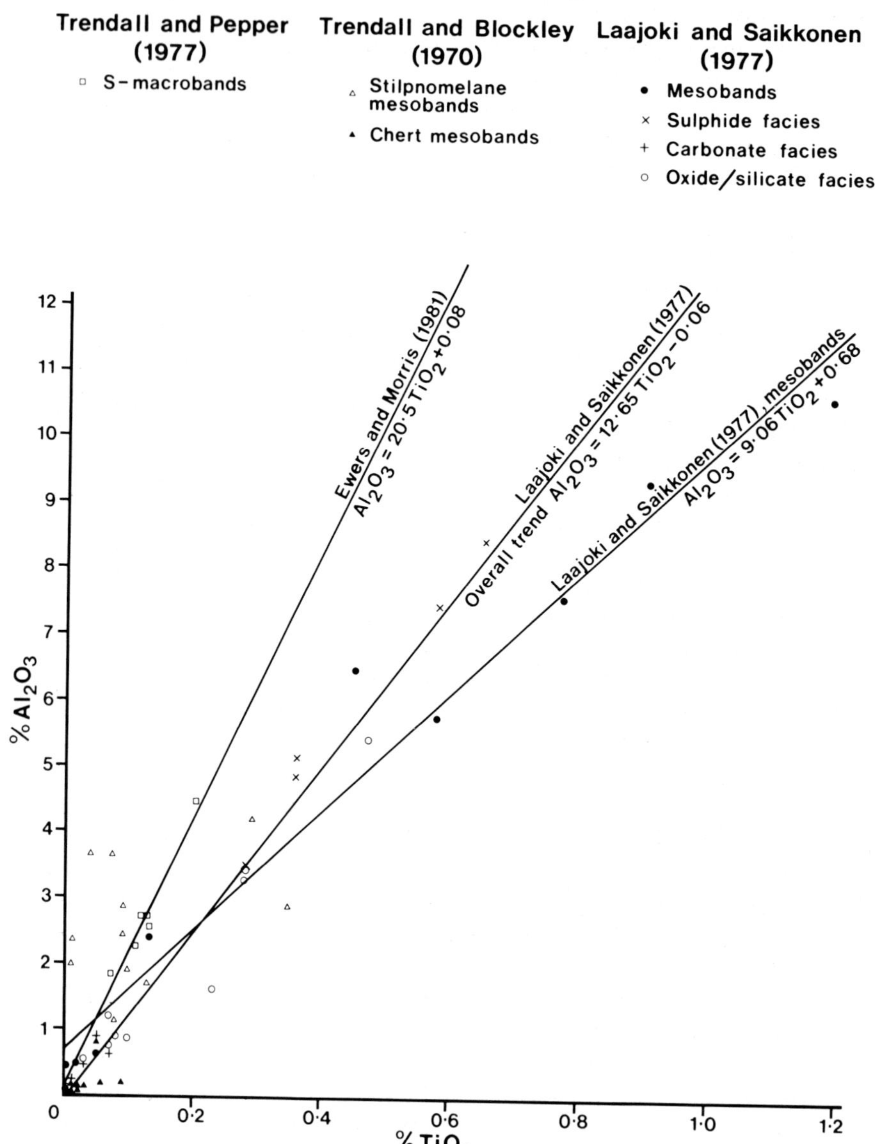

Fig. 8A-1. Plot of Al_2O_3 against TiO_2 for some published data.

Attempts have been made to use other elemental ratios to suggest origins of iron-formations. Pride and Hagner (1972) compared ionic ratios of major elements (e.g., Ca/Mg, Na/K) of iron-formation at Atlantic City, Wyoming, with equivalent ratios for crustal rocks, modern acidic volcanic waters, Dead Sea brine, and fumarolic gases. Although the ratios had a wide range of values, they concluded that iron-formations may have originated through subaerial weathering of crustal rocks, through primary volcanic "emanations", or through some combination of the two. Short pieces of drill core were used for the analyses, and so the ratios obtained may reflect neither the ratio for an individual band nor be representative of the whole iron-formation. An investigation of the patterns generated by such ratios, by analysis of successive bands, might be more informative.

Temporal variations between iron-formations

The Hamersley Iron Province offers a unique opportunity for the study of temporal changes in the composition of iron-formations. Fresh material (drill core) is available from formally named iron-formations which occur in unequivocal stratigraphic relationships to one another within the Hamersley Group.

Table 8A-III presents results of analyses of these iron-formations in order of relative age. Only the section through the Dales Gorge Member is complete (Ewers and Morris, 1981). The remaining values indicate the composition of substantial lengths of drill core but represent only part of the total sequence. It is uncertain whether the iron-formation of column 2, from the lower part of the Wittenoom Dolomite, represents the upper few metres of the Marra Mamba Iron Formation or whether, as is considered more likely, it represents nearly the whole thickness of a thin iron-formation within the Wittenoom Dolomite. The table includes previously unpublished analyses of fresh material from the Weeli Wolli Formation (Davy, in prep.). The data of Table 8A-III indicate trends for SiO_2 and possibly Al_2O_3, TiO_2 and P_2O_5 (increasing upwards), and for CO_2 (decreasing upwards). Total Fe (as Fe_2O_3) peaks in the Brockman Iron Formation.

The data are not ideal. Nonetheless they suggest that there were evolutionary trends within the Hamersley Group. This is surprising in view of the previously noted lack of any trend between individual macrobands of the Dales Gorge Member, a component of the Group. The possible significance of the trends is not yet established.

The average composition of iron-formations

The problem of establishing and comparing the average compositions of iron-formations is that of lack of data.

Average values for Hamersley Group rocks are presented in Table 8A-III,

TABLE 8A-III

Average values of Hamersley Group iron-formations, Western Australia (Values in percentage)

	1	2	3	4	5
Thickness (m)	33.64	4.05	122.43	20.1	96.1
SiO_2	40.85	42.0	43.51	44.34	49.5
Al_2O_3	0.41	0.3	0.36	0.89	2.4
Fe_2O_3	30.52	22.0	43.83	29.23	19.9
FeO		15.3		13.42	14.4
MgO	4.63	2.88	3.03	2.30	2.9
CaO	10.26	6.70	1.81	1.78	3.5
Na_2O	0.196	0.13	0.03	0.53	0.22
K_2O	0.179	0.14	0.06	1.26	0.61
H_2O+		1.10		0.98	2.0
H_2O-		0.12		0.17	0.60
CO_2	12.09	8.37	5.81	4.62	3.1
TiO_2	0.016	0.06	0.03	0.05	0.14
MnO	0.269	0.14	0.07	0.17	0.07
P_2O_5	0.060	0.06	0.20	0.18	0.18
C		0.03		0.01	0.09
S	0.102	0.14	0.07		
FeS_2				0.12	0.15
Σ Fe as Fe_2O_3	30.52	38.7	43.83	43.99	35.7

1 — Marra Mamba Iron Formation, Paraburdoo (Ewers and Morris, 1980) incomplete sequence — part of upper BIF member.
2 — Iron-formation at base of Wittenoom Dolomite. Millstream No. 9, incomplete sequence, (new data added to original paper, Davy, 1975).
3 – Brockman Iron Formation, Dales Gorge Member, complete sequence, Paraburdoo, (Ewers and Morris, 1981).
4 — Brockman Iron Formation, Joffre Member, based on representative samples (Trendall and Pepper, 1977, pp. 13—14).
5 – Weeli Wolli Formation, incomplete sequence. Weeli Wolli No. 1 (new data).

average values for some other formations in Table 8A-IV. In some cases the latter represent published analyses (irrespective of derivation) or are derived from analyses of substantial lengths of drill core. Analyses of oxidized rocks have been omitted even where substantial lengths of core have been analyzed, but it is still doubtful whether the samples which form the basis for the averages recorded in Table 8A-IV are representative of entire, fresh iron-formations.

The iron-formations are grouped by their dominant facies-types. Thus the first group (Table 8A-IVa) are oxide silicate facies, and the second group (Table 8A-IVb) are carbonate facies. However, the compositions indicate that the iron-formations do not conform to "end-member" facies, but that all iron-formations, despite the possible dominance of one facies, are in fact a mixture of facies. In general, the more highly metamorphosed iron-formations have a greater proportion of silicate facies rocks than do the less metamorphosed formations. There have been few determinations of samples of sulphide facies. Perhaps the most interesting compositional feature is the consistent presence of MgO at the 2—4% level in all three facies.

Carbonate facies sequences are distinctively higher in CO_2, and usually also in Mn, than oxide or silicate facies sequences, even though some CO_2 is almost always reported from the silicate facies. In the carbonate facies, Fe(II) is consistently more abundant, and SiO_2 and total Fe less abundant, than in the other two facies. There are no apparent patterns for other components.

Compared with Archaean iron-formations, sulphide facies rocks are relatively rare, and S (sulphide) and free C are low in Proterozoic iron-formations, but there are few analyses of this facies from iron-formations of either age. Proterozoic sulphide facies rocks are present at Puolanka, Finland (Laajoka and Saikkonen, 1977) and the mean value of S in the whole formation exceeds 1%, with individual values considerably higher. Klein and Gole (1981) indicate that the lowest part of the Marra Mamba Iron Formation of the Hamersley Group contains appreciable sulphide.

Trace elements in iron-formations

The study of trace elements in iron-formations is at a more primitive stage than that of major components. There are not enough data, even at the hand-specimen level, for characterization of the general trace-element range of each iron-formation facies.

Some data published since James' paper of 1966 are broken down approximately according to his facies types and given in Table 8A-V. In this table, in addition to the range, a column indicating unusually high values is provided. The sampling difficulties make these trace-element data even more unreliable than major component data: in many cases the freshness of the sample or the validity of the method of sampling is not known.

Oxide and silicate facies data are grouped together but the values recorded indicate that there is little difference in trace element contents between oxide, silicate and carbonate facies rocks. Most iron-formations have very low values for all trace elements. Base metals, which normally co-precipitate with iron oxides, are commonly low, except at Broken Hill, Australia, where mean values of 290 ppm Cu, 1000 ppm Pb and 1840 ppm Zn are reported for 100 samples (Richards, 1966). These samples from Broken Hill were taken from highly metamorphosed rocks of the upper amphibolite facies, and are probably not representative of iron-formations as a whole, since they occur in the vicinity of a major Pb-Zn orebody.

However, certain formations do contain unusual concentrations of trace elements. De Villiers (1971) reported high values for Li (to 0.18%) in several samples from iron-formations of the Transvaal. In other areas many of the higher values are restricted to single samples.

TABLE 8A-IVa

Mean compositions reported for some dominantly oxide or silicate facies iron-formations

	1	2	3	4	5
SiO_2	46.40	40.71	42.21	44.13*	44.45
Al_2O_3	0.90	2.32	0.80	3.85	2.10
Fe_2O_3	18.70	21.04	13.13	30.75	26.42
FeO	19.71	20.10	23.05	15.49	18.79
MgO	2.98	3.15	4.01	2.80	2.27
CaO	1.60	2.63	4.44	1.31	1.65
Na_2O	0.04	1.05	0.10	1.22	0.18
K_2O	0.13	0.57	0.15	0.36	0.50
H_2O+	1.72	0.91	1.89		1.98
H_2O-		0.07	0.05		
CO_2	6.90	6.56	4.90		1.24
TiO_2	0.04	0.12	0.08		0.14
MnO	0.63	0.29	0.08	0.09	0.10
P_2O_5	0.08	0.25	3.00		0.17
C	0.17		1.00		0.08
S			1.12		
FeS_2					
Σ Fe as Fe_2O_3	40.4	43.2	38.5	47.8	47.1

1 — Biwabik Iron Formation. Lepp, 1972.
2 — Kuruman Iron Formation. De Villiers, 1971, p. 41, cited from Hanekom, 1967, table 32.
3 — Puolanka area (metamorphosed). Laajoki and Saikonnen, 1977, table 9, p. 87.
4 — Precambrian iron-formation, nr. Atlantic City, Wyoming, U.S.A. Pride and Hagner, table 3, p. 333. Metamorphosed. SiO_2 by difference.
5 — Average ferruginous Proterozoic rocks, Russian Platform. Ronov and Migdisov, 1971, table 2.
* = calculated by difference.
Values based on calculations from (limited numbers of) representative samples — with or without appropriate weighting.

Although less data are available for the sulphide facies, the higher absolute amounts of trace elements in rocks of this facies must reflect additional source material or quite different depositional parameters from other facies. Some data for Table 8A-V were obtained from samples collected in the vicinity of gold mines (Fripp, 1976). Au and Cu are both known to occur in high (sub-economic to economic) amounts, particularly in Archaean sulphide-facies rocks (Rozhkov, 1971; Sawkins and Rye, 1974; Fripp, 1976; Kirkham, 1979; Appel, 1980). Less is known of other elements of economic interest.

Trace elements have not been sought in any systematic manner: there are no trace-element determinations of single simple bands; there are few determinations of trace elements in macrobands; and there are no values from the complete sequence of any one iron-formation. Again, however, rocks of the Hamersley Group are the best documented. Trendall and Pepper (1977)

TABLE 8A-IVb

Mean compositions reported for some dominantly carbonate facies iron-formations

	1	2	3	4
SiO_2	31.22	32.2	35.9	33.63
Al_2O_3	0.14	1.5	2.1	0.50
Fe_2O_3	12.58	0.6	3.7	6.72
FeO	25.74	31.6	28.7	32.81
MgO	3.94	2.8	4.0	2.22
CaO	5.17	1.6	2.0	0.30
Na_2O	0.05	0.0	0.01	
K_2O	0.10	0.2	0.09	
H_2O+	0.54	0.0	1.2	
H_2O-	0.06	0.2	0.23	18.32
CO_2	19.69	24.8	20.7	
TiO_2	0.02	0.0	0.14	0.11
MnO	0.68	1.9	1.2	1.12
P_2O_5	0.13	0.8	0.11	0.04
C	n.d.	1.8	0.16	
S			0.05	
FeS_2				
Σ Fe as Fe_2O_3	40.9	35.4	35.3	42.8

1 — Temiscamie Iron Formation. Composite of 163 m diamond drill core. Quirke et al., 1960.
2 — Riverton Iron Formation, U.S.A. Bayley and James, 1973, p. 946; from James 1951, p. 257.
3 — Ironwood Iron Formation, U.S.A., Mean values chert-carbonate facies, weighted according to lengths of core. From Huber, 1959, original data p. 91.
4 — Negaunee Iron Formation, Empire Mine, Michigan. Carbonate-chert facies of iron-formation. Composite of 245 ft of core. Gair, 1975, p. 77.
Values based on calculations from (limited numbers of) representative samples — with or without appropriate weighting.

provided determinations of 6 S and 5 BIF macrobands of the Dales Gorge Member, and for 20.1 m of the Joffre Member of the Brockman Iron Formation. Weighted means of the samples analyzed, together with other analyses of substantial lengths of core from other Hamersley Group iron-formations, are given in Table 8A-VI. Although there are differences in trace-element contents, there is no systematic variation.

DISCUSSION AND CONCLUSIONS

The prime purposes of chemical analysis of iron-formations fall into three main categories: (a) general petrographic and chemical characterization; (b)

TABLE 8A-V

Range of common values, and maximum recorded value, for hand-specimen samples of various facies of iron-formation[1] (values in ppm except where stated)

	Oxide and Silicate facies			Carbonate facies			Sulphide facies		
	Normal range (From — to)	Max.	Source	Normal range (From — to)	Max.	Source	Normal range (From — to)	Max.	Source
Sb	<200 —		4,9	<200 —		9	<200 —		9
As	<10 — 90		1,3,4,9,13,15	25 — 200		5,9	<100 — 210	25.6%	5,9
Ba	15 — 180(c)	1800	2,3,4,9,10,11,13,17	<10 — 690		2,5,9,17	<100 — 300		5,9,17
Be	<1 — 25		1,2,13						
Bi	n.d.		4						
B	0 — 240(c)	1000	1,2,13,17	150(d) — 260(c)		17	1800(c) —		
Cd				<1 — 2		5	3 — 10		5
Ce	n.d. — <5	25	4,6,16	14		16	7 — 36		16
Cl	0% — 0.1%		7						
Cr	5 — 160		1,2,4,6,8,13,15,16,17	<10 — 357(c)		2,16,17	40 — 510		8,9,16,17
Co	4 — 38(c)	600	1,2,4,6,9,11,13,15,16,17	10 — 47		5,9,16,17	<20 — 170		5,9,16,17
Cu	<2 — 150	440	1,2,3,4,8,9,10,11,13,16,17	<2 — 79(c)	140	2,5,9,16,17	<5 — 466(c)	(a)3.9%	5,8,16,17
F	<40 — 0.7%		7,13	96 — 136		5	264 — 728		5
Ga	n.d. — 10		1,4,13				13 — 29		9
Ge	n.d. — 30		1,2,9,10,11,13,14	<10		9	<10		9
Au	<0.1 —	0.7	12,16				<1 —	124	(b)8,16
La	n.d. — 120		2,6,13,16	8		16	3 — 19		16
Pb	n.d. — 30	80	1,3,4,9,10,13	<8 — 70		5,9	22 — 243		5,9
Li	n.d. — 50	1800	4	<5		5			
Mo	<5		1,2,3,4,13,15						
Ni	<5 — 151(c)	170	1,2,3,4,8,11,13,15,17	<10 — 205(c)		2,5,9,17	5 — 380		5,8,9,17
Nb	n.d. — 20		1,2,3,13						
Rb				<13 — 13		5	<5 — 25	78	5
Sc	n.d. — 30		2,4,9,13,17	<30		9	<30		9,17
Ag	n.d. — 14	50	1,4				<1 — 13	17.6	8,16

Sr	n.d. — 130	1000	1,2,3,4,9,11,13,17	8 — 219(c)	5,9,17	10 — 140(c)		5,9,17
Te	n.d.		4					
Th	0.05 — 10		3,6					
Sn	n.d. — 10	"50"	1,2,3,4,9	<3	9	<3		9
W	n.d.		2,4					
U	<1 — 2		9,15			2		9
V	<6 — 124(c)	210	2,3,4,10,11,13,15,17	26 — 140(c)	2,17	153(d) — 310	800	9,17
Y	10 — 84(c)	230	1,13,17	47(d) — 56(c)	17	41 — 95		9,17
Zn	<10 — 250	1270	1,2,8,9,10,13,17	100 — 350(c)	5,9,17	5 —	2640	5,8,9,17
Zr	n.d. — 267(c)		2,3,9,13,17	68(d) — 85(c)	17	<30 — 70		9,17

Sources:
1 — Alexandrov, 1973, p. 1050.
2 — Barbosa and Grossi Sad, 1973, pp. 128—129.
3 — Davy, 1979, no. 42365, p. 34.
4 — De Villiers, 1971, p. 41.
5 — Fripp, 1976, pp. 66, 68.
6 — Fryer, 1977, p. 1601.
7 — Gair, 1975, pp. 49, 54, 61, 63, 69, 71, 101.
8 — Gole, 1979, pp. 48—50, 53—54.
9 — Laajoki and Saikkonen, 1977, p. 97.
10 — Novokhatsky, 1973, pp. 154—155.
11 — Plaksenko et al., 1973, pp. 92—93.
12 — Rozhkov, 1971, pp. 92—94.
13 — Schmidt, 1963, pp. 16—17.
14 — Semenenko, 1973, p. 142.
15 — Tugarinov et al., 1973, pp. 37—38.
16 — Appel, 1980, p. 82.
17 — Gross and McLeod, 1980, pp. 224, 226.

(a) Values over 2% cited by Appel, 1980; values to 1%+ implied by Kirkham, 1979, p. 18.
(b) Samples from gold mines.
n.d. = not detected (limits not given).
[1], (c), (d) — Gross and McLeod only record mean values for various facies groups. (c) indicates the mean of their groups with the highest value, (d) indicates the mean of their groups having the lowest values. True maxima and minima are higher and lower than the figures indicated.

TABLE 8A-VI

Trace-element contents, Hamersley Group iron-formation, Western Australia (values in parts per million)

	1	2	3	4	5	6	7	8	9
Sb		<1	1	1.0	<1	1.2	1.0	<1	<1
As	<10	3	14	6	7	8	32	4	10
Ba	<20	12	113	90	88	69	263	50	140
Bi		1	3	3	<2	3	4	<2	<2
B		2	25	15	28	18	49	14	47
Cd		0.5	0.3	0.3	0.1	0.3	0.3	0.1	0.1
Ce	<5								
Cl		140	90	170	80	78	130	80	80
Cr		10	14	10	114	9	30	110	120
Co	20	25	2	2	5	1	6	4	6
Cu	10	10	101	80	41	96	120	45	35
F		60	128	110	179	120	157	150	220
Ga	<2	1	<1	1.0	3	<0.5	3.8	2	4
Ge		<5	2.7	3.0	1	3.0	1.6	<1	3
La	20								
Pb	40	<5	42	60	13	38	54	11	15
Li	6	4	6	6	4	6	5	4	4
Mo	<1	<5	<2	1	<1	<1	2	<1	<1
Ni	<5	20	12	10	63	9	21	80	40
Nb	4				8			10	5
Rb	10	5	31	50	37	16	83	20	60
Sc	<5	<5			<5			<5	5
Ag	<2	1	1.0	0.2	<1	0.9	1.3	<1	<1
Sr	10	2	53	55	36	35	115	30	45
Th	<5	10	12	13	<5	12	11	<5	<5
Sn	4	5	22	5	2	22	20	2	2
V	<5	<5	14	5	17	15	12	15	20
Y	6				18			15	24
Zn	11	85	47	50	49	36	83	46	53
Zr	120				38			30	50

In addition W and U were sought, but were below detection (<5 and <2 ppm, respectively).

1 — Iron-formation at base of Wittenoom Dolomite, new and revised data, Millstream No. 9.

2 — Mount Sylvia Formation, Bruno's Band, core Rhodes Ridge 1 (Davy, in press).

3 — Brockman Iron Formation, Dales Gorge Member, weighted mean of columns 6 and 7.

4 — Brockman Iron Formation, Joffre Member (Trendall and Pepper, 1977, p. 18).

5 — Weeli Wolli Formation, weighted mean of columns 8 and 9.

6 — Brockman Iron Formation, Dales Gorge Member, Average BIF macroband (Trendall and Pepper, 1977, p. 18).

7 — Brockman Iron Formation, Dales Gorge Member, Average S-band (Trendall and Pepper, 1977, p. 18).

8 — Weeli Wolli Formation (Upper part; iron-formation, 71.4 m core). New data.

9 — Weeli Wolli Formation (Lower part; black shale and chert, 52.0 m core). New data.

determination of ore potential; and (c) understanding of origins and evolutionary developments.

Enough analyses of hand-specimen samples of distinctive types of iron-formations (variously described as facies, environmental types etc.) have been carried out for the various types to be generally characterized. In addition, there is abundant data on the Fe content and the proportion and distribution of deleterious components within iron-formations. Such data are generated from ore-search programmes.

However, there is still no definitive classification of iron-formations, no conclusive knowledge of the environment in which they were formed and not enough adequate chemical data for more sophisticated studies to be undertaken.

Detailed geochemistry is important in the resolution of many of the problems of iron-formations. However, workers need to formulate clearly the nature of the problem whose answer is sought from the chemical analysis. Sampling should then be directed to this end.

The Hamersley Province of Western Australia, is, to date, the area studied most effectively from the geochemical point of view. Enough information has been acquired about rocks of the Hamersley Group to suggest that a study of geochemical patterns within this basin could prove valuable. So far, it has been established that there are local variations on the mesoband level, regional variations on the macroband level and that there is some variation between formations with time. However, it is not yet possible to explain variations in other than the most general way. Much more systematic work is needed before an adequate interpretation of their significance can be made.

Similar work should be contemplated for other iron-formation basins of the world, following which comparative studies may be attempted. Even so, the total time span for all the iron-formations in the Hamersley Group is only of the order of tens of millions of years, and extension of any temporal trend for the Hamersley region to iron-formations in other parts of the world would need greater precision in geochronology than is possible at present.

These studies should be paralleled by further studies of changes in composition during diagenesis and metamorphism. One possible line of enquiry might be to examine the trace-element content of the various minerals during different grades of metamorphism. Some preliminary studies on trace elements in magnetite have already been initiated by Rakovich et al. (1971) and Bondareva (1978).

A possible second line of enquiry could be to investigate the distribution of minor- and trace-element components within individual bands and layers, especially in those bands which retain some vestiges of primary sedimentological structures. The distribution of those components whose origin might be expected to be detrital would be of particular interest.

ACKNOWLEDGEMENTS

Permission to publish this chapter has been given by the Director of the Geological Survey of Western Australia.

REFERENCES

Alexandrov, E.A., 1973. The Precambrian iron-formation of the Soviet Union. Econ. Geol., 68: 1035—1062.

Appel, P.W.U., 1980. On the early Archaean Isua iron-formation, west Greenland. Precambrian Res., 11: 73—87.

Barbosa, A.L.M. and Grossi Sad, J.H., 1973. Tectonic control of sedimentation and trace-element distribution in iron ores of central Minas Gerais (Brazil). UNESCO Earth Sci., 9: 125—131.

Bayley, R.W. and James, H.L., 1973. Precambrian iron-formations of the United States. Econ. Geol., 68: 934—959.

Bondareva, N.M., 1978. Trace element distribution in magnetites of iron-siliceous formations of the Ukrainian Shield. Geol. J., Kiev, 38 (3): 62—71 (in Russian).

Davy, R., 1975. A geochemical study of a dolomite-BIF transition in the lower part of the Hamersley Group. West. Aust., Geol. Surv., Annu. Rep. 1974: 88—100.

Davy, R., 1979. A study of laterite profiles in relation to bedrock in the Darling Range, near Perth, W.A.. West. Aust., Geol. Surv., Rep. Ser. 8.

Davy, R., in press. A geochemical study of the Mount McRae Shale and the upper part of the Mount Sylvia Formation in core RD1, Rhodes Ridge, Western Australia. West. Aust., Geol. Surv., Rec.

Davy, R., in prep. The mineralogy and composition of the Weeli Wolli Formation in core WW1, Hamersly area, Western Australia. West. Aust., Geol. Surv., Rec.

De Villiers, P.R., 1971. The geology and mineralogy of the Kalahari manganese field north of Sishen, Cape Province. Geol. Surv. S. Afr., Mem. 59. 85 pp.

Ewers, W.E. and Morris, R.C., 1980. Chemical and mineralogical data from the uppermost section of the upper BIF member of the Marra Mamba Iron Formation. Aust. CSIRO, Inst. Earth Resour., Div. Mineral., Rep. FP 23.

Ewers, W.E. and Morris, R.C., 1981. Studies on the Dales Gorge Member of the Brockman Iron Formation, Western Australia. Econ. Geol., 77: 1929—1953.

Fripp, E.P., 1976. Stratabound gold deposits in Archaean banded iron-formation, Rhodesia. Econ. Geol., 71: 58—75.

Fryer, B.J., 1977. Trace element geochemistry of the Sokoman Iron Formation. Can. J. Earth Sci., 14: 1598—1610.

Gair, J.E., 1975. Bedrock geology and ore deposits of the Palmer Quadrangle, Marquette County, Michigan. U.S. Geol. Surv., Prof. Pap. 769.

Gole, M., 1979. Metamorphosed Banded Iron-formations in the Archaean Yilgarn Block, Ph.D. Thesis, Univ. Western Australia, (unpubl.).

Gole, M.J. and Klein, C., 1981. Banded iron-formations through much of Precambrian time. J. Geol., 89: 169—183.

Gross, G.A. and McLeod, C.R., 1980. A preliminary assessment of the chemical composition of iron formations in Canada. Can. Mineral., 18: 223—229.

Gruner, J.W., 1946. The mineralogy and geology of the taconites and iron ores of the Mesabi Range, Minnesota. Iron Range Resources and Rehabilitation Comm. and Minnesota Geol. Survey, St Paul, Minn., 127 p.

Hanekom, H.J., 1966. The Crocidolite Deposits of the Northern Cape Province. Ph.D. Thesis, Univ. Pretoria (unpubl.).

Huber, N.K., 1959. Some aspects of the origin of the Ironwood iron-formation of Michigan and Wisconsin. Econ. Geol., 54: 82—118.

James, H.L., 1951. Iron formation and associated rocks in the Iron River district, Michigan. Bull. Geol. Soc. Am., 62: 251—266.

James, H.L., 1954. Sedimentary facies of iron-formation. Econ. Geol., 49: 235—285.

James, H.L., 1966. Chemistry of the iron-rich sedimentary rocks. U.S. Geol. Surv., Prof. Pap. 440-W, 60 pp.

Kirkham, R.V., 1979, Copper in iron-formation. Geol. Surv. Can., Pap., 79-1B: 17—22.

Klein, C. and Gole, M.J., 1981. Mineralogy and petrology of parts of the Marramamba Iron Formation, Hamersley Basin, Western Australia. Am. Mineral., 66: 507—525.

Laajoki, K. and Saikkonen, R., 1977. On the geology and geochemistry of the Precambrian iron-formations in Vayrylankyla, south Puolanka area, Finland. Geol. Surv. Finl., Bull., 292, 137 pp.

La Berge, G.L., 1966. Altered pyroclastic rocks in iron-formation in the Hamersley Range, Western Australia. Econ. Geol., 61: 147—161.

Lepp, H., 1972. Normative mineral composition of the Biwabik Formation: a first approach. In: B.R. Doe and D.K. Smith (Editors), Studies in Mineralogy and Precambrian Geology. Geol. Soc. Am., Mem., 135: 265—278.

Novokhatsky, I.P., 1973. Precambrian siliceous iron-formations of Kazakhstan, UNESCO Earth Sci., 9: 153—157.

Pearce, J.A. and Cann, J.R., 1971. Ophiolite origin investigated by discriminant analysis using Ti, Zr, Y. Earth Planet. Sci. Lett., 12: 339—349.

Plaksenko, N.A., Koval, I.K. and Shchogolev, I.N., 1973. Precambrian ferruginous-siliceous formations associated with the Kursk Magnetic Anomaly. UNESCO Earth Sci., 9: 89—103.

Pride, D.E. and Hagner, A.F., 1972. Geochemistry and origin of the Precambrian iron formation near Atlantic City, Fremont County, Wyoming. Econ. Geol., 67: 329—338.

Quirke, T.E., Goldich, S.S. and Kreuger, H.W., 1960. Composition and age of the Temiscamie Iron-Formation, Mistassini Territory, Quebec, Canada. Econ. Geol., 55: 311—326.

Rakovich, F.I., Goniondskaja, L.S. and Drozdovskaja, A.A., 1978. Elements-impurities in magnetites from ferruginous-siliceous rocks and iron ores of the Ukranian Shields of different facies of metamorphism. Geol. J., Kiev, 38 (2): 120—129 (in Russian).

Richards, S.M., 1966. The banded iron formations at Broken Hill, Australia, and their relationship to the lead-zinc-ore bodies. Pt I: Econ. Geol., 61: 72—96. Pt II: Econ. Geol., 61: 257—274.

Ronov, A.B. and Migdisov, A.A., 1971. Geochemical history of the crystalline basement and the sedimentary cover of the Russian and North American platforms. Sedimentology, 16: 137—185.

Rosler, J.J. and Lange, H., 1972. Geochemical Tables. Elsevier, Amsterdam, 468 pp.

Rozhkov, I.S., 1971. Gold content of Krivoy Rog iron ore. Dokl. Akad. Nauk SSR, 196: 92—95.

Sawkins, F.J. and Rye, D.M., 1974. Relationship of Homestake-type gold deposits to iron-rich Precambrian sedimentary rocks. Trans. Inst. Min. Metall. London, 83: B56—59.

Schmidt, R.G., 1963. Geology and ore deposits of Cuyuna North Range, Minnesota. U.S. Geol. Surv., Prof. Pap. 407.

Semenenko, N.P., 1973. The iron-chert formations of the Ukrainian Shield. UNESCO Earth Sci., 9: 135—142.

Trendall, A.F. and Blockley, J.G., 1970. The iron formations of the Precambrian Hamersley Group, Western Australia. West. Aust., Geol. Surv., Bull. 119, 366 pp.

Trendall, A.F. and Pepper, R.S., 1977. Chemical composition of the Brockman Iron Formation. West. Aust., Geol. Surv., Rec. 1976/25 (unpubl.).

Tugarinov, A.I., Bergman, I.A. and Gavrilova, L.K., 1973. The facial nature of the Krivoy Rog iron-formation. UNESCO, Earth Sci., 9: 35—39.

Chapter 8

PART B. RARE EARTH ELEMENTS IN IRON-FORMATION

B.J. FRYER

INTRODUCTION

The geochemistry of the rare earth element (REE) group has proven to be one of the most powerful tools in elucidating igneous processes and is being increasingly applied to the study of sedimentary rocks. The usefulness of the REE group lies in their coherent behaviour during most geologic processes with differences between the chemistry of individual REE being due to slight changes in their ionic (+3) radii, which decrease monatomically from La to Lu (Fig. 8B-1) and to changes in oxidation state for the elements Eu (+3, +2) and Ce (+3, +4). Because of these characteristics, changes in the abundance of the REE and in particular their abundances relative to one another are of fundamental importance in interpreting ancient geologic processes.

In the modern marine environment the REE have very low solubilities and mobilities resulting in short residence times in sea water [50—600 years (Goldberg et al., 1963; Piper, 1974a)]. Consequently, the REE composition of sea water can be expected to rapidly respond to variations in the inputs of the REE to the marine environment and, since residence times are less than mix-

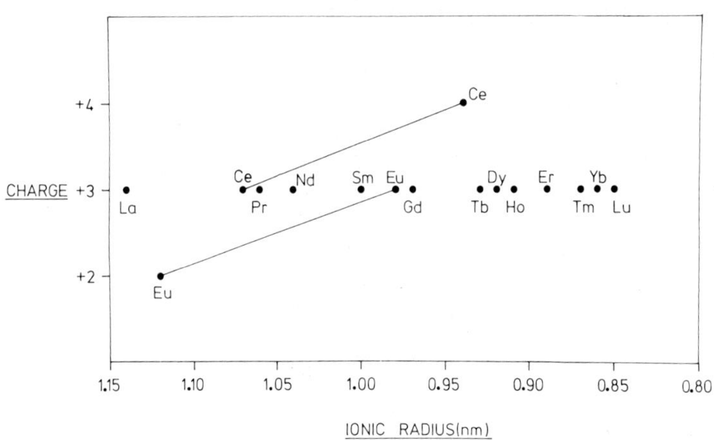

Fig. 8B-1. The relationship of charge to ionic radius for the rare-earth elements.

TABLE 8B-I

REE contents of selected Precambrian iron-formations, modern Fe-rich sediments and ocean water (ppm)

	1*	2	3	4	5	6	7	8	9	10
La	5.09	15.1	253	1.60	2.26	17.2	2.72	1.57	1.62	3.96
Ce	2.84	5.85	803	1.83	3.70	15.4	4.52	2.50	3.05	7.64
Pr										1.12
Nd	3.53	13,6	251	0.96	1.99	22.4	2.37	1.36	1.41	4.91
Sm	0.658	2.94	44	0.265	0.359	3.9	0.434	0.249	0.240	1.20
Eu	0.156	0.695	9.4	0.243	0.202	1.2	0.151	0.080	0.083	0.29
Gd	0.908									1.37
Tb		0.472	7.1			0.77				
Dy	0.984									1.10
Ho				0.111	0.116				0.099	
Er	0.974									0.55
Yb	0.935	2.46	25.5	0.452	0.380	2.0	0.242	0.207	0.389	
Lu	0.165	0.444	4.4	0.082	0.067	0.29	0.032	0.032	0.057	

* Analysis multiplied by 10^6.
1 — Pacific ocean water (Masuda and Ikeuchi, 1979).
2 — average East Pacific Rise crest sediment (Piper and Graef, 1974).
3 — average shallow water manganese nodule (Piper, 1974b).
4 — average Archean jasper-magnetite iron-formation, Temagami, Canada (Fryer, 1977a).
5 — average Archean oxide facies iron-formation, Mary River, Canada (Fryer, 1977a).
6 — average Karelian oxide facies, Finland (Laajoki, 1975).
7 — average Sokoman oxide facies, Labrador (Fryer, 1977b).
8 — average Sokoman silicate-carbonate facies, Labrador (Fryer, 1977b).
9 — average enriched Sokoman oxide facies, Labrador (Fryer, 1977b).
10 — Rapitan pisolitic-cherty iron-formation, Snake River, Canada (Fryer, 1977b).

ing times of the oceans (< 1000 years; Holland, 1978), be capable of maintaining distinctive REE abundances in regions receiving abnormal inputs. This presently is observed as waters draining from the continents (Fig. 8B-2a) and continental shelf waters (Piper, 1974a) are distinctly different from open ocean waters (Fig. 8B-2a) and even open ocean waters vary both laterally and vertically. The reader is referred to Piper (1974a) for a review of REE distributions in the modern sedimentary cycle. Their main characteristics that need concern us here are:

(1) REE abundances in sea water are exceedingly low (Table 8B-I) being measured at the part per trillion level (1E-12 g/g).

(2) Ce becomes oxidized in the oceans to the 4+ state which is exceedingly insoluble and is rapidly incorporated in bottom sediments, particularly Mn nodules (Fig. 8B-2b). This results in depletion of normal sea water in Ce rela-

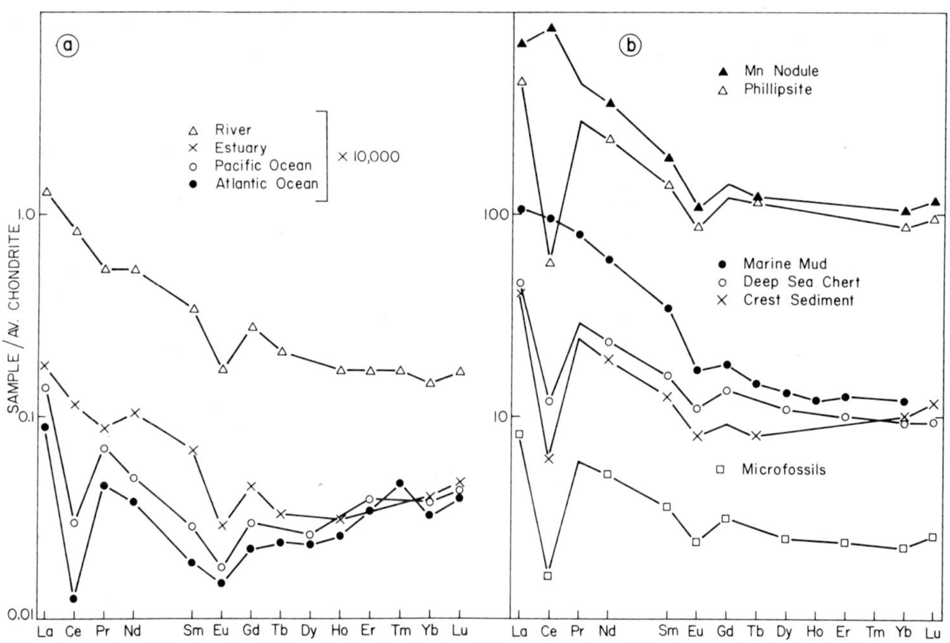

Fig. 8B-2. a. Chondrite-normalized REE abundances in fresh, brackish and marine waters. All concentrations have been multiplied by 10,000 (data for river and estuary water from Martin et al., 1976; for Pacific Ocean from Masuda and Ikeuchi, 1979; for Atlantic Ocean from Hogdahl et al., 1968).
b. Chrondrite-normalized REE abundances in marine phases and muds. Data from Piper, 1974b (average shallow-water manganese nodule and phillipsite); Piper and Graef, 1974 (average East Pacific Rise crest sediment); Shimizu and Masuda, 1977 (average deep-sea chert and average deep-sea siliceous marine microfossils); McLennan and Taylor, 1980 (marine mud from the Gulf of Maine — USGS Standard MAG-1).

tive to the other REE. Estuary waters (Fig. 8B-2a) and shallow water, continental shelf-type environments tend to have little or no Ce depletion.

(3) Eu does not appear to undergo any oxidation state changes in the weathering or sedimentary environment although it is depleted by a factor of about 0.7 compared to adjacent REE reflecting a similar depletion in continental clastic detritus (Fig. 8B-2a, b).

(4) Authigenic sediments accurately reflect (Fig. 8B-2b) the relative Ce and Eu abundances of the waters from which they formed (Piper, 1974a; Shimizu and Masuda, 1977).

(5) Although the residence times of all the REE are very short, those of the heavy REE (Gd-Lu) are longer than those of the light REE (La-Eu) due to the greater stability of heavy REE complexes. As a consequence sea water is relatively enriched in the heavy REE compared to marine clastic sediments.

Anomalous behaviour of Ce and Eu is readily observed in rocks when the analyses are plotted normalized, element by element, to a convenient reference sample, which for the purpose of the following discussion will be taken as chondritic meteorites. Anomalous behaviour in a particular REE is suggest-

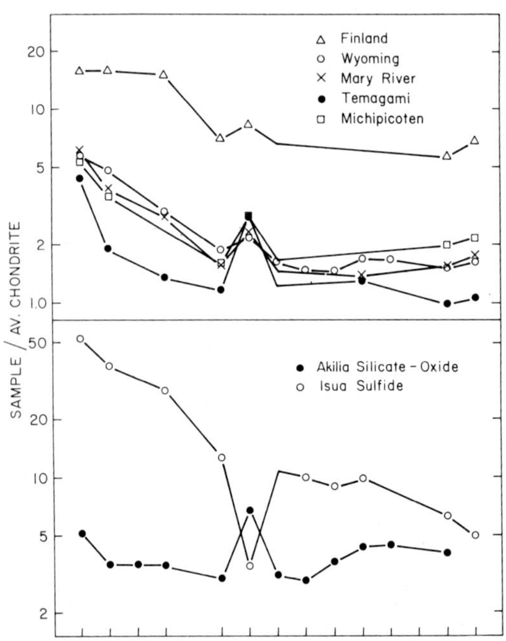

Fig. 8B-3. Chondrite-normalized REE abundances in Archean iron formations. Data from Laajoki and Lavikainen, 1977 (△ Finland, analysis #2 of Ukkolanvaara iron-formation); Wildeman and Haskin, 1973 (○ Wyoming, banded quartz-rich iron-formation); Fryer, 1977 (● Temagami — average of 2, □ Michipicoten — average of 2, X Mary River — average of 6); McGregor and Mason, 1977 (● sample 119232 of iron-formation in Akilia association, West Greenland); Appel, 1980 (○ sample D, sulfide facies Isua iron-formation).

ed when its normalized abundance differs significantly from the predicted abundance obtained by extrapolating from adjacent REE. This procedure is straightforward for Eu but for Ce the identification of anomalous behaviour is more difficult. This results from the very large difference in the +3 ionic radii of La and Ce (Fig. 8B-1) which means that when plotting the REE on a chondrite-normalized diagram with REE atomic number as abscissa, as is normal, some variation from smooth trends can be expected for La as an artifact of the plotting procedure. Consequently small deviations from normal behaviour are difficult to distinguish for Ce. The convention adopted for the purposes of the following discussion will be that Ce anomalies will only be deemed significant if both the extrapolated normalized Ce abundance from La to Pr or Nd and that obtained by backward extrapolation from heavier REE such as Sm through Nd/Pr to Ce deviate beyond experimental error. The application of this convention is best seen on Fig. 8B-3 for the Temagami average, where Ce would be considered anomalous by the former but not the latter extrapolation.

REE DISTRIBUTIONS IN IRON-FORMATION

The distributions of the REE in iron-formations of various ages has recently received increased attention (Laajoki, 1975; Fryer, 1977a, b; Graf, 1977, 1978; Laajoki and Lavikainen, 1977; Fryer et al., 1979; Appel, 1980) the results of which may provide insight into certain aspects of iron-formation genesis.

Most Archean iron-formations are characterized by a relative enrichment in Eu compared to the other REE (Fig. 8B-3) although normal and even depleted Eu abundances have been found. Absolute REE contents in Archean iron-formations are usually lower than normal crustal abundances although the generally impure nature of these sediments makes generalization difficult. No significant anomalous behaviour of Ce has been observed in Archean iron-formations.

In Proterozoic iron-formations Eu abundances range from slight relative depletions to enrichments. Ce is distinctly anomalous in several localities (Figs. 8B-4, 5): see also Fryer, 1977a), although in most iron-formations, average analyses, as shown in Fig. 8B-4, do not show significant Ce anomalies except for the Karelian Paakko Iron Formation. Absolute abundances are generally low and similar to the purest Archean iron-formations although the Paakko Iron Formation has high contents.

Evidence for some mineralogical control in primary REE abundances and patterns in iron-formations is scanty although several studies suggest that it may be important. The apatite-bearing Paakko Iron Formation is enriched in all the REE (Laajoki, 1975) and allanite-bearing samples of Archean iron-formations in Finland are enriched in all, but primarily the light (La-Eu) REE.

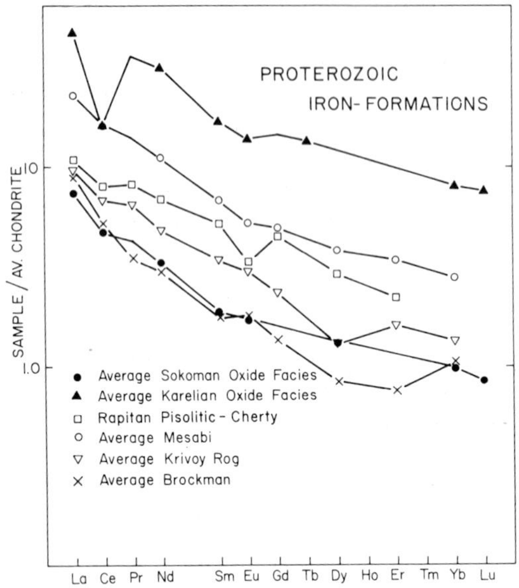

Fig. 8B-4. Chondrite-normalized REE abundances in Proterozoic iron-formations. Data from Laajoki, 1975 (average Karelian oxide facies, Finland); Fryer, 1977a (average Sokoman oxide facies, Rapitan pisolitic-cherty iron-formation, cherty iron-formation, Mesabi Range — part of average); Tugarinov et al., 1973 (average iron-rich zone, Krivoy Rog). Wildeman and Haskin, 1973 (cherty iron-formation, Mesabi Range — part of average); Fryer, 1977b (Brockman Iron Formation — average of 2; jaspilitic iron-formation. Mesabi Range — part of average).

Within the Sokoman Iron Formation of Labrador, Fryer (1977b) found distinct differences between oxide and silicate-carbonate facies of iron-formation (Fig. 8B-5) which substantiated earlier work by Balashov and Goryainov (1966) on iron-formations of the Imandra region, Russia. With this background we will now discuss the implications of the existing REE data for the genesis of Precambrian banded iron-formations.

It must be emphasized at this point that REE studies of iron-formations, completed to date, have all been of what might be termed reconnaissance nature and the existing data is scanty. Because of this the following discussion is necessarily biased by the available data and the author's interpretation of it.

EVOLUTION OF PRECAMBRIAN OXIDATION STATES

At present, the chemical sediments deposited from sea water, such as the Fe-Mn deposits near the mid-ocean ridges and authigenic phases (except for Mn-nodules) accurately reflect the relative sea water depletions in Eu and Ce

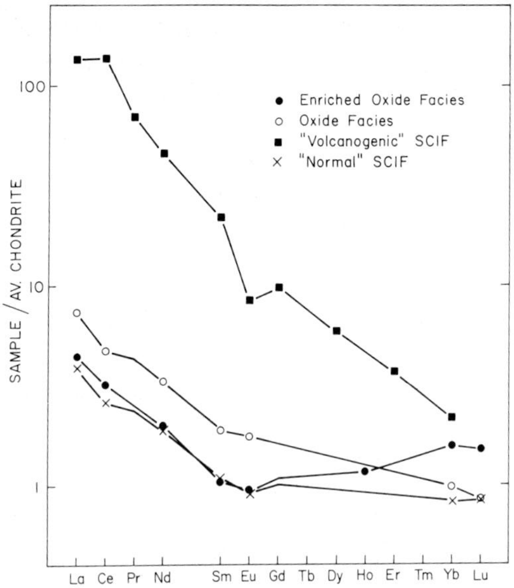

Fig. 8B-5. Chondrite-normalized REE abundances in different facies of the Sokoman iron-formation, Labrador. Data from Fryer 1977b; in prep.

(Fig. 8B-2a, b). Fryer (1977a) therefore suggested that chemical sediments were the best materials to trace changes in the behaviour of the REE in the sedimentary environment through time, particularly the banded iron-formations as they are typically devoid of clastic debris and were widely distributed in space and time.

Fryer (1977a) pointed out that the available data showed that the behaviour of Eu appeared to be time dependent. Archean chemical sediments were consistently enriched in Eu compared to the adjacent REE. Early Proterozoic chemical sediments showed only slightly positive to slightly negative Eu anomalies whereas the late Proterozoic and Phanerozoic iron-rich chemical sediments showed the same large Eu depletions as their contemporaneous clastic sediments. The REE content of clastic sediments, however, has evolved through time due to changing REE compositions of the exposed continental crust (Jakes and Taylor, 1974; Taylor and McLennan, 1981). Thus Fryer (1977a) suggested that compared to their associated clastic sediments, the Early Proterozoic iron-formations were as enriched in Eu as their Archean counterparts. This Eu enrichment was interpreted as indicating that a considerable proportion of the Eu must have been present as Eu^{2+} at some time during the weathering, transportation and/or deposition of the REE in these chemical sediments.

Graf (1978) criticized Fryer's interpretation of the Eu anomalies as being due to surficial processes and hence an indicator of changing oxidation states. Graf (1977) had previously documented that Paleozoic volcanogenic iron-formations from New Brunswick commonly displayed strong positive Eu anomalies and interpreted this Eu enrichment as being due to a hydrothermal input. He suggested that the hydrothermal solution was enriched in Eu due to preferential alteration of feldspar which is anomalously enriched in Eu. These results led Graf (1978) to conclude that as Archean iron-formations of the Algoma type had similar REE characteristics to Ordovician ones, they must both have formed from hydrothermal solutions that interacted with felsic volcanic rocks.

Graf's (1977) results and studies by Kerrich and Fryer (1979) on hydrothermal systems in Archean gold deposits of both epigenetic and syngenetic type which all showed strong Eu enrichment in hydrothermal deposits led Fryer et al. (1979) to reinterpret Eu anomalies in chemical sediments. They suggested that the Eu-enriched nature of Archean and Early Proterozoic sea water was due to a major contribution of strongly reducing hydrothermal fluid discharging onto the sea floor during these times. The hydrothermal regimes involved would be in equilibrium with the quartz — magnetite — fayalite f_{O_2} buffer curve. Under such conditions a significant proportion of the Eu would be present as Eu^{2+} which would be preferentially transported in the hydrothermal fluids. Experimental results (Cullers et al., 1973; Flynn and Burnham, 1978) show that this does take place and furthermore that it is not necessary to have the hydrothermal solution interact with felsic rocks as suggested by Graf (1977). Fryer et al. (1979) therefore suggested that the change to the present situation wherein the Eu content of sea water is determined primarily by the continental river flux to the oceans was a function of the decreasing thermal budget of the earth and the increasing importance of continental weathering.

The model of a seawater composition dominated by volcanogenic sources proposed by Fryer et al. (1979) for the Archean and Early Proterozoic for elements like Eu would explain the available data (except for the earliest iron-formations from West Greenland, see below) if at least some of the iron-formations studied represent deposits in equilibrium with normal sea water of that time. On the other hand, if the iron-formations studied were not deposited from "normal" marine water but were, as Graf (1978) suggests of direct hydrothermal origin, then arguments based on this assumption may not be valid. In either case a hydrothermal signature is present in the REE composition of iron-formations be it direct or indirect. This has implications for possible sources or iron for these deposits, as will be discussed later.

The data for high metamorphic grade iron-formations from West Greenland is equivocal (Fig. 8-4; McGregor and Mason, 1977; Appel, 1980). Both strongly Eu-enriched and depleted REE patterns have been found but no plausible explanation for this behaviour has been advanced. Appel's (1980) explanation that the negative Eu anomalies are due to hydrothermal leaching of basalts is contra-indicated by the available experimental and empirical data. It is more likely that the Eu depletion in some of the samples may be due to preferential removal of Eu in fluids released and expelled during prograde metamorphic reactions particularly in those samples containing high sulphide and organic contents (Isua samples). The redox potential of such rocks would be low enough to ensure a significant proportion of the Eu would be in the 2+ state at the prevailing metamorphic temperatures and hence available for incorporation in the fluid phase upon its release. As Kerrich and Fryer (1979) point out such metamorphic fluids are strongly enriched in Eu relative to the other REE.

Ce behaviour may also show time dependence. No examples of "significantly" anomalous Ce abundances are known for Archean iron-formations. In Proterozoic ones Ce is definitely anomalous with examples of both enrichments and depletions being observed. However the only consistently and strongly Ce-anomalous iron formation is the Paakko Iron Formation with its strong Ce depletion (Fig. 8B-4). Individual samples from the Sokoman, Mesabi, Rapitan and Krivoy Rog iron-formations are also anomalous although the formations as a whole are not. This suggests that Ce was being separated from the rest of the REE by oxidation to the 4+ state at least back to the early Proterozoic which would require that areas of moderately to strongly oxidizing conditions must have existed in Proterozoic seas. Addy (1979) has suggested that Mn nodules, which are thought to be the cause of modern sea water depletion in Ce, would only be capable of removing cerium in quiet oxidized sediments. At a typical marine pH (about 8) Eh values above 0.15—0.45 volts would be required to oxidize Ce.

It is interesting to note that Early Proterozoic iron-formations that exhibit anomalous Ce, at least in part (Sokoman, Mesabi, Paakko, Krivoy Rog), are also enriched in Eu compared to their contemporaneous clastic sediments. If Fryer et al.'s (1979) interpretations are correct, this implies that there was a significant hydrothermal input to the seas from which these deposits were formed, at least for the REE.

MINERALOGICAL FACIES

Although, to this point, the REE characteristics of iron-formations have been interpreted without reference to their mineralogical facies, this is an oversimplication (Fig. 8B-5). Fryer (1977b) has documented small but significant differences in REE abundances and patterns within the Sokoman Iron Formation (Table 8B-I). He found that the silicate-carbonate-facies rocks showed relatively constant REE and other trace-element distributions, compatible with an origin as crystalline precipitates in equilibrium with sea water (Dimroth and Chauvel, 1973). Oxide-facies rocks exhibited widely variable trace-element and REE contents as was expected for rocks whose original trace-element contents were controlled by absorption processes and which must, therefore, subsequently have undergone complex diagenetic changes. Unfortunately, similar studies on REE in other iron-formations to substantiate these findings are lacking.

In addition to the above effects on REE abundances in iron-formations, other as yet unstudied, mineralogical effects must exit. This has been alluded to for the Paakko Iron Formation which contains much higher REE abundances than other iron-formations. This can be directly correlated with its high apatite content, a phase which is known to contain high REE abundances. The apatite may be responsible for its originally high REE contents or have incorporated, and hence caused the retention of, REE released during diagenetic recrystallization of original iron minerals. Other REE-bearing minerals than apatite are known to exist in iron-formations with likely similar affects. Monazite and xenotime have been reported from the Hamersley iron-formation and ore (see Morris, 1983, this volume) and samples including these minerals must have significantly higher REE contents than samples studied to date. If such minerals are erratically distributed in iron-formation then this will give rise to difficulties in sampling for REE studies. Such sampling problems have so far not been recognized. The REE-bearing minerals outlined above would not be expected to impart either Ce or Eu anomalies to otherwise normal REE distributions, however, as studies to date in igneous rocks (Fryer, unpubl. data) show that these minerals accurately reflect the liquids from which they crystallized in terms of the presence or absence of Ce/Eu anomalies. They would, however, drastically affect both the absolute and relative REE abundances of samples containing them.

In addition to the presence or absence of REE-bearing minerals other precautions must be exercised when interpreting the REE data on iron-formations. The REE contents of iron-formation samples are generally very low. As a direct consequence of this any significant addition of either terrigenous or volcaniclastic detritus may severely affect the abundances and hence the interpretation of the REE data.

DIAGENESIS

Diagenetic changes of the REE in iron-formations are undoubtably complex and are currently poorly understood. They may be responsible for the generally very low REE (and other trace-element) contents of Precambrian iron-formations. Recent iron-oxide-rich sediments contain relatively high REE and other trace-element contents (Table 8B-I). If Precambrian oxide-facies rocks originally contained similar high abundances, they may have been subsequently lost as their original absorption control changed to a crystal chemical control, accompanying the diagenetic change from original precipitated iron-oxide hydrate phase to hematite and/or magnetite. Support for this interpretation comes from changes in recent recrystallized siliceous chemical sedimentary components (Shimizu and Masuda, 1977) where greater than an order of magnitude decreases in REE abundances accompany increasing crystallinity.

Direct evidence of REE mobility and fractionation during diagenetic changes is available for a number of iron-formations (Fryer, 1977b). Iron-formation samples that have been significantly enriched in iron oxides subsequent to deposition exhibit strong relative enrichment of the heaviest REE. This is interpreted (Fryer, 1977b) as being due to selective mobilization of the heavy REE by REE carbonate complexes accompanying reduction of ferric iron by organic matter in relatively impermeable beds. The heavy REE and iron were later co-precipitated in more permeable beds through oxidation of the iron. This type of behaviour has been observed in the Brockman Iron Formation and the Rapitan Iron Formation in addition to the Sokoman.

VOLCANIC INPUT TO IRON-FORMATIONS

There is a certain amount of evidence in the REE in iron-formations that may be interpreted in terms of a significant "volcanogenic" source. As discussed earlier, Early Proterozoic and Archean iron-formations are characterized by a relative enrichment in Eu compared to their contemporaneous clastic sedimentary and/or volcanic rocks. This is interpreted (Graf, 1978; Fryer et al., 1979) as being due to a hydrothermal source which may also supply iron for these deposits.

Graf (1978) suggested that Algoma-type iron-formations were deposited from hydrothermal solutions because of their REE signature (positive Eu anomalies) but if this interpretation is correct it may have to be extended to Early Proterozoic deposits as well, at least in a modified way. If the sea water composition from which iron-formations were deposited was influenced by hydrothermal inputs, as evidenced by the REE, then this must have been accompanied by massive inputs of other elements including iron (Fryer et al., 1979). In the Sokoman Iron Formation, this has been documented by Evans (1978) for the Dyke Lake area where the chemistry of the iron-formation has been significantly modified in the vicinity of an alkaline volcanic complex. The iron-formation in this region is strongly enriched in elements characteristic of fluids related to alkaline volcanic rocks, such as Zr, Nb and Y which Evans (1978) related to carbonate complexing. It is also very strongly enriched in the light REE (Fig. 8B-5) and possesses a distinctive negative Eu anomaly which is characteristic of the unusual alkaline fluids associated with these types of volcanics.

SUMMARY

Although the available REE data for iron-formations are not abundant, they do provide significant information regarding the genesis of these rocks. The enrichment in Eu characteristic of most Early Proterozoic and Archean iron-formations can be interpreted as implying a significant hydrothermal input into the water from which they precipitated. Whether or not this was a local input directly related to their deposition or is an aspect of overall marine chemistry of these times is not certain. A local "volcanogenic" input into the Sokoman Iron Formation in one area has been established because of its unusual alkaline nature and "volcanogenic" inputs related to more normal and widespread volcanic activity may yet be established.

Anomalous Ce behaviour in iron-formations of Early Proterozoic age suggests the existence of strongly oxidizing conditions in the marine environment, at least locally, at this time with obvious implications for iron transport and oxygen budgets in the oceans. "Significant" anomalous Ce behaviour in the Archean has not been found to date. Whether or not this is important in terms of changing oxidation states of the oceans remains to be documented by much more extensive sampling, particularly in view of the scattered occurrence of Ce anomalies in Proterozoic iron-formations.

It appears that REE data may be used to distinguish different primary modes of incorporation of trace elements in iron-formations (crystal lattice substitutions vs absorption) and subsequent diagenetic changes they have undergone. This information for the Sokoman Iron Formation substantiates earlier work (Dimroth and Chauvel, 1973) based on textural observations.

Caution must be exercised in interpreting the REE in iron-formations, however, due to the very low REE abundances of these rocks. REE-bearing minerals are known to exist in iron-formations and could cause severe sampling problems. Even minor amounts of extraneous clastic input into iron formation could drastically affect their REE contents and patterns.

Despite the utility of the REE in potentially providing insight into aspects of iron-formation genesis the data remain sparse and more detailed studies on well documented sequences are required before major advances can be made. Work to date can be summarized as reconnaissance in nature and very unevenly distributed in terms of the major iron-formations. In particular, additional work on Australian BIF would provide much useful information on whether or not there has been an evolutionary change in Eu and Ce through time because of the considerable spread in the stratigraphic distribution of the various iron-formations. In addition the effects of REE-bearing phases on REE distributions could be easily evaluated.

Finally, the available data, scanty though it may be, suggests that the study of REE in iron-formations has the potential to provide significant input into the genesis of BIF and in understanding processes that affect them subsequent to their deposition. In addition it may provide insight into changing interactions of the atmosphere, hydrosphere and lithosphere during the Precambrian.

REFERENCES

Addy, S.K., 1979. Rare earth element patterns in manganese nodules and micronodules from northwest Atlantic. Geochim. Cosmochim. Acta, 43: 1105—1115.

Appel, P.W.U., 1980. On the Early Archaean Isua iron-formation, West Greenland. Precambrian Res., 11: 73—87.

Balashov, Y.A. and Goryainov, P.M., 1966. Rare earth elements in the Precambrian iron-bearing rocks of the Imandra Region. Geochemistry (USSR) (English transl.), pp. 951—969.

Cullers, R.L., Medaris, L.G. and Haskin, L.A., 1973. Experimental studies of the distribution of rare earths as trace elements among silicate minerals and liquids and water. Geochim. Cosmochim. Acta, 37: 1499—1512.

Dimroth, E. and Chauvel, J.-J., 1973. Petrography of the Sokoman Iron Formation in part of the central Labrador Trough, Quebec, Canada. Geol. Soc. Am. Bull., 84: 111—134.

Evans, J.L., 1978. The geology and geochemistry of the Dyke Lake area (parts of 23J/8, 9), Labrador. Mineral Development Division, Newfoundland Department of Mines and Energy, Rep. 78-4, 39 pp.

Evenson, N.M., Hamilton, P.J. and O'Nions, R.K., 1978. Rare-earth abundances in chondritic meteorites. Geochim. Cosmochim. Acta, 42: 1119—1212.

Flynn, R.T. and Burnham, C.W., 1978. An experimental determination of rare earth partition coefficients between a chloride containing vapor phase and silicate melts. Geochim. Cosmochim. Acta, 42: 685—701.

Fryer, B.J., 1977a. Rare earth evidence in iron-formations for changing Precambrian oxidation states. Geochim. Cosmochim. Acta, 41: 361—367.

Fryer, B.J., 1977b. Trace element geochemistry of the Sokoman Iron Formation. Can. J. Earth Sci., 14: 1598—1610.

Fryer, B.J., in prep. Geochemical evidence for a hydrothermal input into the Sokoman Iron Formation, Labrador. Can. J. Earth Sci.

Fryer, B.J., Fyfe, W.S. and Kerrich, R., 1979. Archaean volcanogenic oceans. Chem. Geol., 24: 25—33.

Goldberg, E.D., Koide, M., Schmitt, R.A. and Smith, R.H., 1963. Rare earth distributions in the marine environment. J. Geophys. Res., 68: 4209—4217.

Graf, J.L., 1977. Rare earth elements as hydrothermal tracers during the formation of massive sulfide deposits in volcanic rocks. Econ. Geol., 72: 527—548.

Graf, J.L., 1978. Rare earth elements, iron formations and sea water. Geochim. Cosmochim. Acta, 42: 1845—1850.

Hogdahl, O.T., Melsom, S. and Bowen, V.T., 1968. Neutron activation analysis of lanthanide elements in sea water. Adv. Chem. Ser., 73: 308—325.

Holland, H.D., 1978. The Chemistry of the Atmosphere and Oceans. Wiley-Interscience, New York, N.Y., 351 pp.

Jakes, P. and Taylor, S.R., 1974. Excess europium content in Precambrian sedimentary rocks and continental evolution. Geochim. Cosmochim. Acta, 38: 739—746.

Kerrich, R. and Fryer, B.J., 1979. Archaean precious-metal hydrothermal systems, Dome Mine, Abitibi Greenstone Belt: II REE and oxygen isotope relations. Can. J. Earth Sci., 16: 440—458.

Laajoki, K., 1975. Rare-earth elements in Precambrian iron formations in Vayrylankyla, South Puolanka area, Finland. Bull. Geol. Soc. Finl., 47: 93—107.

Laajoki, K. and Lavikainen, S., 1977. Rare-earth elements in the Archean iron formation and associated schists in Ukkolanvaara, Ilomantsi, SE Finland. Bull. Geol. Soc. Finl., 49: 105—123.

Martin, J.-M., Hogdahl, O. and Philippot, J.C., 1976. Rare earth element supply to the ocean. J. Geophys. Res., 81: 3119—3124.

Mason, B., 1979. Meteorites. In: M. Fleischer (Editor), Data of Geochemistry, Chapter B. Cosmochemistry. U.S. Geol. Surv., Prof. Pap., 440-B-1, 132 pp.

Masuda, A. and Ikeuchi, Y., 1979. Lanthanide tetrad effect observed in marine environment. Geochem. J., 13: 19—22.

McGregor, V.R. and Mason, B., 1977. Petrogenesis and geochemistry of metabasaltic and metasedimentary enclaves in the Amitsoq gneiss, West Greenland. Am. Mineral., 62: 887—904.

McLennan, S.M. and Taylor, S.R., 1980. Geochemical standards for sedimentary rocks: trace element data for U.S.G.S. standards SCo-1, MAG-1 and SGR-1. Chem. Geol., 29: 333—343.

Morris, R.C., 1983. Supergene alteration of banded iron-formation. In: A.F. Trendall and R.C. Morris (Editors), Iron-Formation: Facts and Problems. Elsevier, Amsterdam, pp. 513—534.

Piper, D.Z., 1974a. Rare earth elements in the sedimentary cycle: a summary. Chem. Geol., 14: 285—304.

Piper, D.Z., 1974b. Rare earth elements in ferromanganese nodules and other marine phases. Geochim. Cosmochim. Acta, 38: 1007—1022.

Piper, D.Z. and Graef, P.A., 1974. Gold and rare earth elements in sediments from the East Pacific rise. Geochim. Cosmochim. Acta, 17: 287—297.

Shimizu, H. and Masuda, A., 1977. Cerium in chert as an indication of marine environment of its formation. Nature, 266: 346—348.

Taylor, S.R. and McLennan, S.M., 1981. The composition and evolution of the continental crust: rare earth evidence from sedimentary rocks. Philos. Trans. R. Soc. London, Ser. A, 301: 381—399.

Tugarinov, A.I., Balashov, Y.A. and Gavrilova, L.K., 1973. Abundances of the rare earth elements in the lower Proterozoic Krivoy Rog Series. Geokhimiya (English transl.), 1: 28—34.

Wildeman, T.R. and Haskin, L.A., 1973. Rare earths in Precambrian sediments. Geochim. Cosmochim. Acta, 37: 419—438.

Chapter 8

PART C. OXYGEN ISOTOPE GEOCHEMISTRY OF IRON-FORMATION

E.C. PERRY, JR.

INTRODUCTION

In terrestrial rocks three stable oxygen isotopes, ^{16}O, ^{17}O, ^{18}O, occur in approximately the proportions 99.76:0.04:0.20. Terrestrial anomalies of ^{17}O have not been discovered, i.e. all reported $^{17}O/^{16}O$ ratios in terrestrial samples equal the square root of corresponding $^{18}O/^{16}O$ ratios within experimental error. Thus, since ^{17}O abundance is lower and more difficult to determine than ^{18}O abundance, almost all published oxygen isotope data on terrestrial materials are of the abundance of ^{18}O with respect to ^{16}O.

According to convention, oxygen isotope variations are reported as $\delta^{18}O$ values, defined by the relation:

$$\delta^{18}O = 1000x \left[\frac{(^{18}O/^{16}O) \text{ SAMPLE}}{(^{18}O/^{16}O) \text{ SMOW}} - 1 \right]$$

In this equation SMOW is the acronym for Standard Mean Ocean Water which has approximately the oxygen isotope composition of modern ocean water (Craig, 1961; Gonfiantini, 1978).

Differences in isotopic composition between chemical phases A and B are expressed as:

$$\Delta_{AB} = 1000 \ln \frac{(^{18}O/^{16}O)_A}{(^{18}O/^{16}O)_B} \simeq \delta_A - \delta_B$$

Δ_{AB} is, in general, a temperature-dependent function that is virtually independent of pressure (Bottinga and Javoy, 1973; Clayton et al., 1975). In this paper I shall discuss:

Δ_{QM} = ^{18}O fractionation between quartz and magnetite (Becker and Clayton, 1976)

Δ_{SM} = ^{18}O fractionation between siderite and magnetite (Becker and Clayton, 1976)

Δ_{QW} = ^{18}O fractionation between quartz and water (Becker and Clayton, 1976; Knauth and Epstein, 1976).

Oxygen isotopic studies of iron-formation are useful in defining the metamorphic conditions to which these rocks have been subjected. The ^{18}O com-

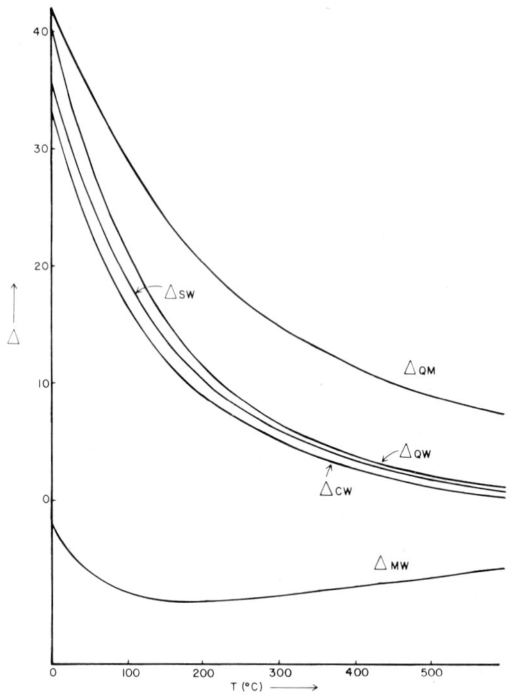

Fig. 8C-1. Fractionation, $\Delta = 1000 \ln\alpha$, vs. Temperature (°C) for quartz-water (Δ_{QW}), siderite-water (Δ_{SW}), calcite-water (Δ_{CW}), magnetite-water (Δ_{MW}), and quartz-magnetite (Δ_{QM}): quartz-water, magnetite-water, and siderite-water from Becker and Clayton (1976); quartz-calcite from O'Neil et al. (1969).

position of two or more phases in a metamorphosed rock often records an equilibrium fractionation that, when compared to experimental or semi-empirical data on temperature vs. isotope fractionation, provides an estimate of the temperature at which the rock equilibrated. This temperature can be, but need not be, the maximum temperature of metamorphism. In particular, iron-formation frequently contains alternating bands in which quartz and/or magnetite are major constituents. These minerals, which have simple stoichiometry and which are resistant to postmetamorphic exchange, differ from each other in equilibrium ^{18}O enrichment to a greater extent than any other pair of common rock-forming minerals (Fig. 8C-1).

Additional important information is recorded by the oxygen isotope composition of iron-formations. Metamorphic decomposition of carbonates results in loss of CO_2 accompanied by a decrease in ^{18}O content of the rocks. The spatial variation in oxygen isotope composition within a rock unit is controlled by diffusion of oxygen isotopes and thus provides information about movement of fluids (H_2O and CO_2) through the system of interest during metamorphism. Information about the Precambrian hydrosphere is preserved in some iron-formations and other chemical precipitates that retain their primary or diagenetic oxygen isotope composition.

PROTEROZOIC IRON-FORMATION

The first major oxygen isotope study of iron-formation (James and Clayton, 1962) concentrated on the geochemistry and metamorphic history of the Proterozoic Negaunee and Vulcan Iron Formations of the Marquette Peninsula of Michigan. All of the important aspects of oxygen isotope geothermometry were considered in this study, but lack of experimental calibration data for isotope fractionation at that time resulted in unreasonably low estimates of temperature, particularly in the staurolite and sillimanite zones of metamorphism. For example, the Vicar mine in the Gogebic district of Michigan is located in an area of chlorite zone metamorphism, Δ_{QM} = 17.8, the temperature of equilibration estimated by James and Clayton (1962) is 150°C, whereas the temperature indicated by Fig. 8C-1 is 240°C. In the Magnetic mine, Republic district, Michigan, in sillimanite zone rocks, Δ_{QM} = 9.2; James and Clayton estimated 320°C, and the revised estimate based on data of Fig. 8C-1 is 485°C.

In discussing the metamorphism of iron-formation it is useful to plot $\delta^{18}O$ values of quartz and of magnetite vs. Δ_{QM} (Fig. 8C-2). Both δ_Q vs. Δ_{QM} and δ_M vs. Δ_{QM} plots contain essentially the same information. δ_Q vs. Δ_{QM} is more directly useful since pure chert rocks associated with iron-formation exist as end members for comparison, whereas rocks containing over 50% magnetite are rare.

Quartz-magnetite fractionation, Δ_{QM}, in an equilibrium assemblage is a function of temperature. As I shall discuss, it need not be a simple record of the maximum temperature of metamorphism but may represent the temperature at which, for example, porosity became low enough to prevent further reaction. In a closed, homogenous system consisting of quartz, enriched in ^{18}O, and magnetite, depleted in ^{18}O, increasing temperature results in a decrease in equilibrium oxygen isotope fractionation between quartz and magnetite. To maintain isotope equilibrium ^{18}O must be transferred from quartz to magnetite. For such a system, a plot of Δ_{QM} vs. δ_Q or Δ_{QM} vs. δ_M is a straight line whose extrapolation to Δ_{QM} = 0 (infinite temperature) is a function of the initial isotopic compositions and the relative abundance of quartz and magnetite. Perry et al. (1978) show that the same general trends should be valid for iron-formations that have more complicated mineral assemblages. For iron-formations having a high bulk ^{18}O composition, i.e. one with a high content of siderite or quartz, the intercept at Δ_{QM} = 0 is high (for example, 19). For an iron-formation with a high content of magnetite, the intercept is low, e.g. 10.

The main assumptions in a $\Delta_{QM} - \delta$ plot are that the system is closed and

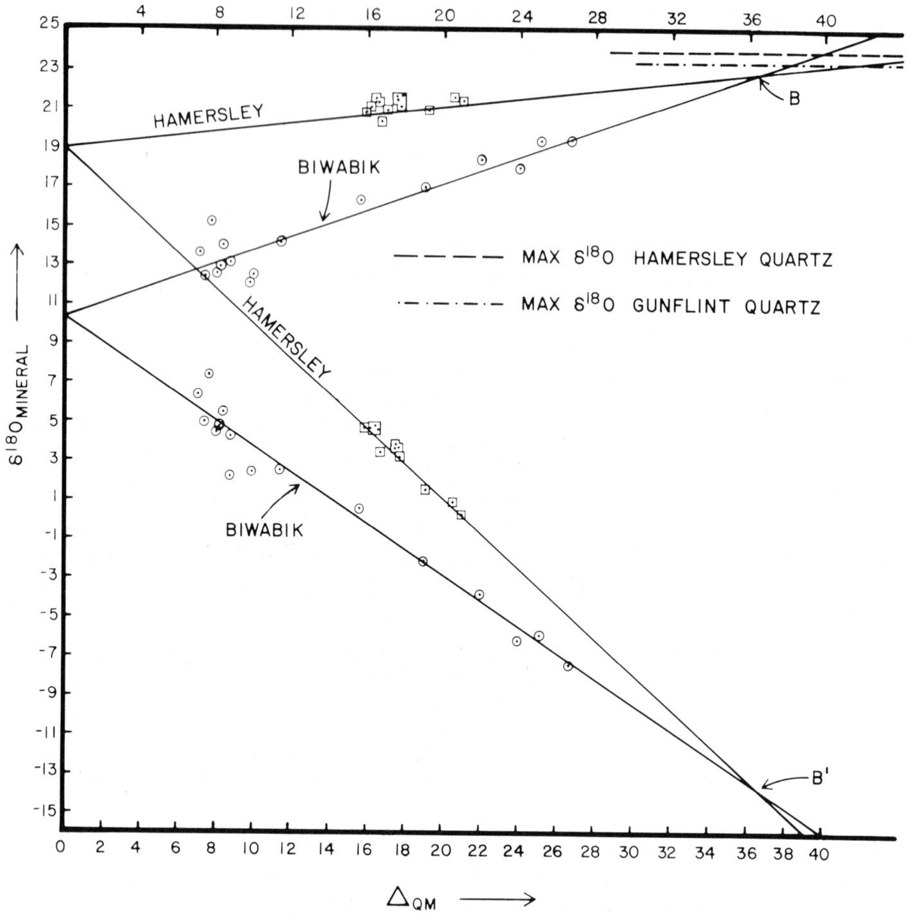

Fig. 8C-2. Trend lines for quartz and magnetite from the Biwabik and Hamersley iron formations. Hamersley data from Becker and Clayton (1976), Biwabik data from Perry et al. (1973). For Hamersley $\delta^{18}O_Q = 0.112\ \Delta_{QM} + 19.2$, correlation coefficient 0.44; $\delta^{18}O_M =$ $= -0.892\ \Delta_{QM} + 19.1$, correlation coefficient −0.97. For the Biwabik data plotted (cf. Perry et al., 1978, fig. 3), $\delta^{18}O_Q = 0.344\ \Delta_{QM} + 10.3$, correlation coefficient 0.94.

that it is homogeneous on a gross scale. Somewhat surprisingly, it appears probable that many Proterozoic iron-formations meet these conditions. Data from two Proterozoic iron-formations, the Biwabik Iron Formation of Minnesota and the Brockman Iron Formation of the Hamersley Group, Western Australia, approximate straight lines for $\Delta_{QM} - \delta$ plots as shown in Fig. 8C-2. Because the oxygen isotope geochemistry of these two iron-formations has been studied extensively and can be used as a model, they are discussed in some detail in the following paragraphs.

When the two iron-formations were originally modelled (Perry et al., 1978) it seemed probable that both were about 2000 m.y. old and thus that they were probably precipitated in ocean water or restricted basin water (Trendall and Blockley, 1970) of quite similar isotopic composition. Age determinations by Compston et al. (1981) indicate that the Brockman Iron Formation may be as old as 2500 m.y. (as much as 500 m.y. older than the Biwabik Iron Formation). I shall, nevertheless, assume that both iron-formations were precipitated from water of similar isotope composition. Some justification comes from the consistency of the results. A more satisfying justification comes later from the good correlation between Brockman and Kuruman Iron Formations, two formations of closely similar ages.

The Biwabik Iron Formation and the Brockman Iron Formation have quite different metamorphic histories. The Biwabik Iron Formation was metamorphosed by the Duluth Complex about 1100 m.y. ago (Perry et al., 1973), whereas the Brockman Iron Formation has been subjected only to burial metamorphisms (Trendall and Blockley, 1970).

Except for a contact metamorphic aureole adjacent to the Duluth Igneous Complex, where temperatures of 600°C or more obtained, the Biwabik Iron Formation (as well as the correlative Gunflint Iron Formation of Minnesota and Ontario) is of exceptionally low metamorphic grade. Perry et al. (1973) report $\delta^{18}O_M$ values as low as −7.5‰ and Δ_{QM} values as large as 26.7 for material from the Biwabik Iron Formation 18 km east of Virginia, Minnesota (Core 2, Table 8C-I). This corresponds to an equilibration "temperature" of less than 150°C, perhaps the lowest measured temperature estimate that has been published for any Precambrian rock.

Proceeding from east to west for more than 50 km across the Mesabi Range, there is a slight increase in regional metamorphic grade as registered by Δ_{QM} (~ 140°C for core 2, east Mesabi, 170°C for core 5, and 160°C for core 7, west Mesabi, Table 8C-I). $\delta^{18}O$ of quartz is uniform for each core but increases from east to west:

Core	$\delta^{18}O_Q$
2 (east)	18.9‰
5 (main)	19.7‰
7 (west)	20.4‰

The increase in $\delta^{18}O_Q$ indicates a change in the bulk composition of the Biwabik Iron Formation between the east and main Mesabi Range (increasing quartz and/or carbonate relative to magnetite in the western cores). This interpretation, that a change in composition of the iron-formation is responsible for the increased $\delta^{18}O_Q$ of the western cores, is supported by partial chemical analyses plotted as fig. 8, 9 and 10 of Perry et al. (1973). Because chemical homogeneity on a broad scale is a fundamental assumption that must be met for the relationships in Fig. 8C-2 to be valid, I have omitted data from cores 5 and 7 as part of the Biwabik Iron Formation Trend in that figure.

It is difficult to determine to what extent quartz and magnetite establish mutual isotopic equilibrium during low-temperature metamorphism. Grain size, fluids present, and ongoing reactions that produce or consume minerals

TABLE 8C-I

Oxygen isotope data and "temperatures" calculated from quartz magnetite fractionation factors, Biwabik Iron Formation, Minnesota

Sample	$\delta^{18}O_Q$	$\delta^{18}O_M$	Δ_{QM}	"T" (°C)
*Core 2**				
2-1714	18.0	−6.2	24.0	150
2-2140-2145	19.4	−5.9	25.1	140
2-2167	19.4	−7.5	26.7	120
Avg. core 2	18.9	−6.5	25.3	140
*Core 5**				
5-632	19.6	−2.0	21.4	190
5-755-760	19.8	−4.6	24.2	150
Avg. core 5	19.7	−3.3	22.8	170
*Core 7**				
7-924	20.4	−4.0	24.2	150
7-1175-1180	20.8	−3.0	23.6	160
7-1200-1205	21.0	−2.7	23.4	160
7-1294	20.5	−3.6	23.8	155
7-1302	19.4	−3.5	22.7	170
Avg. core 7	20.4	−3.4	23.5	160
Outcrop and core samples metamorphosed by the Duluth Complex				
32-64	13.6	6.4	7.1	600
5-64	12.1	2.2	9.8	460
7-64	16.3	0.6	15.6	287
2-64	17.0	−2.1	19.0	224
25-64	18.4	−3.8	22.0	180
MI2056	15.2	7.4	7.7	560
MI2022	12.9	4.7	8.2	540
MI2133	12.5	5.0	7.4	580
MI2067	12.6	4.5	8.0	540
MI2167	13.1	4.2	8.8	500
MI2232	14.0	5.5	8.4	520
37-64 (DH17700)*	12.5	2.5	9.9	450
38-64 (DH17700)*	14.1	2.5	11.5	390

* Location given in Perry et al., (1978) fig. 1.

all are important. Available data from the Biwabik Iron Formation suggest, but do not prove, that:

(1) Quartz and magnetite exchanged ^{18}O readily within the iron-formation at about 120°C, perhaps during recrystallization of fine-grained chert;

(2) At a slightly higher temperature, iron carbonate exchanged ^{18}O with magnetite, probably resulting in nonequilibrium quartz-magnetite oxygen isotope pairs in this temperature interval; since carbonate and magnetite appear to have exchanged while quartz remained inert, Δ_{QM} probably is anomalously high and inferred temperature is anomalously low.

(3) At 200°C or less, isotopic equilibrium was reestablished, and the system remained in approximate isotopic equilibrium to a temperature of over 400°C (in the contact metamorphic aureole).

Evidence for these interpretations comes primarily from a comparison of the oxygen isotope composition of carbonate minerals and quartz. In cores 5 and 7, $\delta^{18}O$ of carbonates is about 3‰ lower in magnetite horizons than in magnetite-free horizons, strongly suggesting exchange with magnetite. Quartz isotope composition is independent of magnetite content (Perry et al., 1973; figs. 9, 10, 12). In core 2 the ^{18}O depletion of carbonate is less, even though the magnetite content is greater (Perry et al., 1973, fig. 8). It is perhaps most reasonable to interpret these data to indicate establishment of quartz-magnetite-carbonate equilibrium at a temperature slightly below that registered in core 2. Subsequent mild regional metamorphism resulted in carbonate-magnetite exchange, particularly in cores 5 and 7, (Table 8C-I), but quartz, after crystallization was isolated from further low-temperature reaction. Contact metamorphism by the Duluth Complex later heated the underlying iron-formation, apparently under closed-system conditions. This produced a re-equilibration of oxygen isotopes in response to decreasing Δ_{QM} [$= f(T)$] at higher temperature. The result was a lowering of $\delta^{18}O$ in quartz and a corresponding increase in $\delta^{18}O$ of magnetite (Fig. 8C-2). The close fit of measured points to the trend lines at low temperature (high Δ_{QM}) suggests that isotopic equilibrium was re-established at 200°C or less. This interpretation, which varies somewhat from that given by Perry et al. (1973), warrants further study.

The Brockman Iron Formation of Western Australia has been deeply buried but subjected to little structural deformation (Trendall and Blockley, 1970). Smith et al. (1982) have indicated variable metamorphism of these rocks. For samples from the Dales Gorge Member in the Wittenoom area, temperature of metamorphism estimated from Δ_{QM} was between 270 and 310°C (Becker and Clayton, 1976). The quartz-magnetite oxygen isotope fractionation varies considerably over short distances (Becker and Clayton, 1976). Becker and Clayton suggest several possible explanations for this variability, favoring localized differences in porosity and permeability that have caused reaction to stop in adjacent subsystems at different stages of metamorphism. It is obvious that under these conditions differences in Δ_{QM} do not represent actual differences in the maximum temperature to which closely spaced units were exposed.

Following Becker and Clayton (1976), Perry et al. (1978) assumed Δ_{QM} values for the Brockman Iron Formation to represent the temperature at which particular subsystems ceased reacting with a large, homogeneous system. Hamersley data of Becker and Clayton (1976), plotted according to this assumption, are incorporated into Fig. 8C-2. $\delta^{18}O_M$ vs. Δ_{QM} intersects the $\delta^{18}O$ axis at about 19‰, probably indicating a higher content of carbonate and quartz (minerals with greater $\delta^{18}O$ contents) than is present in the Biwabik Iron Formation. Perry et al. (1978) attached particular significance to the fact that the Biwabik and Hamersley (Brockman) trend lines (Fig. 8C-2) intersect at B and B′ at Δ_{QM} values of 36.6 and 37.4; the average of these, Δ_{QM} = = 37, corresponds to a temperature of 36°C ± 25°C (Becker and Clayton, 1976). Furthermore, the $\delta^{18}O_Q$ vs. Δ_{QM} lines intersect each other at 23‰, within 1‰, of the 24‰ found by Becker and Clayton to be the most ^{18}O enriched quartz value in cherts associated with the Brockman Iron Formation and the 23.5‰ reported by Perry et al. (1973) as the most ^{18}O enriched chert they found in the Biwabik-Gunflint Iron Formation. Perry et al. (1978) concluded that it was likely that trend lines for both the Brockman and Biwabik Iron Formations represent essentially closed system conditions and that points B, B′ correspond to isotope ratios at the time of sedimentation or early diagenesis. If so, 36°C ± 25°C represents an upper limit to ocean temperature at the time of deposition of these iron-formations. The error quoted is that assigned by Becker and Clayton (1976) to the uncertainty in Δ_{QM} vs. temperature. The association of Proterozoic iron formations with cherts having $\delta^{18}O$ values up to 24‰ sets a further limit. The least-squares linear-regression equation (correlation coefficient = 0.94) for the Biwabik Iron Formation:

$$\Delta_{QM} = \frac{\delta_Q - 10.34}{0.344}$$

indicates a Δ_{QM} of 39.7 and a maximum temperature of deposition of 18°C for this iron formation if δ_Q is assigned a value of 24‰. This temperature estimate is subject to rather large uncertainties of: (1) about ± 25°C for temperature calibration; and (2) additional uncertainty resulting from extrapolation of δ-Δ plots through the temperature of dehydration of chert. Although this uncertainty limits the usefulness of these data in interpreting Precambrian surface temperature, the reasonableness of this estimate lends credence to the assumption that the Biwabik Iron Formation has acted as a closed system for oxygen isotopes.

Recent studies of the Ironwood Iron Formation of Wisconsin and Michigan (Maroney, 1978) and of the Kuruman Iron Formation, Transvaal, South Africa (Perry and Ahmad, 1980) substantiate this pattern. Riebeckite-bearing rocks from crocidolite mines and prospect cores around Kuruman are of greatest interest. These rocks have a low magnetite content and are of low metamorphic grade. Consequently, $\delta^{18}O$ of quartz is high (+24‰ maximum) and varies by only 1.5‰ for a suite of 9 samples. These rocks, then, approximate an end member assemblage of quartz, siderite, and iron silicate containing relatively little magnetite. They form a pair of lines that project through points B, B′ of Fig. 8C-2. Note that the age of the Kuruman Iron Formation is quite similar to that of the Brockman Iron Formation (Beukes, 1983, this volume).

If iron-formations were to exchange readily with surrounding rocks during moderate grades of metamorphism, it would be useless to attempt to interpret their primary characteristics. In this context, it is worth reviewing some of the evidence for limited exchange in the Dale's Gorge Member of the Brockman Iron Formation. Becker and Clayton (1976, p. 1160) noted that final equilibrium must have occurred with a water (pore fluid) of $\delta^{18}O \simeq 13‰$. They conclude from this that:

> "... the final equilibration in the iron formation is likely to have occurred in an essentially closed system at depth, with a very low water—rock ratio, rather than through the exchange of isotopes with a large external oxygen reservoir, such as might be provided by meteoric or ground waters.
>
> Such an equilibration would probably have been restricted to scales of the order of centimeters or perhaps meters ... The overall uniformity of these values ($\delta^{18}O$ of quartz and ankerite) can be explained by the large-scale chemical homogeneity of the original deposit, whereby each locally equilibrating system had more or less the same chemical and isotopic composition."

Hangari et al. (1980) describe an unmetamorphosed magnetite-bearing Paleozoic ironstone in which Δ_{SM} is about 30 (corresponding to an equilibrium siderite-magnetite isotope exchange temperature of about 60°C). This confirms an assumption made by Perry et al. (1978) that magnetite can form diagenetically at temperatures of 60°C or less (cf. Becker and Clayton, 1976).

Study of progressive metamorphism of the Biwabik, Brockman, Negaunee, and Kuruman Iron Formations provides a basis for comparison of other iron-formations with more complicated metamorphic histories. In the case of the Proterozoic Krivoy Rog Formation of the Ukranian SSR, abundant siderite is present in rocks metamorphosed to temperatures (based on quartz-magnetite oxygen isotope fractionation) of up to 475°C (Perry and Ahmad, 1978, 1981). Under these conditions partial breakdown of siderite can occur at system pressures up to several kilobars (French, 1971; Mel'nik, 1973; Perry and Ahmad, 1978, 1981). Siderite decomposition results in loss of CO_2 from the solid system. This can be traced isotopically since ^{18}O is preferentially concentrated in CO_2 and thus is removed from the system. The result observed for Krivoy Rog is shown schematically as a δ-Δ diagram in Fig. 8C-3.

Another interesting aspect of metasomatism seems to be unique, at least among reported occurrences, to iron formation of Krivoy Rog metamorphosed to very high metamorphic grade (granulite facies), in close proximity to marble. Under these conditions, $\delta^{18}O$ values as high as +25‰ are reported for magnetite (Belevtsev et al., 1983, this volume; Sherbak et al., 1981).

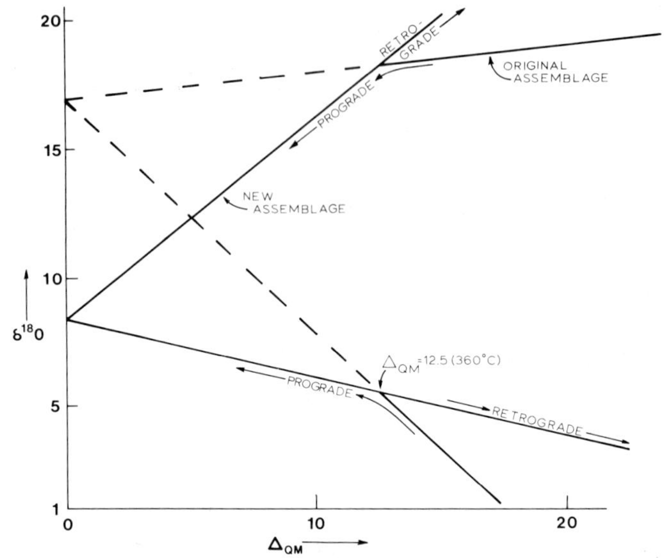

Fig. 8C-3. Schematic diagram showing the probable response of $\delta^{18}O_Q$ to increasing temperature of metamorphism in the Krivoy Rog Iron Formation, USSR. With partial decomposition of $FeCO_3$ to form grunerite at about 360°C the bulk oxygen isotope composition of the rock changes. Further temperature increase gives rise to a new trend line that does not pass through point B of Fig. 8C-2 (from Perry and Ahmad, 1981).

ARCHEAN IRON-FORMATION

Oxygen isotope data are available for few Archean iron-formations; and, of these, most are comprised of metamorphosed lenticular units that have had ample opportunity to exchange ^{18}O with country rocks. A notable exception is the 3800 m.y.-old iron-formation at Isukasia (Isua), west Greenland (Perry et al., 1978). Thick, nearly pure metamorphosed chert layers are associated with this iron-formation; Perry et al. (1978) have proposed that the maximum $\delta^{18}O$ of 20.4‰ determined for quartz in these metacherts approximates the primary $\delta^{18}O$ of SiO_2 deposited from Archean sea water at the time of sedimentation.

Isukasia rocks have undergone metamorphism to amphibolite facies and strong structural deformation (Allaart, 1976). Oxygen isotope fractionation between quartz and magnetite, Δ_{QM}, in these rocks is very uniform, averaging 11.6 (Perry et al., 1978). This corresponds to a temperature of about 390°C. The most cogent argument for preservation of primary $^{18}O/^{16}O$ ratios in the Isukasia rocks is that the metamorphosed chert units are very thick (up to 1 km) and that sharp, compositionally controlled gradients occur at the border of these units. Presumably, diffusion of ^{18}O during metamorphism took place over distances that are short compared to the thickness of nearly pure quartz zones. These rocks contain 60 mole percent magnetite oxygen or less. The average is probably 30% or less of magnetite oxygen in any stratigraphic unit. Thus, all magnetite has been considerably altered in isotopic composition. Perry et al. (1978) found (their fig. 2) that the ^{18}O composition of Isukasia iron-formation is an approximately linear function of magnetite content and that magnetite-rich rocks are depleted in ^{18}O as would be expected for closed-system metamorphism of rocks with a major low ^{18}O component such as magnetite. Under ideal closed-system conditions a plot of δ vs. % magnetite for such a system could be extrapolated to pure magnetite to estimate the diagenetic ^{18}O content of magnetite. For Isukasia, this leads to an estimate of the initial $\delta^{18}O_M$ of $-9‰$. Corresponding Δ_{QM} is 29 (using quartz with an initial $\delta^{18}O$ of 20.4‰); the temperature of diagenesis or low-grade metamorphism estimated from this Δ_{QM} is 100°C. Another, similar attempt to estimate $\delta^{18}O_M$ from Isukasia mine assay samples has lead to an essentially identical result (Ahmad and Perry, unpubl.). An uncertain extrapolation of this sort for a single occurrence isolated in time is almost impossible to interpret.

If the maximum $\delta^{18}O$ of +24‰ for quartz from Isukasia iron-formation and metachert is approximately correct, this value is about 4‰ lower than the values reported by Becker and Clayton (1976) and Perry et al. (1973) for Proterozoic iron-formation. It is about 14‰ lower in ^{18}O than values for modern chert (Knauth and Lowe, 1979).

DISCUSSION

It is possible that the oxygen isotopic composition of iron-formation records the temperature and oxygen isotopic composition of the Precambrian ocean. At present the interpretation of data on the primary oxygen isotope composition of chemical sediments (including iron-formation) is ambiguous. Knauth and Lowe (1979) favor a higher temperature of the Precambrian ocean to explain the observation that the most ^{18}O enriched Proterozoic and Archean cherts are depleted in ^{18}O by 10‰ and 14‰, respectively, with respect to modern deep-sea cherts, whereas Perry et al. (1978) consider a change in the isotopic composition of seawater to be the most important factor. In principle, this question could be resolved by determining the primary isotopic composition of magnetite or its precursor mineral in Precambrian iron-formation since this would permit calculation of Δ_{QM} and, hence, of temperature. Several practical problems restrict this approach at present. The first of these is the uncertainty in the assignment of temperatures to Δ_{QM} values. This uncertainty has been set at ± 25°C by Becker and Clayton (1976). Secondly, magnetite makes up less than 50% by weight of most rocks except in very localized bands. Furthermore, magnetite contains only about half as much oxygen per gram as quartz. Although it is possible to find essentially pure chert bands in which there has been little metamorphic alteration of $\delta^{18}O_Q$, similar bands of magnetite do not exist. Ultimately, the extrapolation of δ vs. Δ_{QM}, demonstrated for the Brockman and Biwabik Iron Formations may help to resolve the question of whether the Proterozoic ocean was hotter or depleted in ^{18}O compared to the modern ocean. For this to be so, the relationship must be demonstrated to hold precisely for other iron-formations and for Paleozoic ironstones; and quartz-magnetite fractionation must be determined with considerable accuracy.

ACKNOWLEDGEMENT

This work was supported by NSF EAR 75-20809 and by Northern Illinois University. I thank my colleague, Dr. Ahmad, for permission to cite work in progress on Transvaal iron-formation.

REFERENCES

Allaart, J.H., 1976. The pre-3760 m.y. old supracrustal rocks of the Isua area, central West Greenland, and associated occurrence of quartz-banded ironstone, In: B.F. Windley (Editor), The Early History of the Earth. John Wiley and Sons, New York, N.Y., pp. 177—189.

Becker, R.H. and Clayton, R.N., 1976. Oxygen isotope study of Precambrian banded iron formation, Hamersley Range, Western Australia. Geochim. Cosmochim. Acta, 40: 1153—1165.

Belevtsev, Ya.N., Belevtsev, R.Ya. and Siroshtan, R.I., 1983. The Krivoy Rog basin. In: A.F. Trendall and R.C. Morris (Editors), Iron-Formation: Facts and Problems. Elsevier, Amsterdam, pp. 211—251.

Beukes, N.J., 1983. Paleoenvironmental setting of iron-formations in the depositional basin of the Transvaal Supergroup, South Africa. In: A.F. Trendall and R.C. Morris (Editors), Iron-Formation: Facts and Problems. Elsevier, Amsterdam, pp. 131—209.

Bottinga, Y. and Javoy, M., 1973. Comments in oxygen isotope geothermometry. Earth Planet. Sci. Lett., 20: 250—265.

Clayton, R.M., Goldsmith, J.R., Karel, K.J., Mayeda, T.K. and Newton, R.C., 1975. Limits on the effect of pressure on isotopic fractionation. Geochim. Cosmochim. Acta, 39: 1197—1201.

Compston, W., Williams, I.S., McCulloch, M.T., Foster, J.J., Arriens, P.A. and Trendall, A.F., 1981. A revised age for the Hamersley Group. Geol. Soc. Aust., Abstr. 3: 40.

Craig, H., 1961. Standards for reporting concentrations of deuterium and oxygen-18 in natural waters. Science, 133: 1833—1834.

French, B.M., 1971. Stability relations of siderite ($FeCO_3$) in the system Fe-C-O. Am. J. Sci., 271: 37—78.

Gonfiantini, R., 1978. Standards for stable isotope measurements in natural compounds. Nature, 271: 534—536.

Hangari, K., Ahmad, S.D and Perry, E.C., Jr., 1980. Carbon and oxygen isotope ratios in diagenetic siderite and magnetite from Upper Devonian Ironstone, Wadi Shatti District, Libya. Econ. Geol., 75: 538—545.

James, H.L. and Clayton, R.N., 1962. Oxygen isotope fractionation in metamorphosed iron formations of the Lake Superior region and in other iron-rich rocks, In: H.L. James, A.E.G. Engel and B.F. Leonard (Editors), Petrologic Studies, volume to honor A.F. Buddington. Geol. Soc. Am., Denver, pp. 217—239.

Knauth, L.P. and Epstein, S., 1976. Hydrogen and oxygen isotope ratios in nodular and bedded cherts. Geochim. Cosmochim. Acta, 40: 1095—1108.

Knauth, L.P. and Lowe, D.R., 1979. Oxygen isotope geochemistry of cherts from the Onverwacht group (3.4 billion years), Transvaal, South Africa, with implications for secular variations in the isotopic composition of cherts. Earth Planet. Sci. Lett., 41: 209—222.

Maroney, M., 1978. Oxygen Isotope Study of the Ironwood Iron Formation, Gogebic Range, Wisconsin and Michigan. M.S. Thesis, Northern Illinois University, DeKalb, Ill.

Mel'nik, Y.P., 1973. Fiziko-khimicheskie uslovia obrazovania dokembriiskikh zhelezistykh kvartsitov. Kiev, USSR.

O'Neil, J.R., Clayton, R.N. and Mayeda, T.K., 1969. Oxygen isotope fractionation in divalent metal carbonates. J. Chem. Phys., 51: 5547—5558.

Perry, E.C., Jr. and Ahmad, S.N., 1978. Oxygen isotope determinations of quartz and magnetite from Krivoy Rog, USSR. Presented at VIIth National Isotope Symposium, Moscow, 23—27 October, 1978.

Perry, E.C., Jr. and Ahmad, S.N., 1980. Oxygen isotope study of Transvaal System iron formation from the vicinity of Kuruman, Cape Province, South Africa. Geol. Soc. Am., Abstr. Programs, 12: 497.

Perry, E.C., Jr. and Ahmad, S.N., 1981. Oxygen and carbon isotope geochemistry of the Krivoy Rog iron formation, Ukrainian SSR. Lithos, 14: 83—92.

Perry, E.C., Jr., Ahmad, S.N. and Swulius, T.M., 1978. The oxygen isotope composition of 3,800 m.y. old metamorphosed chert and iron formation from Isukasia, West Greenland. J. Geol., 86: 223—239.

Perry, E.C., Jr., Tan, F.C. and Morey, G.B., 1973. Geology and stable isotope geochemistry of the Biwabik Iron Formation, northern Minnesota. Econ. Geol., 68: 1110—1125.

Sherbak, N.P., Bartnitski, E.N. and Lugovaya, I.P., 1981. Isotope Geology of the Ukraine. Naukova, Kiev, 246 pp. (in Russian).

Smith, R.E., Perdrix, J.L. and Parks, T.C., 1982. Burial metamorphism in the Hamersley Basin, Western Australia. J. Petrol., 23: 75—102.

Trendall, A.F. and Blockley, J.G., 1970. The iron formations of the Precambrian Hamersley Group, Western Australia, with special reference to the associated crocidolite. West. Aust., Geol. Surv., Bull., 119, 366 pp.

Chapter 9

THE PALAEONTOLOGY AND PALAEOECOLOGY OF PRECAMBRIAN IRON-FORMATIONS

M.R. WALTER and H.J. HOFMANN

INTRODUCTION

All Proterozoic and Archaean iron-formations, except those rare examples from the latest Proterozoic, date from before the origin of any megascopic forms of life (which may be one reason why some are finely laminated, as there were no animals to disturb the sediment). A knowledge of the palaeontology of iron-formations requires a search for evidence of microbial activity. Such evidence takes many forms, and its interpretation must be interdisciplinary. It is no longer sufficient to confine discussions to morphological fossils alone. Microfossils, and sedimentary structures produced by benthic mats of microbes (bacteria, cyanobacteria, and algal protists), i.e. stromatolites, are major sources of information. But in addition, there are the degraded chemical remnants of cells, i.e. kerogen, from which can be derived information on the chemical composition of the original cells, and on the microbial degradational processes. And there are the carbon and sulphur isotopic compositions of minerals that were affected or produced by microbial activity. For instance, autotrophy, the utilization of carbon dioxide as the source of carbon for cell construction, fractionates carbon isotopes, concentrating ^{12}C in the cell material and leaving ^{13}C to accumulate in the environment (and be sampled by precipitating carbonate minerals at or below the sediment—water interface). So the carbon isotopic composition of kerogen and carbonates is a source of palaeobiological information. In addition, the kerogen can be heated to release organic compounds that some workers relate to original microbial populations — this approach is unpopular because of the problems caused by contamination, both natural and in sample handling. The following discussion ranges across all these approaches but it is apparent that very little effort has yet been spent on studying the palaeontology of iron-formations. There is wide scope for further work and a considerable chance such research would solve some of the more significant problems of iron-formation deposition.

Perhaps the earliest report of possible fossils in Precambrian iron-formations is that by W.S. Gresley (1896, 1897), who described megascopic remains from iron ore slabs at the docks at Erie, Pennsylvania, presumably derived from the Vulcan iron-formation at the Chapin Mine near Iron Mountain,

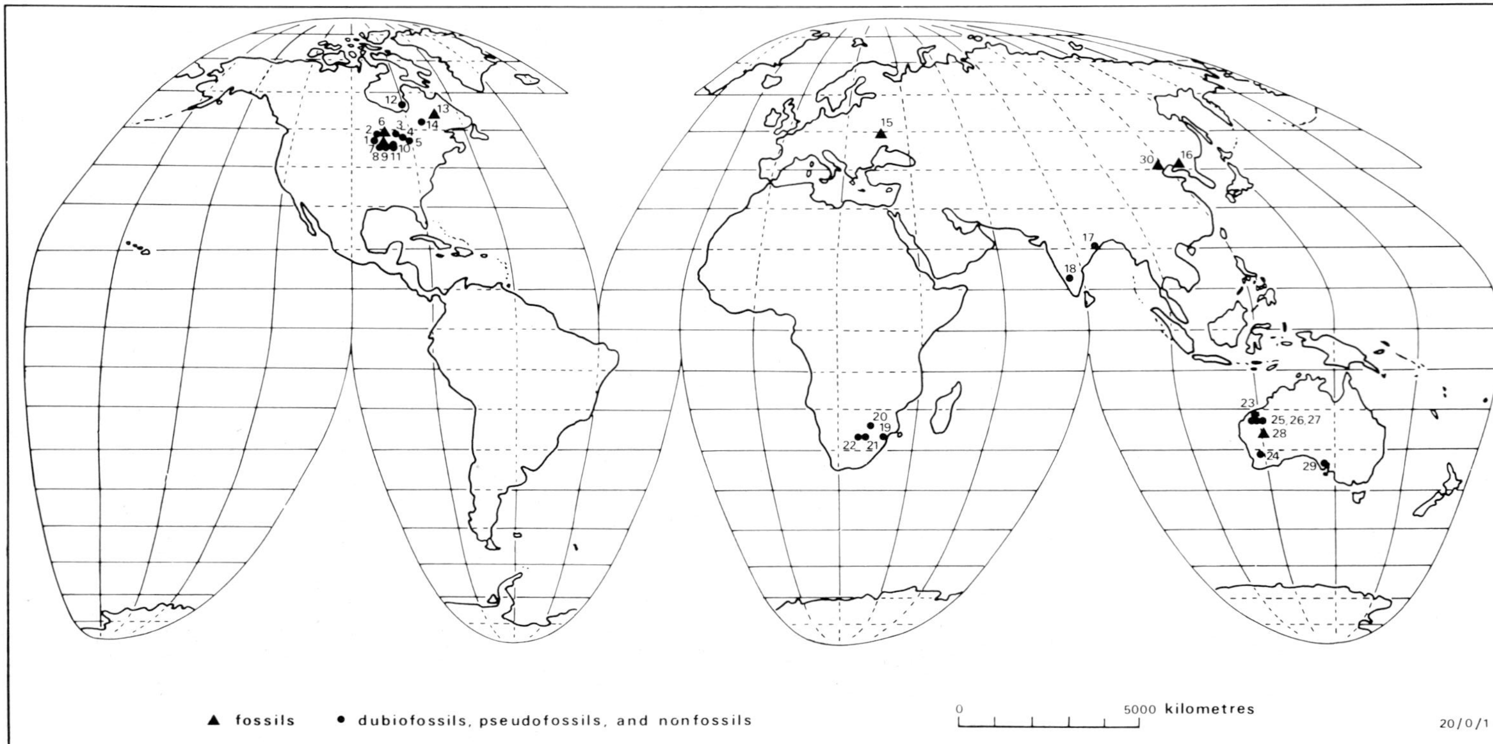

Fig. 9-1. Map showing reported fossil occurrences associated with Precambrian iron-formations.

TABLE 9-I

Reported fossil occurrences associated with Precambrian iron-formations

Type of remains / Occurrence		Filaments					Selected references
	Coccoids	Septate, unbranched	Tubular, unbranched	Branched	Bizarre forms	Dubiofossils, nonfossil	Stromatolites	
	A	B	C	D	E	F	G	
1 Soudan						F		Gruner, 1923; Cloud et al., 1965; Cloud, 1973; Cloud and Licari, 1968; La Berge, 1973
2 Kaministikwia						F		LaBerge, 1973
3 Michipicoten						F		LaBerge, 1973; Goodwin et al., 1976 Karkhanis et al., 1980
4 Woman River						F		LaBerge, 1973
5 Timagami						F		LaBerge, 1973
6 Gunflint-Biwabik	A	B	C	D	E	F	G	Leith, 1903; Grout and Broderick, 1919; Gruner, 1922, 1924; Tyler and Barghoorn, 1954; Moorhouse and Beales, 1962; Barghoorn and Tyler, 1965; Cloud, 1965; Hofmann, 1969, 1971; Walter and Awramik, 1979
7 Tyler	A	B	C		E	F	G	LaBerge, 1967; Cloud and Morrison, 1979, 1980; Schmidt, 1980
8 Baraga (Michigamme)						F		Cloud and Morrison, 1979, 1980
9 Negaunee						F		Mancuso et al., 1971; Lougheed and Mancuso, 1973
10 Riverton						F		LaBerge, 1973
11 Vulcan						F		Gresley, 1897
12 Kipalu						F		Moore, 1918; LaBerge, 1967, 1973; Hofmann, 1971
13 Sokoman	A	B	C		E	F	G	LaBerge, 1967, 1973; Zajac, 1974; Knoll and Simonson, 1981
14 Temiscamie						F		LaBerge, 1967, 1973
15 Krivoi Rog	A					F		Piatnitski, 1924; Timofeev, 1969, 1973, 1979; Mel'nik, 1982
16 Anshan	A					F		Ouyang, 1979; Xu and Chu, 1979; Yin, 1979
17 Singhbum						F		Spencer and Percival, 1952; LaBerge, 1973
18 Bababudan						F		Viswanathiah and Venkatachalapathy, 1980
19 Bomvu Ridge						F		LaBerge, 1973
20 Penge						F		LaBerge, 1973
21 Kuruman						F		LaBerge, 1973; Klemm, 1979
22 Pretoria						F		Cloud and Licari, 1968
23 Gorge Creek						F		LaBerge, 1967, 1973; Hamilton, 1976; Muir, 1978
24 Southern Cross						F		Marshall, 1964; Cloud, 1976; Muir, 1978
25 Marra Mamba						F		LaBerge, 1967, 1973
26 Brockman						F		Edgell, 1964; LaBerge, 1967; Muir and Plumb, 1976
27 Whaleback						F		Karkhanis, 1976
28 Frere	A	B	C		E		G	Hall and Goode, 1978; Walter 1975a,b; Walter et al., 1976
29 Wilgena Hill						F		LaBerge, 1973
30 Chanlingou							G	Zhang, 1981

in the Menominee Range of Michigan. Although some of these were said to have been accepted as biogenic by such authorities as C.D. Walcott, and Charles Schuchert (Gresley, 1897, pp. 532—537), no modern study has confirmed their biologic nature. Furthermore, Walcott (1899, p. 230) suggested that these structures were derived from ferruginous Cambrian sandstone. Other early reports of stromatolites and of microfossils are those of Leith (1903), Cayeux (1911), Moore (1918), Grout and Broderick (1919), and Gruner (1922, 1923, 1924, 1925). In this connection we may also cite here the pioneering chemical and biochemical work of Harder (1919) on the geologic importance of modern iron-depositing bacteria.

This early period was succeeded by a lull in activity in Precambrian palaeontology in general. Interest was rekindled in 1953 with S.A. Tyler's discovery of the Gunflint microbiota on the north shore of Lake Superior (Tyler and Barghoorn, 1954). Since then a great many reports from many parts of the world have been published on various types on microfossils, stromatolites, and dubiofossils in iron-formations, or in rocks associated with them. Reviews of microbiotas associated with Precambrian iron-formations are those of LaBerge (1967, 1973), Cloud and Licari (1968), and Cloud (1973). For a detailed review and discussion of Archaean microbiotas see Schopf and Walter (1983), and for Early Proterozoic microbiotas see Hofmann and Schopf (1983).

Of the literally hundreds of Precambrian iron-formations known around the globe, about thirty are reported as containing morphological fossils (Fig. 9-1; Table 9-I); however, only about half a dozen of these have remains that can be unequivocally called biogenic. The following discussion concentrates on the evidence provided by way of stromatolites and of structurally preserved microorganisms in petrographic thin sections of chert from these few but highly significant occurrences, namely the Gunflint-Biwabik, Tyler, Sokoman, and Frere Iron Formations; the three other occurrences (Krivoy Rog, Chuanlingou and Anshan) are not treated because this material is poorly known (chiefly as HF acid macerates) and was not available to us for study. The Gunflint fossils are the best known, exhibit the greatest morphologic diversity, and are the best preserved of the iron-formation biotas.

Numerous objects now interpreted as pseudofossils have been described from iron-formations. Two groups are especially significant and therefore are briefly discussed here. Large numbers of spheroidal structures 5—40 μm wide (with a modal value near 30 μm) have been illustrated from Archaean and Proterozoic iron-formations by LaBerge (1967, 1973) and from the Early Proterozoic Transvaal iron-formations by Klemm (1979). These were interpreted as microfossils, but in no example has their biogenicity been demonstrated. This is part of a general problem associated particularly with Archaean microspheroids, but in addition only some of the iron-formation spheroids are carbonaceous, and even these could have formed as spherulites in gelatinous sediment (as described by Oehler, 1976). Furthermore, some of them have morphological features unlike those both of known Precambrian microfossils, and of extant bacteria and protists (particularly the linkage to form arrays of carbonaceous spheroids in the example of those from the Hamersley Basin, first described by La Berge, 1967, 1973). All of these structures, Archaean and Proterozoic, are thus regarded at best as dubiofossils, and are likely to be pseudofossils (see discussions in Hofmann, 1971; Hofmann and Schopf, 1983; Schopf and Walter, 1983; and references cited therein). Other microscopic pseudofossils reported from iron-formations are listed in table 9-1 of Schopf and Walter (1983) and table 14-2 of Hofmann and Schopf (1983).

"Stromatolites are organosedimentary structures produced by sediment trapping, binding and/or precipitation as a result of the growth and metabolic activity of micro-organisms, principally [cyanobacteria]" (Walter, 1976a; note that the term cyanobacteria has now largely replaced "cyanophytes" and "blue-green algae", which are synonymous). Some chert mesobands in the iron-formations of the Hamersley and Transvaal Basins have a wrinkled lamination resembling that of stromatolites. This type of lamination is particularly well developed in certain discontinuous chert mesobands, or pods, which have been formally described as stromatolites (Edgell, 1964). However, these structures are distinguishable from stromatolites (Walter, 1972a); they are diagenetically transformed, originally flat, uniform nonstromatolitic sediments (Trendall and Blockley, 1970). Virtually all laminae within the "stromatolitic" cherts of these iron-formation mesobands can be traced laterally into chert-poor areas of the mesobands that are not suggested to be stromatolitic. No such relationship occurs between true stromatolites and intervening non-stromatolitic sediments. To date this has been demonstrated only in the Hamersley iron-formations but the Transvaal examples mentioned by Beukes (1973) are likely to be the same. Abiogenic, stromatolite-like structures from the Gunflint Iron Formation are discussed later in this paper.

PALAEONTOLOGY

Archaean

Neither stromatolites nor convincing microfossils are known from Archaean iron-formations, although both are known from Archaean sediments of other kinds. Like those of younger age, Archaean iron-formations contain organic matter in the form of kerogen. In the oldest iron-formation, that of the Isua supracrustals in Greenland, most of the non-carbonate carbon is present as graphite. Some authorities consider that this may once have been biogenic organic matter, but because of repeated metamorphism of these rocks such an origin is not at present demonstrable (Oehler et al., 1982). Black shales associated with the 3.3–3.5 Ga-old iron-formations of the Gorge Creek Group at Mt. Goldsworthy in the Pilbara Block of Western Australia contain abundant kerogen (and pyrite, to the extent that the shales spontaneously combust when exposed in the open cut iron ore mine). These, and those of other Archaean iron-formations (Table 9-II), have carbon isotopic compositions interpreted as indicative of autotrophy. Though only a few iron-formations have been examined thus far, in all there is evidence of autotrophic microbial activity in the basins of deposition of the iron-formations. The kerogen of the Mozaan iron-formations has unusually little ^{12}C; the cause of this is unknown, but could result from deposition in an anaerobic environment, from which ^{12}C was being lost in the form of methane (Hayes, 1983).

The sulphur-isotope record in Archaean iron-formations has been interpreted as showing one of the major evolutionary transitions in biological history: the advent of dissimilatory sulphate reduction (e.g., Goodwin et al., 1976). This is an energy-yielding process which operates only in anaerobic environments. The Isua iron-formations lack any evidence of biogenic sulphur isotope fractionaction, but those of late Archaean age (e.g. Woman River and Michipicoten Iron Formations) have a moderate spread of $\delta^{34}S$ values in pyrite, somewhat resembling younger, definitively biogenic patterns. While some authorities accept this as "the oldest presumptive evidence" of bacterial sulphate-reduction (Schidlowski et al., 1983, and references therein), others appeal to abiogenic mechanisms (Cameron, 1982) to explain the isotope patterns. Skyring and Donnelly (1982) have reviewed the evidence in detail and suggested that the late Archaean patterns might be indicative of a combination of

TABLE 9-II

Carbon isotopic compositions (relative to PDB) of kerogen and carbonate from some Precambrian iron-formations

	$\delta^{13}C_{ORG}$	$\delta^{13}C_{CARB}$
Frere Formation	−23.3 (9)	−0.66 (7)°
Gunflint Iron Formation	−31.3 (2)	−3⁺
Biwabik Iron Formation	−33.1*	−2 to −19*
Asbestos Hills Banded Iron Formation	−24.6 (1)	—
Krivoi Rog iron-formation	−21.9 (1)	—
Brockman Iron Formation	−27.7 (11)	−8.94 (2)
Mount McRae Shale	−34.0 (7)	—
Mount Sylvia Formation	−18.6 (1)	−12.00 (1)
Marra Mamba Iron Formation	−38.0 (4)	−9 (1)
Mozaan Group iron-formation	−14.4 (2)	−9.6*
Isua supracrustals iron-formation	−17.2 (5)	−2.8 (8)△

All kerogen analyses except those marked * are from Hayes et al. (1983); the number of analyses averaged is given in parentheses. All carbonate analyses except those marked are from M. Baur, J.M. Hayes, S.A. Studley and M.R. Walter (unpublished); results of additional analyses are shown in Fig. 7. *Perry et al. (1973) ⁺Shegelski (1982) °Schidlowski et al. (1983) △Perry and Ahmad (1977) and Schidlowski et al. (1979).

magmatic sulphide and sulphide derived from sulphite reduction (either hydrothermal or biological). There is much scope for further work, and in particular more effort needs to be made to locate remnant sulphates that can be analysed, and to place the data in a palaeoenvironmental context so that they can be interpreted in more detail.

Early Proterozoic.

Microfossils. Structurally preserved remains of organisms from banded iron-formations are found in stromatolitic microbial mats as well as in nonstromatolitic beds. The organisms are all microscopic, generally ranging between 1 and 10 μm in cell diameter, and the most common types are spheroidal and filamentous.

Their taxonomic affinities are difficult to ascertain, given the lack of biochemical information necessary for evaluating their phylogenetic position. Most are candidates for assignment to the Bacteria, Cyanobacteria, and Archaebacteria, and some could fit in any one of these three, based on morphology and size. Others, particularly the larger spheroids up to 30 μm across, have been assigned by some to eucaryotic algal groups. In the absence of definitive morphologies, and without a knowledge of their metabolisms, they can only be classified on the basis of shape and size. Following Hofmann and Schopf (1983), we recognize the following morphologic categories (Table 9-I, Fig. 9-2): coccoids, unbranched septate filaments, unbranched tubular structures, branched tubular structures, and bizarre forms (fossils exhibiting a type of organization that is of uncommon or unknown occurrence among modern microbes). A sixth category comprises dubio, pseudo-, and non-fossil microscopic remains.

In terms of numbers, the *coccoids* constitute the predominant element of iron-formation microbiotas. Their morphology ranges from simple, generally solitary spheroids 2—15 μm across (*Huroniospora*), to clustered spheroids (*Corymbococcus*), to spheroids with internal objects of irregular form and variable position and problematic significance (*Leptoteichos*), to rare, large, ensheathed spheres (*Megalytrum*). The spheres are generally unornamented except for reticulate surface textures that are attributed to degradation and diagenesis. Of somewhat rare occurrence are spheroids with bud-like projections (Fig. 9-2J, K) suggesting that budding may have been one way of reproduction for coccoids. Even more rarely, coccoids exhibiting pore-like apertures have been illustrated (Licari and Cloud, 1968, figs. 19, 20), but because of their rarity one cannot be confident that they served a biological function.

Unbranched, septate filaments represent the second most abundant component of iron-formation microbiotas (*Gunflintia*, and its synonyms). These are narrow filaments 1—5 μm wide, composed of equant cell remains in uniseriate rows, most commonly occurring in variably degraded conditions in

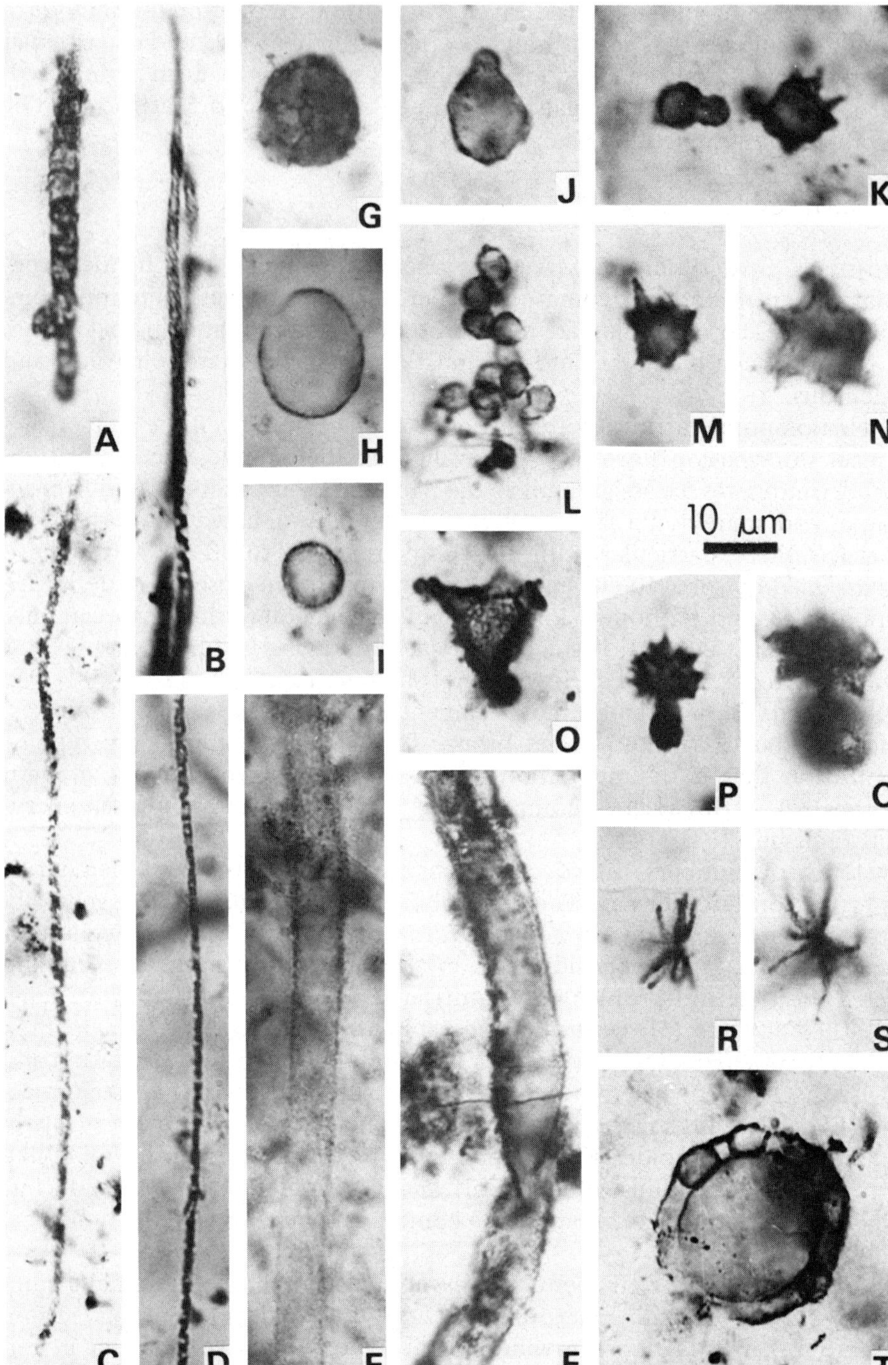
A
B
C
D
E
F
G
H
I
J
K
L
M
N
O
P
Q
R
S
T
10 µm

which cell remains are difficult to recognize (Fig. 9-2A-D). The same filament may exhibit different degradational morphologies (Fig. 9-2B), and some intercalary cells may appear enlarged with respect to others, leading some authors to interpret them as heterocysts (with nitrogen-fixing function), a characteristic of nostocalean cyanobacteria; others interpret them as degradational artifacts.

Unbranched tubular filaments 6—12 μm across are comparatively rare (Fig. 9-2E, F), and include the two genera *Animikiea* and *Entosphaeroides*. Specimens of a third genus, *Siphonophycus*, consisting of 20 μm wide tubes, are exceedingly rare. The tubular elements are generally thought to be evacuated cylindrical sheaths of filamentous cyanobacteria.

Branched tubular filaments up to 10 μm across (*Archaeorestis*), are extremely rare, and only known from the Gunflint Iron Formation, and possibly, the Frere Formation.

The last category, of bizarre forms, includes poorly understood but highly distinctive microfossils characteristically associated with Precambrian banded iron-formations (Figs. 9-2M—T). It consists of *Eoastrion*, which is hardly distinguishable morphologically from the modern iron- and manganese-oxidizing bacterium *Metallogenium;* the acanthomorph acritarchs *Eomicrhystridium*, *Exochobrachium*, *Galoxiopsis*, and *Veryhachium?; Eosphaera*, a large spheroidal microfossil that has been variously interpreted as a green alga (Kazmierczak, 1976, 1979), a red alga (Tappan, 1976, 1980) and a problematic procaryote (Hofmann and Schopf, 1983); *Kakabekia*, a questionable budding bacterium (Awramik and Barghoorn, 1977); and *Thymos* and *Xenothrix*, two genera represented by only one or two specimens each.

Filamentous forms reported from the Whaleback Shale Member and the Michipicoten Group (Archaean) are organic structures that take on a potas-

Fig. 9-2. Microfossils from the Gunflint Formation, north shore of Lake Superior, Ontario. All specimens, except C, are organically preserved; C is replaced by hematite. Photographs with Geological Survey of Canada numbers are the same as those illustrated in Hofmann (1971, p. 121). A—D, unbranched septate filaments. E—F, unbranched tubular filaments, probably sheaths. G—L coccoids. M—T, bizarre forms.

A. *Gunflintia grandis* (GSC photo 140938). B. *Gunflintia minuta*, specimens showing different stages of degradation along length of filament. C. *Gunflintia minuta*, badly degraded specimen preserved by hematite, (type specimen of *Palaeospirulina minuta* Edhorn 1973). D. *Gunflintia minuta*, degraded specimen (GSC photo 140939). E. *Animikiea septata* (GSC photo 140936). F. *Animikiea septata* (GSC photo 140935). G. *Huroniospora macroreticulata* (GSC photo 140942). H. *Huroniospora psilata* (GSC photo 140945). I. *Huroniospora microreticulata* (GSC photo 140943). J. *Huroniospora* sp. showing budding. K. *Huroniospora* sp. showing budding and *Eomicrhystridium barghoorni*, L. *Huroniospora* sp. in a cluster. M. *Eomicrhystridium barghoorni*. N. *Eomicrhystridium barghoorni* (GSC photo 201116-D). O. *Exochobrachium triangulum*. P. *Kakabekia umbellata*. Q. *Kakabekia umbellata* (GSC photo 140941). R. *Eoastrion simplex* (cf. *Metallogenium personatum*) (GSC photo 140940). S. *Eoastrion simplex* (cf. *Metallogenium personatum*) (GSC photo 140946). T. *Eosphaera tyleri*.

sium ferricyanide stain (Karkhanis, 1976; Karkhanis et al., 1980). Reexamination of the Michipicoten structures suggest that these are best regarded as modern contaminants (non-fossil), as they do not penetrate the thin section and contain stainable organic matter.

Stromatolites. These have long been known from the Gunflint and Biwabik Iron Formations (Leith, 1903; Grout and Broderick, 1919; and subsequent authors listed in Table 9-I). In the last decade they have also been reported from a small number of other iron-formations (Table 9-I). The Gunflint and Frere occurrences are good examples, especially as very little has been publish-

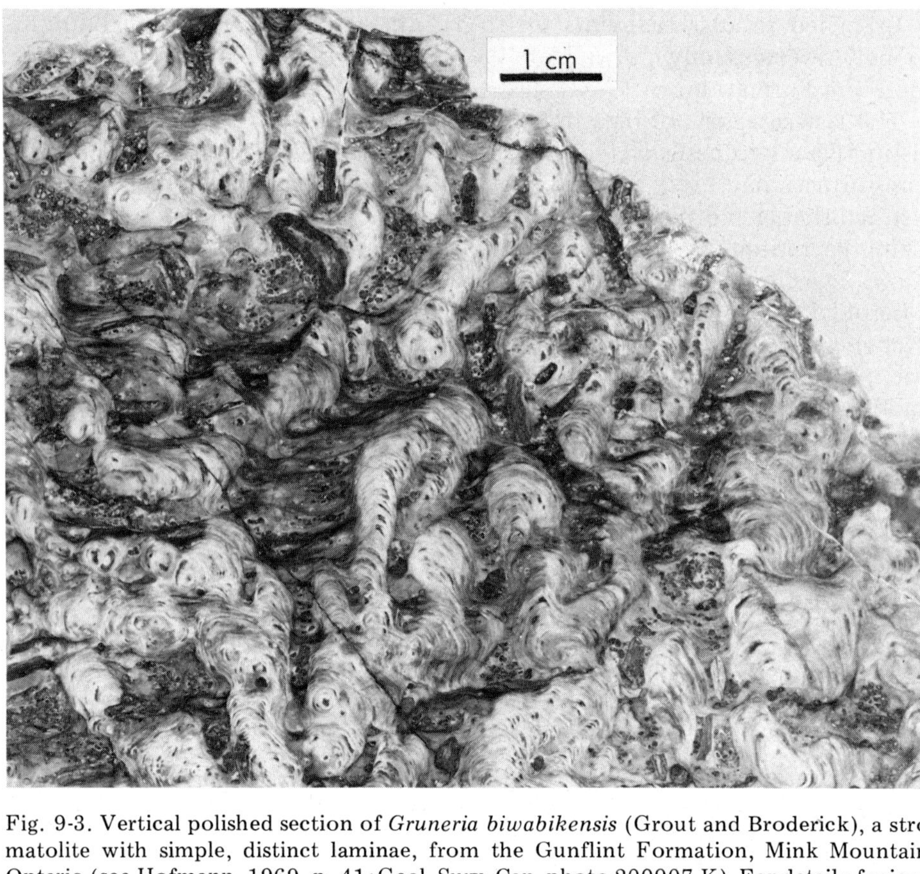

Fig. 9-3. Vertical polished section of *Gruneria biwabikensis* (Grout and Broderick), a stromatolite with simple, distinct laminae, from the Gunflint Formation, Mink Mountain, Ontario (see Hofmann, 1969, p. 41; Geol. Surv. Can. photo 200907-K). For detail of microstructure see Fig. 9-5A.

Fig. 9-4. Longitudinal section (A) and transverse section (B) of *Gruneria ferrata* (Grout and Broderick), a stromatolite with simple, distinct laminae, from the Biwabik Iron Formation, roadside dump of Mary Ellen Mine, Biwabik, Minnesota (see Hofmann, 1969, p. 37; Geol. Surv. Can. photos 200744-C and 200744-F).

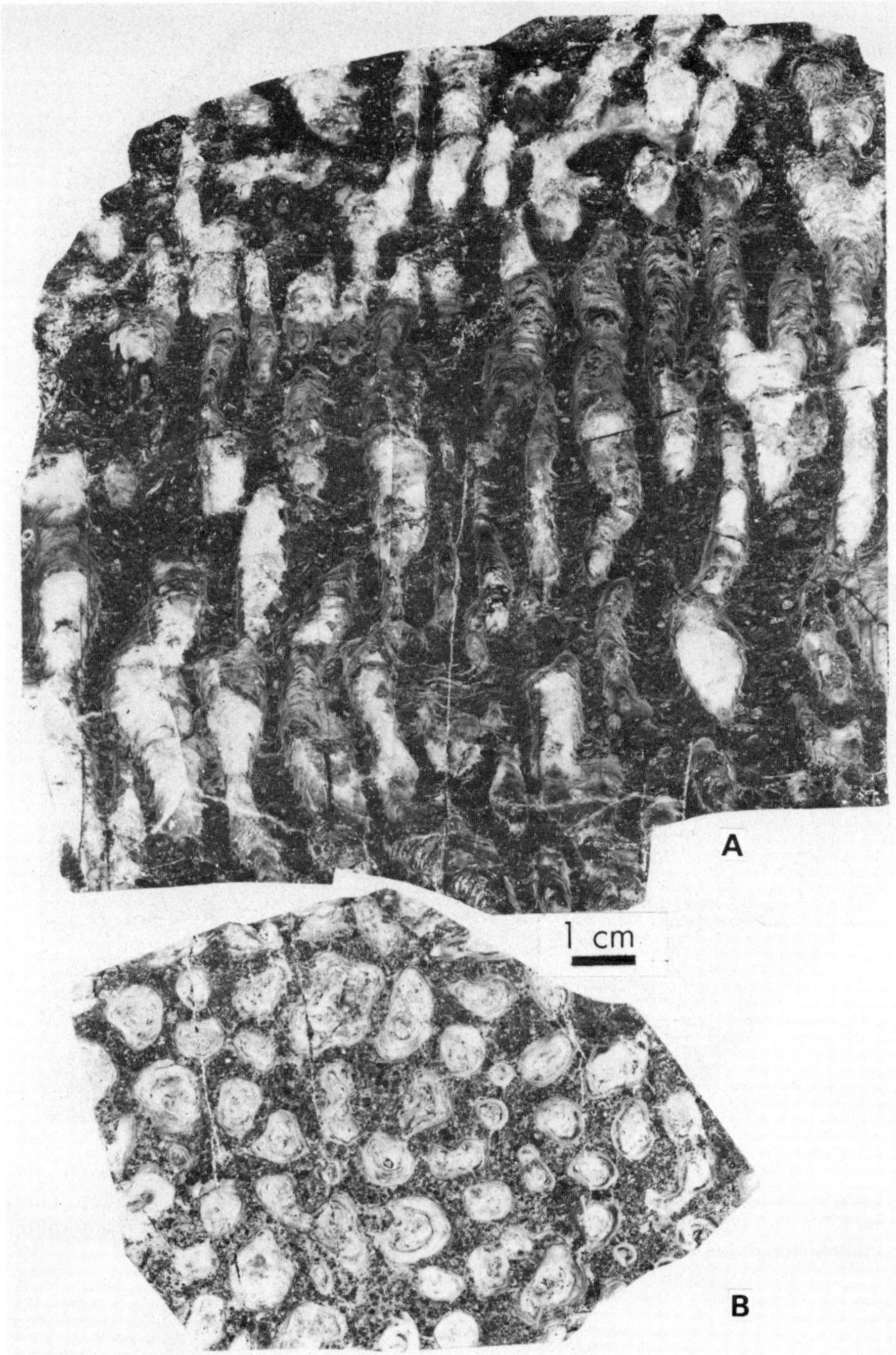
A
1 cm
B

Fig. 9-5. Longitudinal section of columnar stromatolite from the basal part of the Gunflint Formation, Winston Point, Ontario (see Hofmann, 1969, p. 59; Geol. Surv. Can. photo 200907-C). The laminae are simple, diffuse, and composed of permineralized amorphous organic matter and microfossils, such as illustrated in Fig. 9-5B.

ed on the others. The assemblage known from the Frere Formation to date is dominated by stratiform and oncolitic (spheroidal) forms, with none of the columnar forms which are so prominent in the iron-formations of the Animikie Basin. Both the Frere and the Gunflint examples (and those of the Sokoman and Biwabik Iron Formations) are richly microfossiliferous.

Three intergrading kinds of stromatolites can be distinguished by their microstructures: simple-diffuse (or streaky); pillared; and simple-distinct (or banded) (Hofmann, 1969). These microstructures range across a variety of morphologies which have yet to be comprehensively described (Figs. 9-3 to 9-5); those of the Gunflint Iron Formation are the best known (Hofmann, 1969). Microfossils occur only in those with streaky and pillared microstructures, and rarely occur in the latter. Although in many examples it is difficult or impossible to demonstrate that the contained microfossils represent the microbiota that built the stromatolites, rather than entrained plankton or organic detritus (Hofmann, 1969), detailed studies of some of the stromatolites have effectively shown that the bulk of the microfossils are indeed of the constructing organisms (Awramik, 1976, 1977; Walter et al., 1976; Awramik and Semikhatov, 1979; Knoll and Simonson, 1981).

The Gunflint and Biwabik stromatolites with an extremely fine, micrometer thick, banded microstructure, lack microfossils (Hofmann, 1969). In microstructure, morphology, and composition these stromatolites (Figs. 9-3, 9-6A) are indistinguishable from geyserite (Walter, 1972b, 1976b). The same microstructure occurs in Gunflint and Biwabik pisolites; Hofmann (1969) suggested that as these have a somewhat irregular lamination they may be true biogenic oncolites. However, the same degree of irregularity occurs in demonstrably abiogenic pisolites in geysers (Walter, 1976b), and we here accept that the Gunflint pisolites are probably abiogenic. All these evaporative splash deposits have been termed "stiriolites" (Walter, 1976b). Their most distinctive and useful distinguishing feature is their finely banded microstructure (in structures resembling stromatolites in shape).

The carbon in the kerogen in Early Proterozoic iron-formations is isotopically light (Table 9-II) and similar in composition to that in other sediments of that age (Hayes et al., 1983), and for that reason it is considered to have been derived from autotrophic organisms. On the other hand, the carbon in carbonate minerals in the same iron-formations is also light (Table 9-II) and quite different from that in Early Proterozoic dolomites and limestones (Schidlowski et al., 1983). There are two competing interpretations of these observations. Perry et al. (1973) have noted that on a mesoband to macroband scale, those beds that have carbonates with light carbon and oxygen also have the most magnetite. This and other observations lead them to conclude that the magnetite formed by a metamorphic reaction between kerogen and hematite:

$$6\,Fe_2O_3 + C \rightarrow 4Fe_3O_4 + CO_2 \quad (1)$$

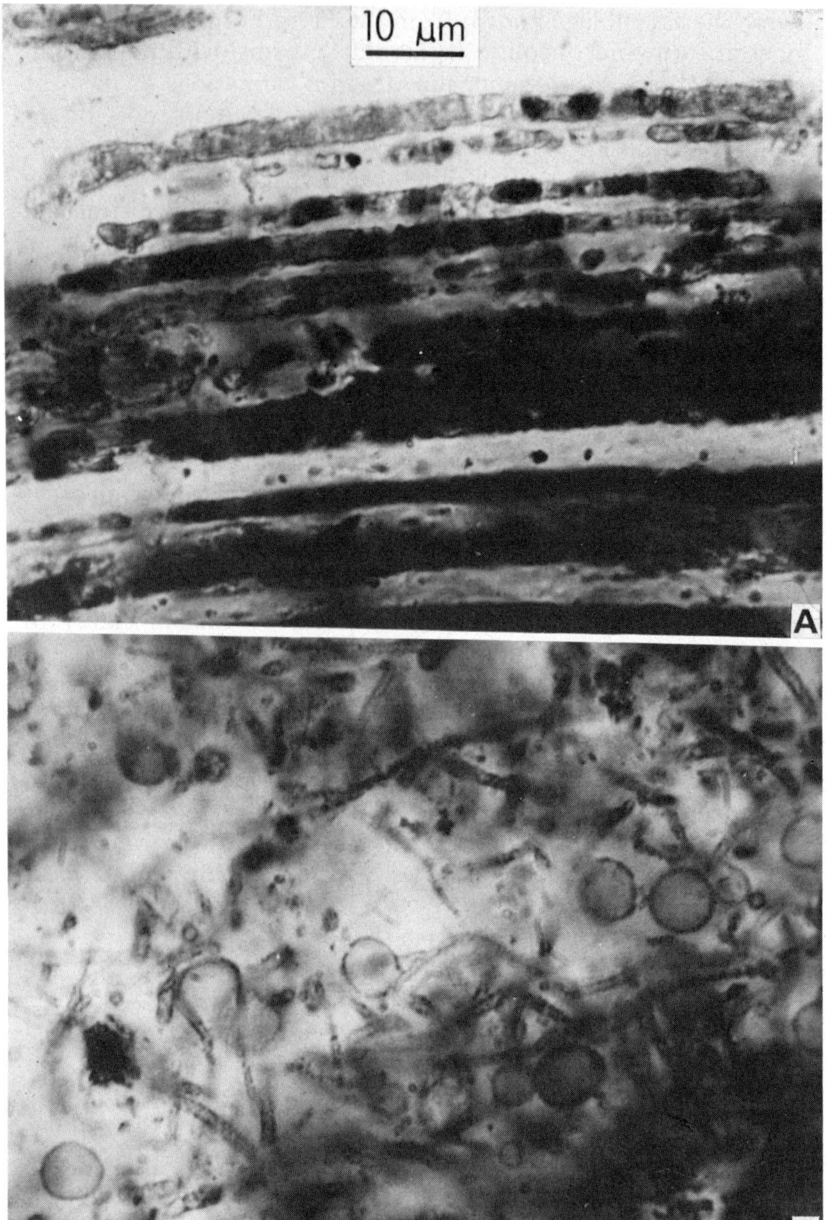
10 μm
A
B

and the carbon dioxide so produced was incorporated in carbonate minerals, taking with it the ^{13}C-depleted carbon of the kerogen. In this interpretation the light oxygen represents thermodynamic equilibrium between the various mineral phases. On the other hand, Becker and Clayton (1972), also working on a mesoband to macroband scale, detected carbonates depleted in ^{13}C in little metamorphosed (Smith et al., 1982) areas of the Hamersley iron-formations, and interpreted their data as indicating release during original sediment deposition of ^{13}C depleted-CO_2 from organic matter and its subsequent incorporation in carbonate minerals. New analytical data by M. Baur, J.M. Hayes, S.A. Studley and M.R. Walter (unpubl.) show that large and regular variations in the carbon and oxygen isotopic compositions of dolomite occur on a *microband* scale (Fig. 9-7) in samples from the Brockman and Marra Mamba Iron Formations of the Hamersley Basin. Both light oxygen and carbon appear to correlate with increased abundance of magnetite. As the compositional variations of up to 10‰ occur on a millimetre scale it is difficult to accept that they could have resulted from the metamorphic reaction given above. While the interpretation of these data is still rather speculative, it does seem likely that they indicate primary and diagenetic environmental fluctuations which perhaps were seasonal, including such reactions as those discussed by Irwin et al. (1977) by which bacteria influence the formation of diagenetic carbonates.

There are few published sulphur isotope data from Early Proterozoic iron-formations, which is surprising considering the potential significance of such information. To the data quoted from the literature by Skyring and Donnelly (1982) can be added those of Shegelski (1982), which are available only in an abstract. All the analyzed samples come from the Gunflint and Temiscamie iron-formations; Shegelski's (1982) data seem to indicate that sulphate-reducing bacteria were active in the iron-formation sediment. This is consistent with other indications that sulphate reducers had evolved by that time, i.e. about 2.0 Ga ago (Cameron, 1982; Skyring and Donnelly, 1982). The sparse but widespread occurrence of pyrite in the Hamersley iron-formations, in the absence of evidence of a hydrothermal source, argues for the occurrence of sulphate-reducing bacteria in the original sediment, but no isotopic analyses are available to support this interpretation; it should be noted that the Hamersley iron-formations are dated at about 2.5 Ga (Trendall, 1983, this volume) and therefore, according to Skyring and Donnelly (1982)

Fig. 9-6. The two main types of microstructure of Gunflint Biwabik stromatolites; scale for A and B identitical.

A. Simple, distinct laminae 2—4 μm thick, of alternating hematite (and limonite) and chert. This type is without microfossils and is of non-cyanobacterial origin, possibly a geyserite like deposit (see Walter, 1972). From Mink Mountain, Ontario (see Hofmann, 1969, p. 45; Geol. Surv. Can. photo 200931-I).

B. Simple, diffuse lamination with abundant structurally and organically preserved microfossils (*Huroniospora* sp. and *Gunflintia minuta*, the two most common taxa. From Winston Point, Ontario (see Hofmann, 1969, p. 45; Geol. Surv. Can. photo 200931-K).

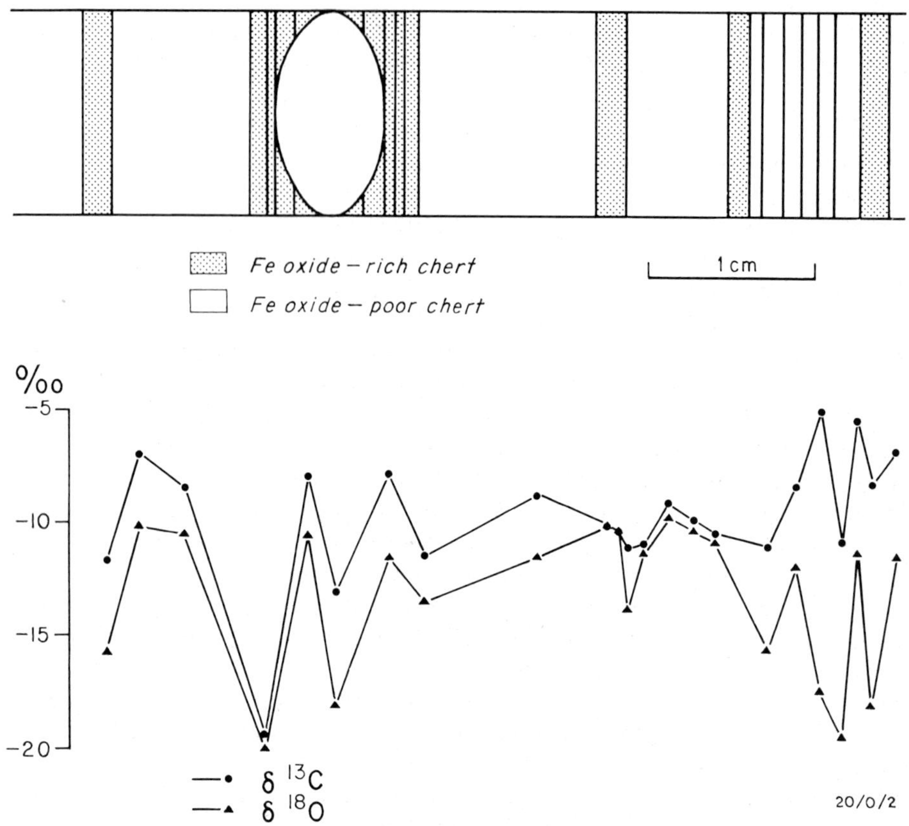

Fig. 9-7. Oxygen and carbon isotopic composition of dolomite from chert-magnetite iron-formation, Marra Mamba Iron Formation, 228.3—388.7 m in drill hole Millstream 9, Hamersley Basin. The upper part of the diagram is a representation of the sampled section of core. From unpublished data of M. Baur, J.M. Hayes, S.A. Studley and M.R. Walter.

and Cameron (1982), whould pre-date bacterial sulphate reduction as a major geochemical process.

Pyrolysis of kerogen from iron-formations from Africa and South America produced *n*-alkane compounds possibly indicative of the former presence of "algae", including cyanobacteria (W. Fiebiger quoted by McKridy and Hahn, 1982). Before these results can be accepted it would be necessary to demonstrate that the presence of such hydrogen-rich compounds is consistent with the grade of metamorphism of the rocks (see discussion in Hayes et al., 1983).

Middle and Late Proterozoic

There is no palaeontological information at all on iron-formations of Late Proterozoic age, and no Middle Proterozoic iron-formations are known. There are oolitic, sideritic and hematitic ironstones of Middle Proterozoic age in the Roper Group of the McArthur Basin, Australia, and these occur with black shales that are rich in kerogen and microfossils (Peat et al., 1981). The microfossil assemblage is very different from that in the older iron-formations.

PALAEOECOLOGY

Archaean

Not until the late Archaean is there any evidence of a benthic microbiota at the sites of deposition of iron-formations, and even then the evidence is restricted to sulphur isotope patterns the interpretation of which is still equivocal. The kerogen in Archaean iron-formations could have originated as a plankton "rain" or could be detrital organic matter derived from distant environments, e.g. adjacent shallow shelves; there is no evidence either way. Whichever it was, isotopic evidence shows that autotrophic organisms were responsible for primary carbon fixation; whether they were chemoautotrophs or photoautotrophs is not apparent. Similarly, if they were photoautotrophs (photosynthesisers), whether they used water as an electron donor (and thus released oxygen), or used H_2S or H_2 or some other electron donor (as do photosynthetic bacteria), is not demonstrable.

While it cannot yet be proven, it is probable that the iron-formation sediments harboured various kinds of bacteria that decomposed buried organic matter, ultimately to CO_2 and CH_4, producing the low concentrations of organic matter found in most sediments. By the late Archaean, this community of recycling microorganisms may have included sulphate or sulphite reducers.

The lack of stromatolites in Archaean iron-formations is consistent with the generally accepted interpretation that the iron-formations were deposited in deep water (e.g., Dimroth, 1977), below the photic zone (the base of which occurs now at about 120 m in the clearest ocean water, although it may have been shallower during the Archaean when the sun was probably less luminous). However, at least one Archaean iron-formation (from the Mozaan Group of the Pongola Supergroup) "probably represents a mud accumulation in a high tidal flat environment" (Von Brunn and Mason, 1977). This also lacks stromatolites, for reasons that can only be speculated upon. Microbial mats generally only develop on a firm substrate, and if the Mozaan iron-formation was deposited as a watery gel like the ferruginous and siliceous sediments in "Iron Ore Bay" on Palea Kameni in the Santorini caldera (MRW, personal observation) no benthic mat could form (as none has formed there).

Early Proterozoic

Here we will take the Gunflint and Brockman Iron Formations as the type examples of two major kinds of iron-formation: peloidal and banded. Most authors would agree with some such distinction, and with Dimroth's (1977) interpretation of the peloidal iron-formations as reworked chemical sediments analogous to limestones formed on a shelf environment comparable to that of the southern side of the Persian Gulf. The Brockman Iron Formation is more problematic, for while it is widely accepted that it was deposited below wave base (Trendall and Blockley, 1970; Trendall, 1976), precise values for this depth have not been established. Many participants in a recent Dahlem Conference considered likely the possibility that the Hamersley iron-formations were deposited on an outer shelf near the junction between deep anoxic water and an overlying oxygenated water body (Button et al., 1982).

There is no known example of a facies transition from typical large-scale peloidal iron-formation to typical large-scale banded iron-formation. Hall and Goode (1977) considered that the peloidal iron-formation of the Frere Formation in the Nabberu Basin is a correlative of banded iron-formations of that same basin and also of the nearby Hamersley Basin, but both interpretations have been challenged (Gee, 1979). The Transvaal iron-formations are very similar to those of the Hamersley Basin (Trendall, 1968), but Beukes (1973) and Button (1976) have described vertical and lateral transitions from stromatolitic dolomites into the iron-formations. No such transitions occur in the Hamersley Basin, where the Wittenoom Dolomite is non-stromatolitic and may well have been deposited below the photic zone. Dimroth (1977) considers that the Transvaal iron-formations were deposited on a shallow shelf environment as were the typically peloidal Sokoman iron-formations of the Labrador Geosyncline, but in the example of the Transvaal iron-formations a more sheltered environment was involved, preventing reworking. It may well be that the Transvaal iron-formations represent an intermediate case between the palaeoenvironmental extremes of the Gunflint and Brockman Iron Formations.

There are striking palaeontological differences between banded and peloidal iron-formations. The former entirely lack stromatolites. In other words, like the Archaean iron-formations, they lack any evidence of the former presence of benthic photoautotrophs, and with their finely banded nature and the lack of evidence of reworking, they can be interpreted as having been deposited below both wave base and the photic zone. The same applies also to the Wittenoom Dolomite of the Hamersley Basin, the lateral equivalent of which, on the shelf to the east of the Hamersley Basin, is the stromatolitic Carawine Dolomite (see Walter, 1983). There may well have been planktonic photoautotrophs in the water body above the site of deposition of the iron-formations and that would explain the presence of ^{13}C-depleted kerogen in the iron-formations; the kerogen could have been derived from detrital organic matter, for instance in the Hamersley Basin from the shallow shelf to the east of the main basin, but the lack of other detritus argues against that possibility. There is no direct palaeontological evidence as to whether any planktonic autotrophs that might have been present were oxygenic. At present that question is best answered by consideration of the mechanism of precipitation of the iron oxides in the iron-formations, and is beyond the scope of this paper.

The presence of ^{13}C-depleted dolomite in the banded iron-formations is interpreted here as indicating microbial decomposition of buried organic matter, and thus, that there was an active microbial population within the sediment. As the isotopic composition of dolomite varies regularly between microbands which are interpreted by Trendall and Blockley (1970) as varves, some response of the microbial population to seasons is suggested. The rich isotopic record in the iron-formations promises to yield an abundance of new information on microbial palaeoecology and associated mineral transformations, but only early results are yet available.

Peloidal iron-formations contain stromatolites associated with sedimentary indicators of a turbulent environment, as many authors have noted (e.g., Hofmann, 1969; Awramik, 1976; Dimroth, 1977). Each of the range of stromatolite morphologies described especially by Hofmann (1969) would record a different palaeoenvironment, as with extant stromatolites, but the differences are likely to have been subtle, as Awramik (1976) has shown. To date, however, no detailed sedimentological study of the stromatolitic facies of any iron-formation has been published, and so what those subtle differences were likely to have been we cannot say. This is another challenge to future workers. It is notable that no high-relief stromatolites are known from iron-formations; all the described stromatolites could have formed in water less than 2m deep.

The probable stiriolites of the Gunflint Iron Formation are here considered to be abiogenic splash deposits marking former shorelines, and, as can be expected, they intergrade with stromatolites built by microbial mats inhabiting those same shorelines.

In the examples of the Gunflint, Biwabik, Frere and Sokoman Iron Formations there is direct evidence of the taxonomic composition of the stromatolite-building microbial mats, in the form of abundant microfossils. Awramik (1976, 1977) and Awramik and Semikhatov (1979) have shown that a specimen of Gunflint Iron Formation with three intergrading but distinctive stromatolite morphologies contains three distinguishable microfossil communities. At present this is more significant for students of stromatolites than those of iron-formations so it is not discussed further here, but it does indicate the potential for distinguishing subtly different palaeoecological niches. The rare oncolites of the Gunflint have a microfossil assemblage that is different again (Awramik and Walter, 1978), but most distinctive is a non-stromatolitic assemblage from carbonaceous cherts in the chert-carbonate facies of the formation; this consists of *Eoastrion* and *Galaxiopsis melanocentra*, with the coccoid *Leptoteichos golubicii* distributed sporadically but abundantly. The latter is considered to have been planktonic (Knoll et al., 1978). There are other possible plankters represented in the Gunflint assemblage but at present too little is known about their facies distribution to allow confident interpretation. Knoll and Simonson (1980) have recognized two assemblages of benthic microfossils within intraclasts in the Sokoman iron-formation.

It is clear that the palaeo-ecologies of banded and peloidal iron-formations are as distinctively different as the rocks, and that a range of microbial niches can be distinguished, but perhaps the most significant question is whether there is anything special about the palaeontology of iron-formations that distinguished them from other rock types and which may contribute to a clearer understanding of their environment of deposition. Here we have to face the problem that with microfossils we are dealing with morphology only, and we know that in extant microbes the same morphologies recur in radically different groups with very different metabolisms and environmental effects. *Gunflintia*-like filaments, for instance, occur in many Proterozoic microfossil assemblages and among many extant bacterial and cyanobacterial groups. *Eoastrion* is more distinctive, but it is still difficult to evaluate the significance of the fact that it is not restricted in occurrence to iron-formations and is also found, for instance, in cherty dolomites (Knoll and Barghoorn, 1976). Some of the rarer forms, such as *Eosphaera*, are known only from iron-formations. One clear distinction can be drawn, however: the upper intertidal microfossil assemblages characteristic of many Early Proterozoic and younger carbonate sequences (e.g., Hofmann, 1976; Muir, 1976; Oehler, 1978) are readily distinguishable from all the iron-formation assemblages.

The stromatolites of the peloidal iron-formations are quite different from those of Early Proterozoic marine open shelf environments such as that of the Athapuscow Aulacogen (Hoffman, 1976) and Transvaal Basin (Truswell and Eriksson, 1973), particularly in lacking any elongate, high-relief domical forms. In addition, among the Early Proterozoic columnar stromatolites that have been formally classified (Bertrand-Sarfati and Walter, 1981), there are none in common between iron-formations and other rock types. Peloidal iron-formations also lack the distinctive and otherwise abundant stromatolite *Conophyton* which seems to characterise calm, marine subtidal, including lagoonal, settings, some of which were very extensive (e.g., Donaldson, 1976).

There is nothing in the palaeontology of Early Proterozoic iron-formations that demonstrates that they were deposited in marine environments, and such comparisons as can presently be drawn show significant differences from marine deposits. This is not to say that the depositional environments were non-marine, although they may have been, but only that the fossils are as distinctive as the rocks themselves.

MICROBIAL DEPOSITION OF IRON-FORMATION

This is a large subject mostly beyond the scope of this paper, but two points must be commented on.

(1) When did oxygenic photosynthesis evolve and thus, when could that have provided the free oxygen needed to precipitate iron oxides?

(2) Is there any evidence of the direct precipitation by bacteria of the minerals of iron-formations?

The origin of oxygenic photosynthesis is discussed at length by several authors in Schopf (1983). It is concluded that it had evolved by the late Archaean, and probably had its origin not long before then. Cloud's (1976) suggestion that iron-formations themselves are evidence of oxygenic photosynthesis, and therefore that this biochemical process had evolved by Isua time (3.8 Ga ago), is not refuted, but is not supported by independent evidence.

Despite intensive study of Precambrian microfossils for the last 20 years no convincing examples of organisms with carbonate or silica skeletons have been discovered. Any models of iron-formation deposition calling for involvement of such organisms must stand without the benefit of palaeontological evidence. On the other hand, there may well have been organisms that indirectly caused mineral precipitation in their extracellular environment, for instance through their influence on pH or alkalinity. The carbonate minerals in stromatolites are at least partly deposited because of such processes (see discussion in Walter, 1983). Suggestions of the direct precipitation (Harder, 1919) or dissolution (Nealson, 1982) of iron minerals by bacteria, as part of the deposition of iron-formations, while they may well be correct, also find little or no palaeontological support. However, one of the more abundant microfossils of peloidal iron-formations, *Eoastrion*, is morphologically closely comparable to the extant iron- and manganese-oxidizing bacterium *Metallogenium* (e.g. Cloud, 1965; Nealson, 1982). Microfossils in peloidal iron-formations are frequently preserved as hematitic moulds and casts, but there is no consistent relationship between the iron minerals and the microfossils.

CONCLUSIONS

The iron-formations of all but the latest Proterozoic were deposited before the origin of animals and of macroscopic plant life. Thus the palaeontology of these iron-formations involves study of traces of microbial life. Such traces consist not only of morphological cellular remnants (microfossils) and sedimentary structures constructed by benthic mats of microorganisms (stromatolites), but also the chemical remnants of cells (kerogen) and patterns of distribution of the stable isotopes of carbon and sulphur that are fractioned during some microbial metabolic processes. All these lines of evidence must be considered together, and in a palaeoenvironmental context, to arrive at convincing interpretations. In fact such an integrated approach is not yet generally possible because of the paucity of palaeontological and geochemical research on Precambrian iron-formations.

Neither stromatolites nor convincing microfossils are known from Archaean iron-formations. These do, however, contain kerogen with a carbon isotopic composition indicative of autotrophy (the use of CO_2 as the source of carbon for cell construction), and so it is apparent that microorganisms were living within the basins of deposition of the iron-formations. Not until the latest Archaean is there any evidence of the former presence of sulphate-reducing bacteria in the iron-formation sediment, and even then the evidence is equivocal. There is no direct evidence of the involvement of microorganisms in the deposition of Archaean iron-formations. The absence of stromatolites is consistent with the interpretation that the iron-formations were deposited in water deeper than about 100m. The palaeontological and geochemical records as a whole seem to indicate that oxygen-producing photosynthesis evolved during the late Archaean, and thus that the iron in older iron-formations must have been oxidised by some other mechanism.

Early Proterozoic iron-formations can be considered to be of two distinct types, peloidal and banded, of which the Gunflint and Brockman iron-formations are the best known examples. The two types are palaeontologically very different. The banded iron-formations of the Early Proterozoic are like those of the Archaean and were probably also deposited in more than 100 m of water; no stromatolites or convincing microfossils are known from these rocks. In contrast, the peloidal iron-formations are richly fossiliferous, with stromatolites and microfossils all consistent with deposition in very shallow, turbulent water (perhaps in some facies less than 2 m deep). Some of the microorganisms are likely to have been photosynthetic organisms that released oxygen. Others may have been iron-oxidizing bacteria, but there is no convincing evidence that microorganisms had a major, direct role in precipitating iron minerals (except perhaps through the photosynthetic production of oxygen). In addition, there is no evidence that silica-secreting organisms contributed to the deposition of the chert of the iron-formations, and in fact no such organisms are known from Precambrian rocks.

It is apparent from the little information already available that palaeontology can reveal much about depositional environments and mechanisms of deposition of Precambrian iron-formations. We hope in this review to have drawn attention to potentially fruitful lines of research that remain as a challenge to future workers.

ACKNOWLEDGEMENTS

M.R. Walter publishes with the permission of the Director, Bureau of Mineral Resources, Geology and Geophysics, Australia.

REFERENCES

Awramik, S.M., 1976. Gunflint stromatolites: microfossil distribution in relation to stromatolite morphology. In: M.R. Walter (Editor), Stromatolites. Elsevier, Amsterdam, pp. 311—320.

Awramik, S.M., 1977. Paleobiology of stromatolites. In: C. Ponnamperuma (Editor) Chemical Evolution of the Early Precambrian. Academic Press, New York, N.Y., pp. 111—131.

Awramik, S.M. and Barghoorn, E.S., 1977. The Gunflint microbiota. Precambrian Res., 5: 121—142.

Awramik, S.M. and Semikhatov, M.A., 1979. The relationship between morphology, microstructure, and microbiota in three vertically intergrading stromatolites from the Gunflint Iron Formation. Can. J. Earth Sci., 16: 484—495.

Awramik, S.M. and Walter, M.R., 1978. Comparative biogeology of the Gunflint Iron Formation (Canada) and Frere Formation (Western Australia). Geol. Soc. Am., Abstr. Prog., 10 (7): 361.

Barghoorn, E.S. and Tyler, S.A., 1965. Microorganisms from the Gunflint chert. Science, 147: 563—577.

Becker, R.H. and Clayton, R.N., 1972. Carbon isotopic evidence for the origin of a banded iron-formation in Western Australia. Geochim. Cosmochim. Acta, 36: 577—595.

Bertrand-Sarfati, J. and Walter, M.R., 1981. Stromatolite biostratigraphy. Precambrian Res., 15: 353—371.

Beukes, N.J., 1973. Precambrian iron-formations of southern Africa. Econ. Geol., 68: 960—1004.

Button, A., 1976. Iron-formation as an end member in carbonate sedimentary cycles in the Transvaal Supergroup, South Africa. Econ. Geol., 71: 193—201.

Button, A. and 10 other authors, 1982. Sedimentary iron deposits, evaporites and phosphorites — state of the art report. In: H.D. Holland and M. Schidlowski (Editors), Mineral Deposits and the Evolution of the Biosphere. Springer-Verlag, Berlin, pp. 259—273.

Cameron, E.M., 1982. Sulphate and sulphate reduction in early Precambrian Oceans. Nature, 296: 145—148.

Cayeux, L., 1911. Existence de restes organiques dan les roches ferrugineuses associées aux minérais de fer huroniens des États-Unis. C.R. Acad. Sci. Paris, 153: 910-912.

Cloud, P.E., Jr., 1965. Significance of the Gunflint (Precambrian) microflora. Science, 148: 27—35.

Cloud, P.E., Jr., 1973. Paleoecological significance of the banded iron-formation. Econ. Geol., 68: 1135—1143.

Cloud, P.E., Jr., 1976. Major features of crustal evolution. Geol. Soc. S. Afr. Annex. Vol. 79, 33 pp.

Cloud, P.E., Jr. and Licari, G.R., 1968. Microbiotas of the banded iron formations. Proc. Natl. Acad. Sci. U.S.A., 61: 779—786.

Cloud, P.E., Jr. and Morrison, K., 1979. New microbiotas from the Animikie and Baraga Groups (∿2 Gyr old), north Michigan. Geol. Soc. Am., Abstr. Progr., 11 (5): 227.

Cloud, P.E., Jr. and Morrison, K., 1980. New microbial fossils from 2 Gyr old rocks in northern Michigan. Geomicrobiol. J., 2: 161—178.

Cloud, P.E., Jr., Gruner, J.W. and Hagen, H., 1965. Carbonaceous rocks of the Soudan Iron Formation (Early Precambrian). Science, 148: 1713—1716.

Dimroth, E., 1977. Facies Models 5. Models of physical sedimentation of iron formations. Geosci. Can., 4: 23—30.

Donaldson, J.A., 1976. Paleoecology of Conophyton and associated stromatolites in the

Precambrian Dismal Lakes and Rae Groups, Canada. In: M.R. Walter (Editor), Stromatolites. Elsevier, Amsterdam, pp. 523—534.

Edgell, H.S., 1964. Precambrian fossils from the Hamersley Range, Western Australia, and their use in stratigraphic correlation. J. Geol. Soc. Aust., 11: 235—261.

Edhorn, A.-S., 1973. Further investigations of fossils from the Animikie, Thunder Bay, Ontario. Proc. Geol. Assoc. Can., 25: 37—66.

Gee, R.D., 1979. Structure and tectonic style of the Western Australian shield. Tectonophysics, 58: 327—369.

Goodwin, A.M., Monster, J. and Thode, H.G., 1976. Carbon and sulfur isotope abundances in Archaean Iron-Formations and early Precambrian life. Econ. Geol., 71: 870—891.

Gresley, W.S., 1896. Organic markings in Lake Superior iron-ores. Science, n.s. 3: 622—623.

Gresley, W.S., 1897. Traces of organic remains from the Huronian(?) Series, at Iron Mountain, Mich., Etc. Trans. Am. Inst. Min. Eng., 26: 527—534.

Grout, F.F. and Broderick, T.M., 1919. Organic structures in the Biwabik iron-bearing formation of the Huronian in Minnesota. Am. J. Sci., 48: 199—205.

Gruner, J.W., 1922. The origin of sedimenatry iron formations: The Biwabik formation of the Mesabi range. Econ. Geol., 17: 407—460.

Gruner, J.W., 1923. Algae, believed to be Archean. J. Geology, 31: 146—148.

Gruner, J.W., 1924. Contributions to the geology of the Mesabi Range. Minn. Geol. Surv., Bull., 19: 1—67.

Gruner, J.W., 1925. Discovery of life in the Archean. J. Geol., 33: 151—152.

Hall, W.D.M. and Goode, A.D.T., 1978. The Early Proterozoic Nabberu Basin and associated iron formations of Western Australia. Precambrian Res., 7: 129—184.

Hamilton, L.H., 1976. Biogenic aspects of Precambrian iron formations from Africa and elsewhere. Int. Geol. Congr., 25th Sess., Sydney, Abstr., 1: 30.

Harder, E.C., 1919. Iron-depositing bacteria and their geologic relations U.S. Geol. Surv., Prof. Pap. 113, 88 pp.

Hayes, J.M., 1983. Geochemical evidence bearing on the origin of aerobiosis, a speculative interpretation. In: J.W. Schopf (Editor), The Earth's Earliest Biosphere: Its Origin and Evolution. Princeton Univ. Press, Princeton, N.J., Chapter 12.

Hayes, J.M., Kaplan, I.R. and Wedeking, K.W., 1983. Precambrian organic geochemistry: preservation of the record. In: J.W. Schopf (Editor), The Earth's Earliest Biosphere: Its Origin and Evolution. Princeton Univ. Press, Princeton, N.J., Chapter 5.

Hoffman, P., 1976. Environmental diversity of Middle Precambrian stromatolites. In: M.R. Walter (Editor), Stromatolites. Elsevier, Amsterdam, pp. 599—611.

Hofmann, H.J., 1969. Stromatolites from the Proterozoic Animikie and Sibley Groups, Ontario. Geol. Surv. Can., Pap. 68—69, 77 pp.

Hofmann, H.J., 1971. Precambrian fossils, pseudofossils, and problematica in Canada. Geol. Surv. Can., Bull. 189, 146 pp.

Hofmann, H.J., 1976. Precambrian microflora, Belcher Islands, Canada. J. Paleontol., 50: 1040—1073.

Hofmann, H.J. and Schopf, J.W., 1983. Early Proterozoic microfossils. In: J.W. Schopf (Editor), The Earth's Earliest Biosphere: Its Origin and Evolution. Princeton Univ. Press. Princeton, N.J., Chapter 14.

Irwin, H., Curtis, C. and Coleman, M., 1977. Isotopic evidence for source of diagenetic carbonates formed during burial of organic-rich sediments. Nature, 269: 209—213.

Karkhanis, S.N., 1976. Fossil iron bacteria may be preserved in Precambrian ferroan carbonate. Nature, 261: 406—407.

Karkhanis, S.N., Goodwin, A.M. and Ponnamperuma, C., 1980. Palaeobiology and organic geochemistry of Archean Helen Iron Formation, Michipicoten area. J. Geol. Soc. India, 21: 1—9.

Kazmierczak, J., 1976. Devonian and modern relatives of the Precambrian *Eosphaera:* possible significance for early eukaryotes. Lethaia, 9: 39—50.

Kazmierczak, J., 1979. The eukaryotic nature of *Eoshpaera*-like ferriferous structures from the Precambrian Gunflint Iron Formation, Canada: a comparative study. Precambrian Res., 9: 1—22.

Klemm, D.D., 1979. A biogenetic model of the formation of the banded iron formation in the Transvaal Supergroup/South Africa. Mineral. Deposita, 14: 381—385.

Knoll, A.H. and Barghoorn, E.S., 1976. A Gunflint-type microbiota from the Duck Creek Dolomite, Western Australia. Origins Life, 7: 417—423.

Knoll, A.H. and Simonson, B., 1981. Early Proterozoic microfossils and penecontemporaneous quartz cementation in the Sokoman Iron Formation, Canada. Science, 211: 478—480.

Knoll, A.H., Barghoorn, E.S. and Awramik, S.M., 1978. New microorganisms from the Aphebian Gunflint Iron Formation, Ontario. J. Paleontol., 52: 976—992.

LaBerge, G.l., 1967. Microfossils in Precambrian iron-formations. Geol. Soc. Am. Bull., 78: 331—342.

LaBerge, G.L., 1973. Possible biological origin of Precambrian iron-formations. Econ. Geol., 68: 1098—1109.

Leith, C.K., 1903. The Mesabi iron-bearing district of Minnesota. U.S. Geol. Surv. Mon., 43, 316 pp.

Licari, G.R. and Cloud, P.E., Jr., 1968. Reproductive structures and taxonomic affinities of some nannofossils from the Gunflint Iron Formation. Proc. Natl. Acad. Sci. U.S.A., 59: 1053—1060.

Lougheed, M.S. and Mancuso, J.J., 1973. Hematite framboids in the Negaunee Formation, Michigan, evidence for their biogenic origin. Econ. Geol., 68: 202—209.

Mancuso, J.J., Lougheed, M.S. and Wygant, T., 1971. Possible biogenic structures from the Precambrian Negaunee (Iron) Formation, Marquette Range, Michigan. Am. J. Sci., 271: 181—186.

McKirdy, D.M. and Hahn, J.H., 1982. The composition of kerogen and hydrocarbons in Precambrian rocks. In: H.D. Holland and M. Schidlowski (Editors), Mineral Deposits and the Evolution of the Biosphere. Springer-Verlag, Berlin, pp. 123—154.

Mel'nik, Y.P., 1982. Precambrian Banded Iron-Formations. Elsevier, Amsterdam, 310 pp.

Moore, E.S., 1918. The iron formation on Belcher Islands, Hudson Bay, with special reference to its origin and associated algal limestones. J. Geol., 26: 412—438.

Moorehouse, W.W. and Beales, F.W., 1962. Fossils from the Animikie, Port Arthur, Ontario. Trans. R. Soc. Can., Sect. 4 (3): 97—110.

Muir, M.D., 1976. Proterozoic microfossils from the Amelia Dolomite, McArthur Basin, Northern Territory. Alcheringa, 1: 143—158.

Muir, M.D., 1978. Microenvironments of some modern and fossil iron- and manganese-oxidizing bacteria. In: W.E. Krumbein (Editor), Environmental Biogeochemistry and Geomicrobiology, vol. 3. Ann Arbor Science Publishers Inc. Ann Arbor, Mich., pp. 937—944.

Muir, M.D. and Plumb, K.A., 1976. Precambrian microfossils in Australia — distribution, significance and problems. Int. Geol. Congr. 25th Sess. Sydney, Abstr., pp. 32—33.

Nealson, K.H., 1982. Microbial oxidation and reduction of iron. In: H.D. Holland and M. Schidlowsky (Editors), Mineral Deposits and the Evolution of the Biosphere. Springer-Verlag, Berlin, pp. 51—66.

Oehler, D.Z., 1978. Microflora of the Middle Proterozoic Balbirini Dolomite (McArthur Group) of Australia. Alcheringa, 2: 269—309.

Oehler, J.H., 1976. Hydrothermal crystallization of silica gel. Geol. Soc. Am. Bull., 87: 1143—1152.

Oehler, J.H. and 12 other authors, 1982. Reduced carbon compounds in sediments — state of the art report. In: H.D. Holland and M. Schidlowski (Editors), Mineral Deposits and the Evolution of the Biosphere. Springer-Verlag, Berlin, pp. 289—306.

Ouyang Shu, 1979. Ultramicro- and micro-fossils from the Anshan Group and the Liaohe Group in eastern Liaoning, northeastern China. In: Selected works for a symposium on iron-geology of China, sponsored by Academia Sinica, 1977. Stratigraphy and Paleontology, Science Press, Beijing, pp. 1—30 (in Chinese, with English abstract).

Peat, C.J., Muir, M.D., Plumb, K.A., McKirdy, D.M. and Norvick, M.S., 1978. Proterozoic microfossils from the Roper Group, Northern Territory, Australia. BMR J. Aust. Geol. Geophys., 3: 1—17.

Perry, E.C. Jr. and Ahmad, S.H., 1977. Carbon isotope composition of graphite and carbonate minerals from 3.8-AE metamorphosed sediments, Isukasia, Greenland. Earth Planet. Sci. Lett. 36: 280—284.

Perry, E.C., Jr., Tan, F.C. and Morey, G.B., 1973. Geology and stable isotope geochemistry of the Biwabik Iron Formation, northern Minnesota. Econ. Geol., 68: 1110—1125.

Piatnitsky, P., 1924. The genetic relations of the deposits of ore in the Krivoy Rog. Trans. Inst. Econ. Mineral. Petrogr. (Lithogaea), 9, 56 pp.

Schidlowski, M., Appel, P.W.U., Eichmann, R. and Junge, C.E., 1979. Carbon isotope geochemistry of the 3.7×10^9-yr-old Isua sediments, west Greenland: implications for the Archaean carbon and oxygen cycles. Geochim. Cosmochim. Acta, 43: 189—199.

Schidlowski, M., Hayes, J.M. and Kaplan, I.R., 1983. Isotopic inferences of ancient biochemistries: carbon, sulfur, hydrogen and nitrogen. In: J.W. Schopf (Editor) The Earth's Earliest Biosphere: Its Origin and Evolution. Princeton Univ. Press, Princeton, N.J., Chapter 7.

Schmidt, R.G., 1980. The Marquette Range Supergroup in the Gogebic Iron District, Michigan and Wisconsin. U.S. Geol. Surv., Bull. 1460, 96 pp.

Schopf, J.W. (Editor), 1983. The Earth's Earliest Biosphere: Its Origin and Evolution. Princeton Univ. Press. Princeton, N.J. in press.

Schopf, J.W. and Walter, M.R., 1983. Archean microfossils: new evidence of ancient microbes. In: J.W. Schopf (Editor), The Earth's Earliest Biosphere: Its Origin and Evolution. Princeton Univ. Press, Princeton, N.J., Chapter 9.

Shegelski, R.J., 1982. Depositional environment of the Gunflint Formation, Ontario Canada. Int. Geol. Corr. Prog. Joint Meeting Proj. 157 and 160, Mexico, 1982, Abstr., p. 37.

Skyring, G.W. and Donnelly, T.H., 1982. Precambrian sulfur isotopes and a possible role for sulfite in the evolution of biological sulfate reduction. Precambrian Res., 17: 41—61.

Smith, R.E., Perdrix, J.L. and Sparks, T.C., 1982. Burial metamorphism in the Hamersley Basin, Western Australia. J. Sediment. Petrol., 23 (1): 75—102.

Spencer, E. and Percival, F.G., 1952. The structure and origin of the banded hematite jaspers of Singhbhum, India. Econ. Geol., 47: 365—383.

Tappan, H., 1976. Possible eucaryotic algae (Bangiophycidae) among early Proterozoic microfossils. Geol. Soc. Am. Bull., 87: 633—639.

Tappan, H., 1980. The Paleobiology of Plant Protists. Freeman, San Francisco, Calif., 1028 pp.

Timofeev, B.V., 1969. Sferomorfidy proterozoya (Proterozoic Sphaeromorphida). Akad. Nauk SSR, Inst. Geol. Geokhron. Dokembriya. Nauka, Leningrad, 146 pp. (in Russian).

Timofeev, B.V., 1973. Microfitofossilii dokembriya Ukrainy (Precambrian microfossils from the Ukraine). Nauka, Leningrad, 58 pp. + 40 plates (in Russian).

Timofeev, B.V., 1979. Mikrofitofossilii Pechengskoi serri (Microfossils from the Pechenga Series). In: B.V. Sokolov (Editor), Paleontologiya dokembriya i rannego kembriya (Paleontology of the Precambrian and Early Cambrian). Nauka, Leningrad, pp. 119—120 (in Russian).

Trendall, A.F., 1968. Three great basins of Precambrian banded iron formation deposition: a systematic comparison. Geol. Soc. Am. Bull., 79: 1527—1544.

Trendall, A.F., 1976. Geology of the Hamersley Basin. 25th Int. Geol. Congr., Sydney, Excursion Guide 43A.

Trendall, A.F., 1983. The Hamersley Basin. In: A.F. Trendall and R.C. Morris (Editors), Iron-Formation: Facts and Problems. Elsevier, Amsterdam, pp. 69—129.

Trendall, A.F. and Blockley, J.G., 1970. The iron formations of the Hamersley Group, Western Australia, with special reference to the associated crocidolite. West. Aust., Geol. Surv., Bull. 119, 353 pp.

Truswell, J.F. and Eriksson, K.A., 1973. Stromatolitic associations and their palaeo-environmental significance: a re-appraisal of a Lower Proterozoic locality from the northern Cape Province, South Africa. Sediment. Geol., 10: 1—23.

Tyler, S.A. and Barghoorn, E.S., 1954. Occurrence of structurally preserved plants in Precambrian rocks of the Canadian Shield. Science, 119: 606—608.

Viswanathiah, M.N.and Venkatachalapathy, V., 1980. Microbiota from the Bababudan Iron Formation, Karnataka. J. Geol. Soc. India, 21: 16—20.

Von Brunn, V. and Mason, T.R., 1977. Siliciclastic-carbonate tidal deposits from the 3000 m.y. old Pongola Supergroup, South Africa. Sediment. Geol., 18: 245—255.

Walcott, C.D., 1899. Pre-Cambrian fossiliferous formations. Bull. Geol. Soc. Am., 10: 199—244.

Walter, M.R., 1972a. Stromatolites and the Biostratigraphy of the Australian Precambrian, with Appendices on Pseudofossils from Australian Precambrian Iron-formation and Greywacke. Ph.D. Thesis, Univ. Adelaide, 466 pp. (unpubl.).

Walter, M.R., 1972b. A hot spring analog for the depositional environment of Precambrian iron formations of the Lake Superior region. Econ. Geol., 67: 965—972.

Walter, M.R., 1975a. Microfossils from a newly discovered Precambrian stromatolitic iron formation in Western Australia. Geol. Soc. Aust., 1st Aust. Geol. Convention, "Proterozoic Geology", Adelaide, May 1975, Abstr., pp. 15—16 (Abstr.).

Walter, M.R., 1975b. The biostratigraphic significance and biological affinities of microfossils from a newly discovered Precambrian stromatolitic iron formation in Western Australia. Int. Geol. Correlation Progr., Symp., Correlation of the Precambrian, Moscow, pp. 41—42 (Abstr.).

Walter, M.R., 1976a. Introduction. In: M.R. Walter (Editor), Stromatolites. Elsevier, Amsterdam, pp. 1—3.

Walter, M.R., 1976b. Geyserites of Yellowstone National Park: an example of abiogenic "stromatolites". In: M.R. Walter (Editor), Stromatolites. Elsevier, Amsterdam, pp. 87—112.

Walter, M.R., 1983. Archean and Early Proterozoic stromatolites: the history of the Earth's earliest benthos. In: J.W. Schopf (Editor), The Earth's Earliest Biosphere: Its Origin and Evolution. Princeton Univ. Press, Princeton, N.J., Chapter 8.

Walter, M.R. and Awramik, S.M., 1979. *Frutexites* from stromatolites of the Gunflint Iron Formation of Canada, and its biological affinities. Precambrian Res., 9: 23—33.

Walter, M.R., Goode, A.D.T. and Hall, W.D.M., 1976. Microfossils from a newly discovered Precambrian stromatolitic iron formation in Western Australia. Nature, 261: 221—223.

Xü Ren and Chu Weiching, 1979. Fossil microbiota of the Anshan Group and preliminary investigation on the genesis of the Anshan iron ore. In: Selected works for a symposium on iron-geology of China, sponsored by Academia Sinica, 1977. Stratigraphy and Paleontology. Science Press, Beijing, pp. 33—38 (in Chinese, with English abstract).

Yin Leiming, 1979. Microflora from the Anshan Group and the Liaohe Group in E. Liaoning with its stratigraphic significance. In: Selected works for a symposium on iron-geology of China, sponsored by Academia Sinica, 1977. Stratigraphy and Paleontology. Science Press, Beijing, pp. 39—58 (in Chinese, with English abstract).

Zajac, I.S., 1974. The stratigraphy and mineralogy of the Sokoman Formation in the Knob Lake area, Quebec and Newfoundland. Geol. Surv. Can., Bull. 220, 159 pp.

Zhang Yun, 1981. Proterozoic stromatolite microfloras of the Gaoyuzhuang Formation (Early Sinian: Riphean), Hebei, China. J. Paleontol., 55: 485—506.

Chapter 10

BANDED IRON-FORMATION — A GRADUALIST'S DILEMMA

PRESTON CLOUD

INTRODUCTION

Banded iron-formation (BIF) is a chemically precipitated sedimentary rock in which iron-rich bands or laminae alternate with iron-poor ones. Although the term is broadly applied, the rock is typically and preponderantly siliceous, consisting of laterally extensive sets of iron-rich and iron-poor laminae or bands of chert, or a metamorphic equivalent (James and Sims, 1973). The following comments are directed to that kind of BIF and not to its peripheral variants, to oolitic or minette-type ores, or to iron-formation in general.

BIF is common among rocks older than about 2×10^9 years and rare among younger rocks. At many localities (e.g., Hamersley Basin, Krivoy Rog, Kuruman, localities in India) submillimeter-scale microbands of such well and extensively laminated BIF cluster in centimeter-scale mesobands (Trendall, 1965). At others one sees only mesobands (e.g., Negaunee, Quadrilatero Ferrifero). A typical unaltered BIF of this nature will contain 25—45% or more Fe-oxide in both the ferric and ferrous states (hematite and magnetite). Such rocks may grade basinward to the sulfide facies of James (1954). The great bulk of the world's economic supply of iron occurs as BIF, either in its pristine state or as the oxidatively enriched product.

Because of its unusual nature, its restriction in time, and its economic interest, possible hypotheses for the origin of BIF have been much discussed. At the request of the editors I here briefly review the history, strengths, weaknesses, and expectable consequences of one such hypothesis — the hypothesis of microbial oxygen balance. This hypothesis is one of a collection of hypotheses which I have offered as a larger model of the interaction among biospheric, atmospheric, and lithospheric evolution on the pre-Phanerozoic earth (Cloud, 1965, 1972, 1974, 1976a, b, 1980). Although during the 1960's this model conflicted with the prevailing views of earlier times, it has since become widely accepted. Yet nothing is gained by simply changing dogmas, and it is also vigorously disputed. In the end the model must stand on its merits, fall on its shortcomings, or undergo revision to accommodate the latter.

A problem with BIF is the absence of convincing modern analogs. Because the most typical BIFs are epicratonal deposits of wide extent, it is unlikely that good modern analogs have simply been overlooked and will subsequently be found. Such a conclusion, however, runs against the grain of gradualistic "uniformitarian" philosophy, especially where exogenous processes and surface conditions different from those now prevailing are proposed to account for this rock.

While sharing the conviction that modern analogs are desirable and should continue to be sought, I take a broader view of uniformitarianism. Unusual rocks may be best accounted for by unusual explanations, even ones without currently operating analogs, provided explanations proposed do not violate natural laws.

Scientific explanation of the records of past events invokes naturalistic processes — either processes observable under prevailing conditions at Earth's surface, or reactions that can be reproduced experimentally under conditions known to exist somewhere in nature, or which follow from established natural laws. The recognitions of that self-evident truth by James Hutton, John Playfair, and Charles Lyell marked the emergence of geology as a science. It came to be known as uniformitarianism, first as a jibe and then as a principle. But the principle has suffered from ambiguity. To some it still implies gradualism and a disinclination to admit circumstances or mechanisms that, although not now observed, can be legitimately invoked as products of the working of demonstrable natural laws. The founders of geology advanced from simpler levels, but they were not simple minded. They knew as well as we do that rates vary and that great earthquakes, tsunamis, typhoons, and hundred-year floods do occur. Strict gradualism has never been a serious alternative. A process need not be observed in nature to conform to naturalistic principles. Neither catastrophes nor conditions different from those now prevailing are excluded from sensible perceptions of "uniformitarianism". Were we living in non-glacial times, glaciation could still be visualized as consistent with the workings of nature. Although we have never witnessed an asteroid impact, we know that they are not forbidden by the principles that normally keep such bodies in orbit.

WHAT NEEDS TO BE EXPLAINED

Any model seeking to explain BIF must account above all for the following:

(1) The rhythmic banding and lamination that occurs and commonly prevails in all true BIF — involving the problem of how to transport the iron to sites of deposition in a soluble state and to precipitate it episodically, and free from conspicuous clastic components, as thin iron-rich bands that extend with little variation in thickness over areas up to hundreds of kilometers across.

(2) The sources and storage sites of important elements, notably iron and silicon.

(3) The essential limitation of BIF to rocks older than $\sim 2 \times 10^9$ years.

(4) Variations in facies of the BIF, including differences in size and depositional sites between typically Archean and typically Proterozoic deposits (regardless of geochronological hair-splitting).

(5) Anomalous occurrences of deposits younger than $\sim 2 \times 10^9$ years.

A more detailed list of characteristics calling for explanation is given in Cloud (1973).

ASSUMPTIONS AND CONSTRAINTS

Geologists, like astrophysicists, deal with signals from the past. To interpret evidence of former events scientifically we must assume that natural laws have remained invariant, or that, if they have changed, they have changed lawfully. Events or conditions that may have occurred or prevailed in distant times need not have current analogs to have been real (or highly probable). The most important ones, however, will have launched signals (verifiable consequences) such as the 3°K general background radiation that records the no-longer seriously doubted occurrence of the Big Bang that initiated the present universe some 18 or 20×10^9 years ago.

Reflection on the nature of signals from the past and on the probable consequences of natural principles suggests the following as self-evident:

(1) Complex signals (e.g., BIF) characteristically imply complex events.

(2) Complex events can rarely be reconstructed from single lines of evidence, even where the record is well preserved.

(3) To reconstruct complex and temporally distant exogenous processes such as those responsible for the BIF, we should assess simultaneously the available evidence bearing on likely interactions between relevant evolving spheres of action — the sedimentary lithosphere, the hydrosphere, the atmosphere, and the biosphere.

(4) Once life appeared on Earth, these interactions could never be quite the same again.

(5) The basic organic molecules of living systems being thermodynamically unstable in the presence of oxygen, chemical evolution from non-living to living systems requires anoxygenous pre-biotic and early biotic conditions, hence an anoxygenous primitive Earth or some plausible oxygen-free haven. [Other reasons for hypothesizing an essentially oxygen-free early Earth are discussed in Mackenzie (1975) and Cloud (1980)].

(6) Because the production of free oxygen involves reversible processes whose products are also thermodynamically unstable and recombine unless sequestered, and because O_2 combines readily with many reduced substances,

it is the current presence of O_2 in hydrosphere and atmosphere, not its former rarity or absence, that most needs explanation.

(7) The appearance of O_2-releasing photosynthesis (with sedimentary burial of C) must eventually affect all other exogenous systems, producing geologically visible results.

THE SEMINAL SOURCES

I came to the problem of the BIF in a roundabout way. Nursall (1959), responding to my suggestion (Cloud, 1948) that the Metazoa arose (probably polyphyletically) and diversified over a relatively short interval in earliest Phanerozoic history, suggested that the timing of that event was a function of O_2 level. Inspired to think about primordial and ancestral atmospheres by the writings of Macgregor (1927), Rubey (1951), Brown (1952), and others, my curiosity about O_2 was aroused by Nursall's comments and came to focus on the BIF with the discovery of James' classic 1954 paper.

The prevalence of large volumes of ferric oxides in sediments having all the characteristics of chemical deposition from open water bodies (summarized by Mackenzie, 1975; Ewers and Morris, 1981) is anomalous, especially if the setting were marine. Modern oxygenous sea water is essentially devoid of iron, as is that of most large lakes above the sill-depth of isolated anaerobic basins. The modern atmosphere and most parts of the hydrosphere are amply oxygenous. How could the iron of the BIF be transported in such quantities under oxidizing conditions, in what states, and from where? And if it was transported in the ferrous state, how was it converted to ferric and ferro-ferric oxides beneath an essentially anoxygenous atmosphere?

The discovery by Tyler and description by Tyler and Barghoorn (1954) of a richly microfossiliferous locality in the 2×10^9 year old Gunflint Iron Formation on the north shore of Lake Superior provided a clue to those puzzles. My own subsequent studies of that apparently wholly procaryotic microbiota strongly implied the presence of photosynthesizing species related to living blue-green algae. Were they producers of free O_2 *before* the evolution of the more familiar eucaryotic photosynthesizers (the true algae and higher plants)? Or could they have depended for energy on some electron source other than H_2O? On Fe^{2+}, for instance? Both would have caused precipitation of ferric iron. Either, given a cyclicity of iron supply or microbial populations, could have accounted for the BIF. Similarly the later appearance of substantial and persistent free atmospheric (thus also hydrospheric) O_2 should have put an end to BIF as a common sedimentary rock.

My change of residence to the Lake Superior region in 1961, followed by two fortuitous conversations in 1964, brought these reflections to a head. The first conversation, with Lloyd Berkner in a cafeteria line in Washington,

D.C., revealed that we held similar views as to the anaerobic state of the primitive earth and a mainly photosynthetic source for its oxygen, although Berkner postulated a much later date for the oxygenous atmosphere than did I. A later discussion with S.S. Goldich, brought out that we two had independently evolved similar views explaining BIF as a by-product of photosynthetic O_2 production — Macgregor (1927) being our mutual seminal source. We even agreed on 1.2×10^9 years ago as the time at which the atmosphere became oxidizing — mistakenly as it turned out. We differed on details of the mechanism and concerning Goldich's views about the secondary origin of the silica in the BIF. The discussions with Berkner led to our jointly organizing a symposium at which the views of Berkner and Marshall (1965) on atmospheric evolution were detailed and updated. The discussion with Goldich led to an agreement on my part that I would defer publication until *after* a paper by Lepp and Goldich (1964) had appeared in print. That, and delay in the description and naming of the Gunflint fossils (Barghoorn and Tyler, 1965), deferred publication of the primitive earth model I first publicly presented to the Colorado Scientific Society in 1962 until the spring of 1965 (Cloud, 1965).

THE ORIGINAL MODEL

That original model (Cloud, 1965), as it relates to BIF, involved the following main elements (quotation marks signify original text):

(1) Free O_2 is not available from primary sources but becomes available only as a product of photolytic dissociation or of photosynthesis involving Photosystem II, with segregation respectively of byproduct hydrogen or carbon.

(2) Photolytic O_2 is and was quantitatively unimportant and evanescent. Although "some oxygen would probably always have been available for a little oxidative weathering" from this source, the hydrosphere and atmosphere could have become truly aerobic only after the evolution of a photosynthetic source of O_2 and the subsequent filling of all potential oxygen sinks.

(3) Banded iron-formation presents a geochemical dilemma — how to transport the iron to its sites of deposition if the hydrosphere and atmosphere were then oxidizing and how to precipitate it if they were reducing.

(4) Nannofossils resembling "blue-green algae or their probable precursors", found in the Gunflint Formation, suggest a possible relation between ferrous iron transported in reducing waters and oxygen-producing but oxygen-intolerant or microaerophilic microbes. The conversion of ferrous iron to the ferric and ferro-ferric oxides of the iron-rich bands could have been simultaneously a product of and a sink for noxious biological O_2.

(5) Eventual filling of such oxygen-sinks as a consequence of increased

biological tolerance to O_2 would have led to saturation of the hydrosphere with O_2, followed by the evasion of photosynthetic O_2 to the atmosphere.

(6) The buildup of O_2 in the atmosphere and its early conversion to O and O_3 by high-energy UV radiation should have produced recognizable geochemical effects. Prominent among them would have been "the retention of ferric iron in the weathering profile, the wide appearance of red beds as an important sediment, and great diminution or disappearance" of the BIF.

At the time this model was proposed my views on a possible source for the silica were ambiguous and my geochronology was mistaken. The oldest extensive redbeds of which I then knew were believed to be only $\sim 1.2 \times 10^9$ years old, and the youngest of the major BIFs were thought to be $\sim 1.7 \times 10^9$ years old. The Gunflint Formation itself was thought to be between 1.7 and 2.1×10^9 years old.

In the discussion that follows I emphasise but do not isolate those parts of the model that bear most specifically on the BIF itself.

SUBSEQUENT MODIFICATION OF THE MODEL

Corrections were soon forthcoming. Correspondents drew attention to redbeds as old as 1.8 to 1.9×10^9 years, as predicted by the model, including some I had thought to be only 1.2×10^9 years old. Others cited BIF-like deposits as young as Devonian and still others the huge late Proterozoic banded iron deposits of Rapitan and Urucum (Jacadigo). The new ages were incorporated in a diagram prepared for a Berkner Memorial Symposium (Cloud, 1968), where I also introduced the subsequently modified idea of the evolution of advanced oxygen-mediating enzymes as the trigger for a near-instantaneous overfilling of O_2 sinks $\sim 2 \times 10^9$ years ago.

A more extensive revision followed (Cloud, 1972), in which, for the first time, I tabulated the geochemical budget for O_2 and C in the sedimentary record, sea, and atmosphere using quantities calculated by Rubey (1951). Based on the near-balance of combining equivalents found, I concluded that aquatic photosynthesis was indeed the preponderant source of combinant O_2 far into the geologic past, before free O_2 could saturate the hydrosphere and begin to accumulate in the atmosphere. Meanwhile Roscoe (1969) and others had emphasized the wide distribution of easily oxidized but extensive detrital uraninite and pyrite as recently as 2.3×10^9 years ago, offering it as supporting evidence for very low levels of atmospheric O_2 until then. And still more recently Mackenzie (1975), has noted that detrital siderite in the $\sim 2 \times 10^9$ year old Animikie sediments of the Lake Superior region carries a similar implication of very low O_2 levels.

By 1970 I had begun to feel dissatisfied with the adequacy of iron from surface weathering and volcanism as continuing sources for the major Proterozoic BIFs — a problem most recently dealt with by Ewers and Morris

(1981). As contemporary sources of times involved, either weathering or volcanism seemed quantitatively inadequate for such deposits; although, cumulatively considered, they are presumably the ultimate sources of the iron. To account for the huge volumes represented in great deposits such as those of the Hamersley Basin, Kuruman, the Labrador Trough and others (Trendall, 1968; Gole and Klein, 1981) ample storage facilities are called for, where iron from whatever source can accumulate in supply centers.

Following a 1972 field trip in southern Ontario with Grant Young and Wm. Church my thoughts turned to what part might have been played by the repeated widespread glaciations of Huronian age, with extension to northern Michigan and Wyoming. Could such glaciation and following frigid climates have resulted in seasonal displacement and upwelling of basinal bottom waters rich in ferrous iron to sites of BIF deposition on epicratonal shelves? When I aired such thoughts at a symposium on BIF later in 1972 (Cloud, 1973), it turned out that Holland (1973) had independently and even more explicitly arrived at similar views on the volume problem and the likelihood of iron storage in Proterozoic ocean-basins. Sources of upwelling other than glacial might be offshore winds, volcanism, or large-scale slumping. Both glaciation and large-scale slumping may be implicated for the late Proterozoic or early Phanerozoic Rapitan deposits of fault basins of the Mackenzie Mts, Yukon Territory, Canada (Eisbacher, 1978).

Seasonal upwelling related to glacial climates would be one way of accounting for microbanding. Seasonal microbial bloom is another. Episodic operation of one or both seasonalities, with local reworking by waves or currents, could explain most of the textural variants described by Gole and Klein (1981). If upwelling is indeed important for the delivery of Fe^{2+}-rich waters to sites of BIF deposition, we might expect to find evidence for contemporaneous or preceding glacial climate, or other potential causes of upwelling associated with BIF at other places and times.

In the same 1973 paper where I addressed basinal storage and upwelling of iron, I also first discussed the problem of silica sources, following with a more extended discussion in 1974. There being no known silica-secreting procaryotes, no probable eucaryotes older than $\sim 1.4 \times 10^9$ years, and no likely eucaryotes older than $\sim 2 \times 10^9$ years old, a biological source for SiO_2 in the BIF is improbable. A physico-chemical source for the silica is much more likely — from a combination of weathering and volcanism, with accumulation in solution to saturation levels in the absence of microbial precipitators. The work of Krauskopf (1956) and others, summarized by Walter et al. (1972), shows that the polymerization of monosilicic acid solutions and the precipitation from them of SiO_2 is favored at near neutral to slightly alkaline states. The abundance of the sodium-amphibole, riebeckite in BIF of the Hamersley Basin and its presence at other localities implies saline, probably marine, waters. Silica- and iron-saturated seawaters having pH from

6.8 to 7.5 seem a not improbable consequence of a high P_{CO_2}, combined with the bicarbonate and silicate buffering systems, a view supported by Mackenzie (1975) who notes that silica-saturated seawater would have precipitated sepiolite at pH >7.5, a sedimentary mineral unknown among pre-Phanerozoic rocks.

Polymerization of H_4SiO_4 and the precipitation of silica gel in the same places to which ferric hydroxides were intermittently added as a product of the combination of Fe^{2+} with biologically generated oxygen could account for the BIF via reactions such as those outlined by Mackenzie (1975), Ewers and Morris (1981), and others. In such a system the potential cations of carbonate sedimentation would remain in solution except where locally elevated pH levels facilitated the conversion of HCO_3^- to CO_3^{2-} and the joining of CO_3^{2-} anions to available Ca^{2+} and Mg^{2+} cations to produce the ferrocarbonate observed in some BIF microbands.

Other elaborations (Cloud, 1976a, b, 1980) have dealt with possible modern biological analogs of the hypothesized BIF precipitators, the question of the carbon that had to be sequestered to yield all the supposedly photosynthetic O_2, refinements in the nature of the interaction visualized between photosynthetic O_2 and BIF, and new records of ancient sulfates and redbeds.

TESTABLE COROLLARIES OF THE MODEL

Hypotheses and models that may be considered fruitful make predictions that can be tested and validated or falsified. The hypothesis or model gains credence with each validation and loses it with each falsification. Eventually, it may either be disproved or survive enough opportunities for disproof to warrant higher levels of confidence. We can test aspects of the general model relevant to BIF against the following six verifiable consequences:

(1) If the iron in BIF is primarily a product of the combination of Fe^{2+} with photosynthetic O_2, then biochemically analogous modern microbes may exist.

(2) Plausible analogs of such modern microbes, moreover, should be found associated with BIF at favorable sites.

(3) Inasmuch as each mole of photosynthetic O_2 is represented somewhere by a mole of C, stoichiometrically equivalent quantities of coeval C must be accounted for. One might also expect traces of phosphate to be associated with carbon of biological origin or iron precipitated as a product of biochemical activity.

(4) BIF and redbeds, being respectively the products of generally anaerobic and generally aerobic environments, should show limited overlap in geologic time.

(5) Because O_2 varies inversely as CO_2, which controls the carbonate equilibria, carbonate rocks should be less common before the oldest redbeds

than after (and more of those present would likely be ferrocarbonates).

(6) Gypsum and other sulfates, although expectable products of the sulfide and sulfite oxygen sinks, should be uncommon sedimentary rocks before the general prevalence of atmospheric O_2 signalled by widespread redbeds.

FINDINGS RELEVANT TO A TEST OF THE MODEL

Consider now the results of testing these six corollaries:

(1) Biochemically analogous modern microbes, unknown to me before the 1970 paper of Stewart and Pearson (although described by Spruit in 1962) are now known to occur among six or seven different genera of blue-green algae and even some eucaryotic algae (e.g.,Oren et al., 1977). They are analogous to the hypothesized microbial flora responsible for the BIF in that, although O_2 releasing photosynthesizers, they prefer low levels of ambient O_2, like to live in the presence of oxygen-depressing compounds (such as H_2S), and are capable of switching from aerobic photosynthesis utilizing H_2O as an electron donor to anaerobic photosynthesis utilizing H_2S as an electron donor. For a good parallel between such microbes and those called upon to generate the conditions that led to precipitation of the BIF, we need only hypothesize that iron, so usual in oxygen-carrying biological pigments, was about as good an ambient O_2 depressor in the ferrous state as H_2S and that perhaps Fe^{2+} could serve as an alternative electron donor. The details of the relationship proposed are discussed in p. 1138 of my 1973 paper on the palaeoecology of the BIF.

(2) Microbial remains of the sort that we should expect to find associated with the BIF consist exclusively of soft carbonaceous matter that is easily oxidized and thus rarely preserved. Abundant microbial remains have, nevertheless, been hermetically sealed and preserved within primary or early diagenetic cherts in or associated with various BIFs. Examples are the Gunflint (Barghoorn and Tyler, 1965), Biwabik (Cloud and Licari, 1968), Frere (Walter et al., 1976), Tyler (Cloud and Morrison, 1980), and Sokoman (Knoll and Simonson, 1981) Formations, as well as strata of the Belcher Group (Hofmann, 1976). The size and morphology of these nannofossils strongly implies a procaryotic nature, and some of them are morphologically almost identical to living oscillatoriacean and nostocacean blue green algae — both of which include living oxygen-sensitive forms that require ambient O_2 depression and can switch from aerobic to anaerobic photosynthesis below about one per cent present atmospheric level of O_2 (the Pasteur Point).

(3) Concerning carbon, it must be admitted at the outset that precious little of it is associated with most BIF. Thin bands of graphitic coal, however, are interbedded with some BIFs — the Archean Soudan BIF for instance

(Cloud et al., 1965). More significant is the C that forms up to 10 or 12 per cent of the black pyritic shales that occur in the deeper parts of some BIF basins (H,L. James, personal communication, Nov.,1962) — significant because the buoyant carbonaceous residues of contemporary microbial biotas may well have drifted away from sites of BIF sedimentation to basin centers where not embedded in coeval silica gel. A third possible explanation for the missing carbon is the magnetite in the BIF itself, a perception we owe to Perry et al. (1973):

$$6Fe_2O_3 + C \rightarrow 4Fe_3O_4 + CO_2, \qquad (1)$$

with evasion of CO_2. (Following the sedimentation of ferric hydroxides, magnetite may also form under limited O_2 availability by the process:

$$2Fe(OH)_3 + Fe^{2+} \rightarrow Fe_3O_4 + 2H^+ + 2H_2O, \qquad (2)$$

a possible explanation for the prevalence of magnetite in Archaean BIFs).

As for phosphate that might logically be associated with iron where the latter precipitated as a result of the biological addition of O_2, interesting tests have been made by Morris (1973). Results are equivocal, however, showing 0.03 to 0.15% P as traces of apatite on a mesoband scale with some but erratic lateral continuity of phosphate and with no consistent relationship to iron.

(4) *BIF and redbeds do overlap* in time, a definite weakness of the model but by no means fatal at the presently known scale and frequency, and considering difference in type. Two large late Proterozoic or early Paleozoic deposits that fit the definition of BIF in most respects, as well as several small but clearly Paleozoic ones, are known (summarized with references in Cloud, 1973). Dimroth (e.g.,1970) has also called attention to significant redbeds in the Chakonipau Formation, beneath the Sokoman BIF.

The downward extension of redbeds, to be sure, is less troublesome than it might seem, because it takes so little O_2 to make redbeds under favorable circumstances, and the admittedly diagenetic origin of the Chakonipau redbeds allows time and alternate sources for oxygenation. It is the relative abundance of redbeds among rocks younger than $\sim 2 \times 10^9$ years that merits emphasis. Nevertheless, Roscoe (1969) has long argued that the beginnings of oxidation might date back as far as 2.3×10^9 years ago, and those older redbeds imply that it may now be necessary to invoke a transition interval from perhaps 2.3 to 2×10^9 years ago, during which temporary local oxidation occurred and trends reversed as new oxygen sinks appeared and became filled.

The greater age-spread of the older Proterozoic BIF suggested by new radiometric ages is not surprising. A nearly continuous range in time before 2×10^9 years ago is to be expected from the general model. The younger BIF-like rocks, however, remain troublesome despite the prominent clastic components that differentiate them so sharply from the typical older BIFs. It is not hard to explain the trivial and local Palaeozoic BIFs as the product of volcanic sources of Fe^{2+} welling upward from anaerobic basins into oxygenous surface waters. But the large late Proterozoic or early Phanerozoic iron deposits of the Jacadigo beds at Urucum, Mato Grosso, along the Brazil — Bolivian border and those of the Rapitan Group, in Yukon Territory, Canada, are more difficult. And the difficulty is compounded for me by the fact that I have seen neither of these deposits in the field. The sedimentary and physical relations of these younger iron-formations, however, constrast with those of the typical older BIF. Those of the Rapitan are associated with lenticular basinal shales and siltstones in the Sayunei Formation, containing authentic glacial dropstones and features of turbidite deposition (Young, 1976; Eisbacher, 1978). That suggests that the iron may have been stored in solution in anaerobic basins before slumping and density currents caused it to well up into oxygenated surface waters and precipitate. Although reported to be large, these deposits seem to comprise but a minor fraction of the Sayunei sequence (Eisbacher, 1978, figs. 10—11). The Jacadigo deposits, on the other hand, are described as erosional remnants of a once extensive sheet of silica-banded hematite that is cut by fluviatile channel fills and associated with rich manganese deposits and possible evidences of glaciation (Dorr, 1973) — a very different picture from the early Proterozoic deposits except for the potential for glacial upwelling. Earlier accounts described the Jacadigo deposits as lenticular (Dorr, 1945). Could they be lake deposits of bacterial origin, perhaps analogous to those of Karelian lakes where bacterial Fe and Mn deposits are forming today, or glacial, like the Rapitan?

The Jacadigo and Rapitan deposits seem no less enigmatic to me now than they did in 1973. Both deserve further study, although it is unlikely that the findings of such a study would invalidate a hypothesis that was developed to explain deposits as different as the classic BIFs.

(5) *As for the distribution in time of carbonate rocks*, it is clear both that such rocks do occur among the oldest sediments known (thin ferrocarbonates among the Isua supracrustals of SW Greenland) and that some later pre-redbed dolomites and limestones are very thick and extensive. It does seem that carbonate rocks are less prevalent and more frequently ferrous among the older rocks, but I see no abrupt transitions and have made no quantitative tabulations. The matter is interesting but not crucial for or against the general model

(6) *Sedimentary sulphates* at one time appeared to be rare before the Phanerozoic and common after the Proterozoic. I related that primarily to a similarly provisional quantum increase in free O_2, seen as a likely trigger for early metazoan evolution. It has always been clear, however, from pseudomorphs and molds in various host rocks, that some gypsum and anhydrite had been deposited in older sediments as isolated crystals or local accumulations. Now increasing evidence for older stratified sulphates has been reported following my earlier suggestion that they would be rare and limited in extent and thickness. It is clear that sedimentary sulphate was

present even in early Archean time (e.g. Lambert et al., 1978) and that it has been regularly deposited at favorable sites from then until now. It is also clear that sedimentary SO_4^{2-} can be produced by bacterial oxidation of sulphides in the absence of any free O_2 and that it does not speak convincingly to questions of atmospheric O_2. I should add that, consistent with the broader model, it does seem to be more abundant in younger rocks. It is, to be sure, also possible that its seeming rarity in pre-redbed rocks may be in part a function of removal by solution.

CURRENT STATUS OF THE MODEL AS IT RELATES TO BIF

One can see from the foregoing that, of six predictions made by the model that are also relevant to the BIF, three are substantially validated and three are not. The model, therefore, needs either to be modified to eliminate or reduce its weaknesses or to be abandoned. As, in its main elements, it has displayed some heuristic merit and may even be valid, it seems more to the point to ask what changes are needed and what tests of the revised model can be devised.

I have already suggested changes of the model in recent discussions (Cloud, in press). One involved shifting emphasis previously placed on internal regulation of microbially produced O_2 to its external regulation, responding to findings relevant to the first test proposed above. The iron of the BIF, in fact, has never been visualized by me as a direct biological precipitate (although were Fe^{2+} used as an electron donor it could be). Instead I think of it as a chemical byproduct of the local and temporary presence of photosynthetic O_2 in the ambient water column. A second modification recognizes that the transition from a generally anaerobic to an aerobic state of hydrosphere and atmosphere was probably not as abrupt as I once supposed but occurred over a range of perhaps 300 million years (2.3 to 2×10^9 years ago with reversals and local episodes of oxidation). For all that, 2×10^9 years ago still seems a reasonable round number to apply to the beginning of a persistent, generally oxidized atmosphere.

In addition to these changes, and as outlined above, I have suggested possible explanations for younger Proterozoic and Phanerozoic banded iron deposits and reduced emphasis on the significance of possible temporal variations in the prevalence of sedimentary carbonates and sulphates.

I have added the likelihood of storage of Fe^{2+} in basinal bottom waters, with mechanisms for inducing upwelling that could lead to precipitation from surface waters above shelves or basins. Archean upwelling may well have been mainly a product of volcanism and basin-margin slumping, leading to precipitation of magnetite in the absence of much free O_2 and at low sulphate and carbonate activity. And some or much Proterozoic upwelling may have been climatically induced by cold sinking surface waters resulting from contemporaneous or preceding glaciation. A verifiable consequence of this would be the occurrence of tillites and dropstones in coeval deposits. In addition to a long succession of glacial deposits of Huronian and Animikian age over a wide area in North America (Church and Young, 1970; Houston et al., 1968), facetted and striated boulders implying a glacial origin have now also been found above BIF in the Turee Creek Group of Western Australia (Trendall, 1976). Though these episodes may not necessarily be directly related to the formation of BIF, further inquiry into the possibility of climatically induced upwelling is called for.

Apart from this, the tests I visualize are extensions of ones already mentioned. We need, above all, quantitative tabulations of associated sediment types based on reliable geologic mapping; estimates such as those of Button (1979), Grandstaff (1980), and Walker (1977) on geochemically probable specific levels of O_2 at spot times in the past; and a continuing search for and description of new microbial floras at all levels of the pre-Phanerozoic column.

Provisional though it still is, I would not at this time change the mechanisms and interactions suggested in my 1973 presentation of the general model. The processes suggested explain the main characteristics of BIF better than other mechanisms so far proposed. I conclude that the model has satisfied the criterion of testability in several important respects, that inquiries to which it has led have been productive of interesting and useful new information, and that a continuation of such inquiry will satisfy Ben Franklin's admonition to devote first priority to those investigations that are both ornamental and useful. I would, however, modify figure 1 of my 1973 paper to extend the range of BIF downward to the oldest known sedimentary rocks (Isua). The stratigraphic range of sedimentary sulphates should also be extended almost or quite as far downward. And modest downward extensions are needed for red beds and the prevalence of carbonate rocks. Should continued geologic mapping or new radiometric ages, however, come to reveal a temporally extensive and areally widespread and voluminous overlap between significant primary or diagenetic redbeds and typical BIF, the concept of a geologically modest transition interval between oxygen states would have to be abandoned.

An alternative explanation for the BIF that is also consistent with other aspects of the general model is the inverse segregation hypothesis of Cairns-Smith (1978). This hypothesis invokes inorganic photochemical oxidation, activated by UV irradiation prior to the evolution of an ozone screen at about 1 per cent present atmospheric level of O_2. An appealing feature of this concept is that it would have been most effective over the interval just before $\sim 2 \times 10^9$ years ago, when the atmosphere was first beginning to become oxidising. A limitation arises in explaining the observed cyclical banding under such a mechanism. This cyclicity would necessarily become a function of an intermittent availability of iron in waters susceptible to UV penetration. Seasonal microbial blooms could no longer be invoked. I see no advantage in inverse segregation as the main causal mechanism for BIF, although it could well be a contributing process and may explain some of the anomalously old surface oxidation reported.

REFERENCES

Barghoorn, E.S. and Tyler, S.A., 1965. Microorganisms from the Gunflint chert. Science, 147: 563—577.

Berkner, L.V. and Marshall, L.C., 1965. History of major atmospheric components. Proc. Natl. Acad. Sci. (U.S.), 53: 1215—1225.

Brown, H. 1952. Rare gases and the formation of the earth's atmosphere. In: G.P. Kuiper (Editor), The Atmospheres of the Earth and Planets (2nd ed.). Univ. Chicago Press, Chicago. Ill., pp. 258—266.

Button, A., 1979. Early Proterozoic weathering profile on the 2200 m.y. old Hekpoort Basalt, Pretoria Group, South Africa: preliminary results. Univ. Witwatersrand, Econ. Geol. Research Unit, Information Circular 133, 21 pp.

Cairns-Smith, A.G., 1978. Precambrian solution geochemistry, inverse segregation, and banded iron formations. Nature, 148: 27—35.

Church, W.R. and Young, G.M., 1970. Discussion of the progress report of the Federal-Provincial Committee on Huronian stratigraphy. Can. J. Earth Sci., 7: 912—918.

Cloud, P., 1948. Some problems and patterns of evolution exemplified by fossil invertebrates. Evolution, 2: 322—250.

Cloud, P., 1965. Significance of the Gunflint (Precambrian) microflora. Science, 148: 27—45.

Cloud, P., 1968. Atmospheric and hydrospheric evolution on the primitive earth. Science, 160: 729—736.

Cloud, P., 1972. A working model of the primitive earth. Am. J. Sci., 272: 537—548.

Cloud, P., 1973. Paleoecological significance of the banded iron formation. Econ. Geol., 68: 1135—1143.

Cloud, P., 1974. Evolution of ecosystems. Am. Sci., 62: 54—66.

Cloud, P., 1976a. Beginnings of biospheric evolution and their biogeochemical consequences. Paleobiology, 2: 351—387.

Cloud, P., 1976b. Major features of crustal evolution. Geol. Soc. S. Afr., Annexure to vol. 79 (Alexander L. Du Toit Memorial Lecture No. 14), 33 pp.

Cloud, P., 1980. Early biogeochemical systems. In: P.A. Trudinger et al. (Editors), Biogeochemistry of Ancient and Modern Environments. Australian Acad. Sci.-Springer Verlag, pp. 7—27.

Cloud, P., in press. Aspects of Proterozoic biogeology. Geol. Soc. Am., Special Paper.

Cloud, P. and Licari, G.R., 1968. Microbiotas of the banded iron formations. Proc. Natl. Acad. Sci. (U.S.), 61: 779—786.

Cloud, P. and Morrison, K., 1980. New microbial fossils from 2 Gyr old rocks in northern Michigan. Geomicrobiol. J., 2(2): 161:178.

Cloud, P., Gruner, J.W. and Hagen, H., 1965. Carbonaceous rocks of the Soudan Iron Formation (Early Precambrian). Science, 148: 1713—1716.

Dimroth, E., 1970. Evolution of the Labrador Geosyncline. Geol. Soc. Am. Bull., 81: 271—2742.

Dorr, J.V.N., II, 1945. Manganese and iron deposits of Morro do Urucum, Mato Grosso, Brazil. U.S. Geol. Surv. Bull. 976-A, 47 pp.

Dorr, J.V.N., II, 1973. Iron formation in South America. Econ. Geol., 68(7): 1005—1022.

Ewers, W.E. and Morris, R.C., 1981. Studies of the Dales Gorge Member of the Brockman Iron Formation, Western Australia. Econ. Geol., 77: 1929—1953.

Eisbacher, G.H., 1978. Re-definition and subdivision of the Rapitan Group, Mackenzie Mountains. Geol. Surv. Can., Pap. 77—35, 21 p.

Gole, M.J. and Klein, C., 1981. Banded iron-formations through much of Precambrian time. J. Geol., 89: 169—183.

Grandstaff, D.E., 1980. Origin of uraniferous conglomerates at Elliot Lake, Canada and Witwatersrand, South Africa: implications for oxygen in the Precambrian atmosphere. Precambrian Res., 13(1): 1—26.

Hofmann, H.J., 1976. Precambrian microflora, Belcher Islands, Canada: significance and systematics. J. Paleontol., 50: 1040—1073.

Holland, H.D., 1973. The oceans: a possible source of iron in iron-formations. Econ. Geol., 68(7): 1169—1172.

James, H.L., 1954. Sedimentary facies of iron formation. Econ. Geol. 49: 235—293.

James, H.L. and Sims, P.K., 1973. Precambrian banded iron-formations of the world. Econ. Geol., 68: 913—914.

Knoll, A.H. and Simonson, B., 1980. Early Proterozoic microfossils and penecontemporaneous quartz cementation in the Sokoman Iron Formation, Canada. Science, 211: 478—480.

Krauskopf, K.B., 1956. Dissolution and precipitation of silica at low temperatures. Geochim. Cosmochim. Acta, 12: 61—84.

Lambert, I.B., Donnelly, T.H., Dunlop, J.S.R. and Groves, D.I., 1978. Stable isotope compositions of early Archean sulphate deposits of probable evaporitic and volcanogenic origins. Nature, 276: 808—810.

Lepp, H. and Goldich, S.S., 1964. Origin of Precambrian iron formations. Econ. Geol., 59: 1025—1060.

Macgregor, A.M., 1927. The problem of the Precambrian atmosphere. S. Afr. J. Sci., 24: 155—172.

Mackenzie, F.T., 1975. Sedimentary recycling and the evolution of sea water. In: J.P. Riley and G. Skirrow (Editors), Chemical Oceanography, Vol. 1(2nd ed.). Academic Press, London-New York-San Francisco, pp. 209—364.

Morris, R.C., 1973. A pilot study of phosphorus distribution in parts of the Brockman Iron Formation, Hamersley Group, Western Australia. West. Aust., Geol. Surv., Annu., Rep. 1972: 75—81.

Nursall, J.R., 1959. Oxygen as a prerequisite to the origin of the Metazoa. Nature, 183: 1170—1172.

Oren, A., Padan, E. and Avron, M. 1977. Quantum yields for oxygenic photosynthesis of the cyanobacterium *Oscillatoria limnetica.* Proc. Natl. Acad. Sci. (U.S.), 74: 2152—2156.

Perry, E.C., Jr., Tan, F.C. and Morey, G.B., 1973. Geology and stable isotope geochemistry of the Biwabik Iron Formation, northern Minnesota. Econ. Geol., 68: 1110—1125.

Roscoe, S.M., 1969. Huronian rocks and uraniferous conglomerates in the Canadian Shield. Geol. Surv. Can., Pap. 68—40, 205 pp.

Rubey, W.W., 1951. Geologic history of sea water. Bull. Geol. Soc. Am., 62: 1111—1148.

Spruit, C.P.J., 1962. Photoreduction and anaerobiosis. In: R.A. Lewin (Editor), Physiology and Biochemistry of Algae. Academic Press, New York, N.Y., 929 pp.

Stewart, W.D.P. and Pearson, H.W., 1970. Effects of aerobic and anaerobic conditions on growth and metabolism of blue-green algae. Proc. R. Soc. London, 175: 293—311.

Trendall, A.F., 1965. Progress report on the Brockman Iron Formation in the Wittenoom-Yampire area. West. Aust, Geol. Surv., Annu. Rep., 1964: 55—65.

Trendall, A.F., 1968. Three great basins of Precambrian banded iron formation deposition: a systematic comparison. Geol. Soc. Am. Bull., 79: 1527—1544.

Trendall, A.F., 1976. Striated and faceted boulders from the Turee Creek Formation — evidence for a possible Huronian glaciation on the Australian Continent. West. Aust., Geol. Surv., Annu. Rep., 1975: 88—92.

Tyler, S.A. and Barghoorn, E.S., 1954. Occurrence of structurally preserved plants in pre-Cambrian rocks of the Canadian Shield. Science, 119: 606—608.
Walker, J.C.G., 1977. Evolution of the Atmosphere. MacMillan Publ. Co., Inc., 318 pp.
Walter, M.R., Bauld, J. and Brock, T.D., 1972. Siliceous algal and bacterial stromatolites in hot springs and geyser effluents of Yellowstone National Park. Science, 178: 402—405.
Walter, M.R., Goode, A.D.T. and Hall, W.D.M., 1976. Microfossils from a newly discovered Precambrian stromatolitic iron formation in Western Australia. Nature, 261: 221—223.
Young, G.M. 1976. Iron-formation and glaciogenic rocks of the Rapitan Group, Northwest Territories, Canada. Precambrian Res., 3: 137—158.

Chapter 11

DIAGENESIS AND METAMORPHISM OF PRECAMBRIAN BANDED IRON-FORMATIONS

CORNELIS KLEIN

INTRODUCTION

This chapter represents a synopsis of the major-element chemistry, mineralogy and petrology of Precambrian iron-formations that have undergone various degrees of metamorphism. In order to be able to delineate reactions that have taken place during metamorphism, the least metamorphosed or, if available, completely unmetamorphosed iron-formation assemblages must be defined at the onset. As such, the first part of this chapter presents information on iron-formations that have undergone late diagenetic as well as very low-grade metamorphic reactions on account of burial*; subsequent parts will discuss the more easily defined, higher temperature metamorphic changes.

This overview will concentrate, by necessity, on those occurrences for which relatively detailed descriptions and quantitative data are available. This means that much of the information in this chapter will be drawn from a restricted number of Proterozoic iron-formations, with additional input from a few Archean iron-formation occurrences. The Proterozoic deposits include: (1) the Sokoman Iron Formation (and its metamorphosed equivalents south of the Grenville Front) in the Labrador Trough, Canada; (2) the Negaunee Iron Formation of the Menominee Group in the Marquette district of Northern Michigan, U.S.A.; (3) the Biwabik Iron Formation of the Mesabi district in Minnesota, U.S.A., and its continuation in the Gunflint Iron Formation of northeastern Minnesota and Ontario, Canada; and (4) the iron-formations of the Hamersley Group of the northwestern part of Western Australia. Information on several Archean iron-formations in the Yilgarn Block of Western Australia, and in southwestern Montana will be incorporated as well.

---

* Late diagenetic reactions are approximately defined by Winkler (1979) as occurring below about 180°C, whereas very low-grade metamorphic reactions would represent a temperature range of approximately 180° to about 300°C. Because in iron-formations the criteria are inadequate to distinguish between late diagenetic and very low-grade reactions, both ranks are treated together.

MAJOR-ELEMENT CHEMISTRY OF BANDED IRON-FORMATIONS

At the outset of any discussion of metamorphic reactions, in response to changing conditions in pressure and/or temperature, one wishes to know whether the metamorphism has been essentially isochemical or whether it has involved the loss or gain of considerable amounts of chemical components. Furthermore, it would be informative to know what general range of chemical bulk compositions is displayed by Precambrian banded iron-formations that are more than about 1.8 b.y. old.

The question of the possible isochemical nature of the metamorphism of banded iron-formations has been addressed by many authors (French, 1968; Bonnichsen, 1975; Floran and Papike, 1978; and Klein, 1978). Klein compared the chemistry of 22 bulk analyses of metamorphic iron-formations in the southern part of the Labrador Trough with 25 bulk analyses of late diagenetic to very low-grade metamorphic iron-formation samples from the central part of the Labrador Trough. The compositions of these two groups of bulk analyses differ mainly in their H_2O and CO_2, and considerably in their SiO_2 contents. The H_2O- and CO_2-rich analyses are those of the essentially unmetamorphosed materials which leads to the conclusion that such volatile components were, to a large degree, lost during metamorphism. The variability in SiO_2 contents between many of the samples is a reflection of the highly variable scale of the chert- or quartz-rich banding. In short, except for the loss of volatile components the metamorphism, at least in the regionally metamorphosed iron-formations of the Labrador Trough, appears to have been essentially isochemical. Furthermore, the $Fe^{3+}/(Fe^{2+} + Fe^{3+})$ ratio of the iron-formations does not appear to have changed in the transition from very low-grade metamorphism to higher metamorphic grade assemblages. This was also concluded by Gole (1981) for the metamorphism of Archean iron-formations in the Yilgarn Block of Western Australia.

The matter of the range of bulk compositions of fresh (unaltered, unleached and unoxidized by post-diagenetic processes) banded iron-formations (*not* iron ores) was evaluated by Gole and Klein (1981a). The averages and ranges for major elements in the iron-formations were obtained from bulk chemical data for two Archean and three major Proterozoic terrains. These results were recalculated to 100% on an H_2O- and CO_2-free basis in order to allow a comparison of unmetamorphosed and metamorphosed occurrences, assuming that the metamorphism was essentially isochemical except for H_2O and CO_2 loss. In addition to the bulk analyses by Trendall and Pepper (1977) of the Brockman Iron Formation, Ewers and Morris (1981) provide average analyses for very long and continuous sections through the BIF and S macrobands of the Dales Gorge Member in the Paraburdoo area in Western Australia. These data, in addition to those of Gole and Klein (1981a) are given in Table 11-I and Fig. 11-1. The data in Table 11-I can be used only for very general comparisons, because banded iron-formations are very inhomogeneous. Such

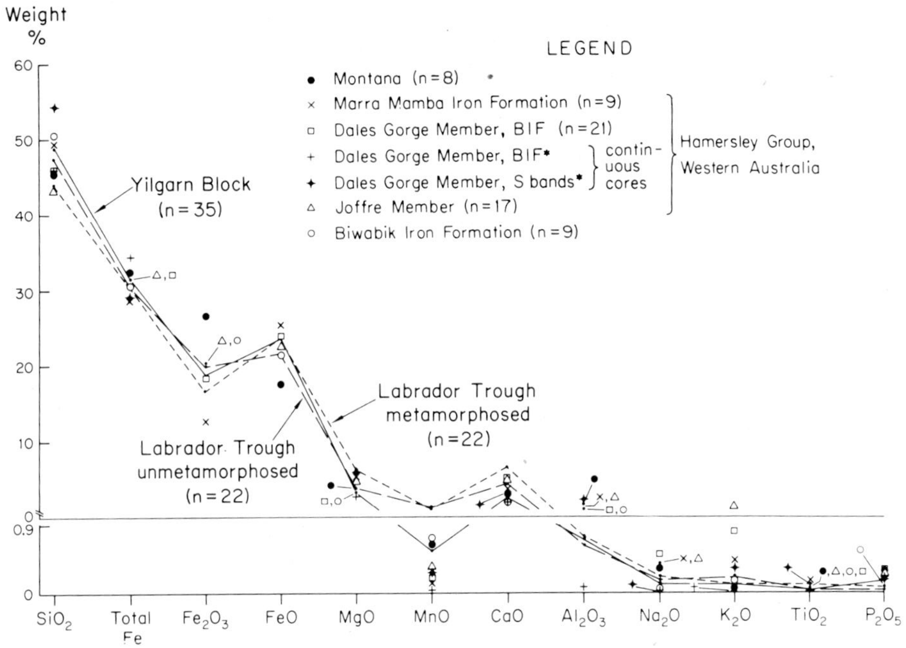

Fig. 11-1. Plot of averaged bulk chemical analyses of banded iron-formations given in Table 11-I. The analyses are recalculated to 100% on an H_2O- and CO_2-free basis. The data points, except for those for BIF and S bands from Ewers and Morris (1981), are from Gole and Klein (1981a). For the two entries marked by *, FeO and Fe_2O_3 data are not shown (only total Fe) because in these analyses total Fe was recalculated as Fe_2O_3 only, without independent FeO determinations. The core lengths sampled by Ewers and Morris (1981) are 103.78 m for the BIF and 18.65 m for the S bands (see also footnote to Table 11-I).

inhomogeneity is the result of various scales of banding and of the diversity of assemblages that can occur in iron-formations with fairly similar bulk compositions, as can be seen from assemblage listings in Klein (1978, table 2) and Gole (1981, table 1). Furthermore, the averages and ranges in Table 11-I (except for columns 5 and 6) are based on a relatively small number of analyses (numbers of samples are listed in the footnote to Table 11-I). Such ranges, particularly those of the Biwabik Iron Formation, do not represent the complete spectrum of compositions that are likely to occur in each area, although some may approximate the actual range. The most accurate averages are given in columns 5 and 6 of Table 11-I as these are based on a very large number of almost continuous samples from drill cores. Column 6 represents S macrobands which compositionally fall within the range of banded iron-formations (James, 1954; Ewers and Morris, 1981).

On the basis of the above it appears that the chemical data in Table 11-I

TABLE 11-I

Averages and ranges of bulk analyses of Precambrian banded iron-formations recalculated to 100% on an H_2O- and CO_2-free basis (after Gole and Klein, 1981a, except for columns 5 and 6)

	Archean		Proterozoic							
	1 Yilgarn Block	2 Montana	3 Hamersley Basin	4	5	6	7	8 Labrador Trough#	9	10 Biwabik
			Marra Mamba	Dales Gorge Member			Joffre Member	unmet.	met.	
				BIF	BIF	S bands				
SiO_2	49.07	45.53	49.16	46.20	45.76	54.01	43.31	47.81	44.33	50.62
	21.5–68.3	33.6–59.5	26.8–67.4	24.8–60.2			7.7–59.1	16.1–87.2	15.2–63.6	46.9–53.4
TiO_2	0.04	0.06	0.18	0.04	0.01	0.11	0.07	0.04	0.10	0.06
	0.00–0.18	0.00–0.18	0.00–0.88	0.00–0.15			0.01–0.28	0.00–0.25	0.00–0.59	0.02–0.13
Al_2O_3	0.70	1.80	1.64	1.03	0.09	2.41	1.72	0.62	0.74	1.13
	0.01–3.51	0.01–5.86	0.00–6.00	0.00–5.12			0.04–7.54	0.05–1.98	0.00–4.23	0.33–2.28
Fe_2O_3	18.98	26.91	12.93	18.40	(49.17)	(32.76)	20.16	19.96	16.87	20.28
	1.7–37.6	3.4–38.3	0.0–39.5	0.1–35.9			3.1–40.1	0.0–65.0	0.0–77.9	12.0–26.3
FeO	23.65	17.51	25.49	23.88	n.d.*	n.d.*	22.53	21.69	23.68	21.43
	14.4–66.6	4.5–27.3	13.6–38.8	13.3–45.6			14.3–33.6	2.1–46.8	1.1–40.9	17.4–26.7
MnO	0.55	0.64	0.13	0.18	0.05	0.28	0.36	1.15	1.01	0.72
	0.04–2.76	0.01–3.26	0.03–0.36	0.00–0.50			0.04–2.68	0.06–2.52	0.12–2.66	0.41–1.02
MgO	3.46	3.82	5.03	3.15	2.85	6.15	4.86	4.00	6.25	3.17
	1.00–9.63	1.25–5.72	2.14–9.42	1.50–6.66			1.78–14.81	0.00–15.04	0.48–13.66	2.59–4.08
CaO	2.68	3.01	3.79	5.22	1.75	3.36	4.97	4.30	6.55	1.98
	0.32–7.61	0.51–11.48	0.00–10.47	0.74–39.6			0.05–37.4	0.00–22.3	0.00–22.4	1.02–3.17
Na_2O	0.11	0.34	0.38	0.50	0.04	0.02	0.39	0.17	0.21	0.06
	0.02–0.57	0.01–0.97	0.01–0.91	0.00–4.97			0.01–2.11	0.01–3.05	0.00–1.49	0.02–0.09

K_2O	0.10	0.07	0.43	0.81	0.02	0.33	1.15	0.20	0.10	0.17
	0.00–0.61	0.01–0.21	0.10–1.19	0.00–4.28			0.00–2.85	0.03–0.46	0.02–0.54	0.07–0.31
P_2O_5	0.16	0.29	0.08	0.31	0.22	0.19	0.25	0.04	0.06	0.09
	0.02–0.45	<0.01–0.55	0.02–0.17	0.17–0.57			0.10–0.60	0.00–0.15	0.00–0.13	0.05–0.17
S	0.81	0.00	0.35	0.14	0.03	0.38	0.11	0.01	0.06	—
	0.00–5.20	0.00–0.01	0.00–2.11	0.00–1.36			0.00–0.84	0.00–0.05	0.00–0.44	—
C	0.00	0.02	0.50	0.22	n.d.*	n.d.*	0.15	0.007	0.06	0.29
	0.00–0.03	0.00–0.10	0.00–1.97	0.00–1.58			0.00–0.55	0.00–0.09	0.00–0.63	0.07–0.58
–O=	0.31	—	0.09	0.08			0.03	—	0.02	—
Total Fe	31.65	32.43	28.85	31.43	34.39	22.91	31.61	30.82	30.21	30.84
	16.2–53.7	22.6–41.0	19.6–46.5	17.7–45.7			25.7–42.1	8.6–52.4	14.3–58.6	29.1–34.2
$\frac{Fe^{3+}}{(Fe^{2+}+Fe^{3+})}$	0.42	0.58	0.31	0.41	n.d.*	n.d.*	0.45	0.45	0.39	0.46
	0.02–0.68	0.10–0.87	0.00–0.59	0.03–0.68			0.09–0.67	0.00–0.99	0.00–0.98	0.29–0.54

0.00 = <0.005%; #Mn-rich banded iron-formations not included in averages; n.d.* = not determined; Fe_2O_3 values in parentheses mean that total iron was recalculated as Fe_2O_3.

1. N = 35, Gole (1981), table 4 no. 1.
2. N = 8, Immega and Klein (1976), table 1 nos. 1 to 8.
3. N = 9, Klein and Gole (1981), table 1 nos. 1–9.
4. N = 21, Trendall and Blockley (1970), table 11 nos. 4 and 6; Trendall and Pepper (1977), table 3 nos. 3, 5, 7, 9 and 11; Klein and Gole (in prep.).
5. Samples from 103.78 m of core of 17 BIF bands, Paraburdoo, Hamersley Range, W.A., from Ewers and Morris (1981), table 4.
6. Samples from 18.65 m of core of 16 S bands, Paraburdoo, Hamersley Range, W.A., from Ewers and Morris (1981), table 4.
7. N = 17, Trendall and Blockley (1970), table 11 nos. 9 and 10; Trendall and Pepper (1977), table 3 no. 12; Klein and Gole (in prep.).
8. N = 22, Klein (1974), table 1 nos. 1 to 7; Klein and Fink (1976), table 5 nos. 1 and 2, table 7 nos. 1 to 5, table 12 nos. 1 and 2; Klein (1978), table 1 no. 1; Lesher (1978), table 8 nos. 2 to 6.
9. N = 22, Klein (1978), table 1 nos. 2 to 23.
10. N = 9, Lepp (1966), table 3 nos. 1 to 9.

and Fig. 11-1 reasonably approximate the CO_2- and H_2O-free composition of many banded Precambrian iron-formations.

DIAGENETIC TO VERY LOW-GRADE METAMORPHIC ASSEMBLAGES

Many of the well-studied Proterozoic iron-formations that have undergone metamorphism preserve sections in which late diagenetic to very low-grade metamorphic lithologies can be evaluated. Parts of the extensive Proterozoic iron-formations in the Hamersley area contain late diagenetic to very low-grade metamorphic assemblages. From a study of associated younger and older rocks Smith et al. (1982), suggest a range of metamorphic conditions for the Hamersley Group from the lowest prehnite-pumpellyite to lowest greenschist facies. A few Archean iron-formations have similarly very low-grade metamorphic assemblages.

Although such iron-formations allow one to study what are considered to be well-preserved, late diagenetic products of iron-formation deposition, it is almost impossible to obtain first-hand information on the primary phases that were originally precipitated in the sedimentary depositional basin. Only through inference, based on laboratory evidence of the crystallization of siliceous gels and the behavior of Fe-oxides and hydroxides at low temperatures and pressures (25°C and atmospheric pressure) and on theoretical evaluations, can one attempt to deduce what the original depositional phases may have been (see chapter 13 for a discussion of this topic). Such deductions are possible only if one accepts the widely held theory that banded iron-formations are of chemical, sedimentary origin (e.g. James, 1954). If, on the other hand, one believes that the presently observed late diagenetic assemblages are the result of some earlier replacement processes (e.g. Dimroth, 1977; Kimberley, 1979, for oolitic iron-formations) such assemblages provide very few, if any, chemical or mineralogical clues about the precursor materials.

Well-preserved but very low-grade metamorphic assemblages make up a large part of the various iron-formations and S macrobands of the Hamersley Basin (Trendall and Blockley, 1970; Klein and Gole, 1981); such materials can be obtained from diamond drill cores that have not been affected by the deep lateritic weathering zone that is ever-present in the Hamersley region. Similarly well-preserved materials occur in the Sokoman Iron Formation of the west-central part of the Labrador Trough (Dimroth, 1968; Klein, 1974; Klein and Fink, 1976; Lesher, 1978). Part of the Negaunee Iron Formation in northern Michigan also contains assemblages that reflect late diagenetic to very low-grade metamorphic conditions. The assemblages therein have many similarities with those of the iron-formations in the Hamersley Basin and the west-central part of the Sokoman Iron Formation. The lowest metamorphic grade rocks in northern Michigan have been assigned to the chlorite zone by James (1955), and assemblages in the Negaunee Iron Formation that are of a very similar

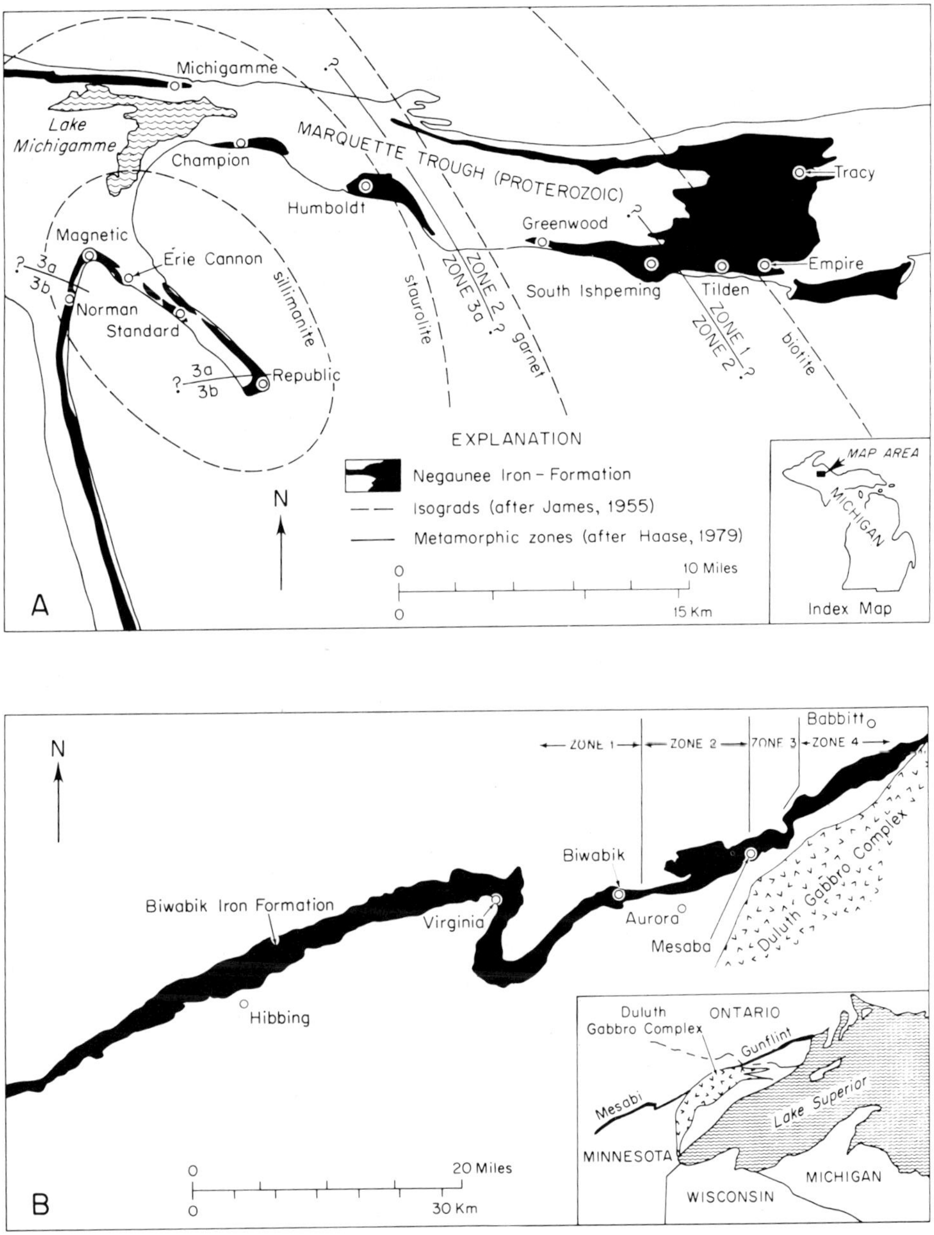

Fig. 11-2. Locations, extent and metamorphic zones of several Proterozoic iron-formations. A. Generalized regional geology of the Marquette Trough which includes the Negaunee Iron Formation. Identified localities are mines or exploration sites (after Haase, 1982a). B. Generalized geologic map of part of the Mesabi Range showing the distribution of the Biwabik Iron Formation and metamorphic zones (from French, 1968).

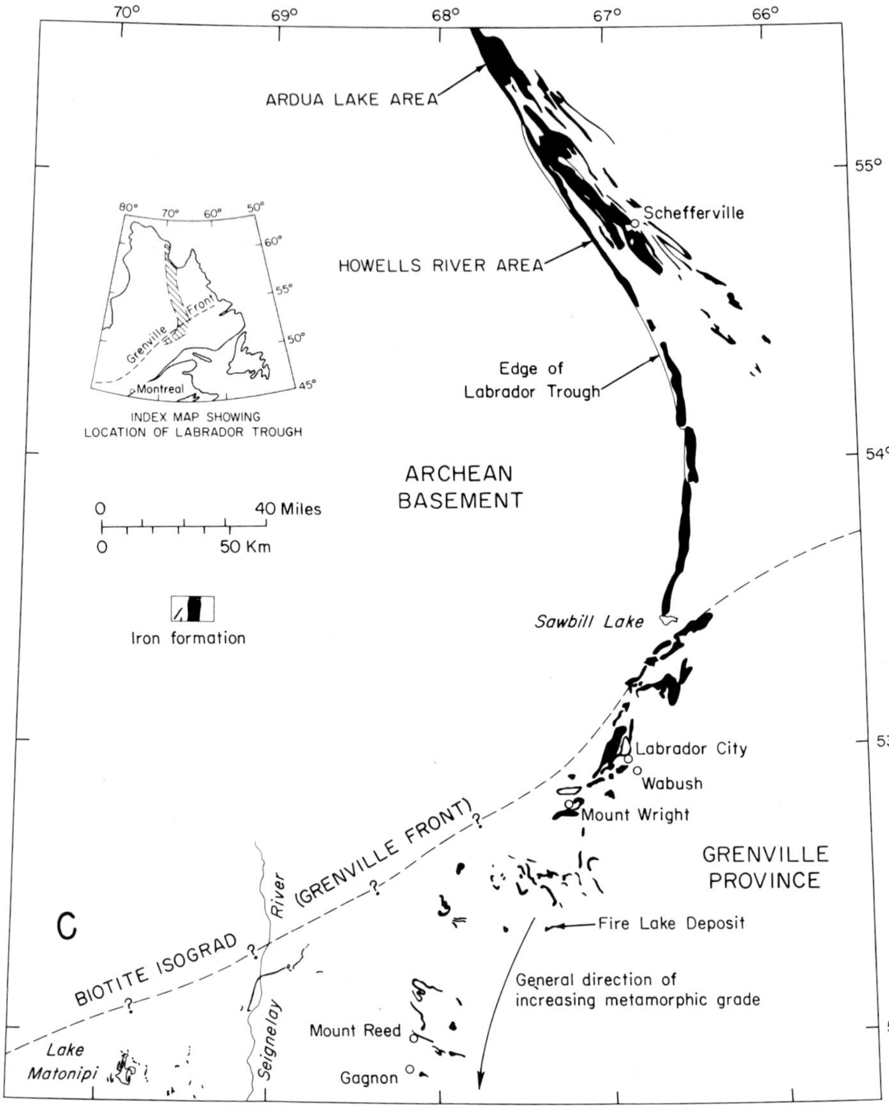

Fig. 11-2. C. Generalized geologic map of the distribution of iron-formation in the central and southwestern parts of the Labrador Trough (after Klein, 1966, 1978). Identified localities refer to mining towns and exploration sites.

metamorphic grade are described as part of "Zone 1" by Haase (1979a, 1982a). The Biwabik Formation in the Mesabi Range of Minnesota similarly contains late diagenetic and/or very low-grade metamorphic assemblages, where the iron-formation has not been affected by thermal metamorphism of the intrusion of the Duluth Gabbro Complex. In French (1968) these assemblages are part of his zone 1. Floran and Papike (1978) have also described assemblages from a zone 1 for the essentially unmetamorphosed part of the Gunflint Iron Formation in Ontario, Canada. As the Biwabik and Gunflint Iron Formations are one and the same, except for their separation by the Duluth Gabbro Complex, the assemblages of French's zone 1 and Floran and Papike's zone 1 are very similar although not identical. Gole (1980a, 1981) describes almost identical, very low grade metamorphic assemblages from Archean iron-formations in the Weld Range, in the northwestern part of the Yilgarn Block in Western Australia.

Location maps, generalized geological maps and distributions of metamorphic zones of several iron-formations that will be discussed in this chapter are given in Fig. 11-2.

The assemblages in late diagenetic and very low-grade metamorphic iron-formations consist of the following minerals: chert (or quartz, as recrystallized chert), magnetite, hematite, siderite, calcite, members of the dolomite-ankerite series, greenalite, stilpnomelane, minnesotaite, riebeckite, ferri-annite, and iron sulfides, mainly pyrite and pyrrhotite. Lesser amounts of chamosite, ripidolite (an Fe-rich chlorite) and ferroan talc are also found. Most of these minerals occur in mesobands (made of alternating chert, or quartz and iron-rich minerals), in fine laminations or in microbands, and in oolitic or granular textures.

Of the above minerals, *greenalite* (approximately $(Fe,Mg)_6Si_4O_{10}(OH)_8$ in composition) tends to exhibit the least crystallized and as such perhaps the most primary textures of any of the minerals in iron-formations. It is most commonly microcrystalline and may occur in laminations, oolites, granules and small irregular masses (see Fig. 11-3). In only very uncommon occurrences such as that of a medium-grained variety close to a contact with a dyke in the Sokoman Iron Formation (Zajac, 1974), or as a recrystallization product under medium grade metamorphic conditions (Floran and Papike, 1978, plate 5A) is greenalite *not* microcrystalline. It almost always has a grain size that is many orders finer than that of any of its coexisting minerals. It is not a common mineral in late diagenetic or very low-grade metamorphic silicate-rich iron-formations*. Indeed it occurs only sporadically in the iron-forma-

* In this discussion the conceptual facies (e.g. silicate-, carbonate-, oxide-, and sulfide facies) formulated by James (1954) are referred to as silicate-rich (or silicate type) and carbonate-rich (or carbonate type) because well-established changes in stratigraphic facies in banded iron-formation are rare. In most iron-formations the various types (facies) generally can not be shown to grade into each other along strike; instead they generally succeed each other in stratigraphic superposition.

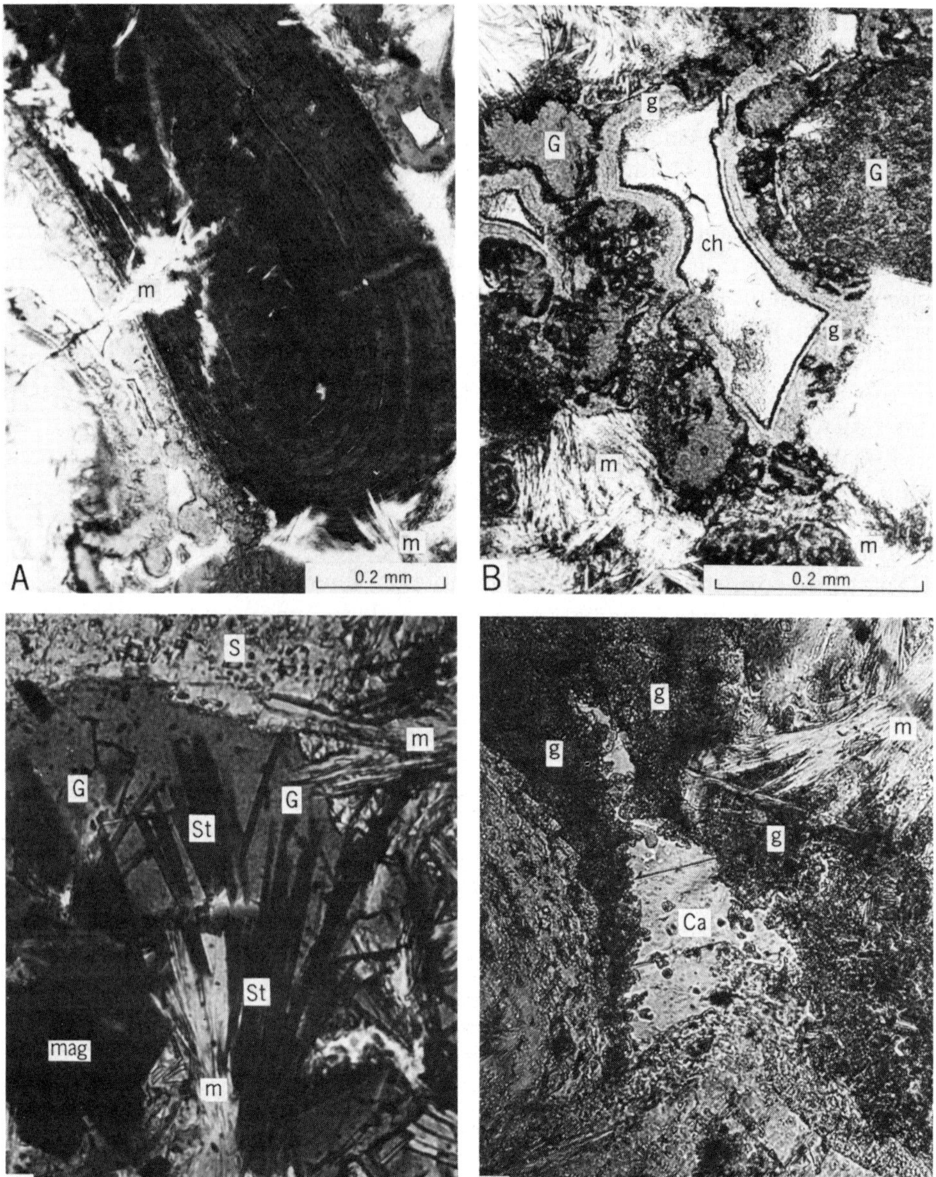
m
m
A
0.2 mm
g
G
G
ch
g
m
m
B
0.2 mm
S
m
G
St
G
St
mag
m
C
0.2 mm
g
g
m
g
Ca
D
0.2 mm

tions of the Hamersley Basin. It is absent from the Negaunee Iron Formation, Michigan, but quite abundant in the Biwabik (French, 1968, 1973) and Gunflint Iron Formations (Floran and Papike, 1975). It is also present in the Howells River (Klein, 1974; Klein and Fink, 1976) and Knob Lake (Zajac, 1974) areas of the Sokoman Iron Formation. Gole (1980a) reports it from Archean iron-formations in the Weld Range of Western Australia as well. Representative analyses of greenalite are given in Table 11-II, and their compositional range is shown graphically in Fig. 11-4. The data show that greenalite has a fairly restricted compositional range even in terms of Fe-Mg substitution. The Al_2O_3 content of greenalite ranges from trace amounts to a maximum of approximately 3 weight percent (Klein, 1974; Floran and Papike, 1975; Gole, 1980a). Some greenalite-like material with as much as $\sim$12% Al_2O_3 (Klein and Fink, 1976) has been found to be berthierine, a 7Å chamosite (Guggenheim et al., 1982). Although in chemical terms, greenalite can be considered the Fe-rich end-member of antigorite ($Mg_6Si_4O_{10}(OH)_8$) its structure is quite different from that of the serpentine minerals. Guggenheim et al. (1982) have determined that the structure of greenalite consists of infinitely extending tetrahedral sheets in which, at regular spacings (around saucer shaped domains) groups of three and four tetrahedra are inverted and coordinate with adjacent octahedral layers. Small amounts of material of chamosite-like composition are sometimes reported from Precambrian banded iron-formations (e.g. LaBerge, 1964; Floran and Papike, 1975; Klein and Fink, 1976; Gole, 1980a; Haase, 1982a). Unless the chemical analyses of such materials are accompanied by careful X-ray determinations it remains uncertain whether such phases are indeed chamosite (7Å), berthierine (7Å), or an iron-rich chlorite (14Å).

Many greenalite occurrences are criss-crossed and transected by two other silicates: stilpnomelane and minnesotaite (see Figs. 11-3B, C and D). *Stilpnomelane* is most commonly a highly pleochroic (yellowish brown to dark brown; sometimes light to dark green) sheaflike silicate which may occur as thin, but quite continuous laminations, as very fine-grained mattes of sheaves and needles, and as coarse-grained sprays and irregularly shaped patches. Stil-

---

Fig. 11-3. Photomicrographs of some textures and mineral assemblages in late diagenetic to very low-grade metamorphic silicate-rich iron-formation. Sokoman Iron Formation in the Howells River area (Klein and Fink, 1976).

A. Greenalite oolite with fine banding; lighter cement surrounding oolite is also greenalite. White patches in oolite are minnesotaite (*m*). Plane polarized light.

B. Greenalite (*g*) cementing greenalite (*G*) granules. Cement interior is chert (*ch*). White, cross-cutting fibers are minnesotaite (*m*). Plane polarized light.

C. Dark brown, highly pleochroic stilpnomelane (*St*) sprays cutting across greenalite (*G*). Minnesotaite (*m*) cuts across greenalite, stilpnomelane and siderite (*S*). Opaque is magnetite (*mag*). Plane polarized light.

D. Calcite (*Ca*) cement inside greenalite (*g*) cement, both of which surround dolomitic oolites. Minnesotaite (*m*) cuts across greenalite patches. Plane polarized light.

TABLE 11-II

Examples of the chemical compositions of greenalite, talc and minnesotaite in banded iron-formations

	Greenalite				Talc-minnesotaite				
	1	2	3	4	5	6	7	8	9
SiO_2	36.0	32.89	35.5	34.5	60.53	59.51	54.40	53.5	50.36
TiO_2	n.d.	0.02	0.02	0.03	0.00	0.04	0.00	0.01	0.04
Al_2O_3	0.1	1.52	0.34	2.43	0.33	0.23	1.09	0.67	0.57
FeO*	48.5	52.51	49.1	47.3	8.69	15.07	20.20	30.7	42.62
MnO	0.1	0.43	0.28	0.50	0.50	0.02	0.08	0.21	0.25
MgO	4.4	1.43	2.10	3.76	25.95	21.36	17.55	10.24	2.07
CaO	trace	0.07	0.11	0.02	0.03	0.01	0.00	0.01	0.02
Na_2O	n.d.	0.00	0.00	0.01	0.00	0.02	0.00	0.00	0.00
K_2O	0.00	0.00	0.04	0.01	0.02	0.01	0.12	0.25	0.35
(H_2O)**	(10.9)	(11.13)	(12.51)	(11.44)	(3.95)	(3.73)	(6.56)	(4.35)	(3.72
	100.00	100.00	100.00	100.00	100.00	100.00	100.00	100.00	100.00
	on basis of 14 oxygens				*on basis of 11 oxygens*				
Si	4.22	4.00	4.28	4.06	3.96	3.99	3.89	3.94	3.93
Al	0.00	0.00	0.00	0.00	0.03	0.01	0.09	0.06	0.05
Σ	4.22†	4.00	4.28†	4.06†	3.99	4.00	3.98	4.00	3.98
Al	0.02	0.22	0.05	0.34	0.00	0.01	0.00	0.00	0.00
Ti	0.00	0.00	0.00	0.00	0.00	0.00	0.00	0.00	0.00
Fe^{2+}	4.76	5.35	4.95	4.65	0.48	0.85	1.21	1.89	2.78
Mn	0.01	0.04	0.03	0.05	0.03	0.00	0.00	0.01	0.02
Mg	0.77	0.26	0.38	0.66	2.53	2.14	1.87	1.12	0.24
Ca	0.00	0.01	0.01	0.00	0.00	0.00	0.00	0.00	0.00
Na	0.00	0.00	0.00	0.00	0.00	0.00	0.00	0.00	0.00
K	0.00	0.00	0.00	0.00	0.00	0.00	0.01	0.02	0.03
Σ	5.56	5.88	5.42	5.70	3.04	3.00	3.09	3.04	3.07

n.d. = none detected.

* total Fe recalculated as FeO only.

** H_2O values obtained by subtracting electron microprobe total from 100%.

† greenalite analyses commonly show excess Si over that required to fill the tetrahedral site occupancy of 4.00 (see Floran and Papike, 1975; Gole 1980a; Guggenheim et al., 1982, for discussion).

1. from Floran and Papike, 1973.

2. from Klein (1974).

3. from Floran and Papike (1975).

4. from Gole (1980a).

5. talc from Lesher (1978).

6. talc in the Dales Gorge Member, Hamersley Group from Miyano (1978a);

7. talc in the Marra Mamba Iron Formation, Hamersley Group, from Klein and Gole (1981).

8. minnesotaite from Gole (1980a).

9. minnesotaite from Klein (1974).

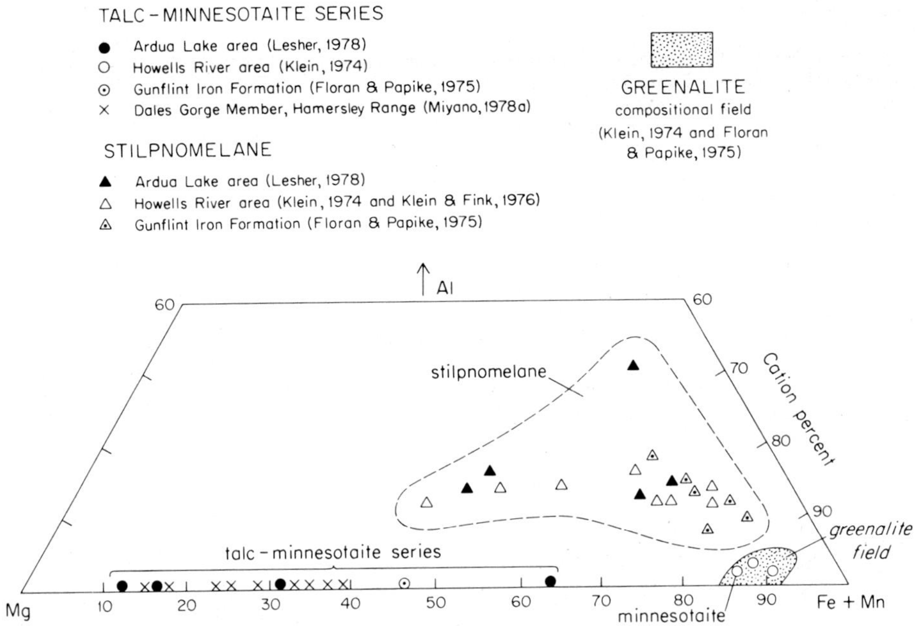

Fig. 11-4. Compositional ranges of stilpnomelane, members of the talc-minnesotaite series, and greenalite (modified after Lesher, 1978) in late diagenetic to very low-grade iron-formations.

pnomelane is probably the most common Fe-silicate in most very low-grade metamorphic iron-formations (Klein and Gole, 1981, table 9, and in prep.; Miyano and Miyano, 1982), and is common in chert (or quartz)-carbonate-magnetite-stilpnomelane associations. It also makes up a major part of the S macrobands of the Dales Gorge Member of the Brockman Iron Formation at Wittenoom in the Hamersley Basin (Trendall and Blockley, 1970; Klein and Gole, in prep). When stilpnomelane is intergrown with greenalite, it commonly has a crosscutting textural relationship with it: fine- to medium-grained sheaves of stilpnomelane criss-crossing microcrystalline, almost amorphous-appearing masses of greenalite (see Fig. 11-3C).

The chemistry of stilpnomelane is much more complex than that of greenalite. Eggleton (1972) determined its structure as well as its average formula, which he gives as $(Ca,Na,K)_4\,(Ti_{0.1}Al_{2.3}Fe_{35.5}Mn_{0.8}Mg_{9.3})(Si_{63}Al_9)$ $(O,OH)_{216} \cdot nH_2O$. Eggleton and Chappell (1978) have suggested $K_{0.6}\,(Mg,Fe^{2+},Fe^{3+})_6Si_8Al(O,OH)_{27} \cdot$ 2—4H_2O as a convenient approximation to the structural formula of stilpnomelane. The main chemical differences between stilpnomelane and greenalite, or minnesotaite, are the presence of small and variable, but essential amounts of alkali cations such as K^+ and Na^+, and relatively large and variable Al_2O_3 contents. Representative analyses of

TABLE 11-III

Examples of the chemical compositions of stilpnomelane (columns 1 through 5) and ferri-annite (columns 6 through 8) in banded iron-formations

	1	2	3	4	5		6	7	8
SiO_2	45.54	44.2	48.15	47.55	46.85		37.60	36.49	35.41
TiO_2	0.26	n.d.	0.00	0.00	0.00		0.16	0.00	0.03
Al_2O_3	4.75	4.5	5.40	4.62	4.20		5.12	1.89	1.38
Fe_2O_3	2.90						7.16△	11.57△	13.34△
FeO	25.38	35.2†	27.59†	31.54†	36.70†		26.26	29.88	30.03
MnO	0.13	0.1	0.07	0.09	0.22		0.10	0.00	0.07
MgO	7.75	3.0	7.55	6.82	2.62		9.60	7.37	7.21
CaO	0.04	0.04	0.29	0.00	0.05		0.00	0.01	0.06
Na_2O	0.82	n.d.	0.16	0.07	0.42		0.00	0.08	0.02
K_2O	1.96	2.0	2.99	2.58	1.06		9.21	7.80	7.87
$H_2O(+)$	8.46	10.6††	7.80††	6.73††	7.88††		4.79††	4.91††	4.50††
$H_2O(-)$	1.56								
Total	99.55**	(100.00)	(100.00)	(100.00)	(100.00)		(100.00)	(100.00)	(100.00)
	Recalculated on basis of 22 oxygens								
Si	7.312	7.364	7.428	7.388	7.520	Si	6.136	6.155	6.018
Al	0.688	0.636	0.572	0.612	0.480	Al	0.985	0.376	0.276
						Fe^{3+}	0.879	1.469	1.706
Σ	8.000	8.000	8.000	8.000	8.000	Σ	8.000	8.000	8.000
Al	0.210	0.248	0.408	0.234	0.314		—	—	—
Ti	0.032	0.000	0.000	0.000	0.000	Ti	0.020	0.000	0.004
Fe^{2+}	3.758	4.904	3.560	4.098	4.928	Fe^{2+}	3.584	4.215	4.268
Mn	0.018	0.014	0.010	0.012	0.030	Mn	0.014	0.000	0.010
Mg	1.854	0.744	1.736	1.579	0.636	Mg	2.336	1.853	1.827
						Σ	5.954	6.068	6.120
Ca	0.006	0.072	0.048	0.000	0.008	Ca	0.000	0.002	0.011
Na	0.256	0.000	0.048	0.021	0.130	Na	0.000	0.026	0.007
K	0.402	0.426	0.590	0.511	0.218	K	1.918	1.679	1.706
Σ	6.536	6.408	6.400	6.455	6.254	Σ	1.918	1.707	1.724

n.d. = none detected.
** additional elements listed: P_2O_5 0.44 and CO_2 0.28 wt. %, making analysis total 100.27 wt. %.
† total Fe recalc. as FeO only.
†† H_2O obtained by subtracting electron microprobe total from 100%.
△ Fe_2O_3 content estimated by allocation of Fe^{3+} to tetrahedral sites, totalling 8.00.
1. Dales Gorge Member, S10 macroband, Hamersley Range (Trendall and Blockley, 1970).
2. Gunflint Iron Formation (Floran and Papike, 1973).
3. Marra Mamba Formation, Hamersley Range (Klein and Gole, 1981).
4. Joffre Member, Hamersley Range (Klein and Gole, in prep.).
5. Sokoman Iron Formation, Canada (Klein, 1974).
6. Joffre Member, Hamersley Range (Klein and Gole, in prep.).
7. and *8.* Dales Gorge Member, Hamersley Range (Miyano and Miyano, 1982).

stilpnomelane, given in Table 11-III, reflect considerable Fe-Mg substitution. Figure 11-4 illustrates the general compositional range in terms of Al, Mg, and (Fe+Mn). If stilpnomelane, on the basis of its major element chemistry, is viewed as an essentially hydrous Fe-Mg-Al silicate (ignoring for the moment the small but essential K content) its formula can be simplified to
$(Fe,Mg,Al)_{2.7}(Si,Al)_4(O,OH)_{12} \cdot xH_2O$.
Such a highly simplified but incorrect formulation facilitates comparison with the chemistry of greenalite $(Fe,Mg)_6Si_4O_{10}(OH)_8$ and minnesotaite $(Fe,Mg)_3Si_4O_{10}(OH)_2$.

It is possible that well-crystallized stilpnomelane sheaves that appear to have formed at the expense of a microcrystalline greenalite-like material (e.g. Klein, 1974; see Fig. 11-3C) are the recrystallization products of an amorphous greenalite-like precursor that locally contained considerable amounts of K^+ and Al_2O_3. Because these elements cannot be incorporated in the greenalite structure, such K- and Al-rich regions would recrystallize, during late diagenetic conditions, to the stilpnomelane structure which has large cavities for the housing of K and minor Na. Gole (1980a) on the basis of textures of stilpnomelane-bearing, Archean iron-formations suggests that stilpnomelane is the result of a reaction between Al-containing greenalite+quartz+siderite+K^+ (aqueous) giving rise to a mixture of stilpnomelane and Al-poor greenalite. In short, where greenalite and stilpnomelane coexist there is considerable evidence for stilpnomelane being a later phase than greenalite. In the majority of iron-formation assemblages, however, stilpnomelane occurs as an abundant, single Fe-silicate and no inferences can be made about its relation to possible precursor materials.

Of the three most common iron-rich silicates in iron-formations, *minnesotaite* is generally not as abundant as stilpnomelane, but it is much more common than greenalite. It commonly occurs as fine- to medium-grained needles that are arranged in sprays, "bow-ties" and irregular patches. Such textures cut across chert, fine- to medium-grained quartz, carbonate, microcrystalline greenalite, as well as medium- to coarse-grained stilpnomelane sheaves. Examples of some of these textural relationships are given in Fig. 11-3. There is ample evidence form the reports of many investigators of late diagenetic to very low-grade metamorphic iron-formations (e.g., French, 1968, 1973; Han, 1971; Klein, 1974; Floran and Papike, 1975; Klein and Fink, 1976; Gole, 1980a; Klein and Gole, 1981; Haase, 1982a) that minnesotaite is a reaction product of greenalite, of stilpnomelane, and of chert (or quartz) + carbonate. Klein (1974) on the basis of well-documented reaction textures has suggested that the following reactions take place in late diagenetic iron-formation assemblages:

$$\underset{\text{greenalite}}{Fe_6Si_4O_{10}(OH)_8} + 4SiO_2 \rightarrow 2\ \underset{\text{minnesotaite}}{Fe_3Si_4O_{10}(OH)_2} + 2\ H_2O \qquad (1)$$

$$2\ Fe_6Si_4O_{10}(OH)_8 + O_2 \rightarrow 2\ Fe_3Si_4O_{10}(OH)_2 + 2\ Fe_3O_4 + 3\ H_2O \quad (2)$$

greenalite — minnesotaite — magnetite

$$FeCO_3 + 4\ SiO_2 + H_2O \rightarrow Fe_3Si_4O_{10}(OH)_2 + 3\ CO_2 \quad (3)$$

siderite — chert — minnesotaite

$$..\ Fe_{2.7}(Si,Al)_4(O,OH)_{12} \cdot xH_2O + 0.33\ Fe^{2+} \rightarrow$$

stilpnomelane-like

$$Fe_3Si_4O_{10}(OH)_2 + H_2O + Al + \text{minor}\ (Na,K) \quad (4)$$

minnesotaite

Unfortunately there is no experimental and only very little theoretical evidence for the delineation of the stability fields, in terms of temperature, pressure, and fluid composition of the above three common silicates in banded iron-formations. However, this will be discussed more fully in a later section.

The ideal end-member composition of minnesotaite is $Fe_3Si_4O_{10}(OH)_2$. Its structure is not identical with that of talc, $Mg_3Si_4O_{10}(OH)_2$ (Guggenheim et al., 1982), but it has not yet been worked out in detail. In iron-formations, it shows a considerable range of Fe-Mg substitutions. Representative analyses are given in Table 11-II, and the compositional range of minnesotaite and ferroan talc is shown in Fig. 11-4. In some assemblages of essentially unmetamorphosed iron-formations such as in the Sokoman Iron Formation at Ardua Lake (Lesher, 1978) and in the Gunflint Iron Formation (Floran and Papike, 1975) and in the very low-grade metamorphic Marra Mamba Iron Formation (Klein and Gole, 1981) and Dales Gorge Member (Miyano, 1978a) of the Hamersley area, ferroan talc is more common than the Fe-rich analogue, minnesotaite. Ewers and Morris (1981) report ferroan talc (identified by X-ray diffractometer patterns of bulk samples) in composite samples of the Dales Gorge Member in the Hamersley area. Lesher (1978) describes talc as thin veinlets or random patches in a chert matrix. It tends to have a more sheaflike habit than the usual minnesotaite needles. Similar to minnesotaite, it appears to be a late diagenetic or very low-grade metamorphic mineral.

Representative examples of greenalite-stilpnomelane-minnesotaite-carbonate-quartz-magnetite associations from late diagenetic to very low-grade metamorphic, Proterozoic and Archean iron-formations are given in Fig. 11-5.

Ferri-annite is a new species of mica with the representative composition of $K_{1.82}(Mg_{1.83}Fe^{2+}_{4.23})\ (Fe^{3+}_{1.69}Al_{0.28}Si_{6.02})O_{20}(OH)_4$ described from the Dales Gorge Member (Miyano and Miyano, 1982) and the Joffre Member (Gole and Klein, in prep.) of the Brockman Iron Formation in the Hamersley Basin. It occurs as flaky or tabular grains and massive aggregates near or within riebeckite-rich zones of banded iron-formation. It commonly coexists with hematite, magnetite, quartz, ankerite, stilpnomelane, and riebeckite. Its pleochroic colors distinguish it from stilpnomelane or mica: X = reddish brown and Y=Z= pale yellow green to greenish brown. Representative electron probe analyses of ferri-annite are given in Table 11-II. Ferri-annite contains much less Al_2O_3 than biotite or phlogopite and because of this, when a formula is recalculated on a basis of 22 oxygens, the mineral shows a cation deficiency

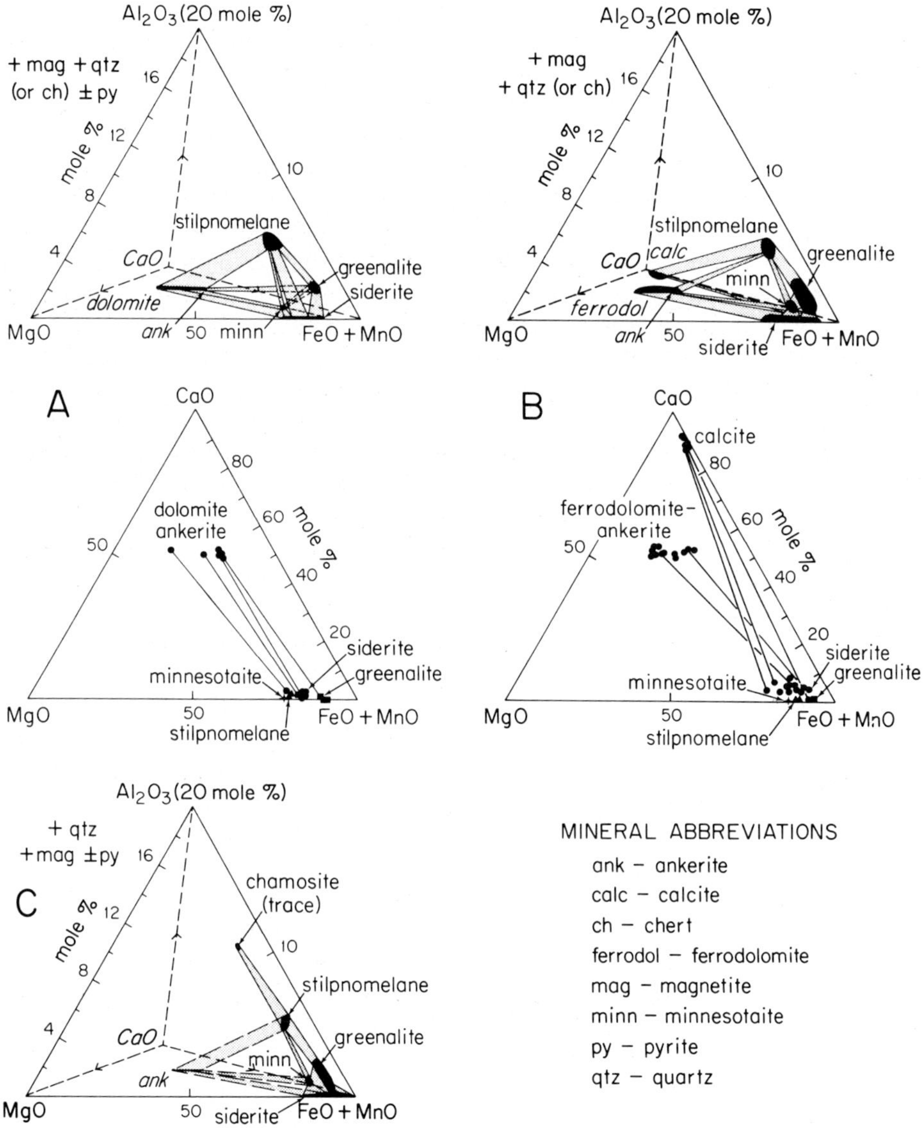

Fig. 11-5. Graphical representation of silicate-carbonate assemblages, and carbonate compositions in late diagenetic to very low-grade metamorphic iron-formation assemblages. Tielines connect coexisting minerals. A and B are from the Sokoman Iron Formation in the Howells River area, west-central Labrador Trough (from Klein and Fink, 1976). C is based upon assemblages reported by Gole (1979) for very low-grade metamorphic Archean iron-formations in the Yilgarn Block of Western Australia.

in the tetrahedral sites (less than 8.00 by 1 to 2 cations). This implies that the tetrahedral site is occupied by variable amounts of ferric iron. In the electron probe analyses of Table 11-III the minimum amount of Fe^{3+} was estimated by assuming that all Fe^{3+}, Al and Si are located in tetrahedral sites.

Chert (or its recrystallized equivalent, *quartz*) is a major constituent of many types of iron-formations. Much of the SiO_2 in iron-formations that have undergone only late diagenetic reactions tends to be very fine-grained, and when it has a grain size range of <0.05 mm this may be referred to as chert (Klein and Fink, 1976). However, in the same chert-rich iron-formation there may be well-recrystallized SiO_2 that would normally be referred to as quartz. There is no clear-cut distinction between the use of chert or quartz in most of the literature. James (1955; table 6) suggested that the gradual increase of the grain size of microcrystalline chert to quartz could be used as an indicator of metamorphic grade. He noted that in low metamorphic grade (chlorite and biotite zones) SiO_2-rich assemblages, the grain size of quartz tended to be less than 0.10 mm, and that with increasing metamorphic grade this gradually enlarged to a maximum of more than 0.20 mm. It appears that such a correlation of SiO_2 grain size with metamorphic grade is possible in silica-rich rocks that have undergone at least chlorite zone metamorphism (Gross, 1961; Dorr, 1964; Klein, 1973). However, severe heterogeneity in quartz (and chert) grain sizes has been noticed in essentially unmetamorphosed rocks (Zajac, 1974; Klein, 1974) and it is possible that SiO_2-rich units which were saturated with H_2O during diagenesis and subsequent very low-grade metamorphism, would have recrystallized more easily and to larger grain sizes than those which behaved as impermeable and less hydrous systems.

Magnetite is the most common Fe-oxide mineral in oxide type iron-formation, but can be abundant in carbonate- or silicate-rich iron-formations as well. It occurs most commonly in well-defined, continuous mesobands, in laminations, within granules, as concentric bands in oolites, and as irregular clusters and patches. Almost universally magnetite tends to be medium-grained, well-crystallized and with a sub- to euhedral habit. It is commonly much coarser grained than co-existing chert (or quartz), hematite and Fe-silicates. Magnetite and carbonates may have rather similar grain sizes because both tend to be strongly recrystallized. The origin of magnetite, in essentially unmetamorphosed banded Precambrian iron-formations, is a much debated subject. Many authors (e.g. Garrels et al., 1973; Gross, 1973; Dimroth and Chauvel, 1973; Perry and Tan, 1973; Klein, 1974; Klein and Bricker, 1977; Mel'nik, 1982) consider it to be, if not a primary precipitate, a diagenetic recrystallization or reaction product of such a precipitate. The types of materials that could have been the precursors of the presently observed magnetite would have been "hydromagnetite" ($Fe_3O_4 \cdot nH_2O$) or mixtures of iron hydroxides [$Fe(OH)_3$ and $Fe(OH)_2$]. However, others (e.g. LaBerge, 1964; French, 1968, 1973; Han, 1971, 1978; Floran and Papike, 1978) interpret much magnetite to be the result of oxidation of iron silicates (e.g. greenalite), and carbonates (e.g. siderite), of reduction of earlier ferric hydroxides, and of metamorphic reactions, as deduced from petrographic textures. Han (1978), Morris (1980) and Morris and Ewers (1981) provide textural evidence for the production of magnetite through a reaction of $Fe(OH)_3$ and diffusing Fe^{2+}. French (1973) gives in his table 1 a listing of the various complex reactions that have been deduced on the basis of textures. As noted above, magnetite grains tend to be sub- to euhedral and relatively coarse in grain size. The edges of such coarse magnetite grains tend to cut across those of coexisting minerals. Klein and Fink (1976) interpret such crosscutting relations to be the result of the recrystallization of an earlier, finer-grained magnetite or "hydromagnetite"-type precursor and not the result of reaction of coexisting minerals (e.g. Fe carbonates or silicates) to form magnetite at their expense.

Hematite may be an abundant Fe oxide in some lithologies of late diagenetic and low-grade metamorphic iron-formations. It commonly is finer in grain size than magnetite, with which it may coexist. It occurs in microbands, in laminations, in concentric bands of oolites, and as granules, or irregular bleb-like masses. In very chert-rich occurrences, it may be so finely interspersed and so fine-grained as to give the chert a red color (jasper). Hematite in unaltered, unleached and essentially unmetamorphosed iron-formations is generally considered to be of primary, sedimentary or perhaps diagenetic origin (e.g. James, 1954; French, 1973; Dimroth and Chauvel, 1973; Klein and Fink, 1976). It appears likely that the sedimentary precursor to the presently observed crystalline hematite is $Fe(OH)_3$ (e.g. French, 1973; Klein and Bricker, 1977; Mel'nik, 1982). Floran and Papike (1978) report fine-grained hematite that has recrystallized and reacted to form coarser grained magnetite in the minnesotaite-rich zone (their zone 2) of the Biwabik Iron Formation.

Of the *carbonates* in unmetamorphosed iron-formation, *siderite* and members of the *dolomite-ankerite* series are probably most common, with *calcite* a lesser constituent. Carbonates may make up a large percentage (50 vol. % or more) of the carbonate-type iron-formation, but they may also be major constituents of silicate-rich and oxide-rich iron-formation. Quartz-magnetite (± hematite) assemblages may contain as much as ten volume percent carbonate. It is very difficult to distinguish various intergrown carbonate species in petrographic sections on account of their similar optical properties, grain size and habit. However, detailed optical and X-ray diffraction studies by French (1968) and subsequent electron microprobe studies of carbonates in very low-grade metamorphic iron-formations (e.g. Klein, 1974; Floran and Papike, 1975; Klein and Fink, 1976; Haase, 1979a, 1982a; Klein and Gole, 1981) have revealed the common presence of calcite-ankerite, ankerite-siderite, and calcite-siderite pairs, as well as calcite-ankerite-siderite coexistences. Extensive Fe-Mg substitution is found in members of the dolomite-ankerite series and in siderite. Figure 11-5 gives some representative compositional ranges of carbonates.

Because carbonates in iron-formation commonly show a medium grain size, often with euhedral, rhombohedral habit, the edges of their grains tend to cut across the grains of coexisting minerals. On the basis of such apparently crosscutting petrographic relations several authors (see French, 1973, table 1) have interpreted such textures as the result of reactions of precursor materials or as replacements. Klein and Fink (1976) conclude that the presently observed medium to coarse grain size of the carbonates is the result of recrystallization and, in places, remobilization (diffusion) of the original sedimentary, very finely crystalline carbonate precipitate during diagenesis and possibly subsequent low-grade metamorphism. As such, the relatively coarse grain size and euhedral habit of the carbonates (as well as of magnetite) may be due mainly to recrystallization of similar, but much finer-grained precursor materials.

Pyrite is a major constituent of sulfide-rich iron-formations (e.g. James, 1954, 1955; French, 1968; Klein and Fink, 1976). Some pyrite-rich black shales, with a considerable carbon content have been referred to as sulfide-rich iron-formations as well. It is unclear if such materials are indeed iron-formations. In most sulfide-rich iron-formations the pyrite grains are coarser grained than the other constituents (e.g. greenalite, or chamosite, carbonates, chert and carbon) and tend to have a euhedral outline (Klein and Fink, 1976); they are commonly concentrated in thin, and well-defined layers. In some parts of the very low-grade metamorphic Marra Mamba Iron Formation of the Hamersley Group (Klein and Gole, 1981) *pyrrhotite* may be present coexisting with pyrite. Gole (1980a) describes coexisting pyrite and pyrrhotite in some very low-grade metamorphic Archean iron-formations as well. French (1968) suggested that pyrrhotite may possibly have been formed by reaction from pyrite in those parts of the Biwabik Iron Formation that had been contact metamorphosed. However, in light of the common association of pyrite + pyrrhotite in several essentially unmetamorphosed iron-formations such a metamorphic origin may not be universal. Pyrrhotite in unmetamorphosed, sedimentary rocks has been reported by Ncube et al. (1978). It is very likely that mackinawite ($Fe_{1+x}S$) is the sedimentary precursor to pyrite (Berner, 1970) and for this reason the stability diagrams for late diagenetic assemblages in iron-formations (Klein and Bricker, 1977) depict a mackinawite field for sulfide-rich iron-formations (see Fig. 11-6).

Carbonaceous material can be present in considerable amounts in some very low-grade metamorphic iron-formations. Gruner (1946) gives values as high as 3.20 wt. % for "organic carbon" for a sample of the Lower Slaty member of the Biwabik Iron Formation. Some of this material is visible under the microscope as dust and black stringers and has been referred to as graphite by Gruner. Chemical analyses of the Ruth shale, a pyritic shaly iron-formation horizon in the lower part of the iron-formation sequence at Howells River, Labrador (Klein and Fink, 1976) show a maximum of 0.9 wt. % C. In such carbon-rich samples the banding is strongly accentuated by very fine carbon laminae. Klein and Gole (1981) report as much as 1.74 wt. % carbon

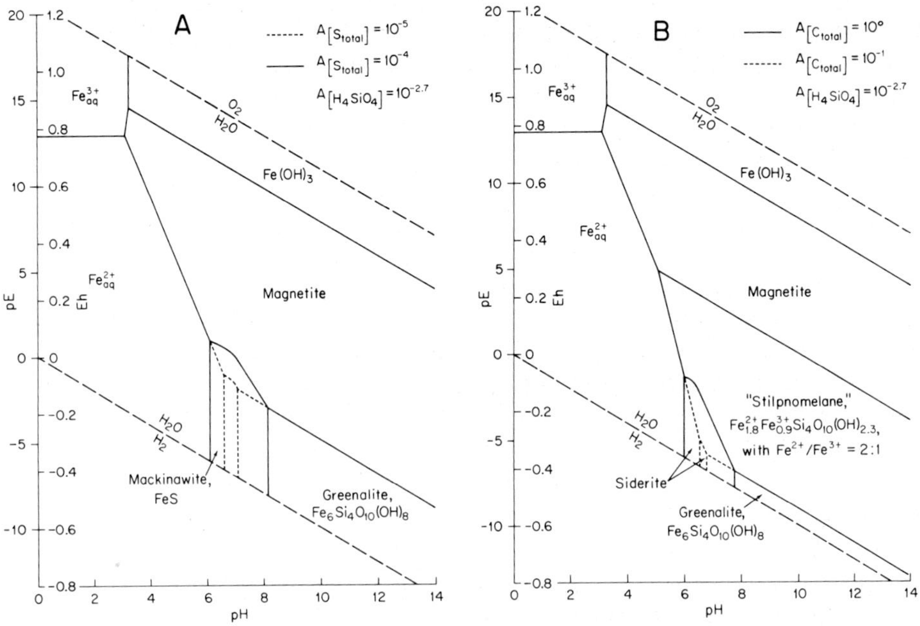

Fig. 11-6. Eh-pH diagrams depicting stability relations among some minerals in diagenetic iron-formation assemblages (from Klein and Bricker, 1977). A. System Fe-H_2O-O_2-S at 25°C, one atmosphere total pressure, saturated with amorphous silica and at two activities of dissolved sulfur. B. System Fe-H_2O-O_2-SiO_2-dissolved carbon, at 25°C and one atmosphere total pressure. Stilpnomelane-like composition with $Fe^{2+}/Fe^{3+} = 2:1$. In both A and B boundaries between aqueous species and solids at $a_{[\text{aqueous species}]} = 10^{-6}$.

in some parts of the Marra Mamba Iron Formation; such parts are pyrite- and pyrrhotite-rich. In short, very fine-grained carbonaceous "graphite-like" material (with poor crystallinity, if any, and poor reflective light properties) is most commonly found in the sulfide-rich part of iron-formations.

Riebeckite (or its fibrous variety crocidolite) is a major constituent of some parts of the Brockman Iron Formation of the Hamersley Basin (Trendall and Blockley, 1970), and in iron-formations of the Griqualand Group of the Transvaal Supergroup of South Africa (Beukes, 1973). These enormously large iron-formation sequences contain large amounts of riebeckite and crocidolite. The Brockman Iron Formation in Western Australia may well contain one of the largest masses of alkali amphibole on earth. Trendall and Blockley (1970) estimated the total reserves and resources of crocidolite (excluding non-fibrous riebeckite) to be approximately 2,410,000 tons in the Brockman Iron Formation. However, in most other late diagenetic to low-grade metamorphic banded iron-formation occurrences (e.g. the Sokoman Iron Formation; Dimroth and Chauvel, 1973; Zajac, 1974; Klein, 1974) it is a very minor and sporadic constituent. Authigenic magnesioarfvedsonite, which is compositionally closely related to riebeckite, has been described from the Green River Formation by Milton et al. (1974). Although riebeckite has a large temperature stability field (Ernst, 1968) it appears to have formed as an authigenic and/or very low-grade metamorphic mineral in many of the iron-formation occurrences. The composition of riebeckite in iron-formations is not greatly variable (see Robinson et al., 1982).

Some iron-rich chlorites may be sporadically present in some of the late diagenetic to very low-grade metamorphic iron-formations. *Ripidolite* (Hey, 1954) is the most common variety. It occurs as irregular, fine-grained patches and stringers (with a grain size similar to that of greenalite; Gole, 1980a), or in somewhat coarser textures with a bladed habit, occasionally forming fans and sheaves (Klein and Fink, 1976; Haase, 1982a). Although textural evidence suggests that it is part of the late diagenetic to very low-grade metamorphic assemblages of iron-formation, it has a considerable stability field with increasing metamorphic grade (see Fig. 11-15). Because of the uncertainty of whether the material is a normal 14Å chlorite or a 7Å septachlorite, the name ripidolite may be given in quotation marks ("ripidolite", see Fig. 11-15).

PHYSICAL AND CHEMICAL CONDITIONS OF IRON-FORMATION DIAGENESIS AND VERY LOW-GRADE METAMORPHISM

The generally accepted view is that the iron-formation minerals discussed above are late diagenetic to very low-grade metamorphic products of chemical sedimentary precursor materials (e.g. James, 1954; Borchert, 1965; Klein and Bricker, 1977; Mel'nik, 1982). If this is so, one should be able, if indeed the required thermodynamic data are available, to calculate low-temperature equilibrium stability diagrams of the diagenetic phases for appropriate temperature, pressure and compositions (see Fig. 11-5). Such diagrams, for 25°C and 1 atmosphere total pressure were calculated by Klein and Bricker (1977), and two of them are shown in Fig. 11-6. The calculated coexistences very closely match the observed late diagenetic assemblages. Miyano and Klein (1983a) have evaluated the stability relations of riebeckite as a function of log f_{O_2} and pH at 130°C, in hematite-magnetite-carbonate associations, similar to those of very low-grade metamorphic iron-formations. Figure 11-7 shows the riebeckite stability field as a function of two different values for $a_{\text{total carbonate}}$.

The pressure and temperature conditions under which the various minerals in late diagenetic assemblages appear to have formed, have been estimated by several authors. French (1973) suggests that greenalite-containing assemblages probably reflect late diagenetic conditions with a temperature range of 100—200°C and a pressure of 1 –2 kbar. Minnesotaite which is pervasively shown to form at the expense of greenalite, as well as other Fe-minerals, is suggested by French to have formed at a higher temperature range, of 200—350°C, at 2—5 kbar. Perry et al. (1973) who determined the oxygen isotopic fractiona-

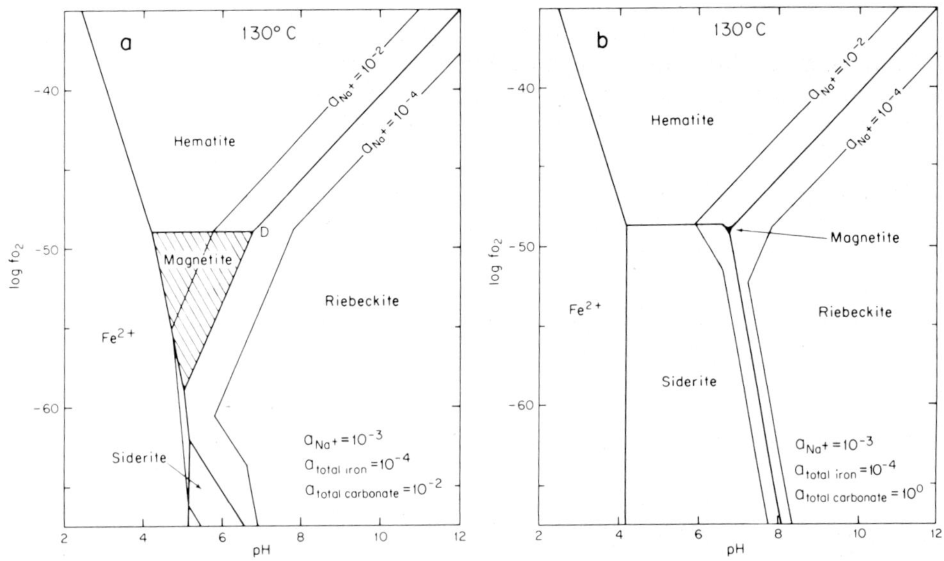

Fig. 11-7. Riebeckite-siderite-Fe-oxide stability relations (after Miyano and Klein, 1983a). In these figures a_{Na^+} ranges from 10^{-2} to 10^{-4}, and $a_{total\ Fe}$ is fixed at 10^{-4}.
a. Log f_{O_2}-pH diagram at $a_{total\ CO_2} = 10^{-2}$ at 130°C.
b. Log f_{O_2}-pH diagram at $a_{total\ CO_2} = 10^{\circ}$, at 130°C.

tion on quartz-magnetite pairs for zone 1 of French, in the Biwabik Iron Formation, concluded that, if quartz and magnetite are in approximate isotopic equilibrium, the approximate isotope temperature of zone 1 would have been about 150°C or less. The lowest value they obtained indicated a possible value of 110°C. Klein and Fink (1976) conclude that the essentially unmetamorphosed assemblages of the Sokoman Iron Formation in the Howells River area probably represent temperatures of 150°C or less. Such temperature values are in approximate agreement with the experimental growth of talc at 90°C and atmospheric pressure by Bricker et al. (1973). If the structures of talc and minnesotaite are reasonably similar, the Fe end member of this group (minnesotaite) would probably be stable at even lower temperatures than talc. Haase (1982a), on the basis of recalculated $^{18}O/^{16}O$ data from the literature, suggests temperatures of about 300°C for one location in zone 1 of the Negaunee Iron Formation, which corresponds with the chlorite zone of James (1955). Miyano (1976) estimates the temperature of burial metamorphism of the Dales Gorge Member of the Brockman Iron Formation in Western Australia to have ranged from a minimum of 100 ± 20°C to a maximum of 230 ± 20°C, at a pressure range of 700—1300 bar. Oxygen isotope temperature estimates by Becker and Clayton (1976) on the Dales Gorge Member and the Wittenoom Dolomite in the Wittenoom area of Western Australia indicate temperatures of 270—310°C. Gole (1980a) concludes that minnesotaite-quartz-magnetite ± pyrite assemblages in Archean iron-formations are the result of the very low grade metamorphism of earlier stilpnomelane-, siderite- and aluminous greenalite-containing lithologies. On the basis of an arsenopyrite geothermometer (Kretschman and Scott, 1976) and on the evaluation of metamorphic mineral assemblages in associated metabasites, he concludes the peak metamorphic temperature for the minnesotaite assemblages to have been between 300 and 350°C, at low pressures. The above temperature data suggest that minnesotaite may begin to form at temperatures as low as 100°C, and that it has a stability range that extends up to 300—350°C in temperature. Indeed, on the basis of thermodynamic data Miyano and Klein (1983b) evaluate the upper stability limit of minnesotaite to be in the 300—350°C range at low pressures (see Fig. 11-11).

MEDIUM-GRADE METAMORPHIC ASSEMBLAGES

Many iron-formation assemblages of medium-grade metamorphic origin are characterized by the common development of amphiboles, mainly members of the cummingtonite-grunerite series. James (1955) correlates the occurrence of grunerite with the garnet and staurolite zones of pelitic schists. However, Klein (1978) notes the first occurrence of Mg-Fe clinoamphiboles at the onset of the biotite zone (see Fig. 11-8). Haase (1982a) infers the transition from his zone 1 (stilpnomelane-minnesotaite-rich) to his zone 2 (grunerite-rich) in the Negaunee Iron Formation in a region that is fairly close to the biotite isograd as mapped by James (1955); see also Fig. 11-2A. It appears therefore that Mg-Fe clinoamphiboles in metamorphosed banded iron-formations are stable from the biotite isograd on, through the garnet and staurolite isograds of pelitic schists. The extensive iron ore deposits in the Labrador City area, Canada show abundant coarse-grained amphibole assemblages, and these deposits are located in the kyanite-staurolite zone (Klein, 1966)

Although members of the Fe-Mg- and Ca-amphibole series are common in medium grade assemblages, they are not the exclusive silicates because some minnesotaite and stilpnomelane may persist into the biotite and garnet zones, respectively (Klein, 1978). In some assemblages that have undergone medium-grade metamorphism, amphiboles may be absent. Essentially unmetamorphosed oxide type iron-formations which consist of quartz, magnetite and/or hematite, persist with this same assemblage throughout all metamorphic grades. Pronounced recrystallization of the original chert (or quartz), magnetite and hematite has taken place but reaction features among the constituent minerals are absent (Klein, 1973; see also Fig. 11-8B). These observations contradict the conclusion of Gunderson and Schwartz (1962, p. 102) that quartz-magnetite horizons produce silicate assemblages upon metamorphism.

In medium-grade metamorphic *quartz-magnetite-hematite* (or specularite) assemblages there is generally no clear-cut evidence of replacement of one iron oxide by the other. Such occurrences may indicate that there has been

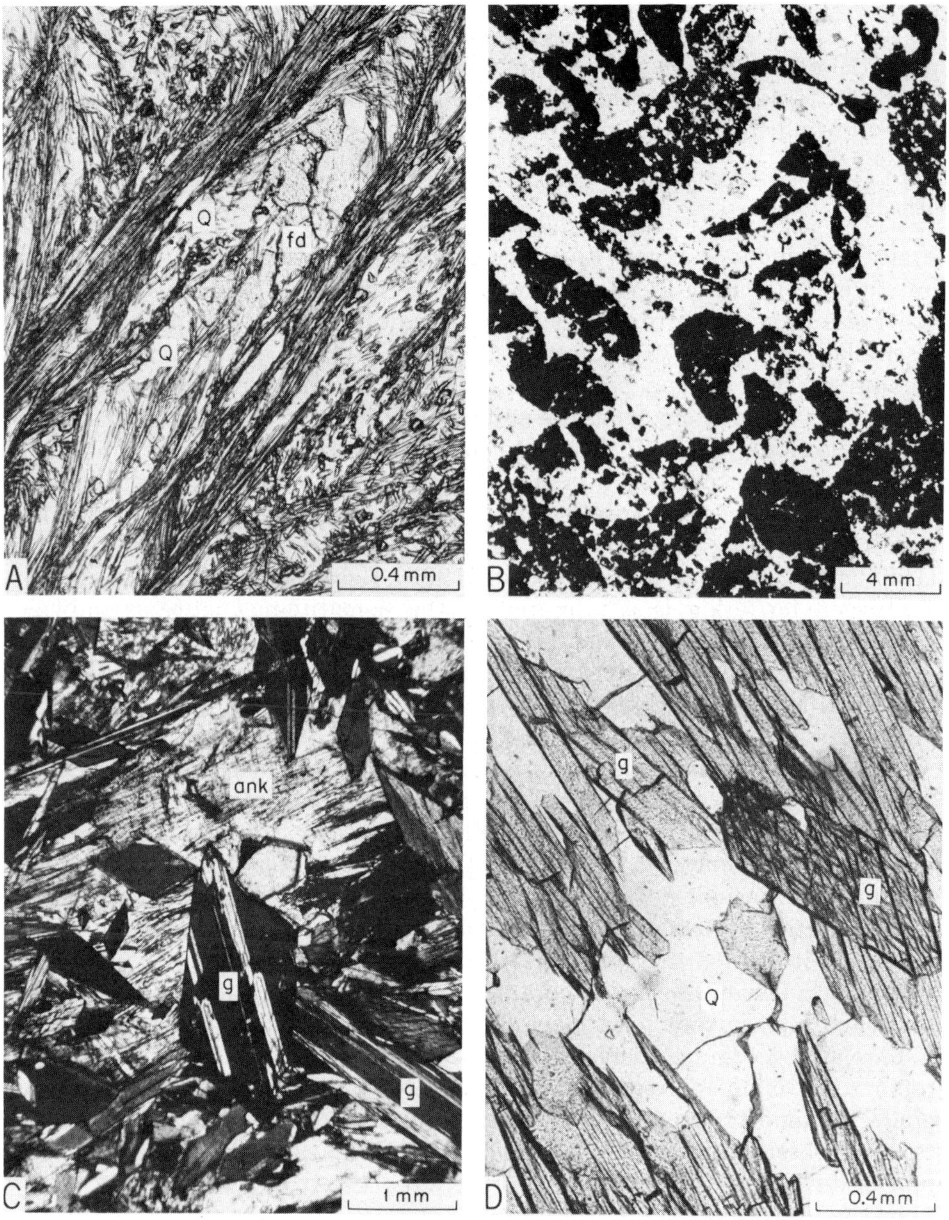

Fig. 11-8. A. Incipient formation of fine needles of grunerite around coarse patches of ferrodolomite (*fd*) and quartz (*Q*). Biotite zone metamorphism, at Pegrum Lake, west-central Labrador Trough. Plane polarized light.

B. Relict granules composed of magnetite and lesser quartz in a quartz matrix. Staurolite-kyanite zone metamorphism, Labrador City area, Newfoundland. Plane polarized light.

C. Coarse-grained grunerite (*g*) schist with patches of ankerite (*ank*). Staurolite-kyanite zone metamorphism, Labrador City area, Newfoundland. Doubly polarized light.

D. Medium-grained grunerite(*g*)-quartz(*Q*) schist. No preexisting carbonate remains. Staurolite-kyanite zone metamorphism. Labrador City area, Newfoundland. Plane polarized light.

no net gain or loss of oxygen as a result of an externally controlled chemical potential of oxygen. Hence oxygen appears, in general, to have behaved as an inert (buffered) component during metamorphism as may be expressed by the following equation:

$$\underset{\text{hematite}}{6\,Fe_2O_3} - \underset{\text{magnetite}}{4\,Fe_3O_4} \rightarrow O_2 \qquad (5)$$

If any movement of oxygen took place, it may have been sufficiently restricted not to exceed the buffering capacity of the magnetite-hematite assemblage; otherwise either hematite or magnetite might have been eliminated from the assemblage. Frost (1979), however, notes that because of the extremely small amount of free oxygen present, f_{O_2} can be buffered over many orders of magnitude without changing the abundance of any of the solid phases or the composition of the fluid phase. As such Frost suggests that the amount of magnetite in an iron-formation should remain constant during metamorphism. Quartz-magnetite-specularite assemblages are common in metamorphosed iron-formations as documented by James (1955), Kranck (1961), Klein (1966 and 1978), Butler (1969), Dorr (1969), Haase (1982a) and Gole (1981).

Oxygen buffering is not limited to the hematite-magnetite assemblages noted above. Oxygen can also be buffered or constrained by complex fluid-solid equilibria involving Fe^{3+} in minerals, and/or redox equilibria due to compositional buffering of the fluid by the mineral assemblage (see Frost, 1979; Haase, 1982b).

Although there is evidence, as indicated above, that there is little to no movement of oxygen between bands of oxide-rich iron-formation during metamorphism, hematite pseudomorphs after magnetite (martite) have been reported from some iron-formations that are concluded to show no evidence of supergene leaching. Gruner (1946) and Gundersen and Schwartz (1962) report martite in hematite-rich bands in the Biwabik Iron Formation. Kalliokoski (1965) notes that in unweathered parts of the iron-formation of El Pao, Venezuela magnetite is commonly replaced by hematite. He suggests that such replacement took place during regional metamorphism and that it is unrelated to later supergene oxidation of the ores. French (1968, p. 83) reports the incipient reduction of hematite to magnetite in relatively low-grade metamorphic conditions in the Biwabik Iron Formation. Based on textural observations Floran and Papike (1978, p. 231) conclude that most magnetite in the Gunflint Iron Formation formed by reduction of earlier hematite.

The above observations suggest that some movement of oxygen has taken place in some instances, but that in most metamorphic assemblages only enough movement took place to allow for the replacement of only a small amount of the original iron oxide.

As mentioned earlier, the presence of members of the *cummingtonite-grunerite* series is diagnostic of iron-formation assemblages having undergone medium-grade metamorphism, ranging from the biotite to the staurolite and kyanite isograds of pelitic schists. These amphiboles which are light beige to light brown in color, have generally a well-developed needlelike habit which overprints earlier (late diagenetic and low metamorphic grade) assemblages (see Fig. 11-8). The types of reactions that give rise to these amphiboles are as follows (Klein, 1973):

$$\underset{\text{ferrodolomite}}{7Ca(Fe,Mg)(CO_3)_2} + \underset{\text{quartz}}{8SiO_2} + H_2O \rightarrow \underset{\text{grunerite}}{(Fe,Mg)_7Si_8O_{22}(OH)_2} + \underset{\text{calcite}}{7CaCO_3} + 7CO_2 \tag{6}$$

$$\underset{\text{siderite}}{8(Fe,Mg)CO_3} + \underset{\text{quartz}}{8SiO_2} + H_2O \rightarrow \underset{\text{grunerite}}{(Fe,Mg)_7Si_8O_{22}(OH)_2} + 7\ CO_2 \tag{7}$$

and

$$\underset{\text{minneostaite}}{7Fe_3Si_4O_{10}(OH)_2} \rightarrow \underset{\text{grunerite}}{3Fe_7Si_8O_{22}(OH)_2} + 4SiO_2 + 4H_2O \tag{8}$$

The first two above reactions have been documented texturally in several studies of metamorphic iron-formations (e.g. French, 1968; Klein, 1973, 1978; Floran and Papike, 1978; Gole, 1981). The third reaction, of the conversion of minnesotaite to grunerite, however, has been documented texturally only twice, by Gair (1975) and Gole (1980b). The occurrence of grunerite at the expense of preexisting minnesotaite reported by Gair is the result of contact metamorphism of iron-formation in the Marquette district of Michigan by a metadiabase. This same reaction is documented by Gole as the result of prograde regional metamorphism of Archean iron-formations in Western Australia. A systematic study of the chemistry and physical properties of members of the cummingtonite-grunerite series in medium-grade metamorphic iron-formation is given by Klein (1964); X-ray parameters are evaluated by Klein and Waldbaum (1967). *Amosite* is a rare asbestiform variety of grunerite which occurs in metamorphosed iron-formations in the Transvaal Province of South Africa (Peacock, 1928; du Toit, 1945; Vermaas, 1952; Hutchison et al., 1975).

Fe-Mg amphiboles are commonly intergrown with Ca-clinoamphiboles such as *actinolite* and *hornblende.* The chemical compositions of such Ca-clinoamphiboles are generally close to the Al-free, tremolite-ferroactinolite join because bulk compositions of iron-formations tend to be low in Al_2O_3 as well as Na_2O. The common occurrence of coexistencies of Ca- and (Fe,Mg)-clinoamphiboles has been documented by Mueller (1960) for the Mount Wright area in Labrador, by Klein (1966, 1968) in the Labrador City area of Canada, by Immega and Klein (1976) in Archean iron-formations of southwestern Montana, and by Haase (1979a, 1982a) in the Negaunee Iron Formation (see Fig. 11-9). Such coexistences are the result of the following type of reaction (Klein, 1966):

$$\underset{\text{ferrodolomite}}{14Ca(Mg_{0.5}Fe_{0.5})(CO_3)_2} + 16SiO_2 + 2H_2O \rightarrow$$
$$\underset{\text{tremolite}}{Ca_2Mg_5Si_8O_{22}(OH)_2} + \underset{\text{grunerite}}{Fe_7Si_8O_{22}(OH)_2} + \underset{\text{calcite}}{14(Ca_{0.9}Mg_{0.1})CO_3} + 14CO_2 \quad (9)$$

The amphiboles, or amphibole pairs commonly show well-developed, thin exsolution textures as described by Bonnichsen (1969), Ross et al. (1969) and Immega and Klein (1976). *Anthophyllite* is not common in metamorphic iron-formation assemblages and tends to be restricted to hematite (or specularite)-rich assemblages with high $Fe^{3+}/(Fe^{2+} + Fe^{3+})$ ratios in the bulk composition. If such Fe^{3+}-rich bulk compositions are essentially devoid of Na_2O, Fe^{3+} is accomodated mainly in the hematite structure (minor magnetite may be present as well) but can not be incorporated in the associated silicates and carbonates, and therefore Mg-rich members of the Mg-Fe amphibole series occur. The most common assemblages are: quartz-specularite-(or hematite)-anthophyllite, and quartz-hematite-anthophyllite-tremolite with minor magnetite (Klein, 1966, 1978; Haase, 1979a). Mn-rich amphiboles such as *tirodite* ($Mn_2Mg_5Si_8O_{22}(OH)_2$), which has been earlier described as manganoan cummingtonite (e.g. Klein, 1964, 1966; Leake, 1978) may be locally abundant in metamorphic iron-formations if their bulk Mn content is appreciable. *Riebeckite-tremolite* and *magnesioriebeckite* have been reported in some relatively Na_2O-rich assemblages in the Labrador City region by Klein (1966).

Members of the *pyroxene* group may be sporadically present in medium-grade metamorphic iron-formation occurrences, but their abundance is generally very minor compared to that of amphiboles. Floran and Papike (1978) describe *hedenbergite* as part of grunerite-rich assemblages in their metamorphic zone 3 of the Biwabik Iron Formation; similar hedenbergite occurrences are reported by Morey et al. (1972). Klein (1966) reports *eulite*, *ferrosalite*, and *aegirine-augite* occurrences in amphibole-rich iron-formations in the Labrador City area, and Haase (1982a) describes members of the ferrosalite-salite series as locally abundant in some parts of the Negaunee Iron Formation. Such relatively rare, pyroxene-containing assemblages in medium metamorphic grade rocks probably do not reflect higher temperatures of origin than the common, associated amphibole assemblages, but may be the result of local reductions in P_{H_2O}, or increases in X_{CO_2}.

Fayalite is a rare mineral in regionally metamorphosed iron-formations but appears to be somewhat more common in iron-formations that have under-

---

Fig. 11-9. Compilation of compositional ranges and major element fractionation data of amphiboles from several medium-grade metamorphic iron-formations. A, B and D are from Immega and Klein (1976); C is from Haase (1979a and 1982a).

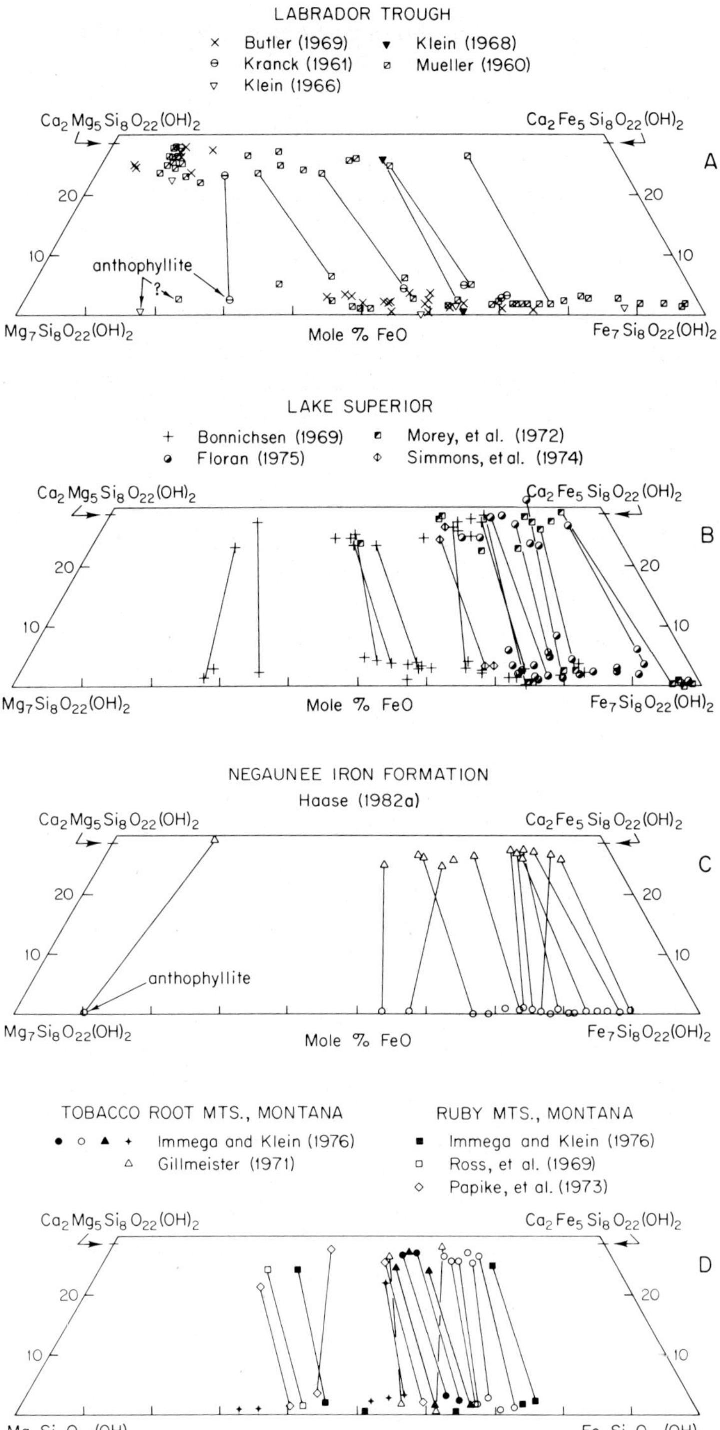

LABRADOR TROUGH
× Butler (1969)
⊖ Kranck (1961)
▽ Klein (1966)
▼ Klein (1968)
⧄ Mueller (1960)
$Ca_2Mg_5Si_8O_{22}(OH)_2$
$Ca_2Fe_5Si_8O_{22}(OH)_2$
A
20
10
anthophyllite
?
$Mg_7Si_8O_{22}(OH)_2$
Mole % FeO
$Fe_7Si_8O_{22}(OH)_2$
LAKE SUPERIOR
+ Bonnichsen (1969)
◒ Floran (1975)
Morey, et al. (1972)
Simmons, et al. (1974)
$Ca_2Mg_5Si_8O_{22}(OH)_2$
$Ca_2Fe_5Si_8O_{22}(OH)_2$
B
20
10
$Mg_7Si_8O_{22}(OH)_2$
Mole % FeO
$Fe_7Si_8O_{22}(OH)_2$
NEGAUNEE IRON FORMATION
Haase (1982a)
$Ca_2Mg_5Si_8O_{22}(OH)_2$
$Ca_2Fe_5Si_8O_{22}(OH)_2$
C
20
10
anthophyllite
$Mg_7Si_8O_{22}(OH)_2$
Mole % FeO
$Fe_7Si_8O_{22}(OH)_2$
TOBACCO ROOT MTS., MONTANA
● ○ ▲ + Immega and Klein (1976)
△ Gillmeister (1971)
RUBY MTS., MONTANA
■ Immega and Klein (1976)
□ Ross, et al. (1969)
◇ Papike, et al. (1973)
$Ca_2Mg_5Si_8O_{22}(OH)_2$
$Ca_2Fe_5Si_8O_{22}(OH)_2$
D
20
10
$Mg_7Si_8O_{22}(OH)_2$
Mole % FeO
$Fe_7Si_8O_{22}(OH)_2$

gone medium-grade conditions of contact metamorphism. Floran and Papike (1978) and Morey et al. (1972) record its occurrence, in association with grunerite and hedenbergite in their metamorphic zone 3 of the contact metamorphosed Biwabik Iron Formation. Haase and Klein (1978) and Haase (1982a) report some fayalite in the Negaunee Iron Formation which may be the result of local contact metamorphism by diabase sills, or due to a localized reduction in P_{H_2O} in portions of the iron-formation adjacent to these sills. Generally, fayalite is considered part of high grade metamorphic iron-formation assemblages (see below).

Rhodonite is a very rare mineral in most iron-formations, because it requires an Mn-rich iron-formation bulk composition. Some of the regionally metamorphosed iron deposits in the Labrador City area are Mn-rich enough to contain considerable rhodonite at kyanite-staurolite zone conditions of pelitic schists (Klein, 1966).

Garnet may be locally present in medium-grade metamorphic iron-formation assemblages but it tends to be relatively rare on account of the inherently low Al_2O_3 contents of the bulk chemistry of iron-formations. The most commonly reported garnets belong to the pyralspite group (pyrope, almandite, and spessartite) but some andradites have also been reported. Garnets which are part of the almandite-spessartite series are quite common in parts of the Negaunee Iron Formation (Haase, 1982a) and are used to define assemblages in the highest metamorphic grade (subzone 3a; Haase, 1982a). Morey et al. (1972) report some almandite-rich and andradite-rich garnets in the contact metamorphosed Biwabik Iron Formation; Immega and Klein (1976) report on the fairly common occurrence of almandite-rich garnets in medium-grade metamorphic Archean iron-formations in southwestern Montana. Klein (1966) reports some almandite-rich assemblages from the iron-formations in the Labrador City region; and Dahl (1979) describes spessartite-rich and andradite-rich garnets in some Archean iron-formations of the Ruby Mountain area in southwestern Montana. When the bulk composition of the iron-formation is manganese-rich some *calderite* garnet ($Mn_3Fe_2^{3+}Si_3O_{12}$) may be present as noted in some parts of the Labrador City region (Klein, 1966).

The *carbonate* species that were present in late diagenetic to very low-grade metamorphic assemblages tend to persist throughout medium grade metamorphic conditions. Calcite, members of the dolomite-ankerite series, and the magnesite-siderite series are present in many medium-grade iron-formation assemblages. The grain size of the carbonates tends to be somewhat coarser than in lower grade occurrences and the textures tend to be more equigranular. Part of the original carbonates has been used up in the production of silicates such as members of the cummingtonite-grunerite series and Ca-clinoamphiboles, as stated by the reaction equations used in the discussion of the amphiboles above (see also Figs. 11-8A and 8C). Indeed, some assemblages, such as quartz-magnetite-grunerite schists may be totally devoid of carbonates, because all of the original carbonate was used up in the formation of the silicate

(Fig. 11-8D). However, much of the bulk of the carbonates may survive in coexistence with the newly formed silicates. The general decomposition reaction of carbonates:

$$\text{carbonate + quartz} \rightarrow \text{silicates} + CO_2 \tag{10}$$

does not always prevail in medium-, or even high-grade metamorphic conditions. This is especially well shown by medium to high grade metamorphic iron-formations in the southern part of the Labrador Trough. In the Labrador City area (Klein, 1966, 1973, 1978) sequences of quartz-ankerite (± siderite)-magnetite are interbanded with quartz-grunerite-magnetite schists. Such occurrences may simply indicate that the chemical potential of CO_2 was high enough locally (in specific carbonate-rich horizons) to prevent the breakdown of the carbonates and the reaction with available quartz to form silicate products, or they may reflect a more complex relationship between $X_{(Fe\text{-}Mg\text{-}Mn)}$ and X_{CO_2} variation in the system during metamorphism (see Haase, 1982b, fig. 6). These interpretations indicate that although many medium grade metamorphic iron-formations are silicate (especially grunerite)-rich, it is not unusual to find quartz-carbonate-rich assemblages closely associated with such silicate assemblages.

James (1955) reports small amounts of carbonate in the garnet-staurolite zone in some of the iron-formations of northern Michigan; French (1968) describes calcite, ankerite and siderite from his zone 3 in the contact metamorphosed Biwabik Iron Formation; Klein (1966, 1978) reports on abundant calcite, dolomite-ankerite, and siderite in iron-formation of medium metamorphic grade in the southern part of the Labrador Trough; Floran and Papike (1978) describe minor amounts of calcite, siderite and ankerite which become relatively scarce in their zone 2 of the Gunflint Iron Formation; Haase (1982a) notes the widespread occurrences of minor amounts of calcite throughout his zone 3 of the Negaunee Iron Formation and the sparse occurrence of some manganoan ankerite in association with the various Fe-silicates (mainly grunerite). A graphical representation of some of the assemblage and mineralogical changes that occur in the prograde sequence of metamorphism of the Negaunee Iron Formation is given in Fig. 11-10.

The occurrence of the above silicate-producing reactions at the expense of quartz (or chert) + carbonates, and of earlier minnesotaite and stilpnomelane, is widely accepted. However there are additional, more complex metamorphic reactions that have been postulated by various iron-formation investigators. LaBerge (1964) concludes that magnetite is not a primary mineral in many iron-formations and suggests that much of it, especially in metamorphic rocks, is the result of the breakdown and oxidation of ferrous-iron minerals (carbonates and silicates). French (1968) and Floran and Papike (1978) suggest that as the result of increasing metamorphism fine-grained hematite and chert (which becomes quartz) are replaced by coarser-grained magnetite. Han (1978) concludes that silicate-rich, carbonate-rich, or mixed types of iron-

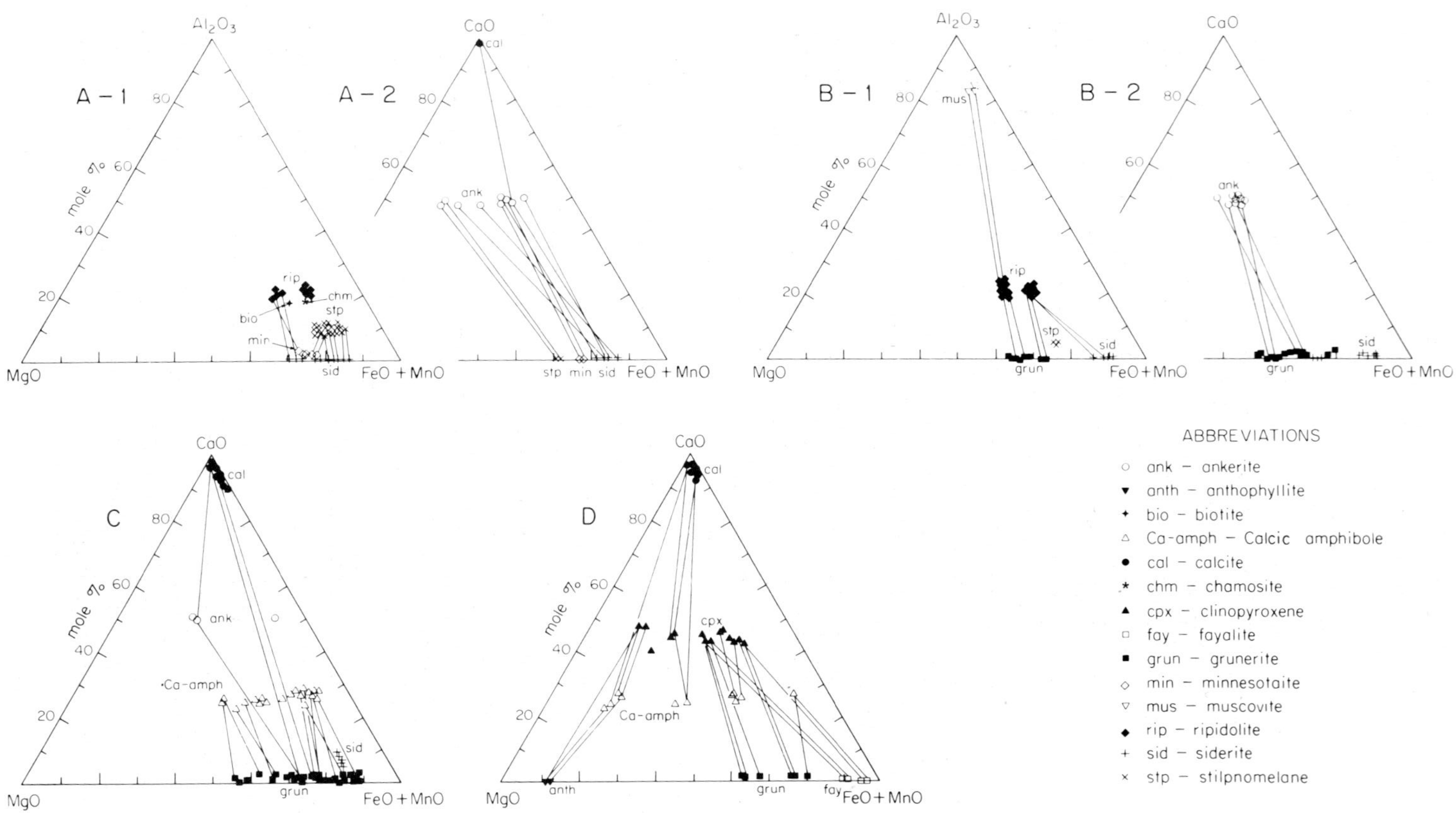

Fig. 11-10. Graphical representation of some assemblages and assemblage changes during progressive metamorphism of the Negaunee Iron Formation (after Haase, 1982a). Tielines connect coexisting minerals.
A-1 and A-2. Phyllosilicate and carbonate compositions in zone 1 (minnesotaite present) of Haase.
B-1 and B-2. Grunerite, phyllosilicate and carbonate compositions in zone 2 (grunerite present) of Haase.
C. Amphibole and carbonate compositions in zones 3a and b (highest metamorphic grade) of Haase.
D. Clinopyroxene, olivine, and clinoamphibole compositions in zone 3b of Haase.

formations might well have been mainly oxide-rich types prior to post-depositional changes that involve silication, carbonatization and magnetitization. Floran and Papike (1978) suggest many complex reactions that involve carbonates, silicates and oxides as reactants as well as reaction products. An example of such a postulated reaction (their reaction no. 15, p. 261) is:

$$\text{greenalite} + SiO_2 + \text{hematite} \rightarrow \text{grunerite} \pm \text{magnetite} + H_2O \quad (11)$$

Many of the above rather complex reactions may well have taken place, however, unambiguous textural evidence in support of such reactions is generally lacking. In essentially unmetamorphosed iron-formation assemblages, late diagenetic textures can often be clearly distinguished because they overprint earlier minerals. However, with increasing metamorphic grade, many of the minerals, except amphiboles, tend to become relatively equigranular in habit, and it becomes very difficult, if not impossible, to distinguish precursor materials from reaction products.

PHYSICAL AND CHEMICAL CONDITIONS OF MEDIUM-GRAINED METAMORPHISM

As stated earlier, the conditions of medium-grade metamorphism of iron-formations range broadly from the biotite grade through the staurolite-kyanite grade as defined on the basis of pelitic schists. Many of the reactions that occur in medium-grade metamorphism involve decarbonation and dehydration.

Several investigators of the metamorphism of iron-formations have evaluated the temperature and pressure of formation of the assemblages, and the possible role of a fluid phase. James (1955) estimates that the first appearance of members of the cummingtonite-grunerite series is reflective of garnet isograd conditions in pelitic schists which, between 2 and 5 kbar pressures, would reflect a temperature range of 450 to 500°C (Winkler, 1979). Klein (1978) notes that Fe-Mg clinoamphiboles appear very close to the biotite isograd in the south-central part of the Labrador Trough. This would imply temperatures somewhat lower than 450°C for the onset of Fe-Mg clinoamphibole formation. The upper limit for abundant amphibole stability is probably close to the high-temperature part of the kyanite-staurolite zone. All of the Labrador City region, in the southwestern part of the Labrador Trough is located in this metamorphic regime. On the basis of the staurolite-kyanite gneisses associated with the iron-formation in that region, Klein (1966) concludes the maximum temperature of metamorphism to have been below 600°C and the pressure between 6 and 10 kbar. Haase (1982a) estimates maximum metamorphic temperatures in the range of 550°C to 615°C for his zone 3 in the Negaunee Iron Formation, on the basis of garnet-biotite geothermometry. He also reports oxygen isotopic temperatures ranging from 400° to 490°C for this same zone. These temperatures are based on isotopic data by James and Clayton (1962) and have been calculated with the quartz-magnetite fractionation curve of Becker and Clayton (1976). Haase (1982a) concludes that the pressure conditions in this zone appear to have been between 2 and 3 kbar. Miyano and Klein (1982b) have evaluated the stability of grunerite on the basis of a combination of thermodynamic data and published pressure-temperature estimates of metamorphism for grunerite-rich assemblages in metamorphosed iron-formations. Figure 11-11 shows their assessment of the stability field of grunerite.

The amphibole-rich assemblages that are diagnostic of medium-grade metamorphism are commonly considered to have originated under high P_{H_2O} conditions. However, Haase (1982b) concludes that with increasing metamorphic grade of the Negaunee Iron Formation the X_{CO_2} of the fluid phase appears to have considerably increased (see last section in this chapter).

HIGH-GRADE METAMORPHIC ASSEMBLAGES

Iron-formations that have undergone the highest grade of metamorphism are characterized by essentially anhydrous assemblages in which variable amounts of ortho- and clinopyroxene predominate. Fayalite may be present, as well as carbonates and garnet, and lesser amounts amphiboles; quartz,

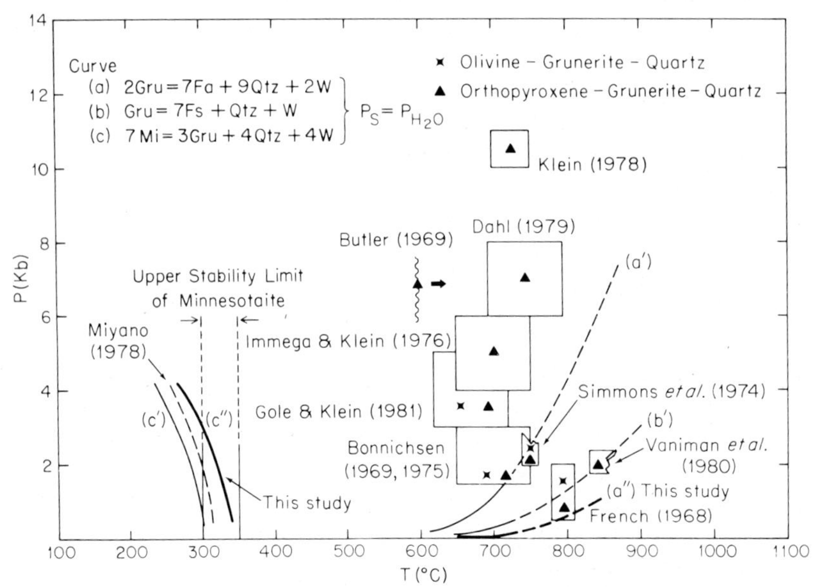

Fig. 11-11. Evaluation of the stability field of grunerite from a combination of published pressure-temperature estimates for various metamorphosed iron-formations and thermodynamic calculations (from Miyano and Klein, 1983b). Curve a'' represents the apparent upper stability limit for grunerite, whereas curve c'' is the adjusted upper stability limit for minnesotaite.

magnetite and/or hematite are still the major constituents of oxide-rich iron formations. Such high-grade assemblages are the result of metamorphic conditions that straddle the sillimanite isograd of pelitic rocks, or that range from upper-amphibolite to granulite facies. Assemblages that reflect the highest temperatures are the result of contact metamorphism, as in the case of the Gunflint and Biwabik Iron Formations near the contact with the Duluth Gabbro Complex. Somewhat lesser temperatures, but higher pressures are evaluated for regionally metamorphosed iron-formations such as those in the southwestern part of the Labrador Trough, and in southwestern Montana.

The oxide-rich iron-formations which consist mainly of mixtures of *quartz*, *magnetite* and *hematite*, even at the highest metamorphic grades, show no reactions among the constituent minerals, only recrystallization. The texture of such lithologies tends to be equigranular, and the grain sizes of quartz and the oxides are often coarse. A very coarse-grained example of a high metamorphic grade quartz-specularite assemblage is from the Gagnon area in Quebec (in the southwestern part of the Labrador Trough) where specularite grains may reach sizes of 1 cm in diameter (Klein, 1978, p. 907). Magnetite, hematite (or specularite) and quartz are also part of the various high grade metamorphic silicate- and silicate-carbonate assemblages. In such occurrences reaction textures between the oxides and silicates or carbonates have not been documented. French (1968) and Floran and Papike (1978), however, state that when hematite-bearing beds of the Biwabik and Gunflint Iron Formations are traced into the contact aureole of the Duluth Gabbro Complex, magnetite occurs in the place of hematite. This implies that the original hematite has been reduced to magnetite by the effects of the gabbro intrusion.

Orthopyroxenes and *clinopyroxenes* make up large percentages of iron-formation lithologies that prior to metamorphism would have been rich in carbonates and hydrous silicates. Bonnichsen (1969) describes abundant ortho- and clinopyroxenes as well as fayalite from the Biwabik Iron Formation, where it has been contact metamorphosed to a pyroxene hornfels by the Duluth Gabbro Complex. A considerable portion of the presently observed, equigranular othopyroxene grains with abundant Ca-pyroxene exsolution lamellae is considered to represent inverted pigeonite. Most of these orthopyroxenes are ferrohyperthene in composition. Commonly such orthopyroxenes coexist with calcic clinopyroxenes. A common iron-formation assemblage as described by Bonnichsen (1969) is: quartz-magnetite-orthopyroxene-Ca clinopyroxene ± fayalite. The presence or absence of fayalite depends on the Fe/(Fe + Mg) ratio of the original bulk composition and total pressure. French (1968) and Gundersen and Schwartz (1962) similarly describe assemblages rich in quartz, magnetite, hedenbergite, ferrohypersthene, fayalite, and pyrrhotite near the contact of the Biwabik Iron Formation with the Duluth Gabbro Complex. Simmons et al. (1974) report very similar overall assemblages from the Gunflint Iron Formation in a zone nearest to the contact. Their orthopyroxene grains, with extensive exsolution lamellae of augite along (001) and (100), are interpreted to have begun crystallization as pigeonite, during the peak temperature of contact metamorphism; upon cooling these grains inverted to the present orthopyroxene with exsolution lamellae. Other minerals associated with such Fe-rich orthopyroxene are augite, fayalite, plagioclase, Fe-Ti oxides, primary cummingtonite, minor quartz, and some retrograde amphiboles. Floran and Papike (1978) describe ferrohypersthene-rich assemblages in the highest contact metamorphic zone (their zone 4) of the Gunflint Iron Formation as well.

Gole and Klein (1981b) report clinopyroxene-orthopyroxene-magnetite ± fayalite-quartz-grunerite assemblages from regionally metamorphosed Archean iron-formations in the Yilgarn Block of Western Australia (see Figs. 11-12B, C and D). As there is no evidence for the orthopyroxene having formed by inversion from pigeonite, their assemblages may well be of somewhat lower temperature origin than those of the highest-temperature zones in the Biwabik and Gunflint Iron Formations; however, the Fe/(Mg + Fe) ratio of the pyroxene and the total pressure during metamorphism are also important parameters in such an evaluation. Figure 11-13 shows the common compositional ranges for ortho- and clinopyroxenes in high-grade assemblages. Similarly pyroxene-rich assemblages are recorded from other regionally metamorphosed iron-formations as well. Kranck (1961), Butler (1969) and Klein (1978) document the extensive occurrence of Fe-rich orthopyroxenes, commonly coexisting with clinopyroxenes from the most highly metamorphosed parts of the iron-formations in southwestern Labrador. Quartz and magnetite are abundant minerals in such associations as well, and some Fe-Mg and Ca-Fe-clinamphiboles appear also to be part of the high-temperature assemblages. The equigranular grain size of the regionally metamorphosed pyroxene-rich assemblages is generally considerably coarser than that of the contact metamorphosed iron-formations discussed above. Immega and Klein (1976) report ferrohypersthene-ferrosalite-magnetite-quartz ± almandite-amphibole assemblages from the high-grade metamorphic terrain of the Tobacco Root Mountains in Montana (see Figs. 11-12A and 11-13). Dahl (1979) records very similar iron-formation occurrences (e.g. quartz-magnetite-ferroaugite-ferro-

Fig. 11-12. Photomicrographs of some high-grade metamorphic assemblages in iron-formations.

A. Granular iron-formation consisting of hypersthene(*hyp*)-ferrosalite(*fsl*)-almandite(*alm*)-magnetite(*mag*)-quartz(*qtz*) and minor hornblende (*hbl*), from Archean terrain in the Tobacco Root Mountains, southwestern Montana (from Immega and Klein, 1976).

B. Coexisting ferroaugite (*cpx*), eulite (*opx*), grunerite (*grun*), quartz (*qtz*) and magnetite (*mag*).

C. Fayalite(*fay*)-eulite(*opx*)-ferroaugite(*cpx*)-grunerite(*grun*)-quartz(*qtz*). Smooth curved grain boundaries occur between all minerals, suggestive of equilibrium crystallization.

D. Fayalite(*fay*)-grunerite(*grun*)-quartz(*qtz*) assemblage in which grunerite mantles fayalite grains so that quartz and fayalite are not in contact. Sampled B, C and D are from Archean iron-formations in the Yilgarn Block, Western Australia (from Gole and Klein, 1981b).

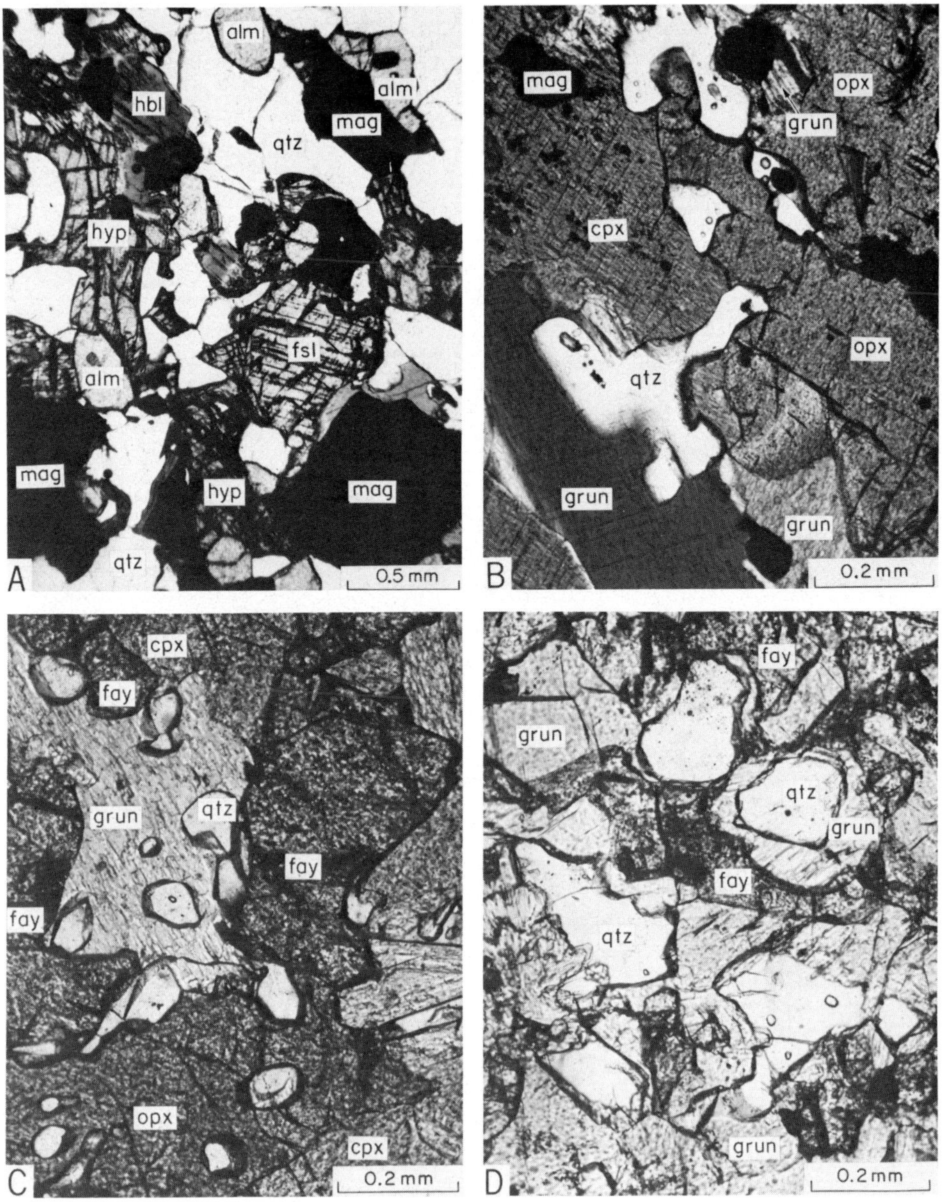
alm
alm
hbl
mag
qtz
hyp
fsl
alm
mag
hyp
mag
qtz
A
0.5 mm
mag
opx
grun
cpx
opx
qtz
grun
grun
B
0.2 mm
cpx
fay
grun
qtz
fay
fay
opx
cpx
C
0.2 mm
fay
grun
qtz
grun
fay
qtz
grun
D
0.2 mm

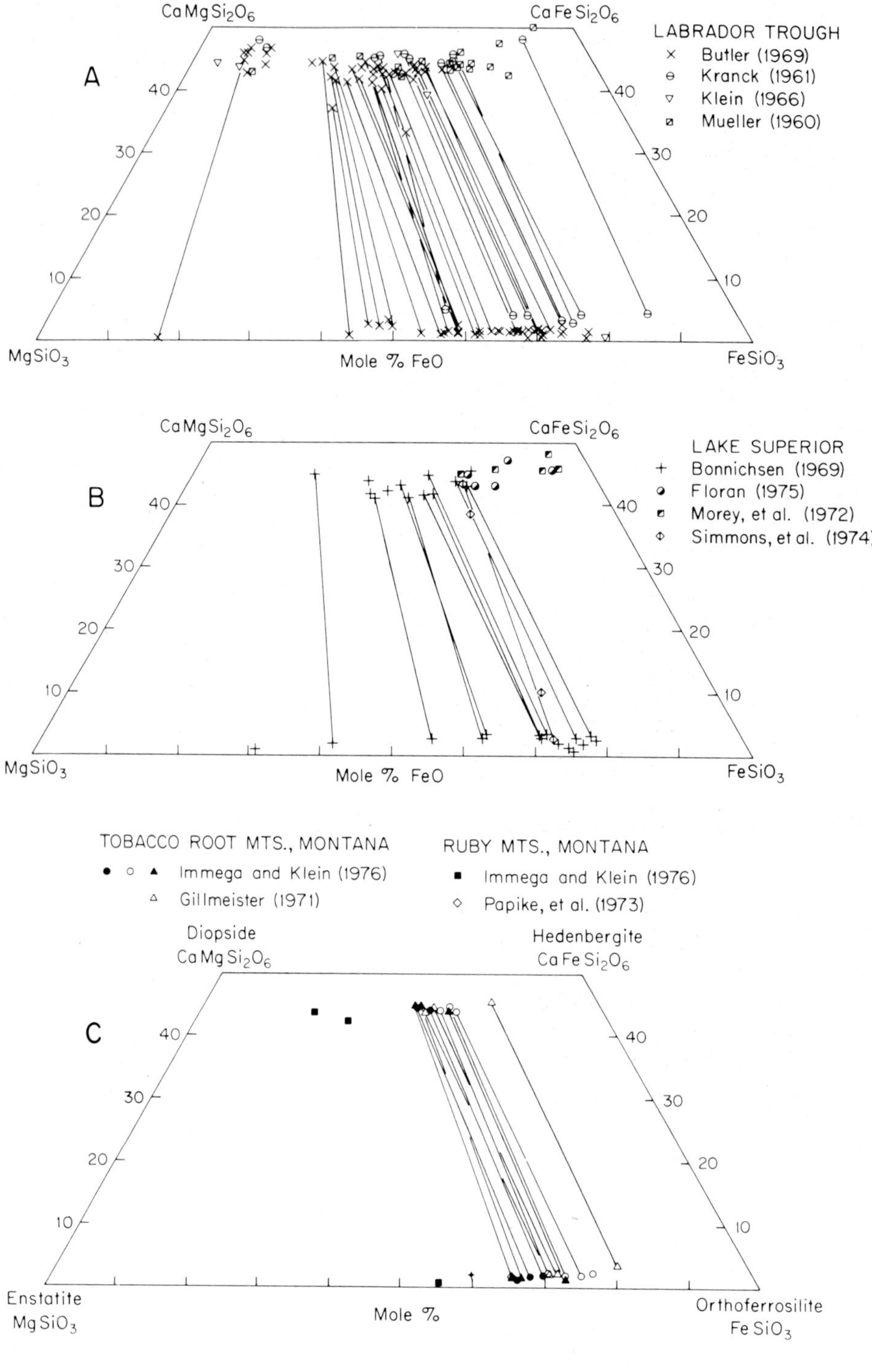
A
CaMgSi2O6
CaFeSi2O6
LABRADOR TROUGH
Butler (1969)
Kranck (1961)
Klein (1966)
Mueller (1960)
MgSiO3
Mole % FeO
FeSiO3
B
CaMgSi2O6
CaFeSi2O6
LAKE SUPERIOR
Bonnichsen (1969)
Floran (1975)
Morey, et al. (1972)
Simmons, et al. (1974)
MgSiO3
Mole % FeO
FeSiO3
TOBACCO ROOT MTS., MONTANA
Immega and Klein (1976)
Gillmeister (1971)
RUBY MTS., MONTANA
Immega and Klein (1976)
Papike, et al. (1973)
C
Diopside
CaMgSi2O6
Hedenbergite
CaFeSi2O6
Enstatite
MgSiO3
Mole %
Orthoferrosilite
FeSiO3

hypersthene-almandite) from the Ruby Range of southwestern Montana as well.

Pyroxenes are the result of the following types of decarbonation reactions (Klein, 1973):

$$\underset{\text{ankerite}}{Ca(Fe,Mg)(CO_3)_2} + 2SiO_2 \rightarrow \underset{\text{clinopyroxene}}{Ca(Fe,Mg)Si_2O_6} + 2CO_2 \qquad (12)$$

and

$$\underset{\text{siderite}}{(Fe,Mg)CO_3} + SiO_2 \rightarrow \underset{\text{orthopyroxene}}{(Fe,Mg)SiO_3} + CO_2 \qquad (13)$$

Amphibole decomposition reactions are also responsible for pyroxene formation, e.g.:

$$\underset{\text{grunerite}}{Fe_7Si_8O_{22}(OH)_2} \rightarrow \underset{\text{orthopyroxene}}{7FeSiO_3} + SiO_2 + H_2O \qquad (14)$$

In high-grade metamorphic rocks with equigranular textures it is generally impossible to distinguish relict minerals or textures, and as such mineral reactions can not be deduced on textural grounds. The above simplified reactions are therefore only suggestive of many of the much more complex reactions that probably have taken place between carbonates, quartz, hydrous silicates and possibly magnetite. Various examples of such complex reactions are given in French (1968) and Floran and Papike (1978).

Fayalite is a major constituent of parts of the Biwabik and Gunflint Iron Formations which have been most highly metamorphosed by the Duluth Gabbro Complex. Such fayalite has a rather restricted composition with an atomic ratio of Fe/(Fe + Mg) of 90% or more (Bonnichsen, 1969). It occurs as anhedral, essentially equidimensional grains. Simmons et al. (1974) report fayalite of composition $Fo_{13}Fa_{87}$. Gole and Klein (1981b) report fayalite compositions with Fe/(Fe + Mg) ranges from 92 to 98 atomic percent, from some Archean iron-formations that have been regionally metamorphosed. Figure 11-14 is a graphical representation of fayalite compositions in pyroxene-rich assemblages.

Amphiboles, such as members of the cummingtonite-grunerite series, and hornblende-like compositions may form a small part of these highest grade metamorphic assemblages. If the amphiboles are relatively coarse-grained, and are not restricted to the grain boundaries of pyroxenes or fayalite, they are considered part of the primary, high-temperature occurrence. Some of the amphiboles, however, in such assemblages are very fine-grained and clearly restricted to the edges of primary silicates such as pyroxenes, and fayalite; these are interpreted to be retrograde in origin (see e.g. Bonnichsen, 1969;

Fig. 11-13. Compilation of compositional ranges and major-element fractionation data of pyroxenes from some high-grade metamorphic iron-formations (from Immega and Klein, 1976).

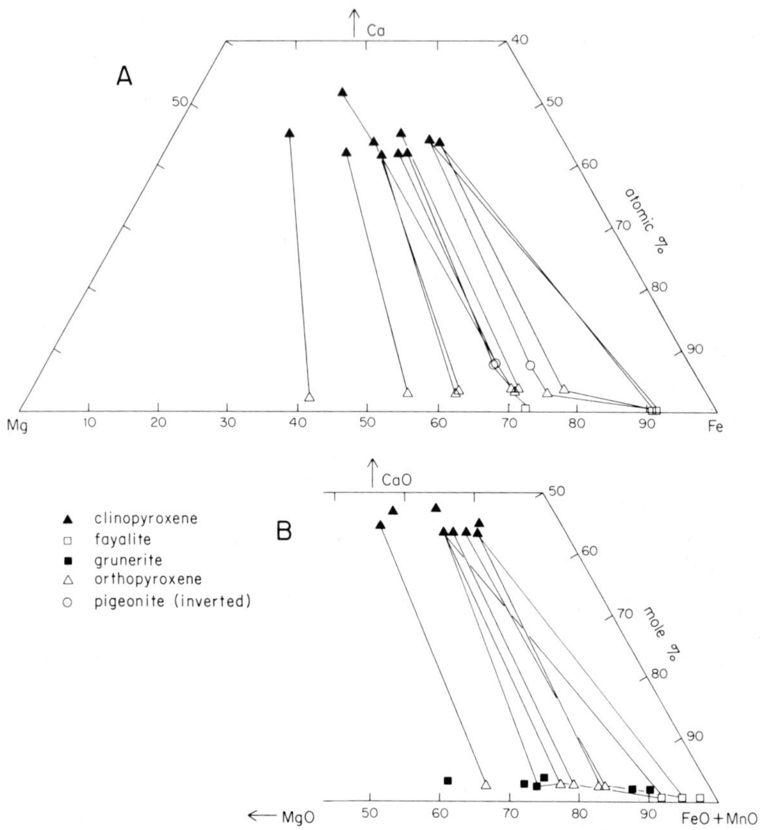

Fig. 11-14. Compositional ranges of orthopyroxene, clinopyroxene and fayalite (and lesser grunerite) in some high-grade metamorphic iron-formation occurrences.
A. Assemblages in the highest-grade metamorphic zone of the Biwabik Iron Formation (from Bonnichsen, 1969);
B. Mineral compositions and assemblages from high-grade Archean iron-formations in the Yilgarn Block of Western Australia (from Gole and Klein, 1981b).

Simmons et al., 1974; Immega and Klein, 1976; Gole and Klein, 1981b).

Carbonates are commonly present in variable amounts in the highest metamorphic grade assemblages. Bonnichsen (1969) reports small to trace amounts of calcite in the highest grade pyroxene-fayalite-rich assemblages. French (1968) similarly reports only calcite in the most highly metamorphosed rocks of the Biwabik Iron Formation, although other species such as siderite and members of the dolomite-ankerite series are present at lower grade. Butler (1969) reports the occurrence of calcite or ankerite or siderite, as well as calcite-ankerite and calcite-siderite pairs in his pyroxene-rich assemblages from the Gagnon region, in the southwestern Labrador Trough. Klein (1978) in his evaluation of the regional metamorphic changes in the iron-formations of the Labrador Trough, notes that calcite and members of the dolomite-ankerite series are abundant throughout the complete metamorphic range exhibited, but that siderite becomes less abundant at the highest grades. This may indicate that the $FeCO_3$ component is more involved in the production of metamorphic silicates than calcite, or dolomite (or ankerite). In these high-grade metamorphic iron-formation assemblages of the southwestern Labrador Trough the carbonates, which commonly are as coarse-grained and equigranular as the coexisting pyroxenes, may make up several volume percent of the assemblage. The range of compositions shown by carbonates as well as other minerals as a function of regional metamorphism from very low to the highest grade in iron-formations of the Labrador Trough, is shown in Fig. 11-15.

Almandite-rich *garnet* may be present locally in some of the highest metamorphic grade assemblages if the bulk composition has an appropriately high Al_2O_3 content. This, however, is not commonly the case, as can be seen in Table 11-I. Almandite garnets are reported by Bonnichsen (1969), Morey et al. (1972), Immega and Klein (1976), Klein (1978) and Dahl (1979).

Other minerals that have been recorded in small amounts as part of the highest metamorphic grade of iron-formation are *graphite* (French, 1968; Butler, 1969), *pyrite* (Butler, 1969) and *pyrrhotite* (French, 1968; Klein, 1978) and *biotite* and *phlogopite* (Klein, 1978).

A schematic diagram of the relative mineral stabilities in iron-formations ranging from very low to the highest metamorphic grade is given in Fig. 11-16.

PHYSICAL AND CHEMICAL CONDITIONS OF HIGH-GRADE METAMORPHISM

The temperature and pressure conditions which produced the highest metamorphic grade iron-formation assemblages have been assessed in almost all of the studies referenced above. The highest metamorphic temperatures have been recorded for the contact metamorphosed Biwabik and Gunflint Iron Formations. French (1968) suggests, on the basis of experimental data, that fayalite which formed at the contact with the Duluth Gabbro probably reflects a temperature range of 780—810°C at a pressure range of 0.5—2 kbar. Bonnichsen (1969), on the basis of $^{18}O/^{16}O$ ratios in coexisting quartz and magnetite, suggests a temperature range of 700—750°C for this same contact zone. Simmons et al. (1974) report a minimum temperature of metamorphism for the most highly contact metamorphosed part of the Gunflint Iron Formation of about 740—760°C, on the basis of magnetite coexisting with primary ilmenite. They evaluate the pressure conditions to have been a minimum of 2 kbar. Floran and Papike (1978) suggest temperatures close to 800°C for this same maximum metamorphism zone of the Gunflint Iron Formation.

Regionally metamorphosed terrains suggest somewhat lower temperatures but considerably higher pressure ranges. Butler (1969) concludes that temperatures in the Gagnon region of the southwestern Labrador Trough exceeded

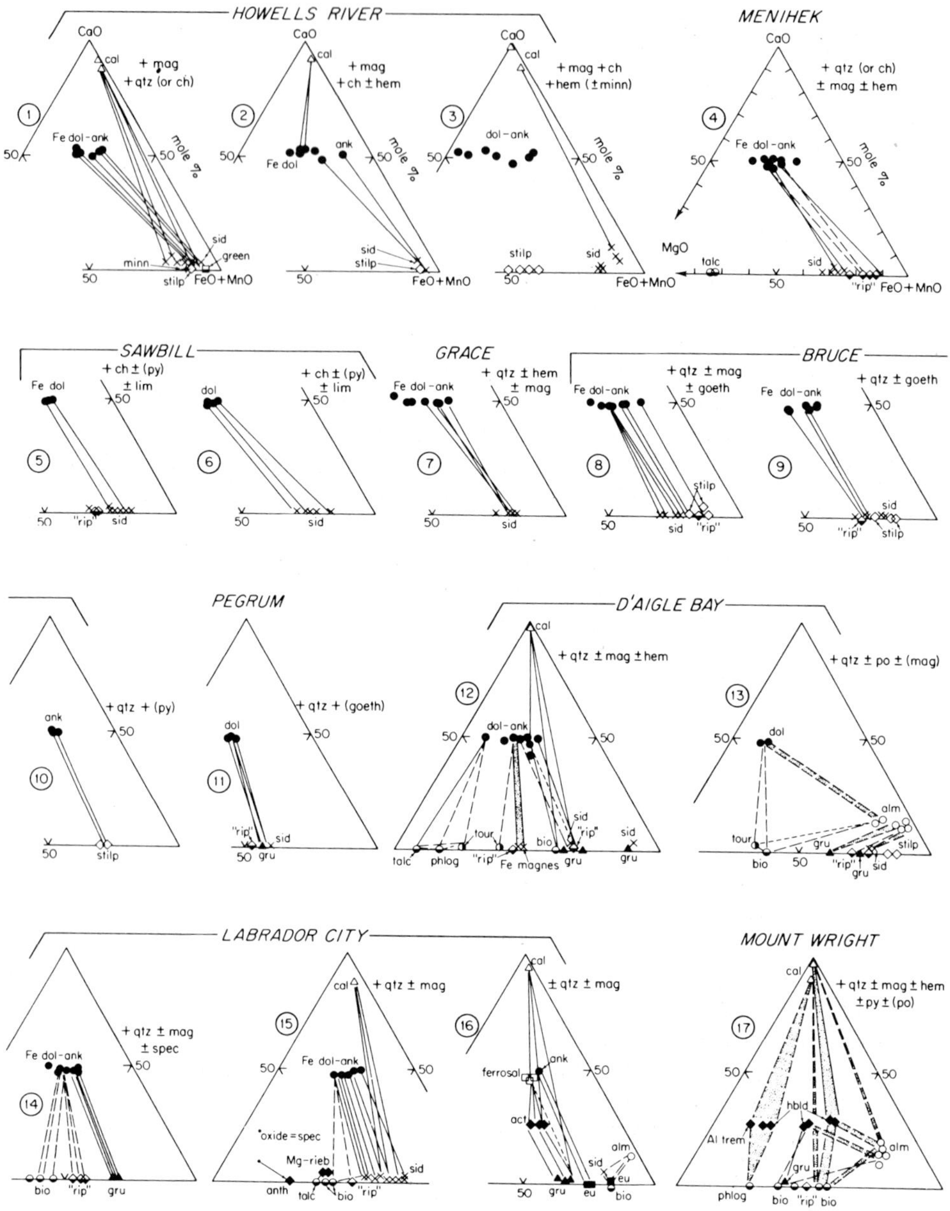
HOWELLS RIVER
CaO
cal
+ mag
+ qtz (or ch)
Fe dol-ank
mole %
minn
sid
green
stilp
FeO+MnO
+ mag
+ ch ± hem
ank
Fe dol
+ mag + ch
+ hem (± minn)
dol-ank
MENIHEK
+ qtz (or ch)
± mag ± hem
MgO
talc
"rip"
SAWBILL
+ ch ± (py)
± lim
Fe dol
dol
GRACE
+ qtz ± hem
± mag
Fe dol-ank
BRUCE
+ qtz ± mag
± goeth
+ qtz ± goeth
PEGRUM
+ qtz + (py)
+ qtz + (goeth)
gru
D'AIGLE BAY
+ qtz ± mag ± hem
+ qtz ± po ± (mag)
tour
phlog
Fe magnes
bio
alm
LABRADOR CITY
+ qtz ± mag
± spec
oxide = spec
Mg-rieb
anth
± qtz ± mag
ferrosal
act
eu
MOUNT WRIGHT
+ qtz ± mag ± hem
± py ± (po)
hbld
Al trem

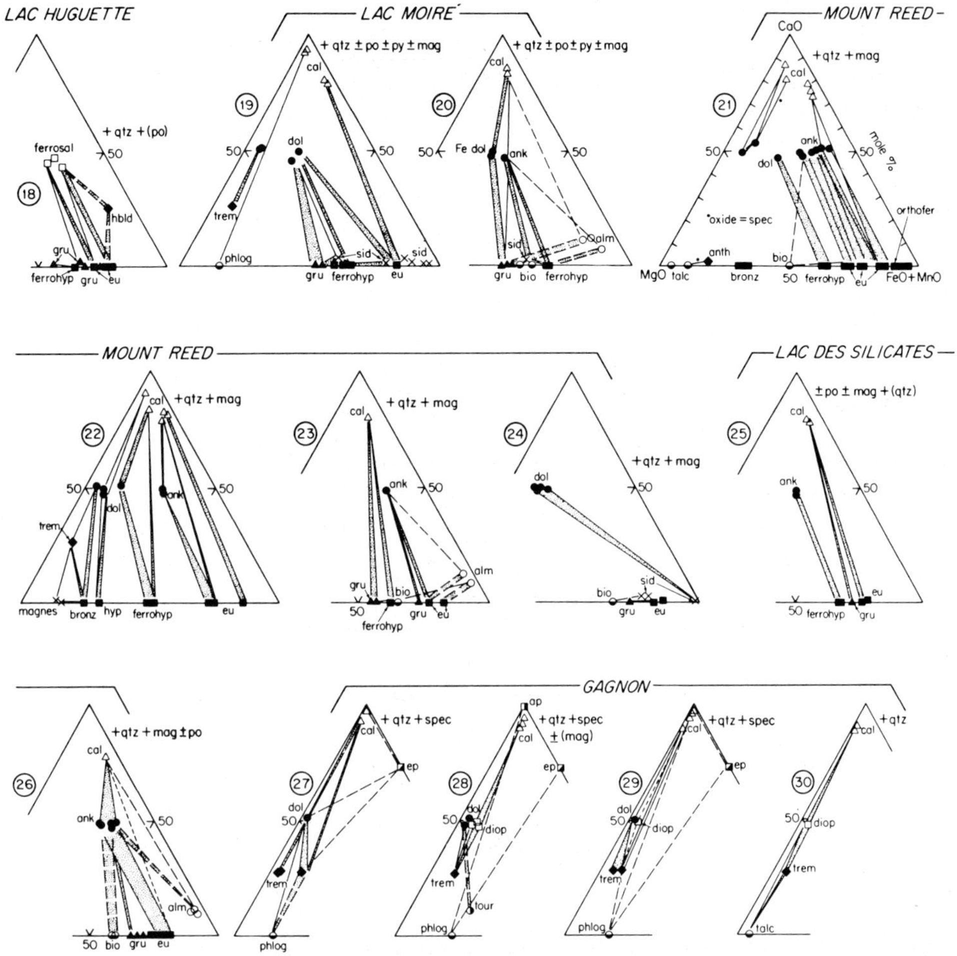

Fig. 11-15. Graphical representation of assemblages in progressively metamorphosed iron-formation in the Labrador Trough and some closely associated rock types. The sequence of assemblage diagrams, numbered from 1 to 30, are arranged in terms of a generally increasing grade of metamorphism. Long-dash tielines are used when a mineral composition does not lie in the plane CaO-MgO-(FeO + MnO). Short-dash tielines are used in case of major tieline crossing. Shaded areas represent regions of many essentially parallel tielines (after Klein, 1978). Some of the locality names can be found in Fig. 11-2C. The name "ripidolite" is put in quotation marks because of the uncertainty of whether it is truly a normal 14Å chlorite, or 7Å septachlorite such as chamosite.

Mineral abbreviations:
act — actinolite
alm — almandite
Altrem — Al tremolite
ank — ankerite
anth — anthophyllite
ap — apatite
bio — biotite
bronz — bronzite
cal — calcite
ch — chert
diop — diopside
dol — dolomite
ep — epidote
eu — eulite
Fedol — ferroan dolomite
Fermagnes — ferroan magnesite
ferrohyp — ferrohypersthene
ferrosal — ferrosalite
goeth — goethite
green — greenalite
gru — grunerite
hbld — hornblende
hem — hematite
hyp — hypersthene
lim — limonite
mag — magnetite
magnes — magnesite
minn — minnesotaite
orthofer — orthoferrosilite
phlog — phlogopite
po — pyrrhotite
py — pyrite
qtz — quartz
"rip" — "ripidolite"
sid — siderite
spec — specularite
stilp — stilpnomelane
tour — tourmaline
trem — tremolite

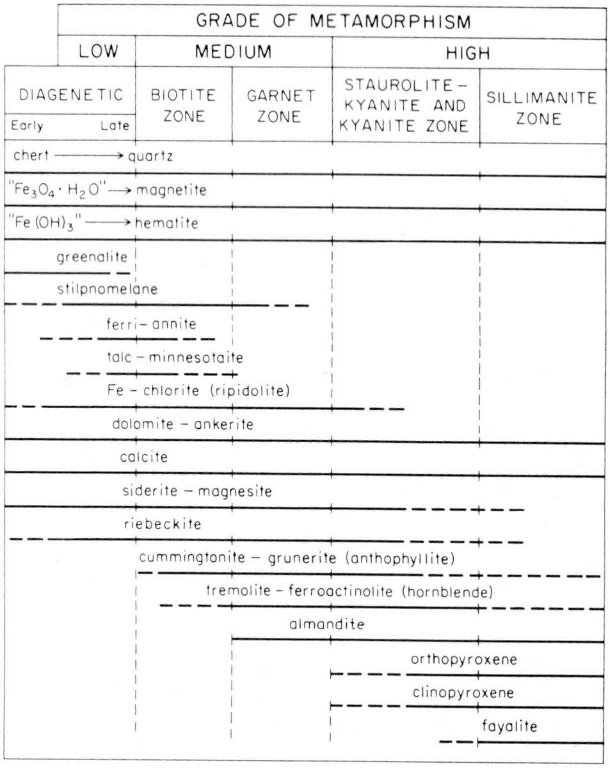

Fig. 11-16. Relative stabilities of minerals in metamorphosed iron-formations as a function of metamorphic zone (compiled from James, 1955; French, 1968; Klein 1978; Floran and Papike, 1978).

600°C. Klein (1978) arrives at a temperature range of 700–750°C and a pressure range of 10–11 kbar for the most highly metamorphosed iron-formation assemblages in the southwestern Labrador Trough. Immega and Klein (1976) suggest temperatures ranging from 650 to 750°C and pressures from 4 to 6 kbar for the pyroxene-rich, Archean iron-formations in the Tobacco Root Mountains of southwestern Montana. Dahl (1979) arrives at a fairly similar range of conditions, 745 ± 50°C and 6–8 kbar, for high-grade Archean iron-formations in the Ruby Mountains of southwestern Montana. Gole and Klein (1981b) interpret fayalite-pyroxene assemblages in Archean iron-formations to reflect 670 ± 50°C at 3–5 kbar.

As in lower-grade metamorphic reactions, many of the reactions responsible for the highest metamorphic grade assemblages involve decarbonation and dehydration. The composition of the fluid phase and the role of volatiles during such metamorphic reactions have been evaluated by several authors. Butler (1969) concludes from his detailed study of high-grade iron-formation assemblages in the Gagnon region of the southwestern Labrador Trough that gradients in the activity of H_2O as well as CO_2 existed during metamorphism. Such gradients appear to have been relatively low along strike, but high across strike. He notes that the direction of such gradients is almost without reversals, which would be consistent with intergranular diffusion of H_2O and CO_2. This implies that CO_2 and H_2O did not behave as externally fixed perfectly mobile components but instead that they behaved as internally fixed initial value components (Zen, 1963). Kranck (1961) concludes that in many parts of the high grade metamorphic iron-formation terrain in the Mount Reed area, of the southwestern part of the Labrador Trough, the rocks were closed either to CO_2 or H_2O, and that the volatile components did not diffuse freely through the rocks. Klein (1978) notes that although there is a general decarbonation and dehydration with increasing metamorphic grade there has not always been a gradual and continual loss of H_2O and CO_2 with increasing metamorphism.

The question of whether oxygen fugacity during metamorphism of iron-formations is controlled by the iron-formation assemblages themselves or by fluids moving in from other sources has been discussed by Frost (1982) and Labotka et al. (1982) as a result of the paper by Vaniman et al. (1980). Miyano and Klein (1983b) have evaluated the relative magnitudes of temperature and f_{O_2} during metamorphism of iron-formation at constant pressure. A summary of their conclusions is given in Fig. 11-17 for various high-grade metamorphic iron-formations. Because the pressures for the various iron-formations are not the same the three numbered regions in Fig. 11-17 reflect different T-f_{O_2} conditions. Iron-formations with similar phase relations are grouped in boxes, marked I, II and III, in Fig. 11-17. Type I represents olivine-free assemblages, type II olivine- and quartz-bearing relations, and type III olivine-bearing and quartz-free relations. Figure 11-11 indicates that olivine is stable at lower f_{O_2} conditions than orthopyroxene at constant temperature. The stability field of orthopyroxene will shrink in the presence of water because grunerite will then become stable. Orthopyroxene in type III assemblages would be replaced by those of type II with increasing water fugacity and temperature (see Miyano and Klein, 1982a).

THEORETICAL EVALUATION OF THE CONDITIONS OF METAMORPHISM OF IRON-FORMATIONS

Theoretical evaluations, by thermodynamic procedures of the physical and chemical conditions during diagenesis and progressive metamorphism of iron-formations have been made by several authors, most notably Mel'nik (1982), Miyano (1978a, b, c), Frost (1979), Haase (1979b, 1982b), and Miyano and Klein (1983b). Considerable data on experimental systems relating to the stability fields or iron-rich minerals are given by Mel'nik (1982), who references many theoretical and experimental works in the Russian literature.

Miyano (1978a) evaluates minnesotaite-carbonate-Fe-oxide (± greenalite)

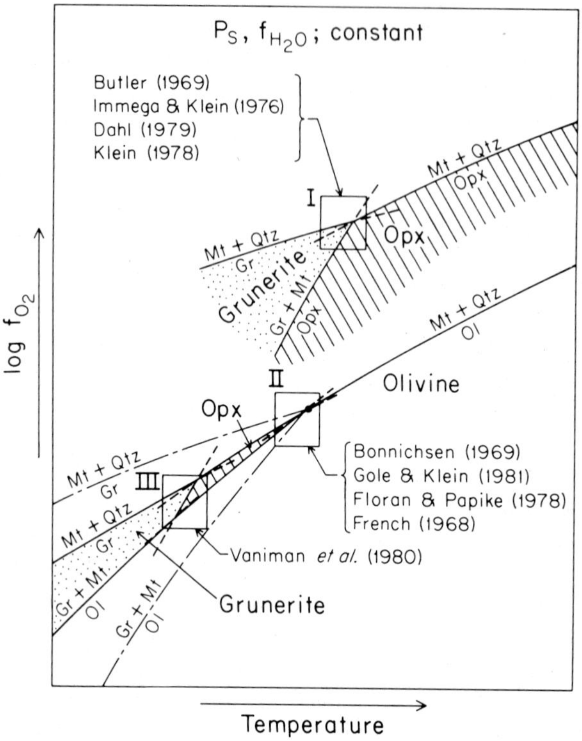

Fig. 11-17. Summary of log f_{O_2}-T conditions for several high-grade metamorphic iron-formations (after Miyano and Klein, 1983b). Type I iron-formations contain no olivine, type II represent olivine- and quartz-bearing assemblages, and type III are olivine-bearing but quartz-free. The phase relations of type III reflect relatively low f_{H_2O} values which occur under dry conditions at high temperature (i.e. contact metamorphism).

coexistences in several very low-grade iron-formations and concludes that the Fe/(Fe + Mg) ratio in minnesotaite is indicative of f_{O_2} as well as P_{CO_2} in the system. He suggests that the very low-grade metamorphic (or late diagenetic) assemblages of the Sokoman (Labrador Trough) and Biwabik Iron Formations reflect a lower P_{CO_2} than those of the Gunflint Iron Formation or the Hamersley Group iron-formations. Miyano (1978b, c) on the basis of the calculated stability field of iron-silicates, carbonates and oxides calculates P_{CO_2} ranges during metamorphism for the same four iron-formations.

Frost (1979) through the use of phase diagrams in the system Fe-Si-C-O-H concludes that hydrous Fe-silicate assemblages, under greenschist facies conditions, are stable only with extremely H_2O-rich fluids. He notes that with prograde metamorphism, the fluid coexisting with siderite-silicate iron-formation will be buffered to moderate or high X_{CO_2} as siderite and quartz react to form silicates. At higher temperatures the fluid, however, would tend to become enriched in H_2O again as grunerite breaks down to form fayalite (or ferrohypersthene) + quartz.

Haase (1979b, 1982b) has calculated the most probable topologies for a petrogenetic grid, depicting metamorphic reactions in iron-formations, on the basis of geometric-chemographic analytical techniques and data from natural occurrences. By calculating the positions of several important metamorphic reactions in temperature-X_{CO_2} space, information can be obtained about the fluid phase coexisting with iron-formation assemblages during metamorphism. Figure 11-18 shows the results of such an approach. Minnesotaite-bearing assemblages are restricted to H_2O-rich fluids. Actinolite-bearing assemblages at moderate temperatures are also indicative of rather H_2O-rich coexisting fluid phases. Within a temperature range that reflects medium metamorphic grade conditions (400—500°C), however, assemblages containing grunerite + garnet are indicative of rather CO_2-rich fluids. This suggests that for localities where actinolite + grunerite and grunerite + garnet assemblages are interbedded on the scale of several meters [as in the Negaunee Iron Formation (Haase, 1982a)] there must have been substantial compositional gradients within the coexisting fluid during metamorphism. Olivine-bearing assemblages within the 450—600°C temperature range are indicative of very CO_2-rich coexisting fluid.

Figure 11-18 also provides a means of following the chemical evolution of a coexisting fluid phase during prograde metamorphism. Many possible T-X_{CO_2}

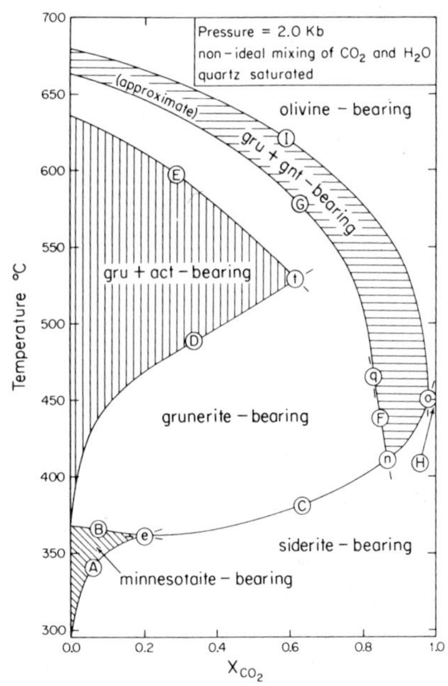

EXPLANATION

- A 3sid + 4qtz + 1H_2O = 1minn + 3CO_2
- B 7minn = 3gru + 4qtz + 4H_2O
- C 7sid + 8qtz + 1H_2O = 1gru + 7CO_2
- D 14cal + 5gru + 16qtz + 2H_2O = 7act + 14CO_2
- E 7act = 14cpx + 3gru + 4qtz + 4H_2O
- F 13ank + 20chl + 72qtz = 6gru + 26gnt + 26CO_2 + 74H_2O
- G 1050chl + 455hbld + 2454qtz = 445gru + 1639gnt + 4392H_2O
- H 2sid + 1qtz = 1olv + 2CO_2
- I 2gru = 7olv + 9qtz + 2H_2O

Abbreviations: act, actinolite; ank, ankerite; cal, calcite; chl = 7 and 14 Å chlorite; cpx, clinopyroxene; gnt, garnet; gru, grunerite; hbld, hornblende; minn, minnesotaite; olv, olivine; qtz, quartz; sid, siderite

Fig. 11-18. Temperature-X_{CO_2} diagram for some common reactions in iron-formation metamorphism (modified after Haase, 1982b). Capital letters refer to specific reactions, whereas small letters refer to invariant points (see text).

paths can be identified, depending on the particular starting mineral assemblage and the initial composition of the coexisting fluid phase. As an example one may consider the common assemblage siderite + quartz ± magnetite. For low values of X_{CO_2} the first reaction encountered would be reaction A, indicated by the production of minnesotaite. At this point, a further temperature increase will cause the composition of the coexisting fluid to be driven to more CO_2-rich values until invariant point [e] is reached. The composition of the fluid will remain at [e], grunerite production will begin and eventually either siderite or minnesotaite will be entirely consumed. If it is the former, the composition of the fluid will now be driven to increasing H_2O values along the trace of reaction B. This is a reversal of the trend started by minnesotaite production. If minnesotaite is entirely consumed by reaction at invariant point [e], a further increase in temperature will now cause the composition of the coexisting fluid to be driven to progressively increasing CO_2-rich values as the trace of reaction C is followed. If siderite is entirely consumed before invariant point [o] is reached, the T-X_{CO_2} path of the assemblage will leave the trace of reaction C and move straight upward. If the temperature increase is sufficient, the trace of reaction I will be encountered, at which point olivine production will begin and the compositon of the coexisting fluid will be driven to more H_2O-rich values. This again marks a reversal of an earlier established trend. If invariant point [o] is reached before siderite is consumed, a situation similar to that for invariant point [e] will obtain but now siderite, grunerite and olivine will be involved. If the initial X_{CO_2} ratio of the fluid phase was very CO_2-rich, reaction H would have been the first to be encountered with increasing temperature and olivine would have been the only Fe-silicate to form. Clearly, many T-X_{CO_2} paths exist and the compositional evolution of metamorphic fluids can be complex. Because of the common occurrence of the assemblage siderite + quartz, however, the above-described trend of pronounced CO_2 enrichment of the coexisting fluid phase is most likely a very important feature of iron-formation metamorphism.

ACKNOWLEDGEMENTS

Research on iron-formations has been made possible by National Science Foundation grants GA-11435, GA-36186, EAR-76-11740 and EAR-80-20377. A Guggenheim Memorial Fellowship in 1978 provided me with an opportunity to do field and laboratory studies of the Hamersley Group iron-formations in Western Australia. C.S. Haase provided me with a synopsis of his findings on the fluid phase during metamorphism of iron-formation. I am grateful to Robert J. Floran, Martin J. Gole, C. Stephen Haase, R.C. Morris and R.E. Smith for their constructive reviews of this paper. I thank Mrs. Thea Brown for the typing of the manuscript and Messrs. W.H. Moran, R.T. Hill and G.R. Ringer for the drafting and photography of the illustrations.

REFERENCES

Becker, R.H. and Clayton, R.N., 1976. Oxygen isotope study of a Precambrian iron-formation, Hamersley Range, Western Australia. Geochim. Cosmochim. Acta, 40: 1153—1165.

Berner, R.A., 1970. Sedimentary pyrite formation. Am. J. Sci., 268: 1—23.

Beukes, N.J., 1973. Precambrian iron-formations of southern Africa. Econ. Geol., 68: 960—1005.

Bonnichsen, B., 1969. Metamorphic pyroxenes and amphiboles in the Biwabik Iron Formation, Dunka River area, Minnesota. Mineral. Soc. Am., Spec. Pap., 2: 217—241.

Bonnichsen, B., 1975. Geology of the Biwabik Iron Formation, Dunka River area, Minnesota. Econ. Geol., 70: 319—340.

Borchert, H., 1965. Formation of marine sedimentary iron ores. In: J.P. Riley and G. Skirrow (Editors), chemical Oceanography, vol. 2. Academic Press, New York, N.Y., pp. 159—204.

Bricker, O.P., Nesbitt, H.W. and Gunter, W.D., 1973. The stability of talc. Am. Mineral., 58: 64—72.

Butler, P., Jr., 1969. Mineral compositions and equilibria in the metamorphosed iron-formation of the Gagnon region, Quebec, Canada. J. Petrol., 10: 56—101.

Dahl, P.S., 1979. Comparative geothermometry based on major-element and oxygen isotope distributions in Precambrian metamorphic rocks from southwestern Montana. Am. Mineral., 64: 1280—1293.

Dimroth, E., 1968. Sedimentary textures, diagenesis and sedimentary environment of certain Precambrian ironstones. Neues Jahrb. Geol. Paläontol., Abh., 130: 247—274.

Dimroth, E., 1977. Facies Models 6. Diagenetic facies of iron-formation. Geosci. Can., 4: 83—88.

Dimroth, E. and Chauvel, J.J., 1973. Petrography of the Sokoman Iron Formation in part of the central Labrador Trough, Quebec, Canada. Geol. Soc. Am. Bull., 84: 111—134.

Dorr, J.V.N., 1964. Supergene ores of Minas Gerais, Brazil. Econ. Geol., 59: 1203—1240.

Dorr, J.V.N., 1969. Physiographic, stratigraphic, and structural development of the Quadrilatero Ferrifero Minas Gerais, Brazil. U.S. Geol. Surv., Prof. Pap. 641-A, 110 pp.

du Toit, A.L., 1945. The origin of the amphibole asbestos deposits of South Africa. Geol. Soc. S. Afr., 48: 161—206.

Eggleton, R.A., 1972. The crystal structure of stilpnomelane. Part II. The full cell. Mineral. Mag., 38: 693—711.

Eggleton, R.A. and Chappel, B.W., 1978. The crystal structure of stilpnomelane. Part III. Chemistry and physical properties. Mineral. Mag., 42: 361—368.

Ernst, W.G., 1968. Amphiboles: Crystal Chemistry, Phase Relations and Occurrence. Springer-Verlag, Inc., New York, N.Y., 125 pp.

Ewers, W.E. and Morris, R.C., 1981. Studies of the Dales Gorge Member of the Brockman Iron Formation, Western Australia. Econ. Geol., 76: 1929—1953.

Floran, R.J., 1975. Mineralogy and Petrology of the Sedimentary and Contact Metamorphosed Gunflint Iron Formation, Ontario—Minnesota. Ph.D. Thesis, State University of New York, Stony Brook, N.Y., 342 pp.

Floran, R.J. and Papike, J.J., 1973. Silicate and carbonate relationships in unmetamorphosed iron-formation. Geol. Soc. Am., Abstr. Programs, Meeting, Dallas, 5: 623.

Floran, R.J. and Papike, J.J., 1975. Petrology of the Gunflint Iron Formation, Ontario—Minnesota: The low grade rocks. Geol. Soc. Am. Bull., 86: 1169—1190.

Floran, R.J. and Papike, J.J., 1978. Mineralogy and petrology of the Gunflint Iron Formation, Minnesota—Ontario: Correlation of compositional and assemblage variations at low to moderate grade. J. Petrol., 19: 215—288.

French, B.M., 1968. Progressive contact metamorphism of the Biwabik Iron Formation, Mesabi Range, Minnesota. Minn., Geol. Surv. Bull., 45, 103 pp.

French, B.M., 1973. Mineral assemblages in diagenetic and low-grade metamorphic iron-formations. Econ. Geol., 68: 1063—1075.

Frost, B.R., 1979. Metamorphism of iron-formation: parageneses in the system Fe-Si-C-O-H. Econ. Geol., 74: 775—785.

Frost, B.R., 1982. Contact metamorphic effects of the Stillwater Complex, Montana: the concordant iron-formation: a discussion of the role of buffering in metamorphism of iron-formation. Am. Miner., 67: 142—148.

Gair, J.E., 1975. Bedrock geology and ore deposits of the Palmer Quadrangle, Marquette County, Michigan. U.S. Geol. Surv., Prof. Pap. 769, 159 pp.

Garrels, R.M., Perry, E.A. and MacKenzie, F.T., 1973. Genesis of Precambrian iron-formations and the development of atmospheric oxygen. Econ. Geol., 68: 1173—1179.

Gillmeister, N., 1971. The Petrology, Stratigraphy and Structure of the Precambrian Metamorphic Rocks of the Central Tobacco Root Mountains with Special Emphasis on Electron Microprobe Analysis of High Grade Mineral Assemblages. Ph.D. Thesis, Harvard University, Cambridge, Mass., 201 pp.

Gole, M.J., 1979. Metamorphosed Banded Iron Formation in the Archean Yilgarn Block, Western Australia. Ph.D. Thesis, University of Western Australia, Nedlands, W.A., 254 pp.

Gole, M.J., 1980a. Mineralogy and petrology of very-low metamorphic grade Archean banded iron-formations, Weld Range, Western Australia. Am. Mineral., 65: 8—25.

Gole, M.J., 1980b. Formation of grunerite in banded iron-formation (BIF), Yilgarn Block, Western Australia (abstr.). Geol. Soc. Am., Abstr. Programs, 12: 227.

Gole, M.J., 1981. Archean banded iron-formations, Yilgarn Block, Western Australia. Econ. Geol., 76: 1954—1975.

Gole, M.J. and Klein, C., 1981a. Banded iron-formations through much of Precambrian time. J. Geol., 89: 169—183.

Gole, M.J. and Klein, C., 1981b. High-grade metamorphic Archean banded iron-formations, Western Australia: assemblages with coexisting pyroxenes ± fayalite. Am. Mineral., 66: 87—99.

Gross, G.A., 1961. Metamorphism of iron-formations and its bearing on their beneficiation. Can. Min. Metall. Bull., 54: 30—37.

Gross, G.A., 1973. The depositional environment of principal types of Precambrian iron-formations. In: Genesis of Precambrian Iron and Manganese Deposits. Proceedings of the Kiev Symposium, 1970, Unesco, Paris, pp. 15—21.

Gruner, J.W., 1946. The mineralogy and geology of the taconites and iron ores of the Mesabi range, Minnesota. Iron Range Resources and Rehabilitation Comm. and Minn. Geol. Surv., 127 pp.

Guggenheim, S., Bailey, S.W., Eggleton, R.A. and Wilkes, P., 1982. Structural aspects of greenalite and related minerals. Can. Mineral., 20: 1—18.

Gundersen, J.N. and Schwartz, G.M., 1962. The geology of the metamorphosed Biwabik Iron Formation, eastern Mesabi District, Minnesota, Minn., Geol. Surv. Bull. no. 43, 139 pp.

Haase, C.S., 1979a. Metamorphic Petrology of the Negaunee Iron Formation, Marquette District, Northern Michigan, Ph.D. Thesis, Indiana Univ., 246 p.

Haase, C.S., 1979b. Multisystems analysis of phase relations in the system CaO-(Fe,Mg)O-Al_2O_3-Fe_2O_3-SiO_2-CO_2-H_2O: a petrogenetic grid for metamorphosed iron-formations (abstr.). Geol. Soc. Am., Abstr. Programs, 11: 436.

Haase, C.S., 1982a. Metamorphic petrology of the Negaunee Iron Formation, Marquette district, northern Michigan: Mineralogy, metamorphic reactions and phase equilibria. Econ. Geol., 77: 60—81.

Haase, C.S., 1982b. Phase equilibria in metamorphosed iron-formations: qualitative T-$X_{(CO_2)}$ petrogenetic grids. Am. J. Sci., 282: 1623—1655.

Haase, C.S. and Klein, C., 1978. Paragenesis of olivine in the regionally metamorphosed

Negaunee Iron-Formation (NIF) at Republic Mine, northern Michigan. Geol. Soc. Am., Abstr. Programs, 10: 414.

Han, T.M., 1971. Diagenetic-metamorphic replacement features in the Negaunee Iron-Formation of the Marquette Iron Range, Lake Superior District. Soc. Min. Geol. Jpn. Spec. Iss., 3: 430—438.

Han, T.M., 1978. Microstructures of magnetite as guides to its origin in some Precambrian iron-formations. Fortschr. Mineral., 56: 105—142.

Hey, M.H., 1954. A new review of the chlorites. Mineral. Mag., 30: 277—292.

Hutchison, J.L., Irusteta, M.C. and Whittaker, E.J.W., 1975. High-resolution electron microscopy and diffraction studies of fibrous amphiboles. Acta Crystallogr., A31: 794—801.

Immega, I.P. and Klein, C., 1976. Mineralogy and petrology of some metamorphic Precambrian iron-formations in southwestern Montana. Am. Mineral., 61: 1117—1144.

James, H.L., 1954. Sedimentary facies of iron-formation. Econ. Geol., 49: 235—293.

James, H.L., 1955. Zones of regional metamorphism in the Precambrian of northern Michigan. Bull. Geol. Soc. Am., 66: 1455—1487.

James, H.L. and Clayton, R.N., 1962. Oxygen isotope fractionation in metamorphosed iron-formations of the Lake Superior region and in other iron-rich rocks. In: A.G.J. Engel, H.C. Jones and H.F. Leonard (Editors), Petrologic Studies: A volume in honor of A.F. Buddington. Geol. Soc. Am., Washington, D.C., pp. 217—239.

Kalliokoski, J., 1965. The metamorphosed iron ore of El Pao, Venezuela. Econ. Geol., 60: 100—166.

Kimberley, M.M., 1979. Origin of oolitic iron formations. J. Sediment Petrol., 49: 111—132.

Klein, C., 1964. Cummingtonite-grunerite series: a chemical, optical and X-ray study. Am. Mineral., 49: 963—982.

Klein, C., 1966. Mineralogy and petrology of the metamorphosed Wabush Iron Formation, southwestern Labrador. J. Petrol., 7: 240—305.

Klein, C., 1968. Coexisting amphiboles. J. Petrol., 9: 281—330.

Klein, C., 1973. Changes in mineral assemblages with metamorphism of some banded Precambrian iron-formations. Econ. Geol., 68: 1075—1088.

Klein, C., 1974. Greenalite, stilpnomelane, minnesotaite, crocidolite and carbonates in a very low-grade metamorphic Precambrian iron-formation. Can. Mineral., 12: 475—498.

Klein, C., 1978. Regional metamorphism of Proterozoic iron-formation, Labrador Trough, Canada. Am. Mineral., 63: 898—912.

Klein, C. and Bricker, O.P., 1977. Some aspects of the sedimentary and diagenetic environment of Proterozoic banded iron-formation. Econ. Geol., 72: 1457—1470.

Klein, C. and Fink, R.P., 1976. Petrology of the Sokoman Iron Formation in the Howells River area, at the western edge of the Labrador Trough. Econ. Geol., 71: 453—487.

Klein, C. and Gole, M.J., 1981. Mineralogy and petrology of parts of the Marra Mamba Iron Formation, Hamersley Basin, Western Australia. Am. Mineral., 66: 507—525.

Klein, C. and Gole, M.J., 1983. Mineralogy and petrology of the Joffre Member of the Brockman Iron Formation, Hamersley Basin, Western Australia. Am. Mineral., in prep.

Klein, C. and Waldbaum, D.R., 1967. X-ray crystallographic properties of the cummingtonite-grunerite series. J. Geol., 75: 379—392.

Kranck, S.H., 1961. A study of phase equilibria in a metamorphic iron-formation. J. Petrol., 2: 137—184.

Kretschman, V. and Scott, S.D., 1976. Phase relations involving arsenopyrite in the system Fe-As-S and their applications. Can. Mineral., 14: 364—386.

LaBerge, G.L., 1964. Development of magnetite in iron-formations of the Lake Superior region. Econ. Geol., 59: 1313—1342.

Labotka, T.C., Vaniman, D.T. and Papike, J.J., 1982. Contact metamorphic effects of the Stillwater Complex, Montana: the concordant iron-formation: a reply to the role of buffering in metamorphism of iron-formation. Am. Mineral., 67: 149—152.

Leake, B.E., 1978. Nomenclature of amphiboles. Am. Mineral., 63: 1023—1053.

Lepp, H.S., 1966. Chemical composition of the Biwabik Iron-Formation, Minnesota. Econ. Geol., 61: 243—250.

Lesher, C.M., 1978. Mineralogy and petrology of the Sokoman Iron Formation near Ardua Lake, Quebec. Can. J. Earth Sci., 15: 480—500.

Mel'nik, Yu.P., 1982. Precambrian Banded Iron-formations: Physicochemical Conditions of Formation. Elsevier, Amsterdam, 310 pp.

Milton, C., Ingram, B. and Breger, I., 1974. Authigenic magnesioarfvedsonite from the Green River Formation, Duchesne County, Utah. Am. Mineral., 59: 830—836.

Miyano, T., 1976. Physicochemical environments during burial metamorphism of the Dales Gorge Member, Hamersley Group, Western Australia (in Japanese with English abstract). Min. Geol. Jpn., 26: 311—325.

Miyano, T., 1978a. Effects of CO_2 on mineralogical differences in some low-grade metamorphic iron-formations. Geochem. J., 12: 201—211.

Miyano, T., 1978b. Phase relations in the system Fe-Mg-Si-O-H and environments during low-grade metamorphism of some Precambrian iron-formations. J. Geol. Soc. Jpn., 84: 679—690.

Miyano, T., 1978c. Stability relations of iron-bearing minerals in the H_2O-CO_2 mixed-volatile region at lower temperatures. J. Geol. Soc. Japan, 84: 711—719.

Miyano, T. and Klein, C., 1983a. Conditions of riebeckite formation in the iron-formations of the Dales Gorge Member, Hamersley Group, Western Australia. Am. Mineral., in press.

Miyano, T. and Klein, C., 1983b. Phase relations of orthopyroxene, olivine, and grunerite in high-grade metamorphic iron-formation. Am. Mineral., in press.

Miyano, T. and Miyano, S., 1982. Ferri-annite from the Dales Gorge Member iron-formation, Wittenoom area, Western Australia. Am. Mineral., 67: 1179—1195.

Morey, G.B., Papike, J.J., Smith, R.W. and Weiblen, P.W., 1972. Observations on the contact metamorphism of the Biwabik Iron-Formation, East Mesabi District, Minnesota. Geol. Soc. Am. Mem., 135: 225—264.

Morris, R.C., 1980. A textural and mineralogical study of the relationship of iron ore to banded iron-formation in the Hamersley Iron Province of Western Australia. Econ. Geol., 75: 184—209.

Mueller, R.F., 1960. Compositional characteristics and equilibrium relations in mineral assemblages of a metamorphosed iron-formation. Am. J. Sci., 258: 449—497.

Ncube, A., Zweifel, H. and Amstutz, G.C., 1978. On the occurrence of bravoite, framboidal pyrite, marcasite, pyrrhotite and possible plant remains in the Laisvall Lead and Zinc deposit, Sweden. Neues Jahrb. Mineral. Abh., 132: 264—283.

Papike, J.J., Cameron, K. and Shaw, K.W., 1973. Chemistry of coexisting actinolite-cummingtonite and hornblende-cummingtonite from metamorphosed iron-formation (abstr.). Geol. Soc. Am., Abstr. Programs, 5: 763—764.

Peacock, M.A., 1928. The nature and origin of the amphibole asbestos of South Africa. Am. Mineral., 13: 241—285.

Perry, E.C. and Tan, F.C., 1973. Significance of carbon isotope variations in carbonates from the Biwabik Iron Formation, Minnesota. In: Genesis of Precambrian Iron and Manganese Deposits. Proceedings of the Kiev Symposium, 1970, Unesco, Paris, pp. 299—305.

Perry, E.C., Tan, F.C. and Morey, G.B., 1973. Geology and stable isotope geochemistry of the Biwabik Iron Formation, northern Minnesota. Econ. Geol., 68: 1110—1126.

Robinson, P., Spear, F.S., Schumacher, J.C., Laird, J., Klein, C., Evans, B.W. and Doolan, B.L., 1982. In: D.R. Veblen and P.H. Ribbe (Editors), Amphiboles: Petrology and Experimental Phase Relations, vol. 9B. Min. Soc. Am., Reviews in Mineralogy, pp. 1—211.

Ross, M., Papike, J.J. and Shaw, K.W., 1969. Exsolution textures in amphiboles as indicators of subsolidus thermal histories. Mineral. Soc. Am., Spec. Pap., 2: 275—299.

Simmons, E.C., Lindsley, D.H. and Papike, J.J., 1974. Phase relations and crystallization sequence in a contact-metamorphosed rock from the Gunflint Iron Formation, Minnesota. J. Petrol., 15: 539—565.

Smith, R.E., Perdrix, J.L. and Parks, T.C., 1982. Burial metamorphism in the Hamersley Basin, Western Australia. J. Petrol., 23: 75—102.

Trendall, A.F., 1973a. Iron-formations of the Hamersley Group of Western Australia: type examples of varved Precambrian evaporites. In: Genesis of Precambrian Iron and Manganese Deposits. Proceedings of the Kiev Symposium, 1970, Unesco, Paris, pp. 257—268.

Trendall, A.F., 1973b. Varve cycles in the Weeli Wolli Formation of the Precambrian Hamersley Group, Western Australia. Econ. Geol., 68: 1089—1098.

Trendall, A.F. and Blockley, J.G., 1970. The iron formations of the Precambrian Hamersley Group, Western Australia. West. Aust., Geol. Surv., Bull. 119, 366 pp.

Trendall, A.F. and Pepper, R.S., 1977. Chemical composition of the Brockman Iron Formation. West. Aust., Geol. Surv., Rec. 1976/25 (unpubl.).

Vaniman, D.T., Papike, J.J. and Labotka, T., 1980. Contact metamorphic effects of the Stillwater Complex, Montana; the concordant iron-formation. Am. Mineral., 65: 1087—1102.

Vermaas, F.H.S., 1952. The amphibole asbestos of South Africa. Geol. Soc. S. Afr., Trans., 55: 199—232.

Winkler, H.G.F., 1979. Petrogenesis of Metamorphic Rocks. Springer-Verlag, New York, N.Y., 348 pp.

Zajac, I.S., 1974. The stratigraphy and mineralogy of the Sokoman Formation in the Knob Lake area, Quebec and Newfoundland. Geol. Surv. Can., Bull. 220, 159 pp.

Zen, E-An, 1963. Components, phases, and criteria of chemical equilibrium in rocks. Am. J. Sci., 261: 929—942.

Chapter 12

DISTRIBUTION OF BANDED IRON-FORMATION IN SPACE AND TIME

HAROLD L. JAMES

INTRODUCTION

The term "banded iron-formation," as used in this and other chapters in the present volume, refers to a thinly layered or laminated rock in which chert (or its metamorphic equivalent) alternates with layers that are composed mainly of iron minerals; the iron content (as Fe) typically is in the range 20–35% and the SiO_2 content is in the range 40–50%. Banded iron-formation, or BIF (pronounced "biff"), following the widely used shorthand introduced by Trendall and Blockley (1970), is more restrictive than the more general term "iron-formation," which may include non-cherty facies (James, 1954). Specifically excluded from classification as BIF are the non-cherty ironstones typified by the oolitic ores of Phanerozoic age.

Two general classes of iron-formation have been defined by Gross (1965, 1973): Algoma type, closely associated with, and probably genetically related to, contemporaneous volcanism; and Lake Superior type, deposited in anorogenic marginal troughs and basins and lacking any obvious volcanic relation. The distinction is a useful one, though not all deposits — even some of major size – can be readily assigned to one or the other class, since it serves to separate the (generally) thin and numerically abundant iron-formations that are so common in greenstone belts of the world, notably those of Archean age, from the great basinal or marginal trough deposits such as those of the Labrador geosyncline.

AGE AND TONNAGE ASSESSMENTS

The two principal aspects of iron-formation deposits dealt with in this chapter are age of deposition and initial tonnage (or volume). Neither can be estimated precisely, not even for the most completely studied deposits, and the margin of error for some can be very large indeed. Nevertheless, the results, crude though they may be, do serve to delineate some of the major aspects of iron-formation distribution in time and space.

The number of known occurrences of iron-formation certainly is in the thousands, and many hundreds have been investigated in detail because of

TABLE 12-I

Estimates of age and initial tonnage for selected eposits of banded iron-formation

Area	Initial tonnage[1]	Estimated age[2]	Possible age limits[3]	References
Africa[4]				
Damara Belt, Namibia	10^{11}	650	590—720	Beukes, 1973.
Shushong Group, Botswana	10^{9}	1875	1750—2000	Beukes, 1973.
Ijil Group, Mauritania	10^{11}	2100	1700—2500	Bronner and Chauvil, 1979.
Transvaal—Criquatown, South Africa	10^{14}	2265	2095—2643	Button, 1976; Hamilton, 1977; Coertze et al., 1978; Van Niekerk and Burger, 1978.
Witswatersrand, South Africa	10^{10}	2720	2643—2800	Coertze et al., 1978; Beukes, 1973; Van Niekerk and Burger, 1978.
Liberian Shield, West Africa	10^{13}	3050	2750—3350	Gruss, 1973; Hurley et al., 1976; Beckinsale et al., 1980.
Pongola Supergroup, Swaziland—S. Africa	10^{11}	3100	2850—3350	Beukes, 1973.
Swaziland Supergroup, Swaziland—S. Africa	10^{9}	3200	3000—3400	Sinha, 1972; Allsop et al., 1968; Anhaeusser and Button, 1976; Eriksson, 1980.
Australia[4]				
Nabberu Basin	10^{13}	2050	1630—2475	Walter et al., 1976; Goode et al., 1983, this volume.
Middleback Range	10^{12}	2200	1780—2600	Parkin, 1969; Thomson, 1976; Trendall, 1973a.
Hamersley area	10^{14}	2500	2350—2650	Trendall, 1973a; Trendall, 1983, this volume.
Yilgarn Block greenstone belts	10^{12}	2700	2600—3000	Gee, 1979; Gole and Klein, 1981.
Eurasia[4]				
Altai region, Siberia—Kazakhstan	10^{12}	375	350—400	Kalugin, 1973.
Maly Khinghan-Uda, Far East U.S.S.R.	10^{13}	550	500-800(?)	Egorov and Timofeieva, 1973; Shkolnik, 1973.
Central Finland	10^{11}	2085	± 45	Laajoki and Saikkonen, 1977; Sakko and Laajoki, 1975.
Krivoy Rog—KMA, U.S.S.R.	10^{14}	2250	1900—2600	Semenenko, 1973; Plaksenko et al., 1973.
Bihar—Orissa, India	10^{13}	3025	2900—3150	Pichimuthu, 1974; Sarkar and Saha, 1977; Sakar et al., 1979.
Belozyorsky—Konsky, Ukraine, U.S.S.R.	10^{12}	3250	3100—3400	Semenenko, 1973.

North America[4]				
Rapitan, western Canada	10^{12}	700	550—850	Young, 1976; Gross, 1973.
Yavapai Series, Arizona, U.S.A.	10^{9}	1795	1775—1820	Anderson and Silver, 1976; Bayley and James, 1973.
Lake Superior, U.S.A.	10^{13}	2250	1900—2600	Bayley and James, 1973; Aldrich et al., 1965.
Labrador Trough and extensions, Canada	10^{14}	2175	1850—2500	Gross, 1968; Dimroth, 1970; Fryer, 1972.
Michipicoten, Ontario	10^{12}	2725	2700—2750	Goodwin, 1973; Brooks et al., 1970.
Beartooth Mountains, Montana, U.S.A.	10^{9}	2920	2700—3140	Page, 1977; De Paolo and Wasserburg, 1979.
Isua, Greenland	10^{9}	3760	± 70	Moorbath et al., 1973; Allaart, 1976; Appel, 1980.
South America[4]				
Morro du Urucum, Brazil and Bolivia	10^{12}	600(?)	450—900	Dorr, 1973.
Minas Gerais, Brazil	10^{14}	2250	1800—2700	Almeida, 1978; Dorr, 1973; Eichler, 1976.
Imataca Complex, Venezuela	10^{13}	3400	3100—3700	Alvarado, 1970; Gruss, 1973; Hurley et al., 1976; Montgomery, 1979.
Antarctica				
Prince Charles Mountains	?	2000(?)	?	Ravich et al., 1982.

[1] Tonnes of iron-formation initially deposited; composition about 30 percent Fe, 45 percent SiO_2. Estimate may include more than one stratigraphic unit and deposits in separated areas.

[2] In millions of years. Specific data used where applicable, otherwise assigned age is arithmetic mean of estimated age limits. Rb-Sr ages, when primary data are given, recalculated using Rb decay constant of 1.42×10^{-11} yr^{-1}.

[3] In millions of years.

[4] Significant deposits not tabulated:

Africa: Gabon (Sims, 1973) and many others in central and western Africa (Marelle and Abdullah, 1970); Morocco (Choubert and Faure-Muret, 1973); Egypt (Bishara and Habib, 1973); Rhodesian craton (Beukes, 1973; Martin et al., 1980).

Australia: Non-cherty ironstones of Adelaide Geosyncline (Parkin, 1969), Yampi Sound (Gellatly, 1972), and Roper Bar-Constance Range (Cochrane and Edwards, 1960).

Eurasia: Norway(Frietsch et al. 1979); Sweden (Frietsch, 1973); Kola-Karelia (Chernov, 1973); Kazakhstan, USSR (Novakhatsky, 1973); Lake Baikal region, USSR (Alexandrov, 1973); Aldan Shield (Vorona et al., 1973); China and Korea (Nishiwacki; 1970); Mysore, India (Pichimutha, 1974).

North America: Many Archean deposits of Canadian Shield (Goodwin, 1973); Northern Rocky Mountains, Montana (Bayley and James, 1973); Atlantic City, Wyoming (Bayley et al., 1973); Seminoe Mountains and Owl Creek Range, Wyoming (Bayley and James, 1973); northeastern Canadian Shield (Jackson and Taylor, 1972).

South America: Relun, Chile (Dorr, 1973); Serra dos Carajas, Brazil (Tolbert et al., 1973); Rio das Velhas, Brazil (Dorr, 1973); Amapa, Brazil (Dorr, 1973).

actual or potential economic significance. No complete tabulation, even of explored deposits, is available so far as I am aware, though several partial listings have been published (James, 1966; Eichler, 1976; Kimberley, 1978; Bronner and Chauvel, 1979; James and Trendall, 1981). The list presented in this paper (Table 12-I) is selective; it includes essentially al deposits of known major dimensions exept for those of China, but it also includes smaller deposits that serve to illustrate certain aspects of dimension and temporal spread in the various continental areas.

The true age of any given unit of iron-formation, as with sedimentary rocks in general and those of Precambrian age in particular, is not readily established. As Goldich (1973) noted in an earlier appraisal of iron-formation ages, direct determination using K-Ar on either whole-rock samples or mineral separates has not proved successful; as with most Rb-Sr determinations*, the resultant value generally proves to be the age of post-depositional metamorphism. The only direct method yielding reasonably certain depositional age is that using Pb-U analyses of zircon from interbedded ash-fall tuff, although in the opinion of some (e.g. Sakko and Laajoki, 1975) whole-rock Pb-Pb analysis may also yield values close to that of actual sedimentation. In general, however, most age assignments are based on limits set by the age of underlying basement rocks and the age of post-depositional metamorphism or igneous intrusion. The uncertainty range commonly is of the order of several hundred million years.

Estimates of initial tonnages of iron-formation are subject to uncertainties even greater than those attached to age assignment. Except in those rare situations in which the depositional basin has remained largely intact and undeformed, evaluation of original areal extent must depend on many independent factors, such as the vagaries of deformation and erosion over periods of time that may be in the billions of years, incompleteness of geologic mapping, and fallibility of geologic judgement, particularly concerning the geometry and continuity of the original basin of deposition. Thickness estimates used in tonnage calculations probably are somewhat more reliable than estimates of original area, but the necessary generalization often has involved combining data from two or more closely associated stratigraphic units, each of which in fact may have had a different original areal extent and its own pattern of thickness variation. Most of the resultant tonnage estimates presented in Table 12—I can at best have only an order-of-magnitude validity.

TEMPORAL AND SPATIAL DISTRIBUTION

The assembled age and tonnage estimates for the various iron-formations are further generalized to yield the proximate time-distribution curve present-

* Rb-Sr ages given in this chapter are based on a Rb decay constant of 1.42×10^{-11} yr^{-1}.

ed as Fig. 12-1. The curve is meant to be illustrative rather than quantitative, and undoubtedly it will be significantly modified as new data become available. The changes, however, are likely to involve the shape of particular peaks, each of which may prove to be bimodal or multimodal, rather than in the gross outlines. Despite the weaknesses in basic data, it is apparent that most of the world's iron-formation falls within four general age groupings: *middle Archean* (age 3500–3000 m.y.), *late Archean* (age 2900–2600 m.y.), *early Proterozoic* (age 2500—1900 m.y.), and *late Proterozoic—early Phanerozoic* (age 750–450 m.y.).

In terms of *quantity* of iron-formation initially deposited (also of that preserved) the sedimentation peak of early Proterozoic age is by far the most significant; this is the age of most great deposits of the Lake Superior type, and fully 90 percent of all known iron-formation is assigned to this deposi-

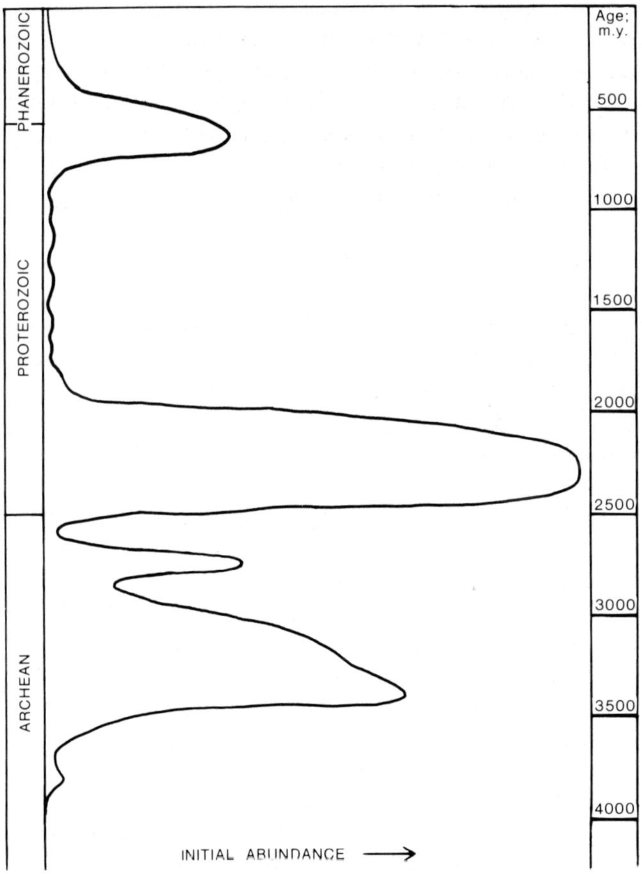

Fig. 12-1. Estimated abundance of iron-formation deposited through geologic time. Horizontal scale is non-linear, approximately logarithmic; range 0—10^{15} tonnes.

tional epoch. In terms of *number* of stratigraphically distinct occurrences, the (quantitatively) minor peak of late Archean age is most notable, with deposits numbered in the thousands; all, or virtually all, of these iron-formations are of the Algoma type, deposited in orogenic environments and closely associated with the volcanic rocks that now define the late Archean greenstone belts in many parts of the world.

The oldest known iron-formation is that at Isua, Greenland, with an age of at least 3750 m.y. The iron-formation is as much as several tens of meters thick, interbedded with quartzite strata that represent the oldest rocks of supracrustal origin yet known (Allaart, 1976). The youngest deposits of significant dimension classed in published reports as banded iron-formation probably are those of the Altai region of western Siberia and eastern Kazakhstan, U.S.S.R. (Kalugin, 1973), with an assigned age of Early or Middle Devonian (about 380 m.y.). These deposits, volcanogenic in association and as much as 65 m thick, are thinly layered, but it is not clear from published descriptions whether or not they contain interlayered chert. Definite chert-banded iron-formation of early Paleozoic age has, however, been reported from Nepal and elsewhere (O'Rourke, 1961), and many volcanic-associated bedded cherts of still younger age, such as those of the mid-Mesozoic part of the Franciscan Complex of California, are physically similar to banded iron-formation, though commonly manganiferous rather than strongly ferruginous.

Deposits of middle Archean age (3500—3000 m.y.)

No broad generalizations concerning deposits of middle Archean age are possible at the present state of knowledge. Some relatively minor deposits, such as those of the Swaziland Supergroup of southern Africa, occur within greenstone belts and can be regarded as Algoma type, similar to those so characteristic of the later Archean greenstone belts around the world, but most are not readily classified.

Probably the largest deposits assigned to this depositional epoch, in terms both of initial abundance and present-day preservation, are those of the Guyana Shield of Venezuela and Guyana, South America, and the probable correlatives in the Liberian Shield of Liberia, Sierra Leone, Guinea, and Ivory Coast, West Africa. Now separated by the relatively recently opened Atlantic Ocean, thick deposits of iron-formation are found scattered over an area that occupied, prior to continental separation, at least 250,000 km^2. The structure of each deposit typically is complex, with highly metamorphosed metasedimentary strata (including iron-formation) infolded in granulitic gneiss. Many individual deposits have been explored — El Pao, San Isidro and Cerro Bolivar in Venezuela; Bomi Hills, Mano River, Bong Range, Goe, Wolgazi, and Nimba in Liberia; Marampa in Sierra Leone; and Simandou in Guinea. Despite the areal and structural separation, the deposits appear to be somewhat similar in stratigraphic associations. The iron-formation units, commonly a hundred to several hundred meters thick, are interbedded with quartzite, quartz-mica schist, and amphibolite. Gruss (1973), noting these common features, suggested approximately contemporaneous deposition on an epicontinental shelf, with localized basins of thick iron-formation accumulation that subsequently became loci for structural synclinoria. Alvarado (1970) estimated the area of the initial basin in which the Venezuelan deposits accumulated to have been about 94,000 km^2. The available age data are not clearcut because of superimposed metamorphism. Gruss (1973) considered the depositional age of the iron-formations to be in the range 2500—3000 m.y. Hurley and others (1976), however, concluded the age of the Imataca Series of the Guyana Shield, which contains the iron-formation, to be at least 3200 m.y., and reported individual Rb-Sr determinations on gneiss from the Liberian Shield in the range 3300—3400 m.y.; values in the range 2700—2800 m.y. are attributed to a superimposed "thermotectonic event". The age assignment of the deposits of the Liberian Shield must, however, be considered still not fully resolved. Recent work using Rb-Sr and Pb-Pb methods, suggests that the Kambui Group, which contains the iron-formation in Sierra Leone, has an age between 2970 and 2750 m.y. (Beckinsale et al. 1980); quite possibly the assumption that the various deposits of West Africa are of equivalent age will prove incorrect.

Two other areas of mid-Archean iron-formation are worthy of note: the Belozyorsky-Konsky synclinal zone in the Ukrainian Shield, about 150 km east-southeast of Krivoi Rog, U.S.S.R.; and the Singhbhum, Bonai, Keonjhar, and Mayurbhanj districts in the States of Bihar and Orissa, India. The Ukrainian deposits, now contained in isolated deep synclinal folds, are of iron-formation in units locally as much as 300 m thick interbedded with metavolcanic rocks and schist. According to Semenenko (1973) the strata were deposited in the "first Precambrian megacycle," marked by "synchronous" granite with ages of 3100—34000 m.y., but no specific data are cited. The Indian deposits are in the Iron Ore Supergroup, with iron-formation (thicknesses ranging up to 350 m) overlain by shale and volcanic rocks and underlain by shale, sandstone, limestone, and volcanic rocks (Krishnan, 1973). The age bracket of the Iron Ore Supergroup appears to be fairly well established by Rb-Sr analyses (Sarkar and Saha, 1977; Sarkar et al. 1979): the underlying Older Metamorphic Group yields age values of about 3150 m.y. and the intrusive Singhbhum Granite yields values about 2900 m.y.

Deposits of late Archean age (2900—2600 m.y.)

The late Archean was an epoch of cratonization in most parts of the world, with widespread volcanism (much of it submarine), contemporaneous volcanogenic sedimentation (including that of iron-formation), and igneous intrusion in localized belts or basins within (presumably) areas of thin sialic crust. Within individual areas the complex events associated with what are now recognized as "greenstone belts" typically occupied no more than about

50 m.y. of time — those of the Vermilion district of northern Minnesota, for example, are tightly bracketed by ages of 2750 m.y. and 2700 m.y. (Sims, 1976). The cratonization process surely was not synchronous even within a single shield, but it does appear to have reached a peak in many parts of the world at about 2700 m.y. ago. However, in some regions the crust was stabilized much earlier — the greenstone belts in the Kaapvaal craton of southern Africa, for example, have ages that approach 3400 m.y. (Anhaeusser and Button, 1976) — and elsewhere much later, as shown by the bracket 1820—1775 m.y. established for the greenstone belt represented by the Yavapai Series of southwestern United States (Anderson and Silver, 1976).

The iron-formations associated with greenstone belts constitute type examples of the Algoma class of deposits. Abundantly represented on every continent, but rarely more than 50 m thick, they are marked by rapid changes in thickness, geochemical facies, and lithic associations. Goodwin (1973) provided a comprehensive outline of the many occurrences in the Canadian Shield, relating them to a series of Archean basins, each characterized by a distinctive pattern of facies variations, sulfide to oxide, in iron-formation units. One of the thickest and most completely studied basins is that of the Michipicoten district of north-central Ontario, the geology of which was reviewed by Goodwin (1973). The district is about 115 km long by about 50 km wide, underlain in large part by felsic to mafic volcanic rocks, now complexly folded. The iron-formation, enclosed in clastic sedimentary rocks in the western part of the basin and in volcanic rocks elsewhere, typically is 100 meters or more thick. In the central and eastern parts of the area, the iron-formation consists of a basal sideritic member overlain by a pyritic member, both yielding to an oxide facies in the western part of the basin.

Iron-formation of late Archean age is abundant in many other cratonic areas, such as the Yilgarn Block of Western Australia (Trendall, 1973a; Gole and Klein, 1981) and the Rhodesian craton of southern Africa (Beukes, 1973). According to Gole and Klein (1981, p. 179) the many separate iron-formations of the Yilgarn Block may have had an original areal extent of 20,000 km^2 or more, with an average aggregate thickness of 30 m.

Examples of iron-formation of late Archean age but not of the Algoma type appear to be scarce. Thin but persistent deposits are present in the Witwatersrand Supergroup of South Africa (see Table 12-I); the beds of iron-formation, generally a few meters thick, are within a sequence of normal clastic rocks, with a probable age of about 2650 m.y. In the northern Rocky Mountains of Montana, U.S.A., highly metamorphosed iron-formation, in units generally less than 25 m thick, is found interbedded with quartzite, schist, and dolomite marble in the structurally complex terrane that makes up the Archean cores of many block uplifts. The extent and initial continuity of these beds are as yet incompletely known (Bayley and James, 1973), but their age is at least 2750 m.y. (James and Hedge, 1980).

Deposits of early Proterozoic age (2500—1900 m.y.)

The early Proterozoic was the major epoch of iron sedimentation in Earth's history, the time of formation of the great deposits that are the principal focus of the present volume. Whether this depositional milieu was relatively brief or in fact existed over the assigned age bracket of 600 m.y. cannot be determined certainly from the data now in hand, but in any case it marks a major milestone in Precambrian history.

In light of other contributions in this volume, no extended discussion of the major basins of iron-formation deposition is offered here, but for completeness, brief reviews are presented of the six most important iron-formation regions (which may or may not coincide with basins of deposition): Lake Superior region, Labrador Trough and extensions, Krivoy Rog—KMA, Transvaal-Griquatown, Minas Gerais, and Hamersley area.

Lake Superior region, U.S.A.

The Lake Superior region includes the Mesabi—Gunflint and Cuyuna ranges of Minnesota, the Gogebic Range of Michigan and Wisconsin, and the Marquette Range, Menominee Range, and Iron River—Crystal Falls district of Michigan (Bayley and James, 1973). Beds of iron-formation, with a maximum thickness of 750 m, occur at several stratigraphic levels in the Marquette Range Supergroup of Michigan and Wisconsin and the Animikie Group of Minnesota, the basal units of which in the Michigan type area are orthoquartzite and dolomite. The major iron-formations, however, such as the Biwabik of the Mesabi Range and the Negaunee of the Marquette Range, are assumed to be approximately of contemporaneous origin, probably deposited in isolated basins in an epicontinental sea (Larue and Sloss, 1980).

Broad limits to the time of deposition of the Marquette Range Supergroup and its iron-formations are set by zircon ages from basement rocks, minimum age 2600 m.y., and from post-supergroup igneous intrusions, maximum age 1900 m.y. (Aldrich et al., 1965; Van Schmus, 1976; Peterman, 1979). Other available data are of uncertain significance. Van Schmus (1976) reported Rb-Sr and zircon ages of 2000—2100 m.y. for certain gneisses in northern Michigan, but the precise relation of these rocks to the supergroup is not known; they lie near the margin of a basement block and are generally assigned to the basement sequence, but quite possibly the isotopic systems were reset during the 1800—1900 m.y. Penokean orogeny. Van Schmus also mentioned in passing (p. 164) a 1950 m.y. Pb-U age for zircon from a volcanic unit in the supergroup, but no specific data are given. Pb-Pb analyses from phosphorite in the Marquette district yield a $^{207}Pb/^{206}Pb$ age of 1890 m.y. (Z.E. Peterman, pers. commun., 1981), which most likely records the time of metamorphism rather than that of deposition.

Labrador Trough and extensions, Canada

Under this heading are grouped the Labrador belt proper, about 1000 km in linear extent, the Cape Smith belt on the north, Belcher Islands on the west, and the Lake Albanel—Temiscamie districts on the southeast — all marginal to the Ungava craton of Archean or very early Proterozoic age. The iron-formation of Labrador, commonly 100 m or more thick, occurs within a sequence dominated by quartzite, arkose, slate, and dolomite (Gross, 1968); its age is between the maximum set by age of basement gneiss, about 2500 m.y. (Baadsgaard et al., 1979) and a minimum set by age of metamorphism, about 1830 m.y., a value thought originally to represent actual age of deposition (Fryer, 1972). If possible correlative formations in fold belts north of the trough are considered, the limits may be narrower: Jackson and Taylor (1972) reported Rb-Sr ages of basement rocks as young as 2395 m.y. and of metamorphosed sediments about 2015 m.y.

Krivoy Rog—KMA, U.S.S.R.

The famous Krivoy Rog district, including its extensions to the north, and the district known as KMA (Kursk Magnetic Anomaly) some 300 km to the northeast, are here grouped together despite lack of proof of age equivalency. It is assumed, however, that like the individual ranges in the Lake Superior region, now separated by even greater distances, these deposits of similar character accumulated during approximately the same interval of time, though possibly or even probably in separate basins.

The Krivoy Rog district (extended) is a narrow belt, about 200 km long, of tight folds in older gneiss (Semenenko, 1973); KMA is structurally similar in an area about 450 km by 140 km (Plaksenko et al., 1973), but entirely mantled by strata of Phanerozoic age. In both areas banded iron-formation, with aggregate thicknesses of several hundred meters, occurs in the middle parts of a sequence of metamorphosed sedimentary strata. The age of the iron-bearing sequence in both districts is between limits set by age of the basement gneiss (about 2600 m.y.) and that of later metamorphism (about 1900 m.y.).

Transvaal—Griquatown, South Africa

The iron-formations of the Transvaal—Griquatown belts of South Africa (Kuruman Iron Formation in the Griquatown belt of Northern Cape Province and Penge Iron Formation of Transvaal) occur in the lower part of the Transvaal Supergroup, overlying thick dolomite (Beukes, 1973). The total length of outcrop, including the gap between the north end of the Griquatown belt and the Transvaal Basin, is about 1200 km. The iron-formation units range in thickness up to 300 m. Shale from the supergroup has yielded a Rb-Sr age of 2263 ± 85 m.y. (Hamilton, 1977), which is close to the mean of the maximum of 2643 ± 80 m.y. determined on zircon from underlying lavas (Van Niekerk and Burger, 1978) and the minimum of 2095 ± 24 m.y. determined by Rb-Sr for a post-Transvaal intrusive (Hamilton, 1977).

Minas Gerais, Brazil

The Caué Itabirite, the principal iron-formation unit in the Quadrilatero Ferrifero of central Minas Gerais, occurs in the middle part of a thick sequence of mostly clastic rocks, the Minas Series. The known areal extent is approximately 400 km by 200 km, with average thickness of iron-formation about 250 m (Dorr, 1973; Eichler, 1976). The age of the series, and its contained iron-formation, is not well established. Regional analysis by Almeida (1978) places deposition in the early stages of the Trans-Amazonian cycle on a basement dated at about 2700 m.y. The strata were then metamorphosed during the later stages of the cycle, which is considered to have ended 1800 m.y. ago.

Possibly of equivalent age to the deposits of the Quadrilatero Ferrifero are those of Serra dos Carajas, some 1600 km north-northwest of Minas Gerais (Tolbert et al., 1973). Information is as yet sparse, but the dimensions and extent of the iron-formation may be comparable to those of the Caué Itabirite, and the general geologic setting appears to be similar.

Hamersley area, Australia

The great deposits of iron-formation of the Hamersley area, unparalleled in extent of exposure and degree of preservation, crop out in an area approximately 550 km by 150 km in Western Australia. They form part of the Mount Bruce Supergroup, which has a maximum thickness of about 15,000 m and accumulated in a basin probably 150,000 km^2 in area (Trendall and Blockley, 1970; Trendall, 1973a, 1983, this volume). Iron-formation forms five separate stratigraphic units in the middle sequence of the supergroup; the aggregate thickness of banded iron-formation is more than 1100 m.

The age of the Hamersley rocks is still under study. As discussed by Trendall (1983, this volume) U-Pb analysis of zircon from contemporaneous airfall tuff has yielded an age of 2490 m.y., whereas limits to possible age assignment are set by a Rb-Sr age of 2350 m.y. for intrusive sills, and by a 2650-m.y. Pb-isotopic model age for galena in underlying basalt.

Deposits of late Proterozoic—early Phanerozoic age (750—450 m.y.)

The late Proterozoic—early Phanerozoic epoch of iron sedimentation is not well understood, perhaps because the assigned deposits may include several genetically unrelated types. Some appear to be of the Algoma class, associated with contemporaneous volcanism; others bear an ill-defined relation to late Proterozoic glaciation. Some of the principal known deposits of this depositional era are discussed below.

Two of the largest deposits of possible Algoma type are in the Maly Khinghan and Uda areas of far eastern U.S.S.R. The Maly Khinghan deposits occupy a belt of tight folds, 150 km long and 30 km wide, that trends north from the Amur River (Chebotarev, 1960; Egorov and Timofeieva, 1973). The iron-formation unit generally is 25–35 m thick, with the lower part characteristically manganiferous. The immediately underlying and overlying strata are shale or slate reportedly of probable volcanic origin. The age assignment is Early Cambrian on the basis of the fossil brachiopod *Modioloides priscus* Walc. (Nalivkin, 1960, p. 135). The deposits of the Uda district are several hundred km north of the Maly Khinghan area but on the same general structural trend; they form a belt 500 km long by 30 km wide that trends northeasterly into the Sea of Okhotsk (Shkolnik, 1973). Descriptive information is sparse, but the deposits evidently consist mainly of banded jasper, in part manganiferous as at Maly Khinghan, in beds of varying thickness (as much as 300 m), interlayered with mafic volcanic flows and pyroclastics. The preferred age assignment appears to be Early Cambrian and/or late Precambrian.

The most notable example of iron-formation spatially and temporally related to deposits of possible glaciogenic origin is that of the Rapitan Group in northwestern Canada (Gross, 1973; Young, 1976). Ferruginous strata — in places, iron-rich maroon shale rather than banded iron-formation — are present in the lower unit of the Rapitan Group, which is exposed in a belt about 600 km long by 50 km wide in the MacKenzie Mountains (Eisbacher, 1977). No precise age is available for the Rapitan, but on regional grounds it can be assigned to the late Proterozoic. The thickest iron-formation is in the Snake River area, near the northwest end of the belt, where it is described as ". . . a succession of jasper and blue hematite beds more than 150 m thick which occur near the base of the Rapitan formation, a crudely stratified, poorly sorted conglomerate at least 1500 meters thick." (Gross, 1973, p. 17). In the North Redstone River area, near the southeast end of the belt, the iron-formation is 20 m thick; it is overlain by 500 m of polymictic conglomerate and underlain by a 400-m thickness of mostly mudstone with scattered large clasts, both units interpreted by Young (1976) as glaciogenic deposits related to late Proterozoic glaciation of North America and elsewhere.

Gross (1973), who apparently did not accept the designation of the coarse clastic rocks as glaciogenic, suggested a possible explanation for the unusual association of chemical sediments (iron-formation) with poorly sorted clastic deposits, namely, that the strata reflect rapid filling of a deep, probably fault-bounded basin or trough, with periodic explosive volcanism and fumarolic discharge of iron- and silica-rich solutions to the seawater. Young (1976), however, pointed to the occurrences of similar deposists of apparent glaciogenic association elsewhere, such as in the Adelaide Geosyncline of South Australia (Trendall, 1973a); he suggested a genetic relation, possibly by reduction in volume of solute (seawater) by freezing, with consequent concentration and precipitation of materials contained in solution.

The fourth major known deposit of iron-formation of probable late Proterozoic age is that of Morro du Urucum, Brazil, and its extensions into Bolivia. The iron-formation, in part manganiferous, is as much as 300 m thick and its lithic associations (Jacadigo Series) are not unlike those of the Rapitan deposits. The full areal extent is not known, and the age is not well established: Dorr (1973) quoted F.F.M. de Almeida, who has studied the deposits, as of the opinion that the Jacodigo Series is "Cambro-Ordovician" in age, but offered his own suggestion that the age is "Eocambrian".

SIGNIFICANCE OF PEAKS IN THE DEPOSITIONAL RECORD

No positive conclusions as to the meaning of the great variation in the rate of iron-formation deposition through time can be drawn as yet, but some speculations can be offered.

The *middle Archean interval* probably is the least sharply defined of all maxima in the depositional record. In fact, the spread of ages for the various units, particularly when coupled with the certainty that geologic record at best is fragmentary and poorly dated, suggests the possibility that no real depositional peak exists — that the time distribution pattern may be related more to accidents of preservation than to a time-controlled cycle of deposition. This, if so, leads to the inference that during this early period of Earth history, seawater at all times contained abundant iron and silica in solution, and that iron-formation was deposited whenever and wherever environmental conditions tilted toward precipitation rather than retention.

The *late Archaean interval* is coincident with the widespread volcanism that produced the great greenstone belts of this age. Individual units of iron-formation are very numerous, but the total quantity is not particularly large. It seems evident that despite undoubted fumarolic contributions of iron and silica on a major scale to a seawater already virtually saturated, the continued crustal instability and the periodic flooding of basins of deposition with volcanogenic debris were not compatible with the quiet environmental conditions necessary for long-continued clean chemical sedimentation. Significant accumulations of iron-formation were possible only during relatively brief periods of quiescence; chemical deposition was halted or heavily diluted by increased input of clastic materials. As can be observed in the greenstone belts, the result was a large number of relatively thin and discontinuous units of iron-formation.

The *early Proterozoic interval* clearly was the period of greatest iron sedimentation in Earth's history — most of the great iron-formation deposits of the world are of this age. The depositional maximum is believed due to coincidence of a number of more-or-less independent variables — some structural, some geochemical, and some biologic. The model here advanced as an explanation for this extraordinary epoch is constructed largely from previously published ideas and concepts, notably those of Button (1976), Cloud (1973), Drever (1974), Eugster and Chou (1973), Holland (1973), James (1954), and Trendall (1973b, 1977). It differs from previous models mainly in emphasis and timing assigned to the various factors, and in their interrelationships. In summary form, it consists of the following elements:

(1) During Archean time, and continuing into early Proterozoic time, the oceans, below a thin surface layer in equilibrium with a weakly oxygenic atmosphere, constituted a major reservoir for dissolved iron and silica derived from diverse sources, including weathering and volcanism.

(2) The almost worldwide era of orogeny and cratonization of late Archean time was followed by a long period of crustal stability — the Eparchean interval of older concepts — during which continental blocks were reduced to surfaces of low relief, with consequent very low levels of clastic input to basins of deposition.

(3) During the early Proterozoic, but probably at different times in different parts of the world, the previously stable cratons were subjected to weak structural disturbances, perhaps in part epeirogenic but mainly extensional, with consequent development of shallow intracontinental troughs and marginal basins.

(4) Encroachment of seawater into the newly formed troughs and basins, with continued replenishment by nutrient-rich deeper ocean waters, triggered a series of events, including rapid growth and evolution of biota, some of which were (or became) oxygen producers, and precipitation of iron and silica to form banded iron-formation. The precipitation of iron and silica was induced in part by greater availability of oxygen (including that locally generated biologically by photosynthesis), in part by increased concentration of iron and silica due to seawater evaporation, and perhaps in part directly by biologic processes. The nature of the iron precipitates would depend upon local conditions: in broad, unrestricted shallow basins, oxidic facies would be produced, whereas in deeper, barred basins and troughs, conditions might favor precipitation of silicate, carbonate, and sulfide facies, none of which require an excess of oxygen.

(5) Geochemical transfer of dissolved iron and silica contained in the constantly upwelling deep ocean waters to marginal basins and troughs, there to be trapped out as chemical precipitates, was an irreversible process, to be halted locally only by change in configuration or filling of the basin of deposition, or ultimately, in a worldwide sense, only when the oceans were essentially depleted of at least their iron content. In essence, the existence of sites appropriate for upwelling and precipitation provided, not just the means whereby large-scale chemical equilibrium could be established between deep oceanic waters and an evolving oxygenic atmosphere, but in fact the necessary milieu for the biologic evolution and biologic processes that were responsible for those changes in the composition of the Earth's atmosphere.

The prime dependence in this model on appropriate structural conditions — a local or regional factor — rather than on independent worldwide evolutionary changes in the composition of the Earth's oceans, atmosphere or biosphere, suggests further that the major iron-formation deposits of early Proterozoic age need not be strictly time-equivalent; conditions favorable for precipitation might well have continued for several hundred million years.

The *late Proterozoic—early Phanerozoic* interval is one of uncertain significance. It followed a gap of more than a billion years during which the only known iron-formation appears to be those relatively minor deposits associated with greenstone belts of mid-Proterozoic age, such as those of southwest U.S.A. Even in the light of the model proposed for early Proterozoic sedimentation, the significance of the gap is not well understood.

As discussed previously, some of the larger deposits assigned to this epoch, notably those of Uda and Maly Khinghan districts of far eastern U.S.S.R., probably are of the Algoma type, directly related to regional orogeny and volcanism. More problematical are those large deposits, typified by the iron-formation of the Rapitan Group of northwestern Canada, that are associated with coarse clastic deposits of possible glaciogenic origin. No ready explanation for this curious association is at hand. Bog iron deposits, precipitated from ground waters in areas of poorly integrated drainage, are of course of common occurrence in regions of Pleistocene glaciation, but these deposits bear little resemblance to chert-banded iron-formation.

In the absence of a convincing hypothesis linking iron-formation to glaciation, the alternative proposed by Gross (1973) for deposits of the Rapitan Group warrants careful consideration. In this hypothesis, the iron and silica are derived from fumarolic emissions in a fault-bounded trough or basin. The structural postulate draws some support from consideration of the late Precambrian history of western North America. According to Stewart (1972), the deposits of the late Precambrian (younger than 850 m.y.) Windermere Supergroup of western Canada and equivalent strata elsewhere reflect a late Precambrian epoch of continental rifting of the North American craton, along a sinuous belt extending from Alaska to northern Mexico. The Rapitan Group, the basal unit of the Windermere sequence, would represent initial deposits in the developing Cordilleran geosyncline. Whether the poorly sorted clastics that make up much of the Rapitan are in fact glacial, as concluded by Young (1976), or are simply the product of rapid erosion of the bounding terrane in a structurally active area, is perhaps not a critical question; glaciation, if it occurred during this period of structural transition, would bear only fortuitous relation to iron-formation sedimentation. Whether the "rift hypothesis" can be applied to other late Proterozoic deposits, such as those of Morro du Urucum, Brazil, remains to be seen, but it is of some interest that one of the few areas in which iron-rich sediments are accumulating today is the rift-dominated Red Sea.

ACKNOWLEDGEMENT

Many of the ideas presented in this chapter were refined during a 1980

Dahlem Conference devoted to biospheric evolution and metallogeny (Holland and Schidlowski, 1982), and I gratefully acknowledge indebtedness to fellow participants.

REFERENCES

Aldrich, L.I., Davis, G.L. and James, H.L., 1965. Ages of minerals from metamorphic and igneous rocks near Iron Mountain, Michigan. J. Petrol., 6: 445—472.

Alexandrov, E.A., 1973. The Precambrian banded iron-formations of the Soviet Union. Econ. Geol., 68: 1035—1062.

Allaart, J.H., 1976. The pre-3760 m.y. old supracrustal rocks of the Isua area, central west Greenland, and the associated occurrence of quartz-banded ironstone, In: B.F. Windley (Editor), The Early History of the Earth. John Wiley and Sons, New York, N.Y., pp. 177—190.

Allsopp, H.L., Ulrych, T.J. and Nicolaysen, L.O., 1968. Dating some significant events in the history of the Swaziland System by the Rb-Sr isochron method. Can. J. Earth Sci., 5: 605—620.

Almeida, F.F.M. de, 1978. Chronotectonic boundaries for Precambrian time division in South America. Acad. Brazil Cienc. An., 50: 527—535.

Alvarado, B., 1970. Iron ore deposits of South America, In: Survey of World Iron Ore Resources. United Nations, New York, N.Y., pp. 302—380.

Anderson, C.A. and Silver, L.T., 1976. Yavapai Series — a greenstone belt. Ariz. Geol. Soc. Digest, 10: 13—25.

Anhaeusser, C.R. and Button, A., 1976. A review of southern African stratiform ore deposits — their position in time and space, In: K.H. Wolf (Editor), Handbook of Strata-bound and Stratiform Ore Deposits. Elsevier, Amsterdam, pp. 257—319.

Appel, P.W.U., 1980. On the early Archaean Isua iron-formation, West Greenland. Precambrian Res., 11: 73—87.

Baadsgaard, H., Collerson, K.D. and Bridgwater, D., 1979. The Archean gneiss complex of northern Labrador. I. Preliminary U-Th-Pb geochronology. Can. J. Earth Sci., 16: 951—961.

Bayley, R.W. and James, H.L., 1973. Precambrian iron-formations of the United States. Econ. Geol., 68: 934—959.

Bayley, R.W., Proctor, P.D. and Condie, K.C., 1973. Geology of the South Pass area, Fremont County, Wyoming. U.S. Geol. Surv., Prof. Pap. 793, 39 pp.

Beckinsale, R.D., Gale, N.H., Pankhurst, R.V., MacFarlane, A., Crow, M.J., Arthurs, V.W. and Wilkinson, A.F., 1980. Discordant Rb-Sr and Pb-Pb whole rock isochron ages for the Archean basement of Sierra Leone. Precambrian Res., 13: 63—76.

Beukes, N.J., 1973. Precambrian iron-formations of southern Africa. Econ. Geol., 68: 960—1004.

Bishara, W.W. and Habib, M.E., 1973. The Precambrian banded iron ore of Semna, Eastern Desert, Egypt. Neues Jahrb. Mineral., 120: 108—118.

Bronner, G. and Chauvel, J.J., 1979. Precambrian banded iron-formations of the Ijil Group (Kediat Ijil, Requibat Shield, Mauritania). Econ. Geol., 74: 77—94.

Brooks, C., Krogh, T.E., Hart, S.R. and Davis, G.L., 1970. The initial Sr^{87}/Sr^{86} ratios of the upper and lower Michipicoten metavolcanics, Ontario, Canada. Carnegie Inst. Washington Yearb., 68: 422—425.

Button, A., 1976. Iron formation as an end member in carbonate sedimentary cycles in the Transvaal Supergroup, South Africa. Econ. Geol., 71: 193—201.

Chebotarev, M.V., 1960 (1958). Geological structure of the South Khingan manganese deposit and essential composition of its ores. Translated by E.A. Alexandrov and Associates. Int. Geol. Rev., 2: 851—866.

Chernov, V.M., 1973. The ferruginous-siliceous formations of the eastern part of the Baltic Shield. In: Genesis of Precambrian Iron and Manganese Deposits. UNESCO Earth Sci. Ser., 9: 85—87.

Choubert, G. and Faure-Muret, A., 1973. The Precambrian iron and manganese deposits of the Anti-Atlas. In: Genesis of Precambrian Iron and Manganese Deposits. UNESCO Earth Sci. Ser., 9: 123—130.

Cloud, P., 1973. Paleoecological significance of the banded iron-formation. Econ. Geol., 68: 1135—1143.

Cochrane, G.W. and Edwards, A.B., 1960. The Roper River oolitic ironstone formations. Aust. Commonwealth Sci. Ind. Res. Org., Mineragraphic Inv. Tech. Paper 1, 28 pp.

Coertze, F.J., Burger, A.J., Walraven, F., Marlow, A.G. and MacCaskie, D.R., 1978. Field relations and age determinations in the Bushveld Complex. Geol. Soc. S. Afr. Trans., 81: 1—11.

De Paolo, D.J. and Wasserburg, G.J., 1979. Sm-Nd age of the Stillwater complex and the mantle evolution curve for neodymium. Geochim. Cosmochim. Acta, 43: 999—1008.

Dimroth, E., 1970. Evolution of the Labrador geosycline. Geol. Soc. Am. Bull., 81: 2717—2742.

Dorr, J.V.N., 1973. Iron-formation in South America. Econ. Geol., 68: 1005—1022.

Drever, J.I., 1974. Geochemical model for the origin of Precambrian banded iron formations. Geol. Soc. Am. Bull., 85: 1099—1106.

Egorov, E.W. and Timofeieva, M.W., 1973. Effusive iron-silica formations of the Maly Khinghan. In: Genesis of Precambrian Iron and Manganese Deposits. UNESCO Earth Sci. Ser., 9: 181—185.

Eichler, J., 1976. Origin of the Precambrian banded iron-formations. In: K.H. Wolf (Editor), Handbook of Strata-bound and Stratiform Ore Deposits. Elsevier, N.Y., 7: 157—201.

Eisbacher, G.H., 1977. Re-definition and subdivision of the Rapitan Group, Mackenzie Mountains. Geol. Surv. Can., Paper 77—35, 21 pp.

Eriksson, K.A., 1980. Transitional sedimentation styles in the Moodies and Fig Tree Groups, Barberton Mountain Land, South Africa: Evidence favoring an Archean continental margin. Precambrian Res., 12: 141—160.

Eugster, H.P. and Chou, I. Ming, 1973. The depositional environments of Precambrian banded iron-formations. Econ. Geol., 68: 1144—1168.

Frietsch, R., 1973. Precambrian iron ores of sedimentary origin in Sweden, In: Genesis of Precambrian Iron and Manganese Deposits. UNESCO Earth Sci. Ser., 9: 77—83.

Frietsch, R., Papunen, H. and Vokes, F.M., 1979. The ore deposits in Finland, Norway, and Sweden — a review. Econ. Geol., 74: 975—1001.

Fryer, B.J., 1972. Age determinations in the Circum-Ungava Geosyncline and the evolution of Precambrian banded iron-formations. Can. J. Earth Sci., 9: 652—663.

Gee, R.D., 1979. Structure and tectonic style of the Western Australian Shield. Tectonophysics, 58: 327—369.

Gellatly, D.C., 1972. Problems of provenance of the Yampi iron ores, West Kimberley region, Western Australia. Aust. Bur. Min. Res. Bull., 125: 191—205.

Goldich, S.S., 1973. Ages of Precambrian banded iron-formations. Econ. Geol., 68: 1126—1134.

Gole, M.J. and Klein, C., 1981. Banded iron-formations through much of Precambrian time. J. Geol., 89: 169—184.

Goode, A.D.T., Hall, M. and Bunting, J.A., 1983. The Nabberu Basin of Western Australia. In: A.F. Trendall and R.C. Morris (Editors), Iron-Formation: Facts and Problems. Elsevier, Amsterdam, pp. 295—323.

Goodwin, A.M., 1973. Archean iron-formations and tectonic basins of the Canadian Shield. Econ. Geol., 68: 915—933.

Gross, G.A., 1955. Geology of iron deposits in Canada, Volume I. General geology and evaluation of iron deposits. Geol. Surv. Can., Econ. Geol. Rep. 22, 181 pp.
Gross, G.A., 1968. Geology of iron deposits in Canada, Volume II, Iron ranges of the Labrador geosyncline. Geol. Surv. Can., Econ. Geol. Rep. 22, 179 pp.
Gross, G.A., 1973. The depositional environments of principal types of Precambrian iron-formations. In: Genesis of Precambrian Iron and Manganese Deposits. UNESCO Earth Sci. Ser., 9: 15—21.
Gruss, H., 1973. Itabirite iron ores of the Liberia and Guyana shields. In: Genesis of Precambrian Iron and Manganese Deposits. UNESCO Earth Sci. Ser. 9: 335—359.
Hamilton, P.J., 1977. Isotope and trace element studies of the Great Dyke and Bushveld mafic phase and their relation to early Proterozoic magma genesis in southern Africa. J. Petrol., 12: 24—52.
Holland, H.D., 1973. The oceans: A possible source of iron in iron-formation. Econ. Geol., 68: 1169—1172.
Holland, H.D. and Schidlowski, M. (Editors), 1982. Mineral Deposits and Evolution of the Biosphere. Springer-Verlag, Berlin/Heidelberg/New York, 332 pp.
Hurley, P.M., Fairbairn, H.W. and Gaudette, H.E., 1976. Progress report on Early Archean rocks of Liberia, Sierra Leone, and Guayana, and their general stratigraphic setting. In: B.F. Windley (Editor), The Early History of the Earth. John Wiley and Sons, New York, N.Y., pp. 511—251.
Jackson, G.D. and Taylor, F.C., 1972. Correlation of major Aphebian units in the northeastern Canadian Shield. Can. J. Earth Sci., 9: 1650—1669.
James, H.L., 1954. Sedimentary facies of iron-formation. Econ. Geol., 49: 235—293.
James, H.L., 1966. Chemistry of the iron-rich sedimentary rocks. U.S. Geol. Surv., Prof. Pap. 440-W, 61 pp.
James, H.L. and Hedge, C.E., 1980. Age of the basement rocks of southwest Montana. Geol. Soc. Am. Bull., Pt. I, 91: 11—15.
James, H.L. and Trendall, A.F., 1982. Banded iron-formation: Distribution in time and paleoenvironmental significance. In: H.D. Holland and M. Schidlowski (Editors), Mineral Deposits and Evolution of the Biosphere. Springer-Verlag, Berlin/Heidelberg/New York, pp. 199—217.
Kalugin, A.S., 1973. Geology and genesis of the Devonian banded iron-formation in Altai, western Siberia and eastern Kazakhstan. In: Genesis of Precambrian Iron and Manganese Deposits. UNESCO Earth Sci. Ser., 9: 159—166.
Kimberley, M.M., 1978. Paleoenvironmental classification of iron-formations. Econ. Geol., 73: 215—229.
Krishnan, M.S., 1973. Occurrence and origin of the iron ores of India, In: Genesis of Precambrian Iron and Manganese Deposits. UNESCO Earth Sci. Ser., 9: p. 69—76.
Laajoki, K. and Saikkonen, R., 1977. On the geology and geochemistry of Precambrian iron formations in Väyrylänkylä, South Puolanka area, Finland. Geol. Surv. Finl. Bull. 292, 137 pp.
Larue, D.K. and Sloss, L.L., 1980. Early Proterozoic sedimentary basins of the Lake Superior region — Summary. Geol. Soc. Am. Bull., Pt. I, 91: 450—452.
Marelle, A. and Abdulla, M.A., 1970. Iron ore deposits of Africa. In: Survey of World Iron Ore Resources. United Nations, New York, N.Y., pp. 62—101.
Martin, A., Nisbet, E.G. and Bickle, M.J., 1980. Archaean stromatolites of the Belingwe Greenstone Belt, Zimbabwe (Rhodesia). Precambrian Res., 13: 337—362.
Montgomery, C.W., 1979. Uranium-lead geochronology of the Archean Imataca Series, Venezuelan Guayana Shield. Contrib. Miner. Petrol.—Beitr. Mineral. Petrol., 69: 167—176.
Moorbath, S., O'Nions, R.K. and Parkhurst, R.V., 1973. Early Archean age for the Isua iron formation, West Greenland. Nature, 245: 138—139.
Nalivkin, D.V., 1960. The geology of the U.S.S.R. Pergamon Press Int. Ser. Monogr. Earth

Sci., v.8, Pergamon Press, New York/Oxford/London/Paris, 170 pp.
Nishiwaki, C., 1970. Iron ore deposits of the Middle East, Asia, and the Far East. In: Survey of World Iron Ore Resources. United Nations, New York, N.Y., pp. 102—206.
Novokhatsky, I.P., 1973. Precambrian ferruginous-siliceous formations of Kazakhstan, In: Genesis of Precambrian Iron and Manganese Deposits. UNESCO Earth Sci. Ser., 9: p. 153—157.
O'Rourke, J.E., 1961. Paleozoic banded iron-formations. Econ. Geol., 56: 331—361.
Page, N.J., 1977. Stillwater Complex, Montana: Rock succession, metamorphism, and structure of the Complex and adjacent rocks. U.S. Geol. Surv., Prof. Pap. 999, 79 pp.
Parkin, L.W. (Editor) 1969. Handbook of South Australian geology. Geol. Surv. S. Aust., Adelaide, 268 pp.
Peterman, Z.E., 1979. Geochronology and the Archean of the United States. Econ. Geol., 74: 1544—1562.
Pichimuthu, C.S., 1974. On the banded iron-formations of Precambrian age in India. Geol. Soc. India, J., 15: 1—30.
Plaksenko, N.A., Koval, I.K. and Schogolev, I.N., 1973. Precambrian ferruginous-siliceous formations associated with the Kursk Magnetic Anomaly. In: Genesis of Precambrian Iron and Manganese Deposits. UNESCO Earth Sci. Ser., 9: 89—94.
Ravich, M.G., Federov, L.V. and Tarutin, O.A., 1982. Iron deposits of the Prince Charles Mountains. In: C. Craddock (Editor), Antarctic Geoscience. Univ. Wisc. Press, pp. 853—858.
Sakko, M. and Laajoki, K., 1975. Whole rock Pb-Pb isochron age for the Pääkkö iron formation in Väyrylänkylä, South Puolanka area, Finland. Geol. Soc. Finl. Bull., 47: 113—116.
Sarkar, S.N. and Saha, A.K., 1977. The present status of the Precambrian stratigraphy, tectonics, and geochronology of the Singhbhum-Keonjar-Mayurbhanj region, eastern India. Ind. J. Earth Sci., S. Ray Vol., pp. 37—65.
Sarkar, S.N., Saha, A.H., Boelrijk, N.A.I.M. and Hebeda, E.H., 1979. New data on the geochronology of the Older Metamorphic Group and the Singhbhum granite of Singhbhum-Keonjar-Mayurbhanj region, eastern India. Ind. J. Earth Sci., 6: 32—51.
Semenenko, N.P., 1973. The iron-chert formation of the Ukrainian Shield. In: Genesis of Precambrian Iron and Manganese Deposits. UNESCO Earth Sci. Ser., 9: 134—142.
Shkolnik, E.L., 1973. Effusive jasper iron-formation and iron ores of the Uda area. In: Genesis of Precambrian Iron and Manganese Deposits. UNESCO Earth Sci. Ser., 9: 187—189.
Sims, P.K., 1976. Early Precambrian tectonic-igneous evolution in the Vermilion district, northeastern Minnesota. Geol. Soc. Am. Bull., 87: 379—389.
Sims, S.J., 1973. The Belinga iron ore deposit (Gabon). In: Genesis of Precambrian Iron and Manganese Deposits. UNESCO Earth Sci. Ser., 9: 323—334.
Sinha, A.K., 1972. U-Th-Pb systematics and the age of the Onverwacht Series, South Africa. Earth Planet. Sci. Lett., 16: 219—227.
Stewart, J.H., 1972. Initial deposits in the Cordilleran geosyncline: evidence of a late Precambrian ($<$ 850 m.y.) continental separation. Geol. Soc. Am. Bull., 83: 1345—1360.
Thomson, B.P., 1976. Precambrian geology and tectonics of the Stuart Shelf and Torrens Hinge Zone. In: B.P. Thomson, B. Daily, R.P. Coats, and B.G. Forbes (Editors), Late Precambrian and Cambrian Geology of the Adelaide "Geosyncline" and Stuart Shelf, South Australia 25th Int. Geol. Cong., Excursion Guide, 33A: 1—11.
Tolbert, G.E., Tremaine, J.W., Melcher, G.C. and Gomes, C.B., 1973. Geology and iron ore deposits of Serra dos Carajás, Pará (Brazil). In: Genesis of Precambrian Iron and Manganese Deposits. UNESCO Earth Sci. Ser. 9: 271—280.
Trendall, A.F., 1973a. Precambrian iron-formations of Australia. Econ. Geol., 68: 1023—1034.
Trendall, A.F., 1973b. Time-distribution and type-distribution of Precambrian iron-for-

mations in Australia. In: Genesis of Precambrian Iron and Manganese Deposits. UNESCO Earth Sci. Ser., 9: 49—57.

Trendall, A.F., 1977. Banded iron-formation in the Archean. In: G.J.H. McCall (Editor), The Archean. Dowden, Hutchinson and Ross, Inc., Stroudsburg, Penn., pp. 270—274.

Trendall, A.F., 1983. The Hamersley Basin. In: A.F. Trendall and R.C. Morris (Editors), Iron-Formation: Facts and Problems. Elsevier, Amsterdam, pp. 69—129.

Trendall, A.F. and Blockley, J.G., 1970. The iron formations of the Hamersley Group, Western Australia, with special reference to the associated crocidolite. W. Aust., Geol. Surv., Bull. 119, 353 pp.

Van Niekerk, C.B. and Burger, A.J., 1978. A new age for the Ventersdorp acidic lavas. Geol. Soc. S. Afr. Trans., 81 (2): 155—163.

Van Schmus, W.R., 1976. Early and Middle Proterozoic history of the Great Lakes area, North America. Philos. Trans. R. Soc. London, 280: 605—628.

Vorona, I.D., Kravchenko, V.M., Pervago, V.A. and Frumkin, I.M., 1973. Precambrian ferruginous formations of the Aldan Shield. In: Genesis of Precambrian Iron and Manganese Deposits. UNESCO Earth Sci. Ser., 9: 243—247.

Walter, M.R., Goode, A.D.T. and Hall, W.D.M., 1976. Microfossils from a newly discovered Precambrian stromatolitic iron formation in Western Australia. Nature, 261: 221—223.

Young, G.M., 1976. Iron-formations and glaciogenic rocks of the Rapitan Group, Northwest Territories, Canada. Precambrian Res., 3: 137—158.

Chapter 13

CHEMICAL FACTORS IN THE DEPOSITION AND DIAGENESIS OF BANDED IRON-FORMATION

W.E. EWERS

INTRODUCTION

Although many questions remain to be answered as to the origins of the Precambrian banded iron-formations, it is generally accepted that they are, or have derived from, chemical sediments. This paper is an attempt to bring together chemical factors involved in the supply of constituents and in the processes of deposition, segregation, consolidation and alteration that have produced these remarkable rocks. It is concerned, therefore, with both quantity and concentration, with equilibrium and non-equilibrium and with processes both in solution and in the primary sediments.

The most comprehensive account of the chemical aspects of this topic is the monograph by Mel'nik (1973). More recent reviews by Eichler (1976) and Dimroth (1976) refer to papers up to 1974. Papers that have appeared since then, and others relevant to the subject are referred to in the text.

Within the James' (1954) definition of banded iron-formation there is a considerable diversity of rocks, a diversity that has been made even greater by the inclusion of rocks altered in varying degrees by metamorphism, metsomatism and weathering. Even in fresh and relatively unmetamorphosed rocks there are marked variations in composition with the four "end-member" facies, oxide, silicate, carbonate and sulphide and many mixed facies. There are also variations which derive from depositional environment, from the shallow-water oolitic and granular forms to the deeper-water finely banded rocks which appear to have escaped the effects of wave or current turbulence. It is these finely banded rocks which might be expected to retain the most information concerning the detailed sequence of deposition, a view that is reinforced by the extraordinary lateral continuity of the fine banding in some of them (Trendall and Blockley, 1970).

This consideration, together with the range of compositional facies represented and the generally low metamorphic grade, has led me to choose one of the classically exposed, finely-banded iron-formations, the Dales Gorge Member of the Brockman Iron Formation, as the main geological reference for this discussion. It is a measure of the confusion on classification of banded iron-formations that this much-studied unit has been classed as Lake-Superior type (Eichler, 1976; Gross, 1980) and as Algoma type (Dimroth, 1976). This unit has much in common with both "types"; much that can be deduced from it will be relevant to iron-formations generally, though clearly not to all of their variations.

THE DALES GORGE MEMBER

The Dales Gorge Member has been extensively studied by the Geological Survey of Western Australia (Trendall and Blockley, 1970; Trendall, 1983, this volume) and more recently in the author's laboratory. From analyses by Trendall and Pepper (1977) it can be seen that the BIF-macrobands contain less iron, more silicate and more carbonate than "end-member" oxide facies iron-formation, and that the intercalated S-macrobands can be regarded as carbonate-silicate facies banded iron-formation. Ewers and Morris (1981) have shown that despite apparently sharp lithological boundaries, the BIF- and S-macrobands grade into one another and that a continuum of chemical precipitation of iron and silica with the change in style of precipitation initiated by periodic additions of volcanic dust can be postulated. LaBerge (1966) recognized volcanic shards in the shaly beds intercalated with the chemically precipitated cherts and chert siderites, typical of the S-macrobands.

Quantities and concentrations

Quantitatively the 17 BIF-macrobands and the 16 S-macrobands have a total thickness of about 142 m, and are thought to have deposited originally over an area of perhaps 10^5 km^2. Based on analyses on drill core representing the whole of the Dales Gorge Member at Paraburdoo, Ewers and Morris (1981) have calculated that the amounts of iron and silica in a column with a cross section of 1 cm^2 are 12.7 kg and 18.1 kg respectively corresponding to 12.7×10^{18} g Fe and 18.1×10^{18} g SiO_2 over the whole depositional area of 10^5 km^2. A somewhat less reliable calculation gives estimates of about 3 times as much iron and silica in the thicker Joffre Member and totals of about 52×10^{18} g Fe and 76×10^{18} g SiO_2 for the two banded iron-formation members of the whole Brockman Iron Formation.

Trendall and Blockley (1970) report an average of 22.5 mg Fe per cm^2 for each microband of 22 analyzed chert mesobands. Individual values range from 8 to 43 mg Fe per cm^2. Using Trendall's assumption that each microband represents a true varve, the average of 22.5 mg Fe per cm^2 corresponds to an annual increment of 22.5×10^{12} g Fe to the whole depositional area. For this amount of iron there is a simple reciprocal relationship between the volume, V (l), of water from which it precipitated and the average decrement in iron concentration, δC (g l^{-1}), in that volume. This average decrement obviously sets a lower limit on the concentration of iron. This relationship is

$$\delta C \times V = 22.5 \times 10^{12}\ \text{g} \tag{1}$$

In the unlikely event that precipitation was from a static body of water of uniform depth and of area 10^5 km^2, a concentration decrement of 1 mg l^{-1} would require a depth of 225 m. In a more realistic flow situation, for example on a shelf or in a basin with intermittent or continuous access to the ocean, the same concentration-volume relationship would apply. It is unlikely that *all* of the iron in a body of water would be precipitated in a given year. On the other hand it is most probable that precipitation (by whatever process) would be initiated in the surface layers only so that, even allowing for circulation, volumes corresponding to depths of tens rather than hundreds of metres would have been involved. These considerations suggest that δC was considerably greater than 1 mg l^{-1} and a realistic minimum concentration is probably about 20 mg l^{-1}.

THE NATURE OF THE SOURCE SOLUTIONS

The discussion to this point implies that iron and silica were introduced in solution. As argued by Trendall and Blockley (1970) the composition of these rocks demands that their iron content, in particular, must have been separated in solution from any feasible geological source. The possibility that the constituents were introduced as solids can be discounted; terrigenous sediments would inevitably have contained more aluminium relative to iron, and cosmic dust would have contained more nickel. No assumption has been made as to the nature of the iron or silica in solution, nor as to whether they were as colloids or in true solution.

Solution chemistry of iron

The oxidized form of iron, iron III, is extremely insoluble in water under naturally occurring conditions; only in very acid volcanic solutions, such as those described by Zelenov (1958) from the Ebeko volcano, does it reach significant concentrations, and above pH 4 its concentration in true solution can be ignored.

Colloidal solutions of iron (III) can be prepared in the laboratory and, in low electrolyte concentrations, may be stable to high concentrations. Such colloids have not been recorded in nature, presumably because they are destabilized by salts. Mel'nik (1973) has reviewed the data on inorganic iron (III) hydroxide sols, pointing out that they are completely precipitated in the presence of sea salts at concentrations of 100 mg l^{-1}, about 1/300 of the present day sea water concentration. Sulphate is a particularly active coagulant. The same author summarized evidence on the stabilization of iron (III) hydroxide sols by silica, the mixed sol being more resistant to coagulation by electrolytes. He concluded, however, that there is no stabilization unless the silica concentration exceeds 200 mg l^{-1}, that is in solutions that are supersaturated with respect to amorphous silica, and which contain polymeric silica hydrates as well as $Si(OH)_4$. Even then stabilization has been demonstrated only at pH values < 4 and between 8 and 11. In summary, it seems very improbable that large bodies of water could have been sustained in a condition conducive to the transport of large quantities of iron as iron (III) hydroxide sols.

Most recent authors have agreed that iron for the banded iron-formations was transported in solution in the iron (II) state as Fe^{2+} or $FeOH^{+}$. Its solubility is controlled by both pH and the redox conditions, as well as by the interdependent activities of several reagents of which the most important, in the natural context, are carbonate, silicate and sulphide.

Considerable uncertainty surrounds the quantitative definition of these controls of solubility. The first precipitates to form when any of the precipitants react with Fe^{2+} are almost certainly ill-defined metastable compounds for which the thermodynamic properties are either unknown or only very approximately known. Thus the oxidation of Fe^{2+} to Fe^{3+} and its subsequent hydrolysis do not produce directly either crystalline hematite or goethite, but a hydrous ferric hydroxide which alters with time to one of a variety of crystalline compounds according to temperature and original solution composition. Values of the free energy of formation, ΔG_f°, of $Fe(OH)_3$ are listed in older tables of thermodynamic properties (e.g. Wagman et al., 1969) but have been omitted from the most recent compilation (Robie et al., 1979). Values of ΔG_f° are available for most of the crystalline solids, hematite, magnetite, siderite, pyrite, greenalite etc., but it is pertinent to ask which, if any, of these are relevant to equilibria which would control the concentration of Fe^{2+}.

Frost (1978) has taken the extreme view that the only relevant materials are the stable crystalline phases such as hematite and magnetite. If this were so it would seem impossible, under credible conditions of pH, Eh, etc., to achieve concentrations of Fe^{2+} in large volumes of water that would satisfy the logistics of supplying iron to the banded iron-formations. Frost expressed his view in criticizing the paper of Klein and Bricker (1977) who constructed their pE-pH diagrams using crystalline magnetite and metastable $Fe(OH)_3$, a device which grossly exaggerated the stability field of magnetite at the expense of the hydrated ferric oxide. Klein and Bricker recognized that if a hydrated magnetite precursor instead of crystalline magnetite had been used in calculating their diagrams, its boundary with $Fe(OH)_3$ would have been at a comparable oxidation potential to the boundary between hematite and magnetite. Such a precursor, as a metastable phase, would also have been more soluble so that, at a given pE and pH, the concentration of Fe^{2+} in "equilibrium" with it would have been greater than if crystalline magnetite were the controlling phase.

A similar argument applies to all of the crystalline phases that are involved in this system. Pyrite is not a direct product of the reduction of SO_4^{2-} in the presence of Fe^{2+}; ill-defined monosulphides are the primary product (Berner, 1970). An iron silicate, whether it be greenalite or minnesotaite, is unlikely to appear initially as a crystalline material; it is much more likely to be an early diagenetic product with an amorphous precursor. Among these phases siderite is perhaps the most likely product of direct precipitation (Mel'nik, 1973), but the in vitro preparation of siderite is characterized by the initial precipitation of amorphous material, which persists for more than a day before crystalline siderite forms (J.E. Wildman, pers. commun., 1979).

Mel'nik (1973) dealt with this problem by using a system of "congruent constants" for oxides, silicates and carbonates of iron which he estimated specifically for "analysis of the conditions of formation of iron ores". In his monograph he does not give the actual values, but he refers to the phases in his stability diagrams as "dispersed". One can deduce that his value for ΔG_f° for Fe_3O_4 (dispersed) is −223.4 kcal mole^{-1}, which compares with −242.4 kcal mole^{-1} quoted by Wagman et al. (1969) for crystalline Fe_3O_4. This calculation assumes Wagman's value, −166.5 kcal mole^{-1}, for ΔG_f° for $Fe(OH)_3$.

Mel'nik concluded that the probable conditions for the chemical precipitation of "ferruginous-siliceous sediments containing iron hydroxide, siderite and secondary iron sulphide were pH about 6, Eh in the range 0.0 to 0.08 V, and Fe^{2+} concentration between 0.1 and 0.4 g l^{-1}".

It is argued above that the minimum iron concentration that would satisfy the logistics of transporting iron in sufficient quantity to form banded iron-formation is of the order of 20 mg l^{-1}. Taking the activity coefficient of Fe^{2+} to be 0.26, this concentration equates to an activity of $10^{-4}M$, which was used by Ewers (1980) to construct a stability diagram displaying the conditions of stable and metastable equilibria in terms of pH and fugacity of oxygen (f_{O_2}). This diagram is given in Fig. 13-1.

The free energies of formation at 298°K used in calculating this diagram are from the self-consistent compilation by Wagman et al. (1969). Minor alterations appear in the most recently updated similar compilation (Robie et al., 1979) but as the value for ΔG_f° for the ill-defined phase $Fe(OH)_3$ is omitted from this version, the older listing is used. In addition the value of ΔG_f° for greenalite is taken from Eugster and Chou (1973) and that for minnesotaite is inferred from Mel'nik (1973). Activities assumed for the dissolved species that may be involved in forming the relevant solid phases are Fe^{2+}, $10^{-4}M$, $H_4SiO_4^0$, $10^{-2.7}M$ (saturated with respect to amorphous silica) and SO_4^{2-}, $10^{-2.76}M$. The small, but significant, amounts of pyrite observed throughout the Dales Gorge Member at Paraburdoo (Ewers and Morris, 1981) suggest an oxygen fugacity under which sulphur would have occurred as sulphate. The solution is assumed to be at equilibrium with an atmosphere in which the fugacity of CO_2, f_{CO_2}, is $10^{-3.5}$ atm.

In Fig. 13-1 the solid phases $Fe(OH)_3$ and $FeCO_3$ are assumed to control

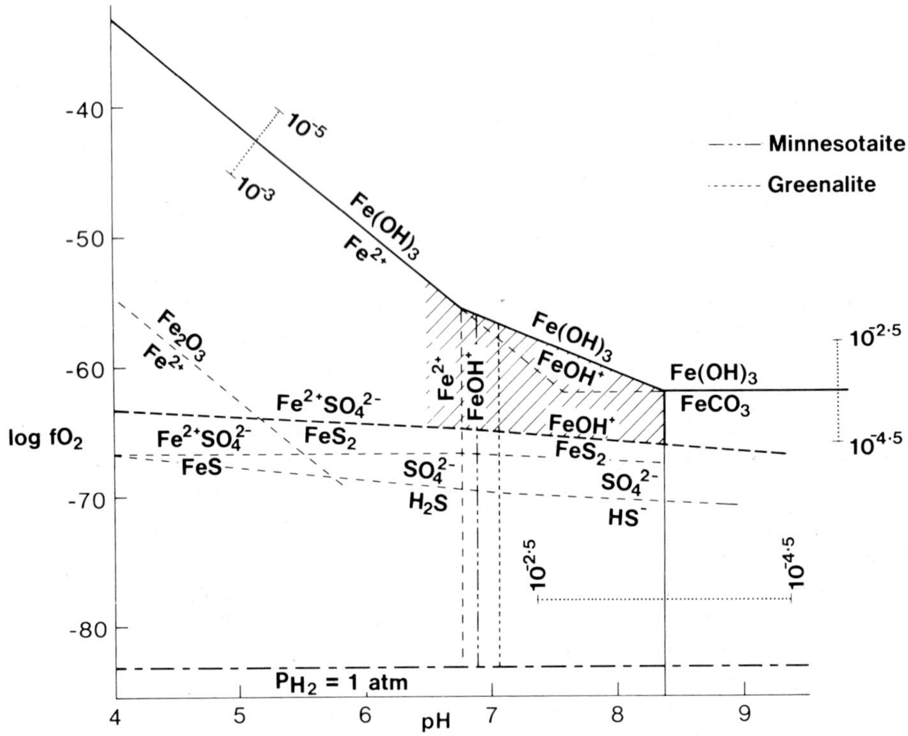

Fig. 13-1. pH, f_{O_2} diagram defining the limits for Fe^{2+} to exist in solution at an activity $\geqslant 10^{-4} M$ at 25°C. Based on data by Wagman et al. (1969), and assuming $a_{SO_4{}^{2-}} = 10^{-2.76} M$, $f_{CO_2} = 10^{-3.5}$ atm. The bar across the boundary Fe^{2+}, $Fe(OH)_3$ indicates shifts when $a_{Fe^{2+}}$ is set at 10^{-3} and $10^{-5} M$. Bars on boundaries of $FeCO_3$ field show shifts when f_{CO_2} is set at $10^{-2.5}$ and $10^{-4.5}$ atm. Reprinted from Ewers (1980) by permission Australian Academy of Science.

the activity of Fe^{2+} or $FeOH^+$. The "kinetic" equilibria involving these phases are shown as solid lines. The dashed lines represent theoretical equilibria between components, which for kinetic reasons appear unlikely to control the activity of dissolved iron. The hatched area represents conditions of f_{O_2} and pH in which the activity of Fe^{2+} or $FeOH^+$ might be expected to be equal to, or greater than, $10^{-4} M$. The area will extend to lower pH and probably to lower f_{O_2}, beyond the indicated equilibrium with pyrite. It could be restricted further on the high pH side if an iron silicate were involved as is shown by the dotted lines for "equilibria" between Fe^{2+} and either minnesotaite or greenalite.

Apart from the uncertainties as to the thermodynamic properties of the metastable phases involved in the primary precipitation of banded iron-formation, there remains some doubt on the ionic equilibrium between Fe^{2+} and

$FeOH^+$. If the data listed by Langmuir (1969) are used, the complex ion, $FeOH^+$, would be eliminated from the diagram and the equilibrium between Fe^{2+} and $FeCO_3$ would shift to a pH of 7.55.

Imperfect as this thermodynamic treatment inevitably is, it serves to demonstrate several conclusions that one can draw as to the factors which caused the precipitation of the predominantly oxide-facies BIF-macrobands, the carbonate-silicate facies S-macrobands and the gradation between them. The precipitation will be dealt with in a later section.

Solution chemistry of silica

The solubility in water of amorphous silica, mainly as $Si(OH)_4{}^0$, is of the order of 120 mg l^{-1} at 25°C. It is affected very little by pH over the range normally encountered in natural waters, 4 to 8.5, or by salinity up to the composition of modern sea water (Krauskopf, 1956; Siever, 1962). In the modern ocean dissolved silica occurs in very low concentrations, usually $\leqslant 1$ mg l^{-1} at the surface and < 10 mg l^{-1} at depth (Spencer, 1975). The very low surface concentrations are attributed to the uptake of silica by organisms, the remains of which settle to the bottom where they may redissolve, may react diagenetically to be incorporated in silicates, or may remain as such in sediments (Mackenzie, 1975).

The silica concentration in ancient seas has been assumed by most authors (Mackenzie, 1975) to have been close to saturation with respect to amorphous silica, $10^{-2.7}M$. This estimate is based on assumptions that silica had been concentrated by evaporation in the sea, that the silica had been stored in the main water column with only limited opportunity to be consumed by the "reverse weathering reactions" proposed by Mackenzie and Garrels (1966), and that biological evolution at that stage had not produced organisms capable of extracting silica from sea water. Stanton (1972) questioned the last of these assumptions, quoting Krauskopf, and also Huber and James, but it has been widely accepted (e.g. Cloud, 1973; Drever 1974). The recent paper by Klemm (1979) has yet to be accepted as convincing evidence to the contrary. The second of the assumptions is perhaps more problematical. Any suspended "degraded silicates" would have been exposed to silica concentrations at least as great as are encountered in the interstitial water of modern ocean floor sediments, though, as they settled, they probably lacked the time for significant reaction to occur. Other difficulties have been discussed by Drever (1974) who reviewed the work of several authors and concluded that magadiite ($NaSi_7O_{13}(OH)_3 \cdot 3H_2O$) is more stable than amorphous silica in contact with modern sea water above pH 5.2, and that the same is true of sepiolite above pH 7.7. Apart from the consequence of these conclusions to the precipitation of silica in banded iron-formations, they call into question the silica concentration assumed in calculating the stability fields of minnesotaite and greenalite in the Figure. Lower concentrations of silica would increase the pH at which these minerals are at equilibrium with Fe^{2+} at activity $10^{-4}M$; with silica activity $2 \times 10^{-4}M$, one tenth of that assumed in Fig. 13-1, the equilibrium for both minnesotaite and greenalite would be close to pH 8.4.

As Stanton (1972) pointed out, one of the features which distinguish the Precambrian banded iron-formations from the younger ironstones is their high ratio of silica to iron. Thus, the supply of silica seems to require concentrations at least as high as those of iron, and perhaps considerably higher. In the Dales Gorge Member, in both BIF- and S-macrobands there are abundant chert mesobands, several centimetres thick, containing very low Fe/SiO_2 ratios. These features might be construed to support the notion that, at least intermittently, high concentrations of silica were available. Mel'nik (1973) summarizes data on the concentrations of silica in hot springs and close to submarine volcanoes. Though concentrations in excess of 200 mg l^{-1} are reported locally, there is no modern instance of sufficient volumes of such solutions being stabilized as colloidal sols capable of distributing silica evenly over the 10^5 km^2 depositional area of the Hamersleys. Quantitatively it is clear that, if a sufficient amount of iron could be supplied at a concentration of 20 mg l^{-1}, a saturated solution of amorphous silica, 120 mg l^{-1}, would be ample to supply the silica content of the banded iron-formations.

THE PRIMARY PRECIPITATION

Environment of deposition

Although this paper is concerned more with chemical factors than with the geological environment of deposition of banded iron-formations, quantitative aspects of the chemistry give some indication as to what the environment might have been. The quantities of iron and silica supplied each year to the Hamersley depositional area have been discussed above. If a concentration of iron of 20 mg l^{-1} is taken as the limiting factor, the minimum volume of water required each year, if all of its iron were deposited, would be 1.125×10^{15} l. A ring of volcanoes capable of replenishing such a volume with iron and silica each year has been discounted by Holland (1973) and his suggestion that the depositional area was replenished from the sea is attractive. It should, however, be pointed out that the volume required is quite large compared with the inflow and outflow of four present-day land-locked seas for which data are given in Table 13-I.

Efficient circulation would also be required to distribute introduced solutes over the whole depositional area. These conditions are best satisfied, among basins listed in Table 13-I, by the Red Sea where the negative water balance ensures vertical circulation although the residence time of water entering at the Straits of Bab el Mandeb is still of the order of 30 years. A shelf situation

TABLE 13-I

	Area 10^5 km^2	Inflow 10^{15} l yr^{-1}	Outflow 10^{15} l yr^{-1}	Reference
Mediterranean	26.0	23.1	21.9	Schink (1967)
Red Sea	4.5	9.1	8.2	Grasshoff (1975)
Baltic Sea	2.02	0.47	0.94	Grasshoff (1975)
Black Sea	4.6	0.190	0.397	Grasshoff (1975)

with unrestricted access to the sea would satisfy the requirements rather better.

Though the oceans can be regarded as a reservoir for iron and silica it is easily demonstrated that, during the 2 to 6 million years in which the Brockman Iron Formation was probably deposited, the reservoir would need to have been replenished. Assuming that the volume of the ocean was the same as the modern world ocean, its total iron content, assuming an average of 20 mg l^{-1}, would have been 25.6×10^{18} g Fe. This is only twice the total iron content of the Dales Gorge Member and about half of that of the Dales Gorge and Joffre Members combined. It would be unreasonable to suppose that a large fraction of the worldwide precipitation of iron was confined to the Hamersley area during the few million years in which the banded iron-formations were deposited. Taking into account the many penecontemporaneous banded iron-formations of the Lower Proterozoic, the notion that these represent a once-off depletion of an ocean reservoir of iron, as free oxygen accumulated in the atmosphere, cannot be sustained. The above argument would be affected, of course, but not invalidated if one could invoke a higher concentration of iron. The highest estimate suggested so far, 400 mg l^{-1} (Mel'nik, 1973), would give an iron content of the world ocean only ten times that of the Brockman Iron Formation.

Any significant concentration of iron in natural waters would demand that those waters were highly reducing. As may be seen from the Figure, at no pH > 4 can the activity of Fe^{2+} approach $10^{-4}M$ (ca 20 mg l^{-1} as a concentration) unless the fugacity of oxygen is less than about 10^{-33} atm. With the present-day oxygen level in the atmosphere, such anoxic conditions occur only where higher salinity near the bottom stabilizes a layer in which organic debris consumes available oxygen. Such conditions are typical of restricted bodies of water with a positive water balance, for example the deeper parts of the Baltic Sea and the Black Sea. In the modern oceans, in which the exchange of water between deeps and surface is estimated to occur over a period of the order of 10^3 years (Broecker, 1971), anoxic deeps are very rare. It would require a very different circulation pattern, with widespread layering as proposed by Degens and Stoffers (1976), for a substantial fraction of the ocean volume to be capable of retaining significant concentrations of Fe^{2+} in the presence of an oxidizing atmosphere.

Clearly the oxygen levels in oceans and other surface waters are linked to the oxygen content of the atmosphere. Cloud (1973) and others have connected the ubiquitous presence of Fe^{2+} in ground waters with the virtual absence of oxygen from the atmosphere. Cloud's argument rests on the view that up to the end of the period which he designated Proterophytic (Cloud, 1972), that is at about 2000 m.y. B.P., the capacity of chemical sinks for oxygen exceeded the available sources. The most serious challenge to this view has come from the proposal by Berkner and Marshall (1964, 1965, 1966) and Brinkman (1969) that substantial quantities of oxygen were produced by photochemical dissociation of water vapour by ultraviolet light. Estimates of the oxygen content of the prebiotic atmosphere by these authors ranged from $\simeq 10^{-3}$ present atmosphere level (PAL) to $\geqslant 0.25$ PAL. Margulis et al. (1976), Walker (1977) and Kasting et al. (1979) have criticized these estimates on the grounds that they failed to account for limitations on the escape of hydrogen from the earth's atmosphere, and therefore underestimated the recombination of oxygen and hydrogen produced by photodissociation. The calculations by Kasting et al. (1979) suggest that the amount of hydrogen escaping from the atmosphere is only of the same order as the amount of hydrogen released into it by volcanoes at present. They estimated that the pre-biotic earth had an atmosphere with a ground level O_2 mixing ratio of the order of 10^{-12} PAL, an estimate which they consider to have been a maximum as they assumed zero flux of oxygen to oxygen sinks in the crust. Kasting et al. concluded that the O_2 mixing ratio increased to a maximum of 10^{-5} PAL at 60 km, whereas the number density of oxygen molecules (number per volume) reached its maximum at about 35 km. A rough calculation based on their graphs of number density indicates that the total inventory of free oxygen in their pre-biotic atmosphere would have been about 0.24×10^{15} g oxygen, more than half of it in the shell between 20 and 40 km above the surface and less than 5% of it in the shell 0 to 20 km above the surface. The total amount of oxygen corresponds to about 2.5×10^{15} g Fe oxidized from FeO to Fe_3O_4 or just over 100 years' precipitation of iron in the Dales Gorge Member. That in the first 20 km shell would correspond to only 5 years precipitation. One cannot make a reasonable estimate as to how much oxygen would be taken up from such a pre-biotic atmosphere by chemical sinks such as the oxidation of dissolved iron. It is clear, however, that the oxygen levels close to the earth's surface would have been depleted below the estimated 10^{-12} PAL. The results of Kasting et al. (1979), suggest therefore that oxygen produced by photodissociation of water could not have contributed significantly to the precipitation of the banded iron-formations; nor could it have interfered significantly with the widespread distribution of dissolved Fe^{2+} in surface waters.

This argument is a reminder that the extensive properties of a system are just as important as the intensive properties in considering geological systems. While the latter may determine what reactions will occur and which phases will form, the extent of the reaction and the logistics of reactant supply are determined by the extensive properties. Drever (1974), in discussing the atmosphere as a source of oxygen to precipitate the iron content of the banded iron-formations, put it very well in his delightful understatement: "an atmospheric oxygen pressure somewhat higher than 10^{-68} atm would be necessary to provide a sufficiently rapid flux of oxygen into the upwelled water". At an oxygen pressure of 10^{-68} atm, one molecule of oxygen would be present, on average, in a volume of a sphere with a diameter more than ten times that of the solar system.

There will have been an atmospheric pressure of oxygen somewhere between that of a pre-biotic atmosphere of $< 10^{-12}$ PAL and the modern atmosphere at which one could conceive of a large proportion of the oceans being sufficiently reducing to carry significant amounts of Fe^{2+} in solution and at the same time for surface waters to be sufficiently oxygenated to oxidize Fe^{2+} to $Fe(OH)_3$. Almost certainly this situation would have resulted from photosynthetic activity presumably, at least initially, in the presence of Fe^{2+} in surface waters.

The chemistry of precipitation

By whatever mechanism a sediment has formed which consolidates into a rock containing magnetite or hematite, if that sediment has drawn its iron content from Fe^{2+}, then an oxidation has been involved. This is probably the strongest *chemical* argument against the suggestion by Kimberley (1974), Dimroth and Kimberley (1976) and Dimroth (1979) that banded iron-formation could have formed by the introduction of iron as an early diagenetic alteration of aragonite beds. Such a replacement would require ferrous iron to be introduced uniformly over a vast area with subsequent intermittent introductions of an oxidizing agent, presumably oxygen, to form the mesobands with varying ratios of hematite and magnetite but also with wide ranging lateral continuity. Even if oxidizing conditions were to alternate repeatedly with the reducing conditions under which iron was introduced, the penetration of bottom sediments by oxygen in the Precambrian would be as unlikely as it is with bottom sediments today. Kimberley's model is considered to have no relevance to banded iron-formations like the Dales Gorge Member and will not be discussed further.

In the Dales Gorge Member the BIF-macrobands are characterized by the presence of iron oxides, both magnetite and hematite. It cannot be seriously questioned that the amount of iron (III) that these represent originates from iron (III) in the primary sediment, and is not the product of a subsequent oxidation. One can assert, therefore, that, for the BIF macrobands at least, the primary precipitation involved an oxidation.

Three mechanisms for oxidation have been proposed:

(1) Atmospheric oxygen at relatively low partial pressures reacting with

Fe^{2+} upwelling from anoxic ocean depths (Holland, 1973; Drever, 1974). As discussed before, oxygen for this mechanism would have been provided by photosynthesis. The presence of oxygen in sufficient quantity would have inhibited the distribution of iron, at least in surface waters, and, with ocean circulation comparable to modern oceans, such a mechanism could probably have operated only over a relatively short time span.

(2) The oxidation of iron as a by-product of photosynthesis (Lepp and Goldich, 1964; Cloud, 1965, 1973). That soluble iron was still available to form banded iron-formations up to about 1.4×10^9 years after the earliest recorded stromatolites were formed (Walter et al., 1980) suggests that efficient chemical sinks for the O_2 released from CO_2 in forming carbohydrates, $(CH_2O)_n$, must have been available throughout this period. The main oxygen sinks were evidently reduced forms of sulphur going to sulphate, most of which accumulated in the ocean, and iron (II) oxidizing to iron (III).

Photosynthesis is favoured by high partial pressures of CO_2 and, as suggested by Ewers and Morris (1981), one question that should be explored is whether the main period of precipitation of the banded iron-formation coincided with a high or low concentration of CO_2 in the atmosphere. There is some evidence (Garrels et al., 1973) that the rate of accumulation of carbon in sediments reached a maximum at that time, and it could be argued that such a maximum corresponded to a minimum in atmospheric CO_2. This circumstance, though unfavourable to photosynthesis, would, for a given pH, allow greater concentrations of Fe^{2+} in surface waters.

(3) The photochemical oxidation of Fe^{2+} has been suggested by Cairns-Smith (1978) and merits careful consideration as a precipitation mechanism. Fe^{2+} in solution has been shown to absorb ultraviolet light in the wavelength range 200—300 nm (Jortner and Stein, 1962). A complex series of photochemically induced electron transfer processes take place, but the resultant overall reaction relating to our present subject can be written as:

$$Fe^{2+}_{aq} + 3H_2O \xrightarrow{h\nu} Fe(OH)_3\downarrow + 2H^+ + \tfrac{1}{2}H_2\uparrow \qquad (2)$$

It will be noted that water is reduced to gaseous hydrogen, just as CO_2 is reduced to $(CH_2O)_n$ in photosynthesis. Experimental evidence for the photochemical oxidation of Fe^{2+} has been restricted to pH values sufficiently low to prevent the precipitation of $Fe(OH)_3$. Jortner and Stein (1962) found that the quantum yield for the reaction decreased initially with increasing pH, but approached a constant at about pH 3. Cairns-Smith (1978) calculated that, in terms of light flux and quantum yield, the mechanism is capable of precipitating each year the quantities of iron measured by Trendall and Blockley (1970). The reaction needs to be confirmed in the pH range 4 to 8 but it is significant that Airey and Dainton (1966) reported the oxidation of ferrocyanide ion $Fe(CN)_6{}^{4-}$ at pH 11.8 in the presence of a suitable electron scavenger.

The photochemical oxidation of iron has several interesting features. It clearly requires an anoxic atmosphere; otherwise ultraviolet light of the appropriate wavelength range would be screened out by oxygen and ozone. The wavelength required is not absorbed by water, being longer than that required for photodissociation of water. The absorption of UV by Fe^{2+} may have special significance in biological evolution, as significant concentrations of iron would act as a protective screen for micro-organisms. Finally, the evolution of hydrogen as a by-product offers a mechanism for maintaining the anoxic conditions on which all of these processes depend.

It does not seem possible to choose between the second and third mechanisms for oxidation of Fe^{2+}; they could have been complementary.

The precipitation of iron in the S-macrobands may have involved oxidation also, but, if it did, the hydroxy-oxides produced must have been reduced by an excess of organic debris incorporated in the primary sediment. It is significant that free carbon occurs only in banded iron-formations that are poor in magnetite and, particularly, hematite. Alternatively, and more probably, as has been pointed out by Ewers (1980) and Ewers and Morris (1981), the S-macrobands may represent a different style of precipitation characterized by higher pH and/or higher concentrations of CO_2. Ewers has suggested that this change in style was triggered by the effect of volcanic dust on the chemical environment, consistent with the evidence produced by LaBerge (1966) and Trendall and Blockley (1970) that the S-macrobands contain evidence of volcanic activity.

The precipitation of iron, whether by oxidation of Fe^{2+} to Fe^{3+} and subsequent hydrolysis to $Fe(OH)_3$, or by forming carbonates or silicates, constitutes a release of acid to the system, as is shown in the following reactions:

$$Fe^{2+} + 0.25\ O_2 + 2.5\ H_2O \rightarrow Fe(OH)_3 + 2H^+ \tag{3}$$

$$Fe^{2+} + HCO_3^- \rightarrow FeCO_3 + H^+ \tag{4}$$

$$Fe^{2+} + \tfrac{2}{3}H_4SiO_4^0 + \tfrac{1}{3}H_2O \rightarrow \tfrac{1}{3}\ Fe_3Si_2O_5(OH)_4 + 2H^+ \tag{5}$$

If any of these reactions are to continue, the acid released will need to be offset by alkali presumably introduced in water buffered by reaction with silicates. Ewers and Morris (1981) have shown that the amount of volcanic ash corresponding to the quantities of Al_2O_3 in the S-macrobands would not have been adequate to buffer the precipitation reactions represented by the S-macrobands. They concluded that reaction of volcanic ash falling outside the depositional zone, or reaction of other silicates with water entering the area, was probably responsible for maintaining the requisite pH during the deposition of S-macrobands. In the sense that oxidation of Fe^{2+} will produce $Fe(OH)_3$ over a wide pH range, the acid released during this process could be seen as maintaining conditions conducive to the deposition of oxide-facies banded iron-formation.

Evaporation has not so far been considered as a mechanism for precipitation of the iron-bearing phases either in the BIF-macrobands or the S-macrobands. In the former case evaporation of a solution, the pH and f_{O_2} of which are described by the line separating the Fe^{2+} and $Fe(OH)_3$ areas in Fig. 13-1, would initially precipitate $Fe(OH)_3$. However, oxidation is necessarily involved according to reaction (3) so that any precipitation will be accompanied by the lowering of both f_{O_2} and pH. Precipitation by this mechanism could not be significant without an addition of oxygen from an external source. For the carbonate and silicate components of banded iron-formation, evaporation could play a more significant role. If water alone were lost during evaporation a solution saturated with respect to siderite or, say, greenalite could precipitate these phases although the reverse of reactions (4) and (5) would be favoured by the resultant lowering of pH. If CO_2 were also lost to the atmosphere during evaporation the precipitation would be described by:

$$Fe^{2+} + 2HCO_2^- \rightarrow FeCO_3 + CO_2 + H_2O \qquad (6)$$

so that the inhibitory role of released acid would be removed.

As noted above, it is the generally held view that organisms capable of building silica into their external skeletons did not exist in the Lower Proterozoic. Accepting this view and that, as a consequence, silica was available in concentrations controlled by the solubility of either amorphous silica or a hydrous silicate such as magadiite or greenalite, then evaporation of such water could have caused the precipitation of silica. There is, however, a problem of quantity. Trendall and Blockley (1970) estimated that the *average* quantity of SiO_2 in each microband, that is for each year, was of the order of 40mg cm^{-2}. At a concentration of 120 mg l^{-1}, this amount corresponds to a surplus of evaporation over precipitation of 3.3 m per year which is close to the upper limit of present day nett evaporation. If the concentration of silica was controlled at a lower level by equilibration with magadiite or greenalite, the amount of evaporation would need to increase correspondingly. This argument is based on the assumption that all evaporation corresponding to precipitation occurred in the depositional area only, probably the best assumption that can be made.

The problem is exacerbated if one takes into account the range of variation in the amounts of silica in microbands. Trendall and Blockley (1970) give a breakdown of the thicknesses of microbands in 300 chert mesobands in the Dales Gorge Member. The maximum thickness of microbands was 1.75 mm, each containing about 400 mg SiO_2 cm^{-2}, and in half of the total sample the microband thickness exceeded 0.6 mm or at least 120 mg SiO_2 cm^{-2}. The amounts of evaporation corresponding to these amounts of silica are 33.3 m and 10 m respectively.

Trendall and Blockley (1970) considered that there had been a massive redistribution of silica during diagenesis, thickening some mesobands (and their component microbands) at the expense of others, including the iron-rich "chert-matrix" and magnetite mesobands. On this hypothesis the average annual deposition of silica would be more relevant to a calculation of evaporation than the present range of silica contents of microbands. And as noted above the average could correspond to a high, though credible, evaporation rate under a temperature regime not too different from that of today. Higher temperatures would correspond to higher silica solubility (about 200 mg l^{-1} at 50°C) and higher evaporation rates.

The alternative viewpoint, that the present composition of mesobands was largely determined by the composition of the primary hydrous deposits (except for H_2O and probably CO_2), as argued by Ewers and Morris (1981), seems to require that other processes, as well as evaporation, were involved in the precipitation of silica, at least for the more silica-rich mesobands. It seems unlikely that the temperature, as it affected true solubility of amorphous silica or rate of evaporation, could have been high enough to explain an annual precipitation of, say 400 mg SiO_2 cm^{-2}.

Freezing of a surface layer as a means of causing precipitation of silica from a saturated solution is also subject to quantitative limitations. Not only is the level of silica solubility lower in water subject to periodic freezing, but the depth of ice formed and remelted on an annual cycle is likely to be insufficient.

An interesting mechanism, suggested by the work of Harder (1965) and Harder and Flehmig (1970), involves the coprecipitation of silica with $Fe(OH)_3$. They demonstrated that if an acid solution of Fe^{3+} containing silica were neutralized with alkali to a pH $\simeq$ 7 the gelatinous hydrolysis product contained up to 67% SiO_2 on a dry weight basis. Oxidation of a solution of Fe^{2+} with H_2O_2 at pH 7 by Ewers and Wildman (unpublished) produced similar results, though the maximum SiO_2 content of the products that they obtained was 40% dry weight. This mechanism is an attractive alternative in that a single process links the precipitation of the two major constituents of banded iron-formation; it is initiated by the oxidation of Fe^{2+} and does not call for high concentrations of silica to account for the amounts contained in most mesoband types. Solution conditions in which coprecipitation could produce the higher ratios of silica to iron in coarsely microbanded chert mesobands have not, however, been discovered.

Thus none of the mechanisms mentioned above appears free from quantitative objection, if the assumed concentration of silica is limited to the true solubility of amorphous silica as monomeric $Si(OH)_4$. For the more silica-rich mesobands it is probably necessary to invoke sols of colloidal silica. Mel'nik (1973) has summarized the literature on sols containing silica, with and without ferric hydroxide, and has also reported experiments on the stability of these materials to variations in pH and electrolyte concentration. The literature is confused and the experiments are difficult to reproduce so that only tentative conclusions can be drawn. In general concentrated sols (> 1500 mg l^{-1}) are most likely to coagulate in the pH range 4—8. The pH range is narrower for less concentrated sols, pH 5—6 for 750 mg SiO_2 l^{-1} but may become wider in the presence of polyvalent cations such as Mg^{2+}. Mel'nik was able to precipitate from a mixed SiO_2-$Fe(OH)_3$ sol a banded "jasperite-like" deposit with concentrations of $MgSO_4 > 1000$ mg l^{-1}.

His summary supports the notion that subtle changes in the history and composition of a solution could bring about dramatic changes in the form and composition of precipitates. Periodic introduction of concentrated silica sols into the depositional area, and changes to their pH or electrolyte concentration as a result of mixing, could initiate the sudden compositional changes so characteristic of mesoband alternation. One process may have had particular significance. Silica sols that were relatively stable in the presence of low concentrations of Fe^{2+} would be destabilized by the oxidation of Fe^{2+} to Fe^{3+} and its subsequent hydrolysis. Very small additions of Al^{3+} or Fe^{3+} to polymeric silica sols are known to coagulate them (Okamoto et al., 1957) and cherts with low iron content could have derived from such a process.

Iron, expressed as its oxides, and silica together constitute about 86% of the Dales Gorge Member (Ewers and Morris, 1981). Most of the remainder is made up of CO_2, MgO and CaO, assumed to have precipitated in carbonates and silicates or their precursors, mainly in response to increased pH. Of the other minor constituents, P_2O_5, occurring as apatite, may have precipitated inorganically although it is probably significant that it is often concentrated in mesobands that are iron-rich or, in S-macrobands, in mesobands that contain appreciable free carbon (Ewers and Morris, unpubl. data). In either case one could speculate on a biological association of phosphorus. Al_2O_3 in S-macrobands was almost certainly introduced in a solid, probably volcanic ash. Significant free carbon occurs only in S-macrobands and probably represents remains of organisms brought down in a relatively rapid accumulation of volcanic ash and in the absence of sufficient iron (III) to oxidize them. It is probable that carbonaceous material was deposited originally in BIF-macrobands also, but has disappeared as a result of oxidation by $Fe(OH)_3$ and is now represented chemically by iron (II) in magnetite and other compounds, with at least some of the carbon remaining in carbonates. The carbon isotope data of Becker and Clayton (1972) and of Perry et al. (1973) on the carbonates of banded iron-formations are consistent with such a mechanism. As noted above, the presence of pyrite throughout the column suggests that sulphate was present but it is unlikely that sulphides were precipitated directly. The sulphur content of the S-macrobands, 0.3% S occurring in pyrite, is consistent with the diagenetic reduction of sulphate in the sediment (Berner, 1970).

The localization of precipitation

The processes that have been proposed for the precipitation of iron by oxidation of Fe^{2+} and hydrolysis of Fe^{3+} to $Fe(OH)_3$ would have been capable of operating wherever surface waters contained sufficient Fe^{2+}, were exposed to sunlight and, for photosynthesis, contained a suitable flora of micro-organisms. Quantitative considerations suggest, however, that the precipitation must have been quite localized. The depositional area of the Brockman Iron Formation, at 10^5 km^2, is less than 0.03% of the area of the modern ocean and yet it is estimated that the amount of iron precipitated in it each year was 22.5×10^{12} g, which is more than 10% of the annual amount of iron at present entering the oceans in the suspended load of streams (Garrels and Mackenzie, 1972). It has been argued above that the oceans could not have acted as a depleting reservoir from which *all* of iron in the banded iron-formations derived, but that they must have been actively replenished during deposition.

An obvious answer to the problem of the localization of deposition is that the sources of replenishment of iron were close to the area of deposition, and yet remote enough to allow the available iron to be distributed sufficiently for the remarkable uniform deposition to occur. Mel'nik (1973) reached this conclusion with respect to the supply of both silica and iron from volcanic sources spatially related but not very close to the area of deposition.

Drever (1974) considered that the localization of deposition was related to areas of upwelling of deep anoxic water which was charged with Fe^{2+} as a result of reduction of the ferric oxide in terrigenous sediments. His model assumed an oxygenated atmosphere but it is probable that upwelling was important, even if it is assumed that the atmosphere contained insignificant amounts of oxygen. Oxidation of Fe^{2+} by photochemical or photosynthetic mechanisms might be expected to have been widespread. If precipitated $Fe(OH)_3$ were mixed in the sediments with organic remains in an area of slow sedimentation, it could be reduced again to Fe^{2+} to be recycled. Grasshoff (1975) has described such a cycle operating at present above and below the halocline in the Baltic, and has shown that phosphorus released from the organic debris is recycled with iron. Phosphorus, as an essential and possibly a limiting nutrient for the photosynthetic mechanism, could be enriched in areas of upwelling, causing much more rapid precipitation of iron in these zones. Thus the combined iron-phosphorus cycle could reinforce the tendency for iron to accumulate in an upwelling situation not too distant from the primary sources of iron and silica.

THE PRIMARY SEDIMENT AND ITS DIAGENESIS

Just as it has been concluded above that the controls on concentration of iron and silica were reactions of soluble species to form ill-defined amorphous material, one must conclude that the primary precipitates consisted of these same materials. The extreme compositional variability of the present rock, at the microband and mesoband level, must have had its origins in some variation within the primary sediments. There is considerable uncertainty, however, as to how much primary differences have been accentuated by subsequent diagenetic and metamorphic processes.

Strakhov, as quoted by Mel'nik (1973), has limited the term diagenesis to those processes which derive their energy essentially from the thermodynamically spontaneous transitions of metastable materials or assemblages towards stable equilibrium phases. Such changes may be accelerated by increased temperature or pressure, but the definition excludes from diagenesis, and categorizes as metamorphism, changes that result in a different set of equilibrium phases that need a higher temperature or pressure for stability.

In using this concept briefly to discuss diagenesis in the banded iron-formations, account should be taken of individual metastable phases, assemblages of them in intimate association, and also assemblages in adjacent mesobands that are close enough to be considered in contact so long as the mobility of soluble components survives the gradual dehydration of the primary precipitate. Examples of metastable phases are: hydrated $Fe(OH)_3$ in relation to goethite and hematite, $SiO_2 \cdot nH_2O$ in relation to opal and quartz, and amorphous carbonates in relation to siderite, ankerite, etc. The primary precipitate of hydrated $Fe(OH)_3$ forms metastable assemblages with several components, carbohydrates $(CH_2O)_n$, Fe^{2+} in solution or coprecipitated with the hydroxide, and compounds of iron (II) such as carbonates or silicates. Magnetite would be a common product in the change of these assemblages towards equilibrium. Furthermore, other individual metastable phases would have varied considerably in composition; $Fe(OH)_3$ precipitates would inevitably have contained some hydrous SiO_2, present in varying concentrations. These amorphous, mixed phases would have been subject to unmixing as such crystalline phases as goethite, hematite, magnetite, quartz or iron silicates began to form.

Trendall and Blockley (1970) consider that microbanding (at the mm level) resulted from annual cycles which, in their simplest form, consisted of an alternation of an iron-rich with a silica-rich layer. They pointed out that the cycles could be more complex and recently Ewers and Morris (1981) have described microbands with up to about 20 subdivisions of the order of tens of μm thick. It seems likely that the original precipitate contained within it subtle variations that have been accentuated diagenetically by unmixing as crystalline phases have formed.

The next scale of banding, mesobanding, with a thickness range from a few mm to a few tens of mm, is characterized by an abrupt change in composition from one mesoband to the next (Trendall and Blockley, 1970). An important type, the chert-mesoband, is often microbanded and within a given mesoband the microbands are of the same style of cyclic variation and are of uniform thickness. This thickness varies considerably from one mesoband to another and, generally, as the microband thickness decreases the ratio iron/silica increases. Starting with the premise that the original precipitate was more or less uniform and undifferentiated, Trendall and Blockley have taken this, and other evidence such as the lateral continuity of microbanding through pods in chert mesobands to mean that there has been a major redistribution of silica normal to the banding during diagenesis. They regard the iron-rich "chert-matrix" and magnetite mesobands as more extreme products of the same segregation process.

Ewers and Morris (1981) have argued that no satisfactory chemical mechanism has been offered for this segregation in terms of the nature of the solution in which silica migrated and of the activity gradients that could have caused the migration. In particular, no chemical explanation has been suggested of a means by which the uniformity of microband thickness in both the thickened and the thinned mesobands was maintained. They propose instead that the compositional variation between mesobands is in the main a product of the primary deposition caused by variations in climatic conditions, in supply of major constituents and of factors, such as phosphorus, affecting the microflora. Thickness variations and podding in particular mesobands are ascribed to lateral flow of silica during distortion under load while the mesobands remained in a plastic state.

There is evidence, however, of some migration of components under activity gradients set up in the processes of diagenetic equilibration, and even where evidence is lacking, it may be fairly deduced that some migration occurred. Several authors (Perry et al., 1973; Dimroth and Chauvel, 1973; Han, 1978; Ewers and Morris, 1981) have discussed the role of such reactions as:

$$4Fe(OH)_3 + C(as(CH_2O)_n) + 8H^+ \rightarrow 4Fe^{2+} + CO_2 + 10H_2O \qquad (7)$$

Fe^{2+} produced in this way may react in the immediate vicinity to form $FeCO_3$ or Fe_3O_4 or it may be available for migration to adjacent mesobands, where $Fe(OH)_3$, in the absence of reactive carbon, may react with it as follows:

$$2Fe(OH)_3 + Fe^{2+} \rightarrow Fe_3O_4 + 2H^+ + 2H_2O \qquad (8)$$

It will be noted that reactions (7) and (8) would produce gradients in Fe^{2+} activity and pH, and diffusion of Fe^{2+} in one direction and H^+ in the other would persist until the disequilibrium between $Fe(OH)_3$ and $(CH_2O)_n$ is removed by the exhaustion of one of these components. Evidence for this mechanism is seen in the overgrowths of magnetite on hematite reported by Han (1978) and Morris (1980), reinforced by the observation (Ewers and Morris, 1981) that the euhedral magnetites are not surrounded by a zone of iron depletion. It is not possible to quantify the extent of this intermesoband or intramesoband migration, but it is unlikely to have been a major factor overall in mesoband differentiation. Activity gradients would also be set up when quartz nucleates in an amorphous proto-chert. Differences in grain size and the phenomenon of "ambient pyrite" (Tyler and Barghoorn, 1963) are probably evidence of this, but there is no indication that such a mechanism has materially affected mesoband composition.

CONCLUSION

In common with many authors who have written on the origins of the banded iron-formations, I would have wished to have transferred more of this topic from areas of opinion and controversy into areas of reasonable certainty. If, however, one takes a realistic view of the uncertainties that still remain in the basic data of geology, of solution chemistry and thermodynamics, of biological evolution and the state of the seas and atmosphere in the early Proterozoic, it is inevitable that subjective opinions still survive. This account is coloured by my opinions, most of them derived from others, but I have attempted to stress the uncertainties mainly in the intensive properties of the system, and the undeniable certainties of some of the extensive properties, in a manner that will stimulate further constructive endeavour.

ACKNOWLEDGEMENTS

I thank many CSIRO colleagues for their helpful discussions leading up to and during the preparation of this chapter. Special thanks are due to R.C. Morris and to the iron ore companies, Hamersley Iron Pty. Ltd., Mt. Newman Mining Co. Pty. Ltd. and Broken Hill Proprietary Co. Ltd., whose sponsorship, through the Australian Mineral Industries Research Association, allowed us to collaborate.

REFERENCES

Airey, P.L. and Dainton, F.S., 1966. The photochemistry of aqueous solutions of Fe (II). 1. Photoelectron detachment from ferrous and ferrocyanide ions. Proc. R. Soc. London, Ser. A, 291: 340—352.

Becker, R.H. and Clayton, R.N., 1972. Carbon isotopic evidence for the origin of a banded iron-formation in Western Australia. Geochim. Cosmochim. Acta, 36: 577—595.

Berkner, L.V. and Marshall, L.C., 1964. The history of oxygenic concentration in the earth's atmosphere. Discuss. Faraday Soc., 37: 122—141.

Berkner, L.V. and Marshall, L.C., 1965. On the origin and rise of oxygen concentration in the earth's atmosphere. J. Atmos. Sci., 22: 225—261.

Berkner, L.V. and Marshall, L.C., 1966. Limitation on oxygen concentration in a primitive planetary atmosphere. J. Atmos. Sci., 23: 133—143.

Berner, R.A., 1970. Sedimentary pyrite formation. Am. J. Sci., 268: 1—23.

Brinkman, R.T., 1969. Dissociation of water vapor and evolution of oxygen in the terrestrial atmosphere. J. Geophys. Res., 74: 5355—5368.

Broecker, W.S., 1971. A kinetic model for the chemical composition of sea water. Quat. Res., 1: 188—207.

Cairns-Smith, A.G., 1978. Precambrian solution photochemistry, inverse segregation, and banded iron-formations. Nature, 276: 807—808.

Cloud, P.E., 1965. Significance of the Gunflint (Precambrian) microflora. Science, 148: 27—35.

Cloud, P.E., 1972. A working model of the primitive earth. Am. J. Sci., 272: 537—548.

Cloud, P.E., 1973. Paleoecological significances of the banded iron-formation. Econ. Geol., 68: 1135—1143.

Degens, E.T. and Stoffers, P., 1976. Stratified waters as a key to the past. Nature, 263: 22—27.

Dimroth, E., 1976. Aspects of the sedimentary petrology of cherty iron-formation. In: K.H. Wolf (Editor), Handbook of Strata-bound and Stratiform Ore Deposits. Elsevier, Amsterdam, 7: 203—254.

Dimroth, E., 1979. Facies models 16. Diagenetic facies of iron-formation. Geosci. Can., 4 (2): 83—88.

Dimroth, E. and Chauvel, J.-J., 1973. Petrography of the Sokoman Iron Formation in part of the central Labrador trough, Quebec, Canada. Geol. Soc. Am. Bull., 84: 111—134.

Dimroth, E. and Kimberley, M.M., 1976. Precambrian atmospheric oxygen: evidence in the sedimentary distributions of carbon, sulfur, uranium, and iron. Can. J. Earth Sci., 13: 1161—1185.

Drever, J.I., 1974. Geochemical model for the origin of Precambrian banded iron formations. Geol. Soc. Am. Bull., 85: 1099—1106.

Eichler, J., 1976. Origin of Precambrian banded iron-formations. In: K.H. Wolf (Editor), Handbook of Strata-bound and Stratiform Ore Deposits. Elsevier, Amsterdam, 7: 157—201.

Eugster, H.P. and Chou, I-Ming, 1973. The depositional environments of Precambrian banded iron-formations. Econ. Geol., 68: 1144—1168.

Ewers, W.E., 1980. Chemical conditions for the precipitation of banded iron-formations. In: P.A. Trudinger, M.R. Walter and B.J. Ralph (Editors), Biochemistry of Ancient and Modern Environments. Australian Academy of Science, Canberra, pp. 83—92.

Ewers, W.E. and Morris, R.C., 1981. Studies of the Dales Gorge Member of the Brockman Iron Formation, Western Australia. Econ. Geol., 76: 1929—1953.

Frost, B.R., 1978. Some aspects of the sedimentary and diagenetic environment of Proterozoic banded iron-formation — A discussion. Econ. Geol., 73: 1369—1371.

Garrels, R.M. and Mackenzie, F.T., 1972. A quantitative model for the sedimentary rock cycle. Mar. Chem., 1: 27—41.

Garrels, R.M., Perry, E.A. Jr. and Mackenzie, F.T., 1973. Genesis of Precambrian iron-formations and the development of atmospheric oxygen. Econ. Geol., 68: 1173—1179.

Grasshoff, K., 1975. The hydrochemistry of landlocked basins and fjords. In: J.P. Riley and G. Skirrow (Editors), Chemical Oceanography. Academic Press, London, 2: 455—597.

Gross, G.A., 1980. A classification of iron formations based on depositional environments. Can. Mineral., 18: 215—222.

Han, Tsu-Ming, 1978. Microstructures of magnetite as guides to its origin in some Precambrian iron formations. Fortschr. Mineral., 56: 105—142.

Harder, H., 1965. Experimente zur "Ausfällung" der Kieselsäure. Geochim. Cosmochim. Acta, 29: 429—442.

Harder, H. and Flehmig, W., 1970. Quartzsynthese bei tiefen Temperaturen. Geochim. Cosmochim. Acta, 34: 295—305.

Holland, H.D., 1973. The oceans: A possible source of iron in iron-formations. Econ. Geol., 68: 1169—1172.

James, H.L., 1954. Sedimentary facies of iron-formation. Econ. Geol., 49: 235—293.

Jortner, J. and Stein, G., 1962. The photochemical evolution of hydrogen from aqueous solutions of ferrous ions. J. Phys. Chem., 66: 1258—1271.

Kasting, J.F., Liu, S.C. and Donahue, T.M., 1979. Oxygen levels in the prebiological atmosphere. J. Geophys. Res., 84: 3097—3107.

Kimberley, M.M., 1974. Origin of iron ore by diagenetic replacement of calcareous oolite Nature, 250: 319—320.

Klein, C. and Bricker, O.P., 1977. Some aspects of the sedimentary and diagenetic environment of Proterozoic banded iron-formation. Econ. Geol., 72: 1457—1470.

Klemm, D.D., 1979. A biogenetic model of the formation of the banded iron-formation in the Transvaal Supergroup/South Africa. Miner. Deposita, 14: 381—385.

Krauskopf, K.B., 1956. Dissolution and precipitation of silica at low temperatures. Geochim. Cosmochim. Acta, 10: 1—26.

LaBerge, G.L., 1966. Altered pyroclastic rocks in iron-formation in the Hamersley Range, Western Australia. Econ. Geol., 61: 147—161.

Langmuir, D., 1969. The Gibbs free energies of substances in the system $Fe\text{-}O_2\text{-}H_2O\text{-}CO_2$ at 25°C. U.S. Geol. Surv., Prof. Paper 650-B: pp. B180—B184.

Lepp, H. and Goldich, S.S., 1964. Origin of Precambrian iron formations. Econ. Geol., 59: 1025—1060.

Mackenzie, F.T., 1975. Sedimentary cycling and the evolution of sea water. In: J.P. Riley and G. Skirrow (Editors), Chemical Oceanography. Academic Press, London, 1: 309—364.

Mackenzie, F.T. and Garrels, R.M., 1966. Silica-bicarbonate balance in the ocean and early diagenesis. J. Sediment Petrol., 36: 1075—1084.

Margulis, L., Walker, J.G.G. and Rambler, M., 1976. Reassessment of roles of oxygen and ultraviolet light in Precambrian evolution. Nature, 264: 620—624.

Mel'nik, IU.P., 1973. Physiochemical conditions of formation of the Precambrian ferruginous quartzites. Kiev, Akad. Nauk Ukrainskoe SSR Institut geokhimii i fiziki mineralov (Izdatel; stvo "Naukova Dumka") 272 pp.

Morris, R.C., 1980. A textural and mineralogical study of the relationship of iron ore to banded iron-formation in the Hamersley Iron Province of Western Australia. Econ. Geol., 75: 184—209.

Okamoto, G., Okura, T. and Katsuma, G., 1957. Properties of silica in water. Geochim. Cosmochim. Acta, 12: 123—132.

Perry, E.C. Jr., Tan, F.C. and Morey, G.B., 1973. Geology and stable isotope geochemistry of the Biwabik Iron Formation, Northern Minnesota. Econ. Geol., 68: 1110—1125.

Robie, R.A., Hemingway, B.S. and Fisher, J.R., 1979. Thermodynamic properties of minerals and related substances at 298.15°K and 1 bar (10^5 pascals) pressure and at higher temperatures. U.S. Geol. Surv., Bull. 1452, 456 pp.

Schink, D.R., 1967. Budget for dissolved silica in the Mediterranean sea. Geochim. Cosmochim. Acta, 31: 987—999.

Siever, R., 1962. Silica solubility 0°—200°C, and the diagenesis of siliceous sediments. J. Geol., 70: 127—150.

Spencer, C.P., 1975. The micronutrient elements. In: J.P. Riley and G. Skirrow (Editors), Chemical Oceanography. Academic Press, London, 2: 245—300.

Stanton, R.L., 1972. Ore Petrology. McGraw-Hill, New York, N.Y., 713 pp.

Trendall, A.F., 1983. The Hamersley Basin. In: A.F. Trendall and R.C. Morris (Editors), Iron-Formation; Facts and Problems. Elsevier, Amsterdam, pp. 69—129.

Trendall, A.F. and Blockley, J.G., 1970. The iron formations of the Precambrian Hamersley Group, Western Australia. West. Austr., Geol. Surv., Bull. 119, 366 pp.

Trendall, A.F. and Pepper, R.S., 1977. Chemical composition of the Brockman Iron Formation. West. Aust., Geol. Surv., Record 1976/25, 25 pp.

Tyler, S.A. and Barghoorn, E.S., 1963. Ambient pyrite grains in Precambrian cherts. Am. J. Sci., 261: 424—432.

Wagman, D.D., Evans, W.H., Parker, V.B., Halow, I., Bailey, S.M. and Schumm, R.H., 1969. Selected values of chemical thermodynamic properties. U.S. National Bureau of Standards, Technical Note 270-4, 152 pp.

Walker, J.C.G., 1977. Evolution of the Atmosphere. MacMillan, New York, N.Y., 318 pp.

Walter, M.R., Buick, R. and Dunlop, J.S.R., 1980. Stromatolites 3,400—3,500 Myr old from the North Pole area, Western Australia. Nature, 284: 443—445.

Zelenov, K.K., 1958. Leaching and transportation of dissolved iron by thermal waters of Ebeko volcano. Akad. Nauk SSSR Dokl., 120: 1089—1092.

Chapter 14

SUPERGENE ALTERATION OF BANDED IRON-FORMATION

R.C. MORRIS

INTRODUCTION

Most definitions of the term supergene either state or imply some relationship with ore-forming processes. In this chapter supergene alteration is used in the wider sense of the chemical and physical processes that affect rocks as a result of meteoric influences, processes which, while most obvious at or near the earth's surface, may produce changes in rocks to depths measured in kilometres. Thus the title embraces weathering — which usually implies decay — as well as the allied processes that result in iron-enrichment, but excludes those more properly related to hydrothermal or other hypogene conditions.

The chapter is not intended as a definitive statement of the effects of weathering on banded iron-formation (BIF). Its main purpose is to draw attention to a generally neglected aspect of the petrography and geochemistry of these rocks, particularly since unsuspected alteration can give rise to misleading chemical data. It is thus meant to complement the discussion by Davy (1983, this volume) on the difficulties inherent in sampling BIF to produce representative information.

The qualitative effects of weathering in banded iron-formation, and the progressive stages that lead to iron enrichment (Morris, 1980) are not hard to follow, but to establish even semi-quantitative data is difficult. In a reasonably homogenous rock such as a gabbro or a granite, weathering profiles can be examined by standard techniques described, for example, by Krauskopf (1967). From such data it is possible to arrive at conclusions regarding the progression of weathering which can be accepted as reasonable approximations, even though absolute certainty of the original composition of each altered specimen cannot be guaranteed. But in any layered sequence, inhomogeneity of the original rock makes interpretation difficult, and of the layered rocks BIF is surely one of the most inhomogeneous.

In general, the range of minerals found in BIF is not large. Even with the variation imposed by metamorphism, it is possible to limit discussion to a handful of mineral groups — silica, iron oxides, silicates, carbonates and apatite. (Sulphides, generally sparse in BIF, will not be considered here). Of these, the alteration of the major components, silica and iron oxides, has been studied for over a century, but there are still areas of controversy on many aspects of the natural systems. Remarkably little information is available on the alteration of iron-rich silicates and carbonates from BIF, even though their more common igneous and sedimentary counterparts have been extensively studied in weathering profiles. Thus much of the discussion that follows will be drawn from non-BIF literature, supplemented mainly from experience with BIF and iron ores of Western Australia.

BIF alteration will be considered here in terms of two broad categories, physical and chemical, with the latter divided into leaching, redox reactions and enrichment. These divisions are not intended as arbitrary pigeon holes but more as convenient points for discussion of the various mineral groups.

PHYSICAL WEATHERING

Though this review is primarily concerned with chemical alteration, physical processes and controls cannot be ignored since these often open the way to the former. The effects from mineral or root swelling, animal burrowing, or fire (Birkeland, 1974; Nicholls, 1976; Blatt et al., 1980) can be locally important, while frost fracturing plays a prominent role in cold climates. Though comparable effects by diurnal temperature changes have been discounted by Blatt et al., (1980, p. 247), Nicholls (1976) suggests that fatigue resulting from repetitive temperature cycling combined with hydrolysis, might result in rock failure. Sheeting and jointing resulting from removal of overburden by erosion is generally more obvious in unlayered rocks such as granites, than in strongly layered rocks, but rapid erosion along susceptible horizons can produce superficially similar features in BIF. Fractures are often hidden until the rocks are weathered. For example, in ore at Paraburdoo (Hamersley Iron Province), fine scale jointing inherited from the original BIF, commonly results in irregular tablets or crude, dice-shaped fragments which are rare in the stratigraphically equivalent ore horizons of the distant mines at Mt. Tom Price and Mt. Whaleback. The joints are not visible in fresh BIF core from the mine, but are selectively affected by alteration to reveal complex patterns in the BIF outcrop. A series of coarser fractures represented by veins of ankerite apparently act as groundwater conduits once the susceptible carbonate has been leached. On a larger scale, crosscutting dolerite dykes in the same area have played a significant role in producing cells of varied supergene-ore types, compared with more uniform ore bodies such as at Mt. Whaleback where dykes are absent (Morris, 1980). On an even larger scale are the joint- or fault-controlled gorges common along the northern scarp of the Hamersley Range proper.

A lesser known physical property of BIF that can lead to accelerated corrosion is the presence of magnetite in sufficient concentration to provide extensive pathways for electron conduction. The mechanisms involved in these electrochemical cells will be discussed under Chemical Weathering.

CHEMICAL WEATHERING

General

As Krauskopf (1967, p. 100) has said, the "chemical reactions of weathering are basically simple. The overall processes involve nothing more abstruse than ionization, addition of water and carbon dioxide, hydrolysis and oxidation". To this we need to add reduction and the formation of complexes, since these often organically mediated processes play an important part in the migration of iron in nature. Indeed, the role of biochemical and organic processes in the breakdown of rocks appears to have been underrated in the geological literature in favour of the traditional appeal to slow abiogenic processes over geological time. For a recent brief review of biochemical weathering readers are referred to Silverman (1979).

Simple though they may be in concept these chemical processes often result in highly complex products, particularly in the intermediate stages, even though the final residues of BIF may consist perhaps only of quartz, hematite, goethite, and kaolinite or gibbsite.

Staining of rocks by iron or manganese oxides is an obvious indication of alteration but subtle effects may sometimes be overlooked. In hand specimen mild oxidation is most easily recognized by the "browning" or limonitization of minerals such as silicates and carbonates. As an example of such weathering, dump samples of originally fresh BIF exposed for just a few decades after mining from the Colonial asbestos mine at Wittenoom, show a marked darkening of the siderite from a pale neutral-grey to a yellow- or reddish-brown. Siderite is particularly susceptible to oxidation and Schaller and Vlisidis (1959) noted the progressive change in a bottled, powdered sample, from 59.42% FeO in 1915, to 0.74% FeO in 1958. The iron-bearing silicates, ferroan talc, minnesotaite, and stilpnomelane appear to be as readily or more easily oxidized, based on a visual comparison of the surfaces of recent, 10-year-old, and 20-year-old saw cuts in BIF core from Wittenoom. Despite storage in dry conditions, the older surfaces show a marked darkening of the silicates to the extent that stilpnomelane can be mistaken for biotite. Such changes are of course largely superficial and though a slight loss of CO_2 and lower Fe^{2+} : Fe^{3+} ratios could be expected, the bulk chemical composition is unlikely to change significantly. However, under natural weathering conditions the process is likely to be accelerated, particularly in the presence of organic agencies (Silverman, 1979), and leaching will lead to a depletion in Mg, Ca, and CO_2 from carbonates, K, Mg, and Si from silicates, and Ca, and P from apatite.

Leaching in BIF, as in other rocks, tends to accentuate structural and textural features. Subtle effects like the enhancement of fine banding, are best seen where the patina of oxidation products is continually removed, as in actively eroding areas. However, even samples from presently forming gorges in the Hamersleys show a wide range of oxidation features and leaching, these variations presumably related to the rapidity of physical erosion. Spalling of large blocks from cliff faces may expose relatively unweathered material, whereas more stable sections are strongly altered. The rate of weathering is such that few visitors to Dales Gorge realise that extensive sections of the gorge floor and northern wall are benches formed during asbestos mining in the late 1930's.

A study of 20 samples taken along a laterally continuous, gently dipping sequence of mesobands at approximately 10 m intervals in from the portal of the abandoned asbestos mine in Colonial Gorge at Wittenoom, shows that minor oxidation effects can be observed to about 40 m in from the present cliff face. Alteration apparently unrelated to the gorge walls is found much deeper than this along susceptible vertical joints, though how much of this is due to natural conditions and how much to exposure by mining cannot be determined. Observations of the critical weathering zone are hampered by the concrete portal, but assuming some 10—15 m of cliff have been removed to make a platform for the mine buildings, it can be stated that some oxidation of carbonates and silicates can be observed to at least 50 m in from the gorge walls, even though significant alteration is probably limited to the first few metres. Thus despite the relatively arid climate of the present weathering cycle, chemical alteration is generally much more rapid than physical erosion in this area.

Supergene alteration can range to great depths. While the crust of altered BIF may extend to many tens of metres below the surface in southern Africa (Beukes, 1983, this volume) or in the Hamersleys, much deeper penetration is commonly associated with the formation of ore throughout the world. Depths of ½ km or more are routinely reported (e.g. MacLeod, 1966; James et al., 1968) while in the Krivoy Rog supergene alteration has been recorded to below 2.5 km (Belevtsev, 1973).

Since redox reactions are electron transfer processes, no free oxygen is required at the site of oxidation of ferrous minerals in the presence of water, even at great depth, provided an electrochemical cell can operate. Such a mechanism has been suggested to explain the formation of deep-seated iron ores of the type mined at Tom Price and Whaleback (Morris et al., 1980). Anodic oxidation at depth can be represented as:

$$Fe^{2+} \rightarrow Fe^{3+} + e^- \qquad (1)$$

with the electrons conducted by magnetite layers to outcrop, where oxygen is reduced at the magnetite cathode:

$$O_2 + 4e^- + 2H_2O \rightarrow 4(OH)^- \qquad (2)$$

while at depth the ferric ion hydrolyses and precipitates:

$$Fe^{3+} + 3H_2O \rightarrow Fe(OH)_3 + 3H^+ \qquad (3)$$

releasing H^+, which reacts with ferrous carbonates and silicates to continue the process. Charge balancing by electrolytic conduction in ground waters completes the cell. Though the mechanism has been specifically applied to deep-seated ore formation, the authors have pointed out that differential corrosion along adjacent laminae in BIF samples, observed during artificial weathering experiments, has its counterpart in outcrop samples, where, for example, certain bands of magnetite, carbonates and silicates may be extensively oxidised or corroded while adjacent bands are relatively fresh.

The pH, Eh conditions and the volume of solution involved obviously play major roles in weathering, with different effects on various minerals. That solution of vast quantities of what are normally considered highly resistant components, such as quartz, does happen under supergene conditions, has been argued by iron ore geologists for at least 70 years (Van Hise and Leith, 1911). For example, to produce an iron ore body such as Mt. Whaleback with ore reserves of more than 1300×10^6 tonnes with Fe $> 60\%$, over 60 per cent of the original volume of BIF must have been dissolved from the present site of the ore. In addition, solution must have affected the upward extension of the strata that has been eroded from above the present surface, to provide the significant additional amounts of iron which are now part of the ore (Morris, 1980).

Silica

Solution of silica is normally considered to be independent of Eh and, within the normal range of ground waters, of pH. Nevertheless there is an apparently complex interplay between the solubility of quartz of about 10 ppm (at 25°C), and the chemical conditions which result in the breakdown of ferrous silicates such as minnesotaite or stilpnomelane, to give silica concentrations in ground waters traversing BIF several times this value (Dorr and Barbosa, 1963; unpubl. company data). Theoretically, amorphous silica (solubility approximately 120 ppm at 25°C) derived from the breakdown of silicates should dissolve under these conditions, though in the presence of quartz, silica should precipitate from such solutions. However, the growth of quartz at low temperature is so slow that this is unlikely to have much effect on the concentration. Possibly, as suggested by Krauskopf (1959), dilution by rainwater in these areas may be faster than solution of silica. Organic matter may also affect the equilibrium by adsorbtion on the silica surfaces, thus inhibiting solution (Siever, 1962). At best it can be assumed that quartz should not dissolve while residual silicates are present, and indeed petrographic examination of a variety of leached BIF samples suggests this is probably the case.

The assumption that large scale solution of quartz can occur must be implicit in the observation of its essential absence from supergene iron ore derived from BIF, but more direct evidence comes from the work of Dorr and

Barbosa (1963) who have described South American examples in some detail. In the Itabira district of Brazil, for instance, solution of quartz from BIF results in generally soft, disaggregated outcrop and winze material which in places may be dug out with the fingernails. This is in marked contrast to the Hamersley area where hard outcrop is the norm. The difference can be reasonably attributed to the average rainfall which at around 300 mm p.a. in the Hamersleys, is about a fifth of that of the Itabira district. Nevertheless, where channeling occurs, such as in many Hamersley ore bodies (Morris, 1980), BIF is often leached to the extent that it, like the associated "blue dust" of adjacent ore horizons, becomes almost entirely unconsolidated. Dorr and Barbosa (1963) have suggested that the primary control of quartz dissolution is time but that fine grain size and hence surface area of the chert, is important, citing examples in Venezuela and India.

Leaching at depths well below the water table, remote from atmospheric oxygen, has been noted in core from the Newman area. Here the BIF has become extremely friable in places but is notable for the presence of *unoxidized* magnetite, siderite, and pyrite. Thin sections of some of the less leached zones show that originally abundant silicate (probably minnesotaite-ferroan talc), intergrown with chert, has been leached out, and in some cases actually replaced by silica. In more friable specimens it is uncertain whether disaggregation is the result of solution of the contact areas of the quartz grains or merely the removal of the intergrown silicate and carbonate. However, since parts of the core contain unconsolidated residues of little but magnetite, it is a reasonable assumption that quartz itself is dissolving in these zones.

Under some circumstances BIF may be totally silicified. The outcrop specimen shown in Fig. 14-1 has proved instructive in that the transformation from oxidized BIF with ferric minerals intact (martite and primary hematite) to its silicified equivalent, can be traced in a single thin section. The replacement of the magnetite grains tends to follow the pattern of the martite lamellae — suggesting that oxidation of the rock occurred before or during silicification. The conditions under which the ferric iron was dissolved can only be surmised, though chelation, low pH, or reducing conditions, individually or in concert, all suggest organic mediation.

The opposite process where quartz is pseudomorphed by iron oxide has been shown to be an important part of supergene ore formation (James et al., 1968; Morris, 1980) though how this is accomplished has not been established. In enriched samples, it appears that the progress of iron oxide replacement of chert is by a series of generally imperceptible irregular shells from the margins to the interiors of the grains (Morris, 1980, fig. 5), unlike the cleavage-controlled replacement of silicates and carbonates. The presence of numerous minute voids in goethite is not uncommon in iron-enriched excherty samples (Morris, 1980, fig. 5F), indicating a stage where solution of silica sometimes continued without further replacement by iron hydroxyoxides. These observations indicate that the iron oxide shells were not impermeable barriers to silica or iron solutions during the process.

It can be speculated that domains of widely divergent pH could form as a result of reactions of the type shown in eq. 1—3. These could be on a large scale, as implied by the electrochemical model or on a grain to grain basis. Where high alkalinity results, the solubility of silica should dramatically increase but reprecipitation would remain sluggish even under changed conditions. In any case, groundwater flow and varying concentration of dissolved components could well be controlled by alternate wet and dry seasons. During high flow-through, part of a molecular layer could be dissolved from exposed quarts grains, to be replaced (in some zones) later in the season by iron oxide, as the water became supersaturated following stagnation. Such a process might well take millions (James et al., 1968) or hundreds of millions (Trendall, 1975) of years to complete.

As has been shown by a number of studies (e.g. James et al., 1968) the structural situation plays a major role in channelling groundwaters through BIF. However, Trendall (1975) in describing the wide variety of structures of the Hamersley ore bodies noted equally favourable structures that had not resulted in enrichment. While the missing factors are conjectural, the most likely is that of suitable topography. For example, internal drainage on a high plateau would obviously be more effective in leaching than a situation of dominant surface run-off.

Fig. 14-1. Silicification of BIF. The conversion from oxidised BIF (dark) with narrow "magnetite" mesobands to completely silicified BIF (light), can be traced in a single thin section.

Carbonates

Unlike the pH dependant dissolution of calcite or dolomite, the inorganic solution of carbonates containing ferrous iron (ankerite, siderite and the ferroan calcites and dolomites) is dependent on both pH and Eh. Thus, in mildly acid oxidizing conditions, mineral solution should normally result in rapid oxidation of ferrous iron and its precipitation as hydrous ferric oxide(s). This, as seen in eq. 3, would result in increased acidity, further accelerating the solution of carbonate. In the presence of organic complexing agents, ferric iron could be kept in solution to redeposit elsewhere as the organic component was destroyed by oxidation. However, complete replacement of carbonates by goethite is common both in outcrop BIF and in ore throughout the world. Such pseudomorphing of siderite requires the addition of nearly 50% more iron; if dolomite or ankerite are involved even more iron would be required. Preservation of delicate internal textures such as the shapes of quartz or silicate inclusions (Fig. 14-2), indicates a metasomatic process rather than merely void filling, though this can also happen. Oxidation and replacement tend to follow cleavage patterns and it is not uncommon to find virtually unaltered carbonate and solid goethite on opposite sides of a cleavage plane.

Presumably congruent solution of these carbonates would occur under acid reducing conditions which, if continued for long enough, should leave voids unstained by iron oxides. However, the presence of such voids cannot be taken as evidence of such a process since goethite pseudomorphs may also dissolve later, given the right conditions (Morris, 1980).

Apatite

Apatite, the only significant phosphate mineral reported in fresh BIF, is commonly distributed as micrometre-sized grains in patterns which are related to the bedding and which follow complex distortions with remarkable precision. The distribution can be readily determined with simple peel and stain techniques (Morris, 1973; Morris and Ewers, 1978).

Apatite is slightly less soluble than calcite, dissolving very slowly in the pH range 7—8 and more rapidly with increasing acidity. Because it is generally extremely fine-grained and anhedral, direct microscope evidence of its behaviour in BIF during weathering is scarce. (A comprehensive account of weathering of *rich* phosphate deposits is given by Altschuler (1973).) Follow-

Fig. 14-2a, b. Porous goethite after carbonate and silicate. These areas, originally containing carbonate euhedra with inclusions of the silicate-chert matrix, were pseudomorphed by goethite which was later preferentially leached, removing most of the pseudomorphed quartz and some of the silicate. The dark grey areas represent plastic-filled voids resulting from this selective leaching. (Reflected light).

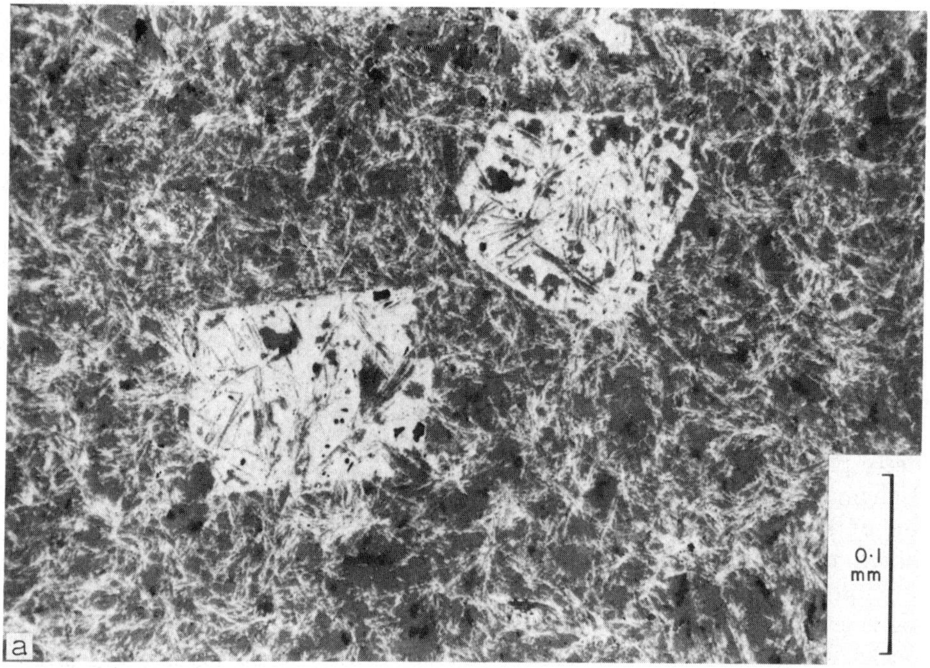
a
0·1
mm

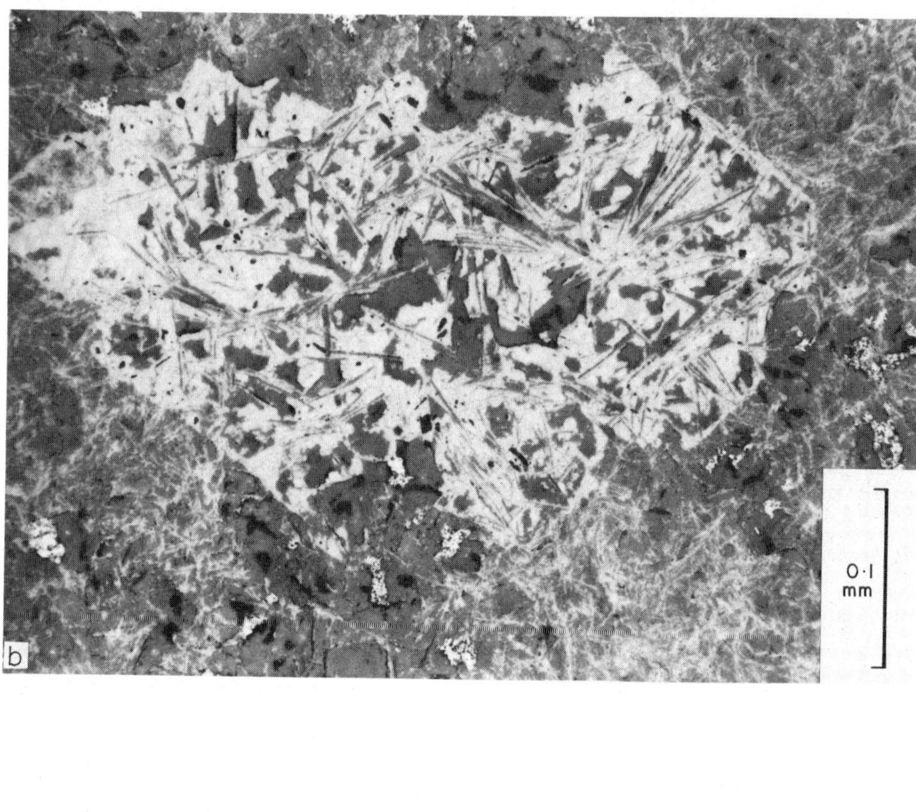
b
0·1
mm

ing a survey of BIF samples from cliffs and gorges in the Hamersley Iron Province, Ewers and Morris (1981) concluded that apatite was too easily leached from surface rocks for such samples to be used for accurate chemical data. One criterion of alteration used for the survey was that phosphate-prints of fresh BIF generally showed sharp patterns, while even mildly leached rocks gave diffuse prints resulting from the loss of fine-grained apatite. Despite such leaching, minute specks of apatite (as well as carbonates) may be preserved within chert grains but, as weathering increases, even these appear to be removed.

Metasomatic replacement of apatite by iron oxides or iron phosphates is presumably possible but no unequivocal evidence of this has been found. In heavily altered or mineralised BIFs, particularly in the silicate-rich members, discrete phosphorus minerals occur as scarce micrometre-sized grains. Xenotime containing traces of many rare-earth elements, as well as submicrometre phosphates containing Ca, Ba, La, Ce, Nd and Sm, were identified by Graham (1973) in samples from Rhodes Ridge (Hamersley area). A wider survey of supergene alteration in BIF (Ewers and Morris, unpubl. data) has shown that these sparse microscopic phosphates often contain significant aluminium with varying proportions of Fe, Ba, REE, Ca, K and Si. The grains are too small for quantitative data but it seems likely that many of the specks represent plumbogummite-type species. Norrish (1968) has shown that crandallite-gorceixite members of this group are widespread in soils as products of weathering. The absence of such compounds from fresh BIF supports the view that they are not resistant residuals, though a monazite-like mineral has been detected in these rocks (Morris, 1973).

Berge (1970) has recorded aluminium phosphates in the weathering profile of BIF from the Goe Range, Liberia and sporadic segregations of iron and aluminium phosphates possibly related to avian guano (Bridge, 1974) have been found in iron ores in Western Australia. Barbour (1973) reported that in the ores and altered BIF of the Itabira District of Brazil, phosphorus is related to the adsorption and iron exchange properties of clay minerals and hydrated iron oxides. In the Hamersleys, the bulk of the residual phosphate in weathered or enriched samples is present within goethite, but not all goethite contains phosphorus. The distribution pattern in these hydrous iron oxides (Graham, 1973; Ewers and Morris, unpubl. data) suggests co-precipitation of iron and phosphorus from solution or adsorption of phosphorus on the growing surfaces of the hydroxyoxides. Secondary apatite, though usually unimportant in ores, may give rise to local phosphorus anomalies.

Silicates

Loughnan (1969) has indicated that three concurrent processes operate during weathering of silicates: release of components by breakdown of mineral structure, removal of certain of these components in solution, and the formation of new minerals from the remnants of the originals. With layer lattice minerals such as biotite, hydration often leads to significant expansion which in turn can cause considerable physical disruption (Birkeland, 1974). Accelerated weathering studies on BIF have shown that stilpnomelane is susceptible to this process of expansion (Morris et al., 1980).

Siever and Woodford (1979) have suggested from dissolution studies on finely ground silicates ($< 53\ \mu m$) that armouring of mineral surfaces by precipitation of ferric hydroxide inhibits further dissolution. However, petrological examination of BIF shows that oxidation of silicates, like that of carbonates, commonly occurs on cleavage surfaces *within* the grains. Production of H^+ (eq. 3) as a result of the hydrolysis reaction and the precipitation of iron hydroxyoxide *in situ* is likely to enhance the dissolution, particularly if concomitant hydration causes expansion of the mineral structure.

Very little has been reported on the breakdown products of silicates in BIF of the Hamersleys apart from a few unpublished reports of the "shale"-rich horizons in the iron ore deposits. In general it would be expected that smectites and chlorites should predominate as a result of the essentially arid climate of the area. Trendall and Blockley (1970, p. 115) report a study by N.L. Marsh in which "montmorillonoid" was a significant component in shale horizons in relatively unaltered core, but most of the sporadic studies by the iron ore companies and by the writer have shown that in weathered samples kaolinite is by far the most common clay component, sometimes accompanied by minor gibbsite, with mica where the original rock is rich in this component. However, where stilpnomelane, in particular, is present as a minor component of BIF proper, it is often replaced by goethite.

Magnesian talc is reported as a remarkably persistent mineral during weathering (Guild, 1953), but there is no doubt that the iron-bearing varieties are prone to rapid attack. Even in rocks considered to be fresh — based on the criterion of the absence of hydrous iron products in siderite — minnesotaite and ferroan talc may be sufficiently deeply coloured to be mistaken for stilpnomelane, whereas their unweathered state is pale yellow-green to colourless. In more altered rocks, the minerals are often replaced by goethite, and indeed ore derived from the Marra Mamba Iron Formation is notable for the high content of goethite-pseudomorphed forms of these minerals (Fig. 14-2). The characteristic yellow staining of the BIF outcrop of this formation is partly due to the easy oxidation of these components. Trendall and Blockley (1970) suggest that the abundance of minnesotaite is possibly one explanation of the susceptibility to weathering of this formation as a whole, compared with the cliff-forming Dales Gorge Member, though the presence of abundant carbonate (Ewers and Morris, 1980) is also significant.

Massive riebeckite, while physically very resistant, is still readily affected by chemical weathering, usually ending as a "*hard yellow-brown aggregate of limonite and secondary silica not unlike weathered chert*" (Trendall and Blockley, 1970 p. 176). Polished sections usually show the original matted or rosetted forms. The asbestiform variety, crocidolite, appears more susceptible to leaching, judging from the peculiar "whitening" effect seen in exposed mine dump samples, but replacement by iron oxide to give griqualandite is common in enriched rocks, while the silica-replaced product, "tiger's eye", is a popular ornamental stone.

Though these examples of supergene alteration are drawn from the Hamersleys, similar features have been found by the writer in BIFs from other areas in and out of Australia. Mann (1953) described goethite and hematite replacement features in North American samples involving siderite, minnesotaite, stilpnomelane and grunerite, but concluded that the process was hydrothermal, whereas James et al. (1968) and Gair (1975) supported a supergene origin.

Iron oxides

Hematite

Hematite is undoubtedly the most stable component of BIF under most conditions of supergene alteration judging from the common presence of totally unaltered *primary* hematite in ores and highly altered BIF (Morris, 1980). Though its direct hydration to goethite under such conditions may be theoretically possible, Langmuir (1971) considers the conversion to be infeasible on kinetic grounds. Observations on ores and mineralized samples of Hamersley rocks support Langmuir's statement on hematite stability, but the often reported "hydration of hematite" requires clarification. There is, for example, ample evidence in the goethite-rich "hydrated" zone of the Mt. Tom Price iron ore mine or the equivalent "hard cap" of the Mt. Whaleback mine at Newman, that the original material was once hematite-rich, and is now commonly very goethitic. But the goethite of these zones occurs as ochreous, colloform, or botryoidal masses, not as a solid-state or metasomatic replacement of individual crystals of hematite. In all probability the original hematite was taken into solution with the aid of organic material and the iron later reprecipitated as goethite or as colloform hematite, leading to destruction of original texture — a process common during lateritization.

Magnetite

Oxidation of magnetite, as commonly reported in the geological literature, is generally considered to produce two end phases — hematite, first called martite or "cubic" Fe_2O_3 by Breithaupt (1828) and, γFe_2O_3, originally described as magnetic peroxide of iron in 1859 (Robbins, reported in Twenhofel, 1927), and later named maghemite (Wagner, 1927). But this simple picture is not supported by the wealth of data from experimental studies, even though no agreement has been reached on the mechanisms or on some of the products of oxidation.

Martite, of course, is the name for hematite pseudomorphs after magnetite (and after pyrite, Deer et al., 1969, p. 409). Since Gruner (1929) showed that hematite replacement preferentially followed the octahedral planes of magnetite, the resultant lattice texture — usually best seen under crossed polars — has been an excellent guide to the original mineral in heavily altered rocks. However, many martites do not show these lattice textures, while others show complex mixtures of lattice and zonal textures, the latter often marking growth patterns (Figs. 14-3 and 4; Han, 1978; Morris, 1980).

It is generally held that the oxygen lattice is essentially immobile during oxidation of magnetite, while the small metal ions diffuse through the structure (e.g. Davis et al., 1968). In the case of isolated grains oxidized in air at temperatures high enough to produce only hematite, the system remains closed to iron and oxidation must proceed by addition of oxygen, represented thus:

$$
\begin{array}{lll} & 4Fe_3O_4 + O_2 & \rightarrow 6\alpha Fe_2O_3 \\ \text{molar volumes} & 178.096 & \rightarrow 181.644 \end{array} \quad (4)
$$

which results in a theoretical 2% volume increase. Davis et al. (1968) suggested this expansion could be accomodated by diffusion of iron through defects to surface sites, where by oxidation it is added as hematite on or between the octahedral parting surfaces, thus producing the lattice martite texture.

Where groundwaters are involved a similar result could be achieved by reaction with water (Garrels and Christ, 1965):

$$
\begin{array}{lll} & 2Fe_3O_4 + H_2O & \rightarrow 3\alpha Fe_2O_3 + 2H^+ + 2e^- \\ \text{molar volumes} & 89.048 & \rightarrow 90.822 \end{array} \quad (5)
$$

which is effectively eq. 4 with oxygen transfer through the reaction:

$$
2H^+ + \tfrac{1}{2}O_2 + 2e^- \rightarrow H_2O \quad (6)
$$

However, loss of iron and attendant porosity could develop thus:

$$
\begin{array}{lll} & 3Fe_3O_4 & \rightarrow 4\alpha Fe_2O_3 + Fe^{2+} + 2e^- \\ \text{molar volumes} & 133.572 & \rightarrow 121.096 \end{array} \quad (7)
$$

The oxidation of magnetite to form defect-structured "*maghemite*" has had a great deal of attention for its importance in the electronic industry and in rock magnetism. In standard mineragraphic texts (Schouten, 1962; Ramdohr, 1969; Uytenbogaard and Burke, 1971) maghemite is described as pale blue, grey-blue, grey or white in reflected light. Observations on thousands of Hamersley samples show that magnetic phases more pink-brown than the original magnetite are very common products of oxidation (Fig. 14-3). Comparable blue or grey phases have been found in only a handful of specimens (Fig. 14-3b). Pink-brown oxidized magnetite phases are also common in the few specimens examined by the writer from BIF localities in Liberia, Brazil, North America, South Australia and the Yilgarn Block of Western Australia. Newhouse and Callahan (1927) reported many examples of a "*brownish magnetic mineral which replaced bluish magnetite*" with a colour difference "*so slight that unless both minerals are in the field it is hard to determine which one is present*", but Ramdohr (1969, p. 977) found this colour data so different from his own observations that he suggested a "*misprint*" was responsible. As the optical properties of these pink-brown phases are similar to unoxidized magnetite their aberrant nature may have been ignored like the observations of Newhouse and Callahan, and the minerals simply classified as magnetite or as an initial stage of oxidation.

The defect spinel structure is metastable but can be stabilised by a variety of foreign ions such as Ti, V, and Zn occupying some of the vacant sites. Protons often fill this role and Swaddle and Oltmann (1980) suggested that oxidation to maghemite should be rapid if sufficient "water protons" were present to preserve the structure, but that the conversion would be inhibited if too many were present, by eliminating the normal lattice vacancies through which the iron diffuses. Smith (1979) pointed out that the literature on synthetic maghemite was not always clear on the presence of structural water, but that possibly two forms existed: a maghemite with a primitive-cubic structure, stabilized by "water", and a tetragonal, anhydrous γFe_2O_3.

For some time it was accepted that cation vacancies in iron-deficient magnetite were entirely in the octahedral sites; such compounds giving diffuse XRD lines, a regular decrease in cell dimension from $Fe_3O_4 \rightarrow Fe_{3-\upsilon}O_4$ and a bluish colour in reflected light. Experimental high-temperature studies reported by Kullerud et al. (1969), however, indicated that a form existed with vacancies in both octahedral and tetrahedral sites, which gave sharp XRD lines, an increased cell edge, and a colour in polished section similar to, or darker than, unoxidized magnetite. They introduced the term keno-tetrahedral magnetite for this phase, keno-octahedral magnetite for the material with only octahedral vacancies (maghemite proper?), and suggested the general term *kenomagnetite* where no specific vacancy sites were established [keno = empty (Greek)]. Following this last suggestion the naturally occurring pink-brown phases thought to be intermediate between magnetite proper and the presumed fully oxidized phase, maghemite (Fig. 14-3b), have been termed kenomagnetite (Morris 1980), but no relationship between these and the high-temperature artificial phases is implied.

Fig. 14-3.a. A typical, uniform pink-brown kenomagnetite phase (*Kmt*) after magnetite. Note the partial cores and rims of hematite (martite (*M*)). White irregular grains are primary hematite (*H*), and the matrix is goethite (dark grey-mottled) after chert, carbonate and silicate. (Oil immersion-reflected light).

b. Oxidized magnetite grains showing the rarely seen transitional stages between the typical pink-brown kenomagnetite phase (*Kmt*), and the blue, presumed end-phase, maghemite (*Mh*) (lighter tones). The "magnetites" show evidence of their original growth pattern by the re-appearance during oxidation, of cores and rims now defined as martite (*M*), whereas the bulk of the grains remain as defect-structured spinel. (Oil immersion-reflected light).

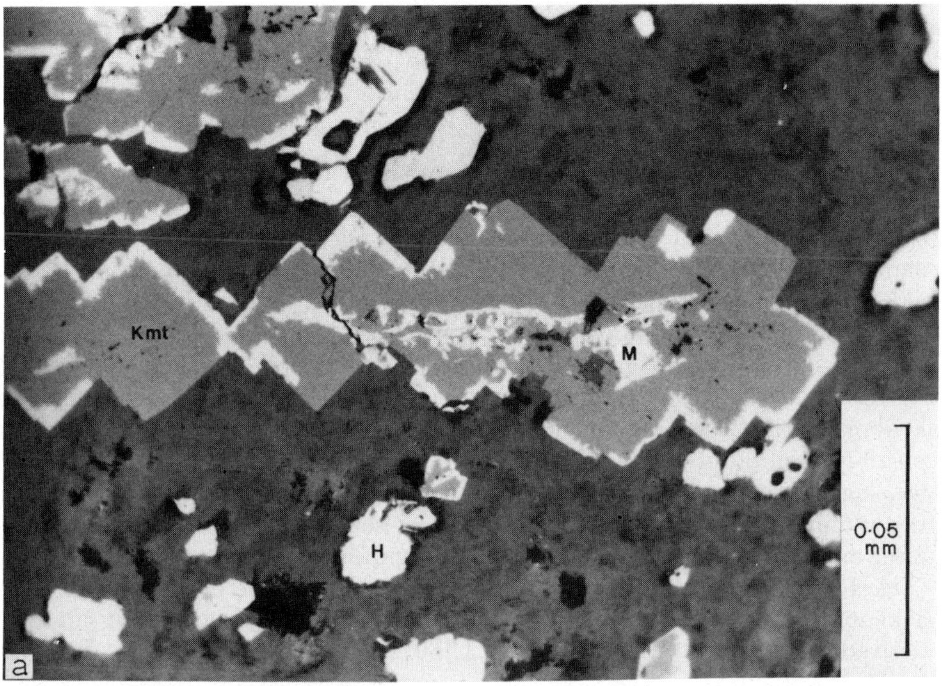
Kmt
M
H
0·05
mm
a

H
Kmt
M
Mh
0·05
mm
b

In forming kenomagnetite from magnetite, eq. 7 can be written:

	$3Fe_3O_4 \rightarrow 3Fe_{8/3}O_4 + Fe^{2+} + 2e^-$		(8)
molar volumes	$133.572 \rightarrow 133.572^+$	(kenotetrahedral magnetite)	
molar volumes	$133.572 \rightarrow 131.03$	(keno-octahedral magnetite)	

involving minimal volume change but a change in density as a result of the removal of up to one iron from each nine in the system. There is seldom any evidence of significant iron-staining around such grains, but the usual abundance of pseudomorphed carbonates and silicates in the vicinity of such oxidized magnetites, suggests a local sink for this iron.

In the Hamersley rocks it is not often that kenomagnetite forms without at least some martite-hematite forming as well, either as trellis lamellae or by following the growth patterns (Han texture) (Fig. 14-3). Very often the kenomagnetite is subsequently replaced by goethite (Fig. 14-4), a process possibly facilitated by the protonization mentioned earlier. Ramdohr (1969) has suggested the direct conversion of unoxidized magnetite to goethite, but this is unlikely. On empirical grounds, the presence of goethite pseudomorphs after magnetite, or the observation of significant porosity within martite (Fig. 14-5), can usually be taken as an indication of the earlier existence of a kenomagnetite phase — the porosity resulting from leaching of goethite derived from kenomagnetite (Morris, 1980).

DISCUSSION

Whether analyses of altered BIF can be considered of value will naturally depend on the use to which they are to be put. While no one would seriously dispute the importance of the two major components, silicon and iron, in understanding the origin of BIF, it could hardly be argued that precise silica-iron ratios of individual specimens have any more than broad significance, in view of the inhomogeneity of these rocks. Thus even severe oxidation would be of little concern in the interpretation of such data. On the other hand, ferrous-ferric ratios might be of interest in say, diagenetic or oxygen isotope studies, and here even mild oxidation could prejudice the results.

It is probably the elements most easily affected by weathering that are

---

Fig. 14-4a. Magnetite in oxidized BIF partly converted to martite (mostly lamellae), and partly to kenomagnetite (*Kmt*) (mid-grey). The kenomagnetite shows minor inversion to hematite (irregular white areas) and is partly hydrated to goethite (*G*) (ramifying dark grey areas). Matrix is chert. (Oil immersion-reflected light).

b. The central, large oxidized magnetite shows the original growth pattern revealed by differential oxidation. White is martite. The kenomagnetite (*Kmt*) (mid-grey) is partly hydrated to goethite (*G*) (dark grey). The other grains in the field show kenomagnetite entirely hydrated to goethite. Matrix is goethite after chert. (Oil immersion-reflected light).

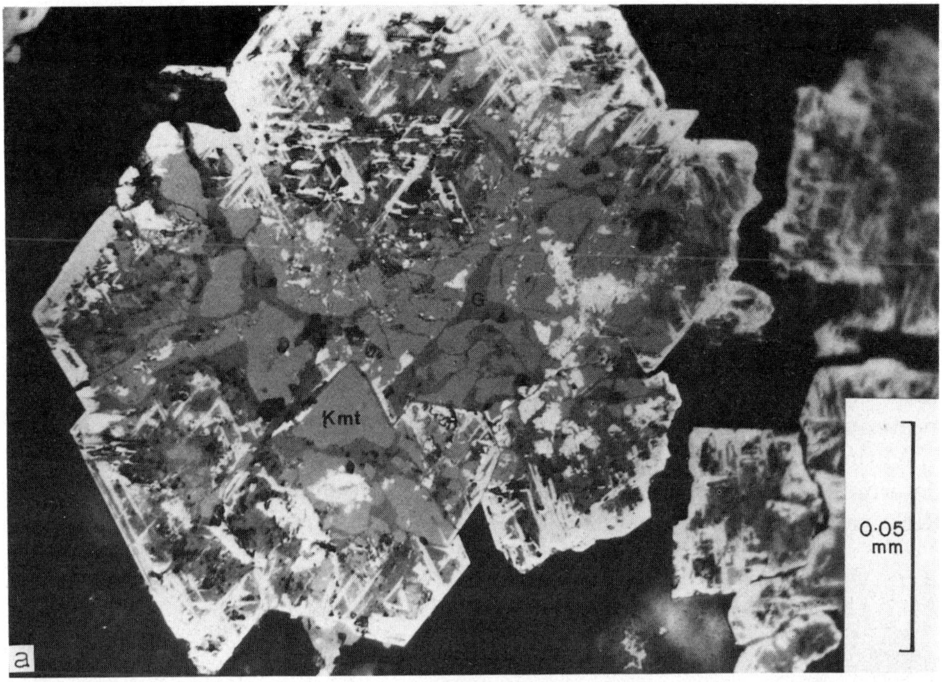
G
Kmt
0·05
mm
a

G
M
Kmt
G
M
G
0·05
mm
b

the most important in BIF research. Phosphorus, for example, is of major economic significance to the Western Australian iron ore industry and there is no question that establishment of base values in BIF requires analysis of totally unleached rocks. Rare earth elements are attracting more interest in BIF studies (Fryer, 1983, this volume), but little is known of their distribution in BIF and even less of how they are affected by weathering. If present entirely in resistant minerals such as non-metamict zircon or monazite, then minor weathering is not likely to significantly affect the data. But what if the hosts are apatite or carbonate which are readily leached from outcrop specimens? A case could be argued for other components, K, and Al from silicates, Ca, Mg, Sr, and CO_2 from carbonates, used by a multitude of writers to discuss various aspects of the origin of BIF. It is important, therefore, to at least note the potential hazard to data that unrecognized alteration can produce.

Various techniques can be used to detect subtle alteration effects, or to determine original mineralogy when alteration is severe. Minor oxidation in BIF can be recognized in hand specimen by "browning" of normally pale

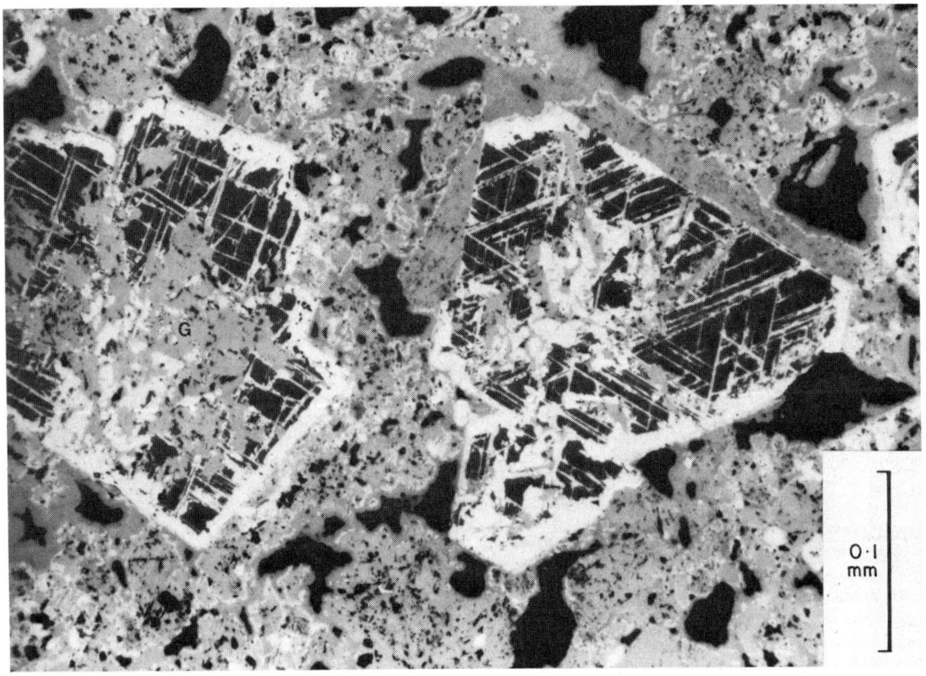

Fig. 14-5. Skeletal "magnetite" in ore. Selective leaching has removed much of the goethite (*G*) (originally kenomagnetite) from the grains leaving the martite lamellae intact (black areas are plastic-filled voids). Repeated solution and deposition of goethite and minor hematite in the matrix has destroyed the original pseudomorphed BIF texture. (Oil immersion-reflected light).

minerals such as siderite or the talc and minnesotaite groups. Under the microscope the incipient stages can be seen as faint mottlings of increased colour. With slightly more advanced alteration, "limonite" or manganese oxide stains and replacement are easily visible on streak-prints (Morris and Ewers, 1978). Leaching may be detected by its effect on apatite, since its loss from BIF results in diffuse phosphate-prints. Petrographic examination of oxidized or iron-enriched specimens is probably best achieved with polished thin sections. With experience, the different goethite-pseudomorphed species can often be recognized, not only by shape but sometimes by differences in the colour of the goethite. Such textural examination will show that unlike the process of lateritization which usually destroys the original rock texture, enrichment of BIF generally tends to preserve the texture by metasomatic replacement. Later processes may even enhance these textures, for example, by selective leaching of certain pseudomorphed species (Fig. 14-2), or by dehydration of specific goethitized components to hematite (Morris, 1980).

CONCLUSION

The susceptibility of BIF to rapid supergene alteration can be attributed to a number of factors. Firstly, to the presence of ferrous iron in easily leached minerals such as carbonates and silicates — oxidation of ferrous iron to ferric followed by precipitation of ferric hydroxyoxides releases hydrogen ion, which accelerates further attack. Secondly, to channelling of groundwaters and hence accelerated alteration along susceptible horizons, as a result of the marked lamination and variability of the rocks. And thirdly, to the possibility that conductive layers of magnetite (or graphite and sulphides) may aid electron transfer during redox reactions by the formation of large electrochemical cells.

There are still many geoscientists who find it difficult to accept that quartz can, in not unreasonable time, be dissolved almost completely from rocks such as BIF. Even harder to accept, apparently, is that quartz may be metasomatically replaced by iron oxides under supergene conditions. But the processes have been well documented. Rainwater in traversing BIF (or any siliceous rock) will dissolve silica, firstly by attacking silicates and when these have been destroyed quartz will follow. There is no question that iron oxides can take their place under certain conditions, since many ore samples show the parent textures as well, if not better, than the original BIF. Given the right structural and topographical situations, groundwaters will move through and out of the system and, particularly if channelled by artesian conditions and aided by organic agents, will in time remove not only silica, but also the more resistant iron oxides, by solution or suspension.

Finally, there can be little doubt that the commonly stated view — that chemical weathering in arid regions is outstripped by physical erosion — is certainly not true for the Hamersley Iron Province or for the equivalent areas in southern Africa.

ACKNOWLEDGEMENTS

I am grateful to many colleagues in the Australian iron ore industry for their active support through the Australian Mineral Industries Research Association Limited over the past years, and my special thanks go to W.E. Ewers who led our research program on BIF and iron ores until his recent retirement. The chapter has benefitted from the critical reading of C.R.M. Butt, W.E. Ewers, M.J. Gole and J. Graham.

REFERENCES

Altschuler, Z.S., 1973. The weathering of phosphate deposits — geochemical and environmental aspects. In: E.J. Griffith, A. Beeton, J.M. Spencer and D.T. Mitchell (Editors), Environmental Phosphorus Handbook. John Wiley, New York, N.Y., pp. 33—96.

Barbour, A.P., 1973. Distribution of phosphorus in the iron ore deposits of Itabira, Minas Gerais, Brazil. Econ. Geol., 68: 52—64.

Belevtsev, Y.N., 1973. Genesis of high-grade iron ores of the Krivoyrog type. In: Genesis of Precambrian Iron and Manganese deposits. Proc. Kiev Symposium 1970. UNESCO, Paris, pp. 167—180.

Berge, J.W., 1970. Implications of phosphorus fixation during chemical weathering in the genesis of supergene iron ores. Geol. Min. Metall. Soc. Liberia, Bull., 4: 33—43.

Beukes, N.J., 1983. Palaeoenvironmental setting of iron-formations in the depositional basin of the Transvaal Supergroup, South Africa. In: A.F. Trendall and R.C. Morris (Editors), Iron-Formation: Facts and Problems. Elsevier, Amsterdam, pp. 131—209.

Birkeland, P.W., 1974. Pedology, Weathering, and Geomorphological Research. Oxford University Press, New York, N.Y., 285 pp.

Blatt, H., Middleton, G. and Murray, R., 1980. Origin of Sedimentary Rocks. 2nd Ed. Prentice-Hall, Englewood Cliffs, N.J., 782 pp.

Breithaupt, A., 1828. Quoted in Dana's System of Mineralogy. C. Palache, H. Berman, C. Frondel. 1: 834 pp.

Bridge, P.J., 1974. Avian derived phosphate from inland Western Australia. West. Aust. Naturalist, 13: 24.

Davis, B.L., Rapp, Jr., G. and Walawender, M.J., 1968. Fabric and structural characteristics of the martitization process. Am. J. Sci., 266: 482—496.

Davy, R., 1983. Chemical composition of BIF. In: A.F. Trendall and R.C. Morris (Editors), Iron-Formation: Facts and Problems. Elsevier, Amsterdam, pp. 325—343.

Deer, W.A., Howie, R.A. and Zussman, J., 1969. An Introduction to the Rock Forming Minerals. Longmans, London, 528 pp.

Dorr, J. van N., II, and Barbosa, A.L.M., 1963. Geology and ore deposits of the Itabira district, Minas Gerais, Brazil. U.S. Geol. Surv., Prof. Pap. 341C.

Ewers, W.E. and Morris, R.C., 1980. Chemical and mineralogical data from the uppermost section of the upper BIF member of the Marra Mamba Iron Formation. CSIRO Inst. Earth Resour., Div. Miner. Rep. No. FP 23.

Ewers, W.E. and Morris, R.C., 1981. Studies of the Dales Gorge Member of the Brockman Iron Formation, Western Australia. Econ. Geol., 76: 1929—1953.

Fryer, B.J., 1983. REE in iron-formation. In: A.F. Trendall and R.C. Morris (Editors), Iron-Formation: Facts and Problems. Elsevier, Amsterdam, pp. 345—358.

Gair, J.E., 1975. Bedrock geology and ore deposits of the Palmer Quadrangle, Marquette County, Michigan. U.S. Geol. Surv., Prof. Pap., 769.

Garrels, R.M. and Christ, C.L., 1965. Solutions, Minerals, and Equilibria. Harper and Row, New York, N.Y., 450 pp.

Graham, J., 1973. Phosphorus in iron ore from the Hamersley iron formations. Aust. Inst. Min. Metall., Proc., 246: 41—42.

Gruner, J., 1929. Structural reasons for oriented intergrowths in some minerals. Am. Mineral., 14: 227—237.

Guild, P.W., 1953. Iron deposits of the Conghonas district, Minas Gerais, Brazil. Econ. Geol., 48: 639—676.

Han, T-M., 1978. Microstructures of magnetite as guides to its origin in some Precambrian iron-formations. Fortschr. Mineral., 56: 105—142.

James, H.L., Dutton, C.E., Pettijohn, F.J. and Wier, K.L., 1968. Geology and Ore Deposits of the Iron River—Crystal Falls District, Iron County, Michigan. U.S. Geol. Surv., Prof. Pap., 570: 134 pp.

Krauskopf, K.B., 1959. The geochemistry of silica in sedimentary environments. In: Silica in Sediments. Soc. Econ. Paleontol. Mineral., Spec. Publ., 7: 4—20.

Krauskopf, K.B., 1967. Introduction to Geochemistry. McGraw-Hill, New York, N.Y., 721 pp.

Kullerud, G., Donnay, G. and Donnay, J.D.H., 1969. Omission solid solution in magnetite: kenotetrahedral magnetite. Z. Kristallog., 128: 1—17.

Langmuir, D., 1971. Particle size effect on the reaction goethite = hematite + water. Am. J. Sci., 271: 147—156.

Loughnan, F.C., 1969. Chemical Weathering of the Silicate Minerals. Elsevier (America), New York, N.Y., 154 pp.

MacLeod, W.N., 1966. The geology and iron deposits of the Hamersley Range area, Western Australia. West. Aust., Geol. Surv., Bull. 117, 170 pp.

Mann, V.I., 1953. The relation of oxidation to the origin of soft iron ores of Michigan. Econ. Geol., 48: 251—281.

Morris, R.C., 1973. A pilot study of phosphorus distribution in parts of the Brockman Iron Formation, Hamersley Group, Western Australia. West. Aust., Geol. Surv., Annu. Rep., 1972: 75—80.

Morris, R.C., 1980. A textural and mineralogical study of the relationship of iron ore to banded iron-formation in the Hamersley Iron Province of Western Australia. Econ. Geol. 75: 184—209.

Morris, R.C. and Ewers, W.E., 1978. A simple streak-print technique for mapping mineral distributions in ores and other rocks. Econ. Geol. 73: 562—566.

Morris, R.C., Thornber, M.R. and Ewers, W.E., 1980. Deep-seated iron ores from banded iron-formation. Nature, 288: 250—252.

Newhouse, W.H. and Callahan, W.H., 1927. Two kinds of magnetite? Econ. Geol., 22: 629—632.

Nicholls, G.D., 1976. Weathering of the earth's crust. In: J.P. Riley and R. Chester (Editors), Chemical Oceanography, 5, 2nd Ed. Academic Press, London, pp. 81—101.

Norrish, K., 1968. Some phosphate minerals of soils. 9th Int. Congr. Soil Sci., Trans., 2: 713—723.

Ramdohr, P., 1969. The Ore Minerals and Their Intergrowths. English translation of the 3rd Ed., Pergamon, Oxford, 1174 pp.

Schaller, W.T. and Vlisidis, A.C., 1959. Spontaneous oxidation of a sample of powdered siderite. Am. Mineral., 44: 433—435.

Schouten, C., 1962. Determination Tables for Ore Microscopy. Elsevier, Amsterdam, 242 pp.

Siever, R., 1962. Silica solubility, 0—200°C, and the diagenesis of siliceous sediments. J. Geol., 70: 127—150.

Siever, R. and Woodford, N., 1979. Dissolution kinetics and weathering of mafic minerals. Geochim. Cosmochim. Acta, 43: 717—724.

Silverman, M.P., 1979. Biological and organic chemical decomposition of silicates. In: P.A. Trudinger and D.J. Swaine (Editors), Biogeochemical Cycling of Mineral-Forming Elements. Elsevier, Amsterdam, pp. 445—465.

Smith, P.P.K., 1979. The observation of enantiomorphous domains in natural maghemite. Contrib. Mineral. Petrol., 69: 249—254.

Swaddle, T.W. and Oltman, P., 1980. Kinetics of the magnetite-maghemite-hematite transformation, with special reference to hydrothermal systems. Canad. J. Chem., 58: 1763—1772.

Trendall, A.F., 1975. Geology of Western Australian iron ore. In: C.L. Knight, (Editor) Economic Geology of Australia and Papua New Guinea. I. Metals. Australas. Inst. Min. Metall., Geophys. Monogr., 5: 883—892.

Trendall, A.F. and Blockley, J.G., 1970. The iron formations of the Precambrian Hamersley Group, Western Australia; with special reference to the associated crocidolite. West. Aust., Geol. Surv., Bull. 119, 336 pp.

Twenhofel, L.H., 1927. Changes in the oxidation of iron in magnetite. Econ. Geol., 22: 180—188.

Uytenbogaard, W. and Burke, E.A.J., 1971. Tables for Microscopic Identification of Ore Minerals. 2nd Ed., Elsevier, Amsterdam 430 pp.

Van Hise, C.R. and Leith, C.K., 1911. The geology of the Lake Superior region. U.S. Geol. Surv., Monogr., 52: 641 pp.

Wagner, P.A., 1927. Changes in the oxidation of iron in magnetite. Econ. Geol., 22: 845—846.

SUBJECT INDEX

This index has been compiled mainly from lists prepared by the authors of each chapter. The entries thus reflect their judgment of the relative importance of the subject in question.

Page numbers shown in italics refer to Figures and Tables or to their captions. The text on these pages should also be consulted.